Big Data Research for Social Sciences and Social Impact

Big Data Research for Social Sciences and Social Impact

Special Issue Editors

Miltiadis D. Lytras
Anna Visvizi
Kwok Tai Chui

MDPI • Basel • Beijing • Wuhan • Barcelona • Belgrade

Special Issue Editors

Miltiadis D. Lytras
The American College of Greece
Greece
Effat University
Saudi Arabia

Anna Visvizi
The American College of Greece
Greece
Effat University
Saudi Arabia

Kwok Tai Chui
The Open University of Hong Kong
Hong Kong

Editorial Office
MDPI
St. Alban-Anlage 66
4052 Basel, Switzerland

This is a reprint of articles from the Special Issue published online in the open access journal *Sustainability* (ISSN 2071-1050) from 2018 to 2020 (available at: https://www.mdpi.com/journal/sustainability/special_issues/big_data_research).

For citation purposes, cite each article independently as indicated on the article page online and as indicated below:

LastName, A.A.; LastName, B.B.; LastName, C.C. Article Title. *Journal Name* **Year**, *Article Number*, Page Range.

ISBN 978-3-03928-220-3 (Pbk)
ISBN 978-3-03928-221-0 (PDF)

Contents

About the Special Issue Editors

Miltiadis D. Lytras Ph.D., is an expert in advanced computer science and management, and an editor, lecturer, and research consultant, with extensive experience in academia and the business sector in Europe and Asia. Dr. Lytras is Research Professor at Deree College—The American College of Greece—and a Distinguished Scientist at the King Abdulaziz University, Jeddah, Kingdom of Saudi Arabia. Dr. Lytras is a world-class expert in the fields of cognitive computing, information systems, technology-enabled innovation, social networks, computers in human behavior, and knowledge management. In his work, Dr. Lytras seeks to bring together and exploit synergies among scholars and experts, and is committed to enhancing the quality of education for all.

Anna Visvizi Ph.D., is a political scientist and economist, editor, and research consultant with extensive experience in academia and the think-tank sector in Europe and the US. As the author of several publications, Dr. Visvizi has presented her work across Europe and the US, including Capitol Hill. Dr. Visvizi's expertise covers issues pertinent to the intersection of politics, economics, and ICT. This is translated in her research and advisory roles in the areas of AI and geopolitics, smart cities and smart villages, innovation promotion, global migration management, and economic integration, especially the EU and BRI. Currently, Dr. Visvizi serves as Associate Professor at Deree College—The American College of Greece. Until December 2018, Dr. Visvizi was Head of Research at the Institute of East-Central Europe (IESW), Poland. In her work, Dr. Visvizi places emphasis on engaging academia, the think-tank sector, and decision-makers in dialogue to ensure well-founded and evidence-driven policy-making.

Kwok Tai Chui received the B.Eng. degree in electronic and communication engineering—Business Intelligence Minor and Ph.D. degree from City University of Hong Kong. He had industry experience as Senior Data Scientist in Internet of Things (IoT) company. He joined the Department of Technology, School of Science and Technology, at The Open University of Hong Kong as Research Assistant Professor. He was the recipient of 2nd Prize Award (Postgraduate Category) of 2014 IEEE Region 10 Student Paper Contest. Also, he received Best Paper Award in IEEE The International Conference on Consumer Electronics-China, in both 2014 and 2015. He has more than 45 research publications including edited books, book chapters, journal papers, and conference papers. He has served as various editorial position in SCI-listed journals including Managing Editor of *International Journal on Semantic Web and Information Systems*, Topic Editor of *Sensors*, Guest Editors of *Sustainability*, *Sensors*, *Energies*, *Applied Sciences*, *Journal Future Generation Computer Systems* and *Journal of Internet Technology*. His research interests include computational intelligence, data science, energy monitoring and management, intelligent transportation, smart metering, healthcare, machine learning algorithms and optimization.

Preface to "Big Data Research for Social Sciences and Social Impact"

A new era of innovation is enabled by the integration of social sciences and information systems research. In this context, the adoption of Big Data and analytics technology brings new insights to social sciences. It also delivers new, flexible responses to crucial social problems and challenges. We are proud to deliver this edited volume on the social impact of Big Data research. It is one of the first initiatives worldwide analyzing of the impact of this kind of research on individuals and social issues. We are grateful to all the contributors to this edition for their intellectual work and their sound propositions and arguments. The relevant debate is aligned around three pillars:

Section A. Big Data For Social Impact

We emphasize an initial assessment of the social impact of Big Data. For this purpose, we communicate the main findings of a preliminary study. We then discuss the missing variable in big data for social sciences: the decision-maker. We analyze the role of (smart) citizens from data providers to decision-makers, using the city of Barcelona as a case study. The integration of sustainability in the context of Big Data research is also discussed though analysis of the impact of big data on the development of sustainable online communities, towards the sustainable development of online communities in the Big Data era: a study of the causes and possible consequence of voting on user reviews.

More emphasis is paid to advanced social mining techniques, such as sentiment analysis and opinion mining, in the context of social networks. Finally, a research study on the impact of big data research on innovation is communicated.

Selected topics in this section include:

- Big Data and Their Social Impact: A Preliminary Study;

- The Missing Variable in Big Data for Social Sciences: The Decision-Maker;

- (Smart) Citizens from Data Providers to Decision-Makers;

- Towards Sustainable Development of Online Communities;

- Sentiment from Online Social Networks;

- Big Data Approach as an Institution.

Section B. Techniques and Methods For Big Data-Driven Research on Social Sciences and Social Impact

In this section, various, sophisticated research and case studies of Big Data-driven research on social impact are presented. Selected topics include:

- Opinion Mining on Social Media Data: Sentiment Analysis of User Preferences;

- Towards Sustainable Urban Communities: A Composite Spatial Accessibility Assessment for Residential Suitability Based on Network Big Data;

- Destination Image Analytics Through Traveller-Generated Content, Spatiotemporal Analysis to Observe Gender Based Check-In Behavior Using Social Media Big Data: A Case Study of Guangzhou, China;

- Sales Prediction by Integrating the Heat and Sentiments of Product Dimensions;

- Assessing Technology Platforms for Sustainability with Web Data-Mining Techniques;

- Semantic Network Analysis of Legacy News Media Perception in South Korea: The Case of PyeongChang.

Section C. Big Data Research Strategies

In the concluding section of this edition, the debate on the social impact of Big Data Research on sustainability is promoted, with an integrative discussion of complementary research strategies including:

- Skill Needs for Early Career Researchers—A Text Mining Approach;

- Identifying Promising Research Frontiers of Pattern Recognition through Bibliometric Analysis;

- A Conceptual Framework for Assessing an Organization's Readiness to Adopt Big Data;

- Application of Machine Learning in Predicting Performance for Computer Engineering Students: A Case Study;

- Analyzing Online Car Reviews Using Text Mining;

- Context–Problem Network and Quantitative Method of Patent Analysis: A Case Study of Wireless Energy Transmission Technology.

We want to thank the professional staff at MDPI for their qualitative work that made this edition possible. If you would like any further information on this edition, we are at your disposal, and we invite you to join us in our next editions, on the topics of *Big Data Research*, *Social Sciences* and *Sustainability*.

Miltiadis D. Lytras, Anna Visvizi, Kwok Tai Chui
Special Issue Editors

Editorial

Big Data Research for Social Science and Social Impact

Miltiadis D. Lytras [1,2,*] and Anna Visvizi [1,3]

[1] School of Business and Economics, Deree College—The American College of Greece, 153-42 Athens, Greece; avisvizi@acg.edu
[2] Effat College of Engineering, Effat University, Jeddah P.O. Box 34689, Saudi Arabia
[3] Effat College of Business, Effat University, Jeddah P.O. Box 34689, Saudi Arabia
* Correspondence: mlytras@acg.edu; Tel.: +30-210-600-9800

Received: 13 December 2019; Accepted: 14 December 2019; Published: 24 December 2019

Abstract: This Special Issue of Sustainability devoted to the topic of "Big Data Research for Social Sciences and Social Impact" attracted significant attention of scholars, practitioners, and policy-makers from all over the world. Locating themselves at the cross-section of advanced information systems and computer science research and insights from social science and engineering, all papers included in this Special Issue contribute to the debate on the use of big data in social sciences and big data social impact. By promoting a debate on the multifaceted challenges that our societies are exposed to today, this Special Issue offers an in-depth, integrative, well-organized, comparative study into the most recent developments shaping the future directions of interdisciplinary research and policymaking.

Keywords: big data research; social and humanistic computing; social sciences; social good; social impact; machine learning; knowledge management; web science; data science; social inclusive economic growth; sustainability; innovation; innovation networks

---

1. Introduction—Overview of the Edited Volume

The evolution of big data research and social media and the contributions of individuals and organizations in social networking data ecosystems resulted in a new, sophisticated context for technology-facilitated social interactions [1]. In parallel, a number of social challenges and problems, and the critical need to enhance the capability of our society to deal with delicate social issues, sets new directions for research [2]. The focus of this edited volume is on the intersection of advanced information systems and social sciences research [3,4]. It is about analyzing the impact of data and their processing to the understanding and addressing of significant social problems. It is also about analyzing data as a social good that must be protected and be aligned with significant rules and ethical principles.

Social good is generally an action or application that benefits society. In the past, governments and nonprofit organizations usually drove it. With the advancement of social media via computer-mediated technologies like WeChat, WhatsApp, Weibo, Twitter, Instagram, Facebook, and YouTube, billions of registered users utilize social interactions through social media. As a result, everyone can contribute to society easily and achieve social good.

Tremendous growth of digital information (from granular data to aggregated data) is available for numerous social sciences and social impact applications, for instance, environmental protection, healthcare and education. Data analytics are ubiquitous and purpose-oriented in different forms: descriptive analytics, diagnostic analytics, predictive analytics and prescriptive analytics. Typical challenges for adopting big data technologies for social sciences and social impact are data handling and storage, data quality, computational power of computers and algorithm customization for special application and security. More importantly, it seems that our societies have an infinite need for utilizing the value of big data for social impact and thus applications on social good [5–7].

This edited volume aims to consolidate recent advances in big data for social good. Topics of interest for this Special Issue include (but are not limited to):

- Innovative applications of data analytics to social sciences and social impact problems like energy, healthcare, education, food, poverty, injustice, and inequalities in society;
- Machine learning algorithms for big data applications for social sciences and social impact;
- Advanced techniques for handling unstructured, unlabeled and/or missing data;
- Data quality control of big data for social good;
- Big data research-driven policy-making for social impact;
- Standardization for big data infrastructure and framework;
- Big data implications for society;
- Big-data-driven KPIs research for international benchmarking;
- Socially inclusive economic development and growth through big data applications;
- Human-centric big data research;
- Ethical issues on social research of big data.

The final selection of papers includes 19 research studies organized in three sections:

1.1. Section A: Big Data for Social Impact

- Lytras, M.D.; Visvizi, A. Big Data and Their Social Impact: Preliminary Study. *Sustainability* **2019**, *11*, 5067.
- Arnaboldi, M. The Missing Variable in Big Data for Social Sciences: The Decision-Maker. *Sustainability* **2018**, *10*, 3415.
- Calzada, I. (Smart) Citizens from Data Providers to Decision-Makers? The Case Study of Barcelona. *Sustainability* **2018**, *10*, 3252.
- Zhao, J.; Wang, J.; Fang, S.; Jin, P. Towards Sustainable Development of Online Communities in the Big Data Era: A Study of the Causes and Possible Consequence of Voting on User Reviews. *Sustainability* **2018**, *10*, 3156.
- Carrera, B.; Jung, J.-Y. SentiFlow: An Information Diffusion Process Discovery Based on Topic and Sentiment from Online Social Networks. *Sustainability* **2018**, *10*, 2731.
- Yau, Y.; Lau, W.K. Big Data Approach as an Institutional Innovation to Tackle Hong Kong's Illegal Subdivided Unit Problem. *Sustainability* **2018**, *10*, 2709.

1.2. Section B: Techniques and Methods for Big Data Driven Research for Social Sciences and Social Impact

- Păvăloaia, V.-D.; Teodor, E.-M.; Fotache, D.; Danileţ, M. Opinion Mining on Social Media Data: Sentiment Analysis of User Preferences. *Sustainability* **2019**, *11*, 4459.
- Zhao, Y.; Zhang, G.; Lin, T.; Liu, X.; Liu, J.; Lin, M.; Ye, H.; Kong, L. Towards Sustainable Urban Communities: A Composite Spatial Accessibility Assessment for Residential Suitability Based on Network Big Data. *Sustainability* **2018**, *10*, 4767.
- Marine-Roig, E. Destination Image Analytics Through Traveller-Generated Content. *Sustainability* **2019**, *11*, 3392.
- Muhammad, R.; Zhao, Y.; Liu, F. Spatiotemporal Analysis to Observe Gender Based Check-In Behavior by Using Social Media Big Data: A Case Study of Guangzhou, China. *Sustainability* **2019**, *11*, 2822.
- Lyu, X.; Jiang, C.; Ding, Y.; Wang, Z.; Liu, Y. Sales Prediction by Integrating the Heat and Sentiments of Product Dimensions. *Sustainability* **2019**, *11*, 913.
- Blazquez, D.; Domenech, J.; Garcia-Alvarez-Coque, J.-M. Assessing Technology Platforms for Sustainability with Web Data Mining Techniques. *Sustainability* **2018**, *10*, 4497.

- Yoon, S.-W.; Chung, S.W. Semantic Network Analysis of Legacy News Media Perception in South Korea: The Case of PyeongChang 2018. *Sustainability* **2018**, *10*, 4027.

1.3. Section C: Big Data Research Strategies

- Maer-Matei, M.M.; Mocanu, C.; Zamfir, A.-M.; Georgescu, T.M. Skill Needs for Early Career Researchers—A Text Mining Approach. *Sustainability* **2019**, *11*, 2789.
- Park, I.; Yoon, B. Identifying Promising Research Frontiers of Pattern Recognition through Bibliometric Analysis. *Sustainability* **2018**, *10*, 4055.
- Olszak, C.M.; Mach-Król, M. A Conceptual Framework for Assessing an Organization's Readiness to Adopt Big Data. *Sustainability* **2018**, *10*, 3734.
- Buenaño-Fernández, D.; Gil, D.; Luján-Mora, S. Application of Machine Learning in Predicting Performance for Computer Engineering Students: A Case Study. *Sustainability* **2019**, *11*, 2833.
- Kim, E.-G.; Chun, S.-H. Analyzing Online Car Reviews Using Text Mining. *Sustainability* **2019**, *11*, 1611.
- Ree, J.J.; Jeong, C.; Park, H.; Kim, K. Context–Problem Network and Quantitative Method of Patent Analysis: A Case Study of Wireless Energy Transmission Technology. *Sustainability* **2019**, *11*, 1484.

2. Conclusions—The Value Added of this Special Issue

This collection of papers provides an integrative discussion on key issues and challenges related to the adoption of big data research and their social impact. Below we provide a list of the key findings and ideas communicated in this Special Issue:

- The key understanding of big data and their social impact requires sophisticated studies for the measurement of value and the associated perception from individuals and groups. It also requires a sophisticated approach for the linkage of social value to key social challenges and problems. In this context, various key performance indicators have to be justified. A number of individual concerns related to data privacy and anonymity are also important to be addressed.
- The promotion of big data research for social impact still emphasizes the role of the missing variable in big data for social sciences: the decision-maker. The sophisticated analysis of social-sensitive data requires decision makers with the capability to analyze and to link these data with significant social problems. Without this human-centric approach in decision, making any effort for social impact will be of limited contribution.
- The evolution of big data research for social impact requires strategic efforts and initiatives towards sustainable development of online communities in the big data era. These communities will adopt rules and will promote the required culture for linking advanced research based on big data and analytics for addressing significant societal problems with actions.
- Social evolution of big data research is also related to sophisticated data processing methods like analysis of sentiment from online social networks.
- Big data research can justify also a new era for the evolution of social innovation. The exploitation of skills and capabilities beyond local boundaries will link social local capabilities to global social challenges.
- Advanced data mining and analytics approaches are required for the revelation of hidden insights regarding big data linked to social problems. In this direction, many more things have to be done. In the current era, limited isolated approaches prove the capacity of these techniques to deal with social issues. At the other extreme, some rules and ethical norms have to applied in initiatives like social rating systems that violate privacy and personal human rights.
- Sustainability concerns are significant in the context of the social impact of big data research. The perception of big data as social good that must promote social value is a basic axiomatic sentence, but there are many grey areas for the provision of this value as a transparent good in the benefit of the global society.

We are pleased to be able to present this collection of papers to the research community. The promotion of socially sensitive research especially that that addresses topics related to the use of technology will be a trend in the years to come [8,9]. The understanding that social impact and social value are the key objective of any technology-driven innovation is the basic step towards sustainable and socially inclusive growth and development.

Author Contributions: All authors contributed evenly to this Editorial. All authors have read and agreed to the published version of the manuscript.

Conflicts of Interest: The authors declare no conflict of interest.

References

1. Lytras, M.D.; Raghavan, V.; Damiani, E. Big data and data analytics research: From metaphors to value space for collective wisdom in human decision making and smart machines. *Int. J. Semant. Web Inf. Syst.* **2017**, *13*, 1–10. [CrossRef]
2. Lytras, M.D.; Mathkour, H.I.; Abdalla, H.; Al-Halabi, W.; Yanez-Marquez, C.; Siqueira, S.W.M. Enabling technologies and business infrastructures for next generation social media: Big data, cloud computing, internet of things and virtual reality. *J. Univers. Comput. Sci.* **2015**, *21*, 1379–1384.
3. Lytras, M.D.; Mathkour, H.I.; Abdalla, H.; Al-Halabi, W.; Yanez-Marquez, C.; Siqueira, S.W.M. An emerging— Social and emerging computing enabled philosophical paradigm for collaborative learning systems: Toward high effective next generation learning systems for the knowledge society. *Comput. Hum. Behav.* **2015**, *5*, 557–561. [CrossRef]
4. Visvizi, A.; Lytras, M.D. Rescaling and refocusing smart cities research: From mega cities to smart villages. *J. Sci. Technol. Policy Mak.* **2018**. [CrossRef]
5. Lytras, M.D.; Aljohani, N.R.; Hussain, A.; Luo, J.; Zhang, X.Z. Cognitive Computing Track Chairs' Welcome & Organization. In Proceedings of the Companion of the Web Conference, Lyon, France, 23–27 April 2018.
6. Visvizi, A.; Mazzucelli, C.; Lytras, M. Irregular migratory flows: Towards an ICT' enabled integrated framework for resilient urban systems. *J. Sci. Technol. Policy Manag.* **2017**, *8*, 227–242. [CrossRef]
7. Crusoe, J.; Ahlin, K. Users' activities for using open government data—A process framework. *Transform. Gov. People Process Policy* **2019**, *13*, 213–236. [CrossRef]
8. Visvizi, A.; Lytras, M.D. *Politics and Technology in the Post-Truth Era*; Emerald Publishing: Bingley, UK, 2019; ISBN 9781787569843.
9. Visvizi, A.; Daniela, L. Technology-Enhanced Learning and the Pursuit of Sustainability. *Sustainability* **2019**, *11*, 4022. [CrossRef]

 *sustainability*

Article

Big Data and Their Social Impact: Preliminary Study

Miltiades D. Lytras [1,2,*] and Anna Visvizi [1,3]

[1] School of Business & Economics, Deree College, The American College of Greece, 153-42 Athens, Greece; avisvizi@acg.edu
[2] Effat College of Engineering, Effat University, Jeddah P.O. Box 34689, Saudi Arabia
[3] Effat College of Business, Effat University, Jeddah P.O. Box 34689, Saudi Arabia
[*] Correspondence: mlytras@acg.edu; Tel.: +30-210-600-9800

Received: 26 July 2019; Accepted: 10 September 2019; Published: 17 September 2019

Abstract: Big data is the buzz-word of today, and yet their specific impact on individuals and societies remains assumed rather than fully understood. Clearly, big data and their use have already given rise to a number of questions, including those of how data can be collected and used in ethical and socially sensitive ways. Building on these points, the objective of this study was to explore how precisely big data and big data based services influence individuals and societies. This paper elaborates on individuals' perceptions of data, especially on how they perceive the actual sharing of their data. In this way, this paper defines a value space for the social impact of big data relevant to three factors, namely the intention to share personal data, individual's concerns, and social impact of big data.The main contribution of this study consists of the insights into the still nascent area of research that unfolds at the cross-section of social science and computer science. We expect that in the next years this area of research will gain prominence.

Keywords: social impact; big data research; information systems; analytics; decision making; social sciences

---

1. Introduction

Recent developments in data-driven information systems, set big data research and business analytics at the core computer science and social science. In computer science research, there is a consensus that big data and data analytics research will foster a new generation of information systems capable of managing the collective wisdom in human decision making and smart machines [1]. Emerging research areas like cognitive computing [2] combined with artificial intelligence and machine learning, permit advanced and sophisticated methods for processing data, including sentiment analysis, image processing, natural speech recognition and text mining. In parallel emerging technologies, including cloud computing, internet of things and virtual reality, the value proposition of application and services that process data in different formats such as text, images, videos, microcontents in social media is further enhanced [3,4]. The development of a huge data ecosystem around the globe, in which providers and users of data promote business value in terms of data and decision making, is a key development of our times. In this context, users of applications and services worldwide participate consciously, or unintendedly, to an integrated data dissemination and aggregation process with critical trust and privacy issues.

A great discussion on the real impact of big data research has been initiated. An interesting study [5] sets the significance of the human decision maker at the center of any type of big data information processing cycle. There is an agreement between different academics that big data can make a big impact [1,6].

Big data research is aligned with the evolution in emerging information technologies research. New information processing paradigms further promote the significance of big data, and have a great impact on its volume and coverage. A number of application domains and industries already adopt big data research with significant success. Consider social networks research and the contribution of social networks to the big data ecosystem [2,3]. Other examples are artificial intelligence and machine learning applications in various domains, such as customers/clients of big data repositories for personalized and targeted services [3,4].

In the recent literature of big data research, an increasing section is dedicated to the capacity of big data to support social sciences research. There is the anticipation that big data is potentially a social good that must be secured and be used for the transparency of services, and for the evolution of a user-centric new culture for sustainable computing. In parallel, several concerns have been documented, mostly related to trust, privacy and the protection of personalities in the new technology-driven domain of services and applications. In Figure 1, below, we provide our initial framework for the investigation of the social impact of big data.

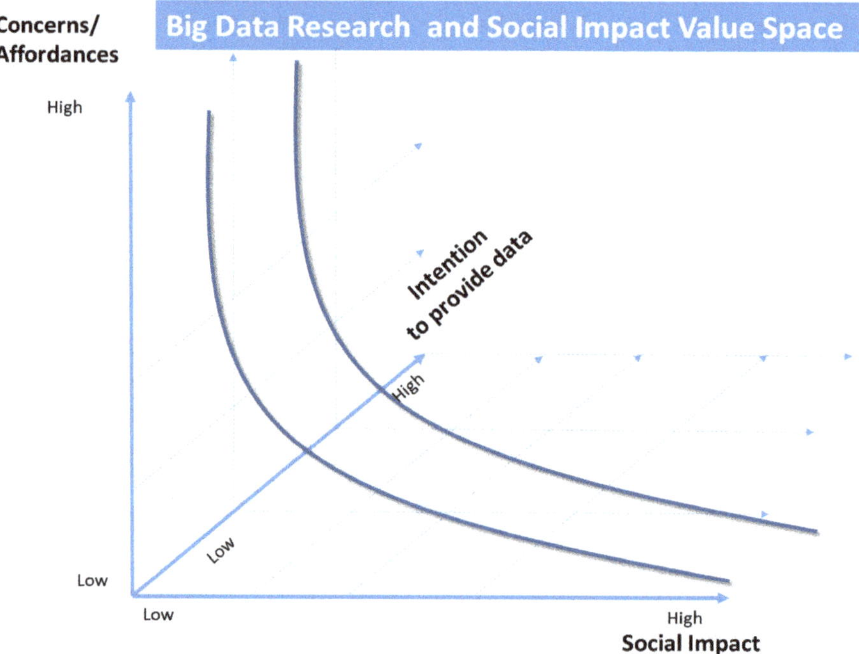

Figure 1. An initial framework for understanding the social impact of big data research. Source: The authors.

In our approach three critical factors need further investigation:

- User concerns/affordances: The first factor, namely user concerns or affordances is related to all the psychological, social, personal or professional concerns of users in relevance to the use of applications and services that generate and share personal data or other kinds of data from individuals within the big data ecosystem
- Intention to share data/informed consent: This aspect of our research problem is related to the conscious agreement or the intrinsic motivation of users to share their data for the purposes of big data application. In our research, we are interested in the connection of social challenges and social problems to the intention of users to share their data. Furthermore, we wanted to understand if in some scenarios, users of applications and services share their data without formal agreement due to their interest in exploiting the added value of the service for themselves or for the society.

- Social impact of big data: The third critical factor also determines the value space in Figure 1. The measurement of the social impact of big data seems to require interdisciplinary approaches and metrics, thus we must deploy heuristics for the attachment of value contribution to the perception of users for the impact of big data research to their lives and to our society. This is the ultimate objective of our research, nevertheless, the requirements and the various research strategies we deployed exceed the length and scope of this paper.

The value space of big data is defined as an aggregation of three factors/forces. At the X-axis, the social impact of big data is presented in a spectrum of low to high value. Users of big data applications develop perceptions and have their own interpretation mechanisms for the impact of big data. On the Y-axis, concerns and fears of users of big data application develop an intrinsic motivation mechanism for the use of such application. They deploy different ways for the use of big data applications and they also express their concerns for various aspects of these applications. Furthermore, on the Z-axis in Figure 1, it is shown that users also execute a different degree of willingness to share their data, for the proper functioning of big data applications. Various studies in the literature mention these factors, and our previous research has tried to investigate these factors. The value space that is defined by these three axes, can be used as a model for discussing big data applications and services and for mapping such services in wider contexts e.g., smart cities research. From a practical point of view, this model can also be exploited by real users of big data applications for the customization of available services or the personalization for added value of such applications. Also, from a policy making view, such a model can guide public consultation and debate on how we protect the data and identity rights of citizens against big data applications without compromise of social value and impact.

Figure 1 is used as a metaphor to communicate the overall idea of our research, that somehow users, with their perceptions and intention to use big data applications, define their personal value space and maybe also a societal value space. We understand that in our approach some key assumptions are integrated. We do, however, believe that it is worthy to investigate this research problem which has many psychological and social aspects. In the next section we provide a critical review of the relevant literature towards the justification of our research model that will be presented in Section 3 of this research study.

2. Literature Review—Understanding the Debate on Big Data and their Social Impact

The agenda of big data research is quite wide and involved various multidisciplinary communities. From a computer science and information systems perspective issues related to standardization, data mining, aggregation of data, interoperability and recommendation systems are at the top of research priorities. From a social science perspective, data as a social construct affecting issues related to identity management, personality, privacy and security are the focus of social research. Furthermore, the concept of the digital self, that combines personal, professional, social, and other features of individuals is gaining more interest [1]. In an evolving way, big data that refer to human entities and communities of people are established with convenient computational methods that permit social analysis and reference.

The connection of big data research to social sciences as well as the big impact of data-intensive applications and processing methods to societal challenges provides a very interesting research challenge. From the one side we have the social actors, humans, decision makers that both provide and consume data available in diverse, interconnected information systems [5]. The quest for impact on big data platforms and big data [6] requires a detailed study of different factors and accordingly new metrics like analytics or KPIs (key performance indicators) [6]. Humans, from this point of view, realize a critical mental shift in their behavior. From data providers they are requested to perform a decision maker role, within the boundaries and across hi-tech socio-technical structures like smart cities [7].

From a different angle, the big data ecosystem requires distribution and aggregation of information in modes that were unforeseen in the past. The sophistication and the huge capacity of big data services to process significant volumes of data, automatically, without human intervention, sets critical questions related to privacy, security and data protection [8].

Especially in the context of social networks and social media [9], the information diffusion has exceeded any prediction. The ease of sharing information as well as the increased openness of such data warehouses permits advanced data processing that leads to critical insights about the data providers. In this situation, big data applications serve as intermediaries, matching the gap between the providers and the consumers of data, allowing several innovative business models to appear [10]. There is a connection that needs further investigation. The power of big data applications as intermediaries and as unique business models for adding value to raw data with data processing data, like sentiment analysis and opinion mining [11]. The capacity of new information processing methods to conclude about sentiments, attitudes or opinions is directly linked to some forms of social impact for such applications [12].

Within this complex big data ecosystem, individuals, organizations as well as governments need to develop frameworks to measure their readiness for the integration of big data research for measurable individual and social objectives [13,14]. One direction for the exploitation of big data research is analytics. The exploitation of value through huge volumes of data, requires the development of big data analytics capabilities [14,15], aiming to provide visualizations and summaries of data that can promote enhanced decision making. From a social science perspective, this connection directly leads to a new era of smart urbanism, where human actors, e.g., citizens, exploit processed data in meaningful visual forms for the improvement of the quality of their lives [16,17].

Another key aspect of big data literature is related to the big data hype. The utilization of big data research for business or social purposes must identify opportunities, myths as well as risks [18]. It is necessary for our societies and for policy making purposes to ask various provocative questions related to the ownership, supervision, consumption and protection of big data [19]. Consider, for example, a system for social rating based on microcontent contributions of citizens on social media, capable of measuring sentiments, political beliefs etc. Smart cities and smart government research [20] must take into consideration, a number of delicate issues related to privacy, security, safety and social responsibility of individuals and groups. Without a focus on sustainability [21,22], social inclusive economic growth and social justice, any isolated, monolithic big data application in the long term will unfortunately fail to promote its social impact. Novel approaches are required in the management of big data and their interoperability, as well as the annotation of data and services for improved social services [23,24]

What seems to be less analyzed, is the social dimension, and the social dynamics of big data, that refer to groups of people, businesses or social constructs. In both cases, the ultimate objective of big data research is to provide useful insights for the personalization of services and the targeting of value adding services.

A key challenge of big data research is to justify and to develop value reference layers to big data. The usability of big data, for various purposes and targeted markets needs to be clarified. In our research, our focus is on the social impact of big data. The key research question is related to the capacity of big data to have a social impact, and to enable bold solutions and responsive actions related to social problems.

In Figure 2, below, we organized in a simple way, some complementary aspects of the big data literature from an information systems/computer science perspective. The list of topics is definitely not exhaustive, but rather representative of significant aspects of the research.

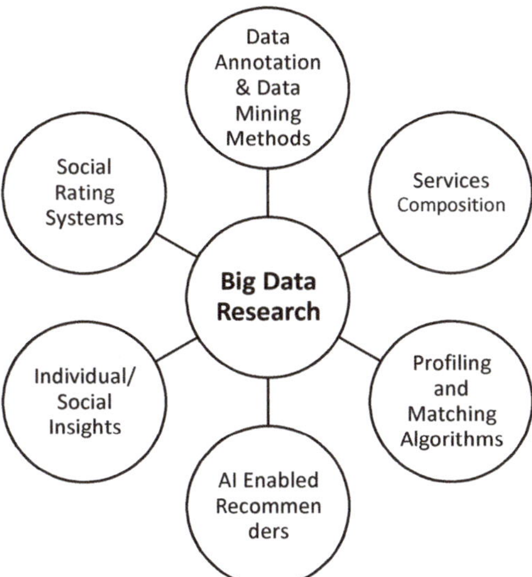

Figure 2. An overview of the key aspects of big data research literature for social sciences. Source: The authors.

Data annotation and packaging, as well as the emerging data mining methods such as sentiment analysis or social mining, set a new domain of research. Especially for social sciences, the capacity of methods like sentiment analysis to understand opinions, to analyze social behavior and the attitudes of individuals can be used extensively for sophisticated social sciences research.

Dynamic big data service composition and selection is also another very interesting area of research and literature. Big data, per se, have limited value without some services (clients) that consume these data for well-defined purposes. The design of socially aware value adding services that consume big data, will soon be a key trend in social sciences. From this perspective, we are going to realize a convergence of social sciences and computer science. Consider real time platforms that provide analysis and collective intelligence, over social media micro-contents (e.g., analysis of sexual harassment, bullying, anti-terrorism detection etc.).

Advanced user profiling [1,5] is also critical for the launch and management of social sensitive applications powered by big data research. The standardization of profiles is the first step toward interoperability of applications and social services. To this direction, latest developments in computer science as well as in policy-awareness frameworks, provide significant contributions. From a social impact perspective, a key question is how, within governmental institutions, and regulations, can we envision trustable, participatory and democratic platforms that exploit big data profiles for social good. Social rating systems, or social filtering platforms are key examples for this emerging area of research. Furthermore, from a social perspective, another key concern is about the ownership of the big data. However, this topic goes beyond the scope of our research articulated in this research study. Smart cities research is another example of critical integration for social sciences and computer sciences research [6–8]. In all these cases several research questions link big data research to critical social impact [11–17].

The availability of big data ecosystems offers numerous options for sophisticated services [6–8]. With the evolution of artificial intelligence e.g., machine learning approaches, we can have systems that are trained by the availability of big data. For special social problems, like poverty, exclusion, migration, we have a brand-new era of information services and recommender systems.

All the previous complementary aspects of research move the quest of our time forward for the provision of sophisticated social insights over individual's data or communities' data. Several ethical issues are involved, but it seems that the next thread of big data services and applications will materialize some of the aspects mentioned in this compact literature review.

In our approach the big data social impact research problem is part of a greater smart cities research approach [25–32]. In the next section, we provide our research methodology for this challenging research problem. From the beginning we have to communicate the limitations of our study and also the complexity of the research phenomenon. This study is based on a pilot survey, in which participants are academics and researchers from computer science and social sciences. The generalization of findings and conclusions should be analyzed within this context.

3. Research Methodology

In our research study we emphasize on an interpretive qualitative research, together with quantitative research. The understanding for the social impact of big data, is a very complex research problem because:

- The technical aspects of big data are complicated and continuously evolving.
- The social aspects of big data research involve human actors with complicated profiles and social references, and thus it is hard to generalize conclusions that are derived from a sample population.

In Figure 3 we provide a synopsis of the research model of our study. There are six critical research objectives:

Research Objective 1. To understand the degree of awareness of big data research and the actual use of big data applications by our responders.

Research Objective 2. To analyze, the perceived value of big data as it is realized by our responders.

Research Objective 3. To understand the main concerns of users of big data services, as well as their affordances towards greater deployment of such applications.

Research Objective 4. To clarify the impact of big data to individual's life and perceptions.

Research Objective 5. To understand the determinants of social impact of big data in our societies in relevance to critical social problems and challenges.

Research Objective 6. To develop guidelines for socially aware big data-enabled services.

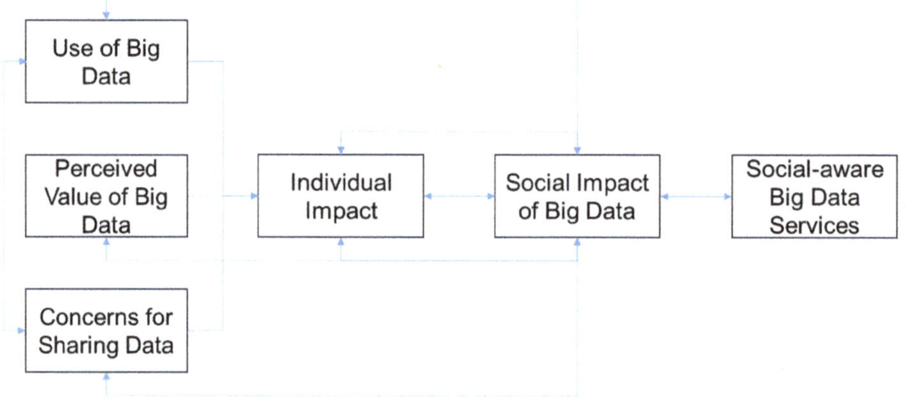

Figure 3. The research model for the study of social impact of big data. Source: The authors.

Two different directions of research have been deployed:

a. A quantitative research based on a sample of responders for some initial insights related to research objectives 1–4. In the next section we present the results of this analysis. Briefly, we designed a research tool for the collection of responses based on a survey, and we collected 108 questionnaires in total from academics, students and researchers.
b. A qualitative research design focused on targeted interviews with five heavy-users of big data applications and five non-users. The aim was to gain an in depth understanding of their mental models for the use and value of big data.

In this research paper we present the main findings of the first research direction, the quantitative analysis. From the beginning of our analysis we communicate the limitations of our study that refer to the limited sample and the complicated aspects of human behavior. We have completed the collection of the questionnaires and we are also currently designing the structured interviews for the heavy and light users of big data applications. Furthermore, we must state that this research on the social impact of big data is part of a greater research project related to smart cities research. In the latter, the objective is to explore end-users perceptions of smart cities' applications, especially as seen from the perspective of personal data sharing. We also understand that this is a complex research problem that needs further investigation and various contributions from different disciplines. Indeed, it is necessary to initiate a sound dialogue on these matters among researchers and decision-makers to create and explore synergies.

4. Analysis and Main Findings

Our effort to elaborate on the social impact of big data research is by default a very challenging research. The critical psychological and personal factors, affecting the adoption of big data applications goes far beyond traditional information systems or computer science research. From the other side, the technical sophistication of advanced applications and services also poses critical challenges to the protection of privacy, identity management and safety on the internet. From the beginning of this analysis, we have to declare the following facts:

- All the findings presented in this section refer to the "biased" population sample of our survey: Academics and researchers that are probably not the average users of big data applications.
- The generalization of findings and their interpretation must consider the previous fact.
- Beyond the previous two statements, the contribution of our study remains important: It is one of the first studies, that integrates social sciences and information systems research in the context of measuring the social impact of big data.
- We do not intend to discuss advanced statistics in this survey, but only descriptive statistics. Our intention is to provide interpretations of the main findings and to use these for the development of a global social impact of big data research study.

In this section we will present the main findings and their interpretation for the context of our research. As it was communicated in the previous section, we focus on the quantitative analysis of our survey. The presentation of key facts of our research will be accompanied by interpretations relevant to the key research objectives, clarified in Section 3. In Section 5, we provide an additional discussion. We start with an overview of the demographics of our survey.

4.1. Demographics

Our survey has a critical objective to understand the social aspects of big data and to interpret and measure the integration of social sciences and big data research.

The questionnaire used is available in Appendix A. We deployed SurveyMonkey software and we targeted users of applications, also familiar with the use of social networks. This survey is used as a pilot survey, since we are planning a global big data and social impact research to run in 2020. We circulated our survey to students and academics in universities of our scientific networks and we received, within three months from July to October 2018, 108 responses. In Tables 1–3 we summarize the main demographics data from our study. In total 108 respondents; academics, researchers and students in management, international business, social sciences and information systems, in age clusters from 18 to 70 years old (Table 2). We admit that the sample is biased, since it is constituted by academics and researchers which are familiar with big data and analytics research. In the preliminary study, this special feature of responders was needed. We also have to declare that the generalization of the findings of this preliminary research study, should be done within the context of this limitation. Our purpose is to use the key findings of this research in order to populate a new research tool that will target broader clusters of users of big data applications.

Table 1. Age of responders.

Answer Choices.	Responses	Numerical
Under 18	0.00%	0
18–24	3.74%	4
25–34	18.69%	20
35–44	38.32%	41
45–54	26.17%	28
55–64	11.21%	12
65+	1.87%	2

Table 2. Discipline of responders.

Answer Choices	Responses	Numerical
Social Sciences	32.67%	33
Sciences (including Computer Science, Engineering)	67.33%	68
Total		101

Table 3. Awareness of big data concept by responders.

Answer Choices	Responses	Numerical
Yes	96.26%	103
No	3.74%	4
Total		107

Two thirds of the respondents were from a science discipline and one third from the social sciences. The balance is achieved since some respondents have joint expertise. Most of them were junior or experienced researchers in domains related to management, international business, social sciences and information systems. In Table 2 we summarize the relevant information.

Another key characteristic of our sample is that the vast majority or participants expressed their awareness about big data. Almost 97% of the responders claimed that they were aware of the big data phenomenon. This is important, also as a finding of our research, because big data for several communities is considered as an information system, or computer science research domain, but it seems that also social scientists are quite aware of it (Table 3).

In the next section of our survey, and in our relevant research model, we are interested in understanding the exposure of our responders to big data applications as well as their concerns and perceived value. A key motivation in our research is to understand the degree to which users of big data applications have concerns or feel ambiguity or danger in terms of trust or privacy.

4.2. Use of Big Data Applications and Perceived Value

The transparency of big data applications and their ubiquitous nature means that several times users of applications, or services that deploy big data, use them as black boxes. They do not care about the computational aspects of the information model of the application, but rather they want to enjoy the service. In our survey we discovered that 55% of applications users, have a deep knowledge of the big data applications that they use, that collect and aggregate data, which are stored in a distant server (Table 4). Also 45% of participants said that they are not users of big data applications, which also proves that currently, there are many people that do not intend to use advanced big data services. The question for sure is, how many of them dislike the use of smart cities services because they're afraid of the violation of their privacy or for other kinds of personal concerns. As a next question, we also focus on this.

Table 4. Do responders use big data services?

Answer Choices	Responses	Numerical
Yes	55.24%	58
No	44.76%	47
Total		105

Given the key findings that most of the participants of our survey are aware of big data applications and phenomenon, and also that more than half of the respondents are using big data applications extensively in their lives for various purposes, it is quite challenging to investigate, the perceived value they attach to the use of these applications and if it is relevant to their "quality of life" or just a "contribution" to their expectations and perceptions. While it is hard, even from a scientific way, or a statistical "correct" approach, to measure this value, we deployed a heuristic rule/approach. We asked our participants on a scale from 0–100 to attach a "numerical value" to the value of big data in their lives.

The result is summarized in Table 5. The average rating of all the responders for the value of big data in their lives is 66 out of 100. This numerical value seems to be overall "positive" in the sense that most responders attach a value greater than the average in the spectrum of low (0) to high (100) values. It is also evident from this value that responders seem to be skeptical about some aspects of big data. Thus, we need to understand the main concerns of users, specifically the features of big data that make them worried or concerned.

Table 5. Value of big data in the lives of the responders.

Answer Choices	Average Number	Total Number
	66	6.964
Total Respondents		106

Table 6, below, shows one of the most interesting findings of our survey, and deserves a more detailed analysis. Given the overall, rather high score (66 out of 100), for the perceived value of big data as provided by our respondents, we tried to understand some qualitative aspects of this positive effect.

Table 6. How responders perceive the value of big data. Personal ratings on the following statements.

	1 (Totally Disagree)	2 (Rather Disagree)	3 (Neutral)	4 (Somehow Agree)	5 (Fully Agree)	Weighted Average
Big Data allow me to enjoy personalized services	2.83%	2.83%	21.7%	49.06%	23.58%	**3.88**
Big Data offer me access to unique services	0.94%	3.77%	29.25%	47.17%	18.87%	**3.79**
Big Data enabled services, save time for me and effort	1.90%	2.86%	20.00%	51.43%	23.81%	**3.92**
Big Data is about social security	13.46%	29.81%	30.77%	17.31%	8.65%	**2.78**
Big data protect my privacy	27.18%	41.75%	25.24%	4.85%	0.97%	**2.11**
Big Data promote the collective intelligence and this has an impact in my daily life	3.81%	2.86%	18.10%	53.33%	21.90%	**3.87**
Big Data promote interoperability of services worldwide—thus i enjoy integrated services	1.90%	6.67%	25.71%	48.57%	17.14%	**3.72**

From the answers of our sample, we found that one of the key features *of big data applications, is that they save time and effort for their users.* Thus, developers of big data applications or designers of smart cities services must know that users would be happier to use their applications if they realized that they would be saving time. The next most important value components, according to our findings are the interoperability that big data applications offer to users, irrelevant of country or place. Users like to enjoy the same services, worldwide, with the same quality and transparency. This is also one more extremely interesting finding. If users, want to enjoy mobility, and in parallel to have access to the same services, then the big data research community and industry must promote this ecosystem of services worldwide. From the other side, this request and wish of users, needs to comply with several local and/or global policy making requirements.

Uniqueness of services is another critical value component of big data applications. Users understand that several services that are big data enabled, are unique, so they are somehow happy to use them. From this perspective, another key characteristic of big data applications is their innovative nature. Users are happy to use innovative services, that save them time in their lives, and can be enjoyed locally or globally with the same, high quality standard. This also means that big data industry should be always in a progressive, evolutionary process for the launch of novel services and innovations to the market.

Personalization, also seems to be valued by our responders. Users understand that most big data applications exploit their data for the enhancement of their personal experience. For this finding, there is also a side effect. Users recognize that big data applications challenge the protection of their personal data. Somehow a compromise between user experience and privacy is understood by our responders.

Another important finding of our survey is also that users recognize that big data applications somehow aggregate the collective intelligence of humans and potentially this could improve the quality of life. A direct interpretation of this finding is that some big data applications must certainly promote collective intelligence, but at the same time they must secure the trust, and the protection of privacy.

Finally, in Table 6, one more user perception is recorded. There is a rather neutral understanding that big data research can potentially promote also social security. With the bold debate on social media, fake news, fake profiles, social networks, analytics scandals, it seems that social security is a big theme for social security and currently users are not convinced for big data contribution.

These key findings permit several interpretations for further investigation in future studies:

- If citizens and users of big data application recognize that several big data applications save time and effort for them, then the next research question is what is the cost they are willing to pay for their use, in terms of money or indirect costs, for example partial loss of their privacy, or agreement from their side to offer their personal data under specific conditions.

- Also, if users are interested in personalization of services, the next research question is which are the clusters that categorize different users to different clusters, and how happy would they be for such personalization if greater openness and access to their personal data is required.
- The finding related to the uniqueness of services is also critical. If users recognize that some big data are unique, then the next question is how can they resist to the necessity to share their personal data with such applications. For the big data industry, how easy it is to keep a limit to the penetration of sensitive personal data for the parametrization of their services to different users features?
- The anticipation of interoperability of big data services across cultures and nations also needs to be understood in a social context. Recent examples of cool applications like FaceApp, prove the capacity of big data applications to generate spontaneously huge data bases of critical personal data e.g., faces.

In the next section we present a third level of analysis for our survey with an emphasis on the potential social impact of big data. One of the key assumptions of our research model is the following: If citizens use big data applications and if they attach a positive value to their behavior, then the next step is to understand if the individual behavior also have some social contracts and positive implications for the society. In our approach this is defined as the social impact of big data.

4.3. Social Value of Big Data

In our modern, complicated social environment, in order to respond to social challenges, we have to understand them. In our survey, with its given limitations and limited reach, we made a first effort to record the key societal challenges of our times. Given the focus of our research on the social impact of the big data, the direct connection between these concepts is the capacity of big data applications to promote bold actions or solutions to societal challenges or problems. In Table 7, we summarize the outcome of our effort. According to our survey, the top five societal challenges of our time are:

- Security
- Socially inclusive economic growth
- Access to education/quality of education
- Equal opportunities for all
- Job opportunities

Table 7. The main societal challenges of our times based on the own beliefs of responders.

Answer Choices	Responses	Numerical
security	74.77%	80
social inclusive economic growth	48.60%	52
equal opportunities to all	42.99%	46
fair justice	34.58%	37
poverty	31.78%	34
depression	14.02%	15
education quality	54.21%	58
Job opportunities	39.25%	42
happiness	27.10%	29
Total		107

If we accept, as a working hypothesis that our responders reflect a greater population, then a key point for our future research is to analyze the social impact of big data research in terms of its capacity to promote sustainable goals related to the societal challenges reflected in Table 7. For example, a direct interpretation could be the following:

- Can big data applications enhance the capacity of people to have access to high quality open education? (Can we develop a big data learning platform to offer free, open, personalized training modules to individuals that will enhance their skills and competencies?)
- Can we design big data enabled, advanced, sophisticated services that promote the feeling of security in modern societies? (e.g., can we build big data enabled antiterrorist detection systems over social media?)
- Is there a way to exploit big data research in order to promote socially inclusive economic growth by defining, for example, new markets, or new data-intensive industries or innovations?
- Can we deploy big data research in order to promote new and better jobs in our societies? Is there, for example, any possibility to "measure talents", to codify skills and competencies and to match job profiles with candidates etc.?
- Can we integrate big data research with sophisticated computational methods like artificial intelligence and machine learning in order to investigate "personalities" and personal habits that are linked to critical social challenges e.g., antiterrorism detection, harassment etc.?

We understand, and we admit that this is an extremely significant objective that goes beyond the scope and the depth of this limited survey. Nevertheless, it offers a very good starting point for further analysis and integration to a forthcoming greater research in terms of scope and coverage. It is also a good context for skepticism and interdisciplinary understanding. Somehow, big data research needs new contributions from the social sciences that have for years developed research tools and theories for understanding human behavior and personality. In our understanding in the context of the virtual world, internet and social media, there are many more things to be done in this direction.

In a next step, we tried to cross-check and to integrate the perception of our responders with one more question, related to the value and the concerns of users about big data use. Our intention, from a research point of view, is to build a theoretical framework to be tested in a future research about the connection between the social impact of big data, concerns and intention of use. This will be presented in Section 5 of our paper.

In Table 8, the findings reflect key aspects of users concerns about big data use and their associated added value.

Table 8. How responders perceive the value of big data.

	1 (Totally Disagree)	2 (Rather Disagree)	3 (Neutral)	4 (Somehow Agree)	5 (Fully Agree)	Weighted Average
Big data applications violate my privacy	3.74%	11.21%	26.17%	45.79%	13.08%	**3.53**
Big data services require advanced computer knowledge	0.93%	21.50%	12.15%	46.73%	18.69%	**3.61**
Big data services promote the gap between computer literated and non computer litarated	0.94%	18.87%	25.47%	39.62%	15.09%	**3.49**
I somehow feel that my data are used for uknown purposes	1.87%	5.61%	8.41%	37.38%	46.73%	**4.21**
Big data services do not use transparent methods for processing my data	1.87%	6.54%	14.02%	41.12%	36.45%	**4.04**
I do not like that companies use my data for getting customer insights about me	0.93%	10.28%	19.63%	28.97%	40.19%	**3.97**

The main concern is related to uncertainty. Users of big data applications have a fear that their data will be used for unknown purposes. From a policy making point of view it is an absolute requirement that regulatory and legislative frameworks provide protection to users. The feeling of our responders is that currently there is a significant gap in this area. In close relevance to this finding, users also believe that big data applications do not use transparent methods for data processing. Some initiatives, like the General Data Protection Regulation (GDPR) are headed in the right direction. Users must have the right to be informed about who uses their data, for which purposes, and under which methods. One more bold finding of our survey is that our responders are also not happy that companies use big data research in order to gain better customer insights about them. This is also another huge theme for further research, on which we will elaborate further in the conclusion of our research study.

Some interpretations of these research findings include the following:

- What is the role of governmental authorities and supervising bodies towards the design, implementation and well-functioning of big data awareness policies related to privacy and data protection?
- How can users and citizens have an increased awareness about the processing methods of their data. Is it possible for them to have access to an IT-service where all the "users" of their personal data appear and are analyzed further?
- Concerning the social impact of the big data research the key question is, is there a fair-justice approach in which social bodies or organizations can supervise and rate "behavioral" oriented big data of individuals or groups. Answers to these questions are not obvious. They need significant social agreement and consultation.
- Concerning the compromise of sharing personal data for the use of unique big data applications, another critical question is how can users resist in using unique services without letting third party organizations gain significant insights into their personalities?
- Also, if we promote it as socially-fair to offer supervising organizations access to personal data, then does the individual level of decision refer to the degree of declining such services? We have to investigate a rather increasing population of users that deny to use big data applications due to this fear.

The respondents answers to our questions related to their reluctance to offer customer insights to companies about themselves guided the next part of our survey. We need to understand how people and users of big data applications interpret the new trend in informatics and social computing about analytics research. For this purpose, we attached 3 more questions that are summarized in Tables 9–11.

Table 9. Are responders aware about the concept of data analytics?

Answer Choices	Responses	Numerical
Not at all familiar	1.87%	2
Not so familiar	9.35%	10
Somewhat familiar	30.84%	33
Very familiar	39.25%	42
Extremely familiar	18.69%	20
Total		107

Nine out of ten respondents in our sample are familiar with the analytics concept and consider analytics to be potentially beneficial for users. In a similar approach, as we did for big data research, our responders attach a value of 76 out of 100 to the value of analytics. Given the limitation of our numerical approach to value measuring, this is an indication that users consider analytics to be of greater value by ten units, than big data (76 versus 66 out of 100). This indicates that users indirectly attach and associate increased value to advance decision-making capabilities.

Table 10. What responders think about data analytics.

Answer Choices	Responses	Numerical
Not at all important	0%	0
Not so important	0.93%	1
Somewhat important	16.82%	18
Very important	50.47%	54
Extremely important	31.76%	34
Total		107

Table 11. How responders rate the value of data analytics for social purposes (related to the capacity of data analytics to provide insights to social problems or challenges).

Answer Choices	Average Number	Total Number
	76	8.172
Total Respondents		107

Given the rather high value attached by our responders, to the impact of data analytics research for social purposes, we asked our participants to clarify the key aspects of analytics research that have increased social impact. In Table 12, we present these responses. The most interesting finding is that our sample states that analytics enhance social aware services. According to our responders, the analysis and organized presentation of analytics (e.g., key performance indicators, or visual overviews of big data or advanced data mining methods) can lead significant socially aware responses to social problems. An interpretation of this finding is that designers of IT/IS services and social scientists have to collaborate to deliver fully functional big data and analytics platforms for social issues and problems.

Table 12. How responders perceive the value of data analytics for social impact.

	1 (Totally Disagree)	2 (Rather Disagree)	3 (Neutral)	4 (Somehow Agree)	5 (Fully Agree)	Weighted Average
Data Analytics provide usefull insights for understanding critical social problems/challenges	0.00%	1.89%	10.38%	58.49%	29.25%	**4.15**
Data Analytics techniques enable social-aware services	0.00%	2.83%	14.15%	59.43%	23.58%	**4.04**
Data Analytics promote solutions to social problems	0.95%	4.76%	28.57%	53.33%	12.38%	**3.71**
Data Analytics enhance social responsibility	1.89%	7.55%	41.51%	38.68%	10.38%	**3.48**
Data Analytics support civic engagement	0.00%	7.55%	40.57%	42.45%	9.43%	**3.54**
Data Analytics help the recognition and monitoring of new social problems	0.94%	3.77%	14.15%	55.66%	25.47%	**4.01**
Data analytics promote Social coherence	0.00%	10.38%	43.40%	37.74%	8.49%	**3.44**

In close relevance, our survey also concludes that based on the responses of participants, analytics can also help to clarify novel social problems and issues, not easily diagnosed with other methods. This finding can initiate, for sure, a social dialogue and a policy-driven open participatory procedure.

While our responders recognize the potential social impact of analytics research, more than one third (36%) of them still state that they are still not ready for sharing their data for social aware services

(Table 13). This proves that there is a great distance to be covered until a new era of social aware services commences. From the other side, the combined big data and analytics research and integration for social purposes increases the perceived value to 70 points (Table 14). Finally, 98% of the responders are confident that in the next few years social aware big data services will be a key trend (Table 15). This proves that most users nowadays feel that somehow their data are processed for such purposes and that soon they will be forced, or they will be happier to use such services.

Table 13. Would respondents be happy to share their data in order to promote social aware big data services?

Answer Choices	Responses	Numerical
Yes	63.21%	67
No	36.79%	39
Total		106

Table 14. Responders rating of the impact of big data and analytics research to society (negative to positive).

Answer Choices	Average Number	Total Number
	70	7.360
Total Respondents		105

Table 15. Do respondents think that big data driven services for social impact will be a trend in the near future?

Answer Choices	Responses	Numerical
Yes	96.11%	104
No	1.89%	2
Total		106

5. Key Implications and Future Research Directions

The research problem under study, is complicated. The contribution of our research, within the given limitations, is significant: It provides interesting insights for the perception of big data users for the added value of these services to their lives and the potential social impact. We summarize, in the next paragraphs, the key response of our study to the research objectives we set out to address. We also elaborate on future research directions and policy making implications related to sustainability.

Research objective 1: To understand the degree of awareness of big data research and the actual use of big data applications by our responders.

- Key finding 1: Based on the qualitative analysis and interpretation of respondents' answers, we conclude that most responders, irrelevant if they are from a sciences or social sciences background, are aware of big data research. Of them, 55% also consciously use big data services due to unique features.
- Further research direction 1: It is interesting to investigate the people that do not use big data services, and under which conditions they would use big data applications for social purposes.
- Policy making and sustainability implication: The development of a policy aware regulation framework to deal with affordances and concerns of users of big data services.

Research objective 2: To analyze, the perceived value of big data as it is realized by our responders.

- Key finding 2: The majority of responders recognize the added value of big data services to their lives. They understand and value that applications save them time, and improve their life in some respects. Furthermore, they recognize that most big data services offer unique services that can be enjoyed with the compromise of sharing personal data. Additionally, they feel that somehow big data applications aggregate collective wisdom and thus they improve the quality of their lives.

- Further research direction 2: The perceptions of individuals about the added value of big data services and big data research needs to be codified. An ontological approach to value components of big data should be promoted further. This will lead to a new measurement approach capable of codifying various aspects of value related to personality, ethics, social responsibility, social justice, open democracy etc.
- Policy making and sustainability implication 2: The evolution of big data research, from a policy making perspective needs to set all the required regulations about trust, transparency, security, anonymity as well as to protect the digital self of individuals. From a social perspective, it is nevessary to integrate social insights in a well-defined context for socially inclusive economic and social growth. We need policies that will enhance the reusability and exploitation of big data for social purposes.

Research objective 3: To understand the main concerns of users of big data services, in regards to their use, as well as their affordances towards greater deployment of such applications.

- Key finding 3: Most of the responders recognize that there are three critical obstacles to big data applications: The unknown information processing methods of their data, and their purposes, as well as the fact that most of the time they consider that the ultimate destination of big data research is to gain more customer insights about them.
- Further research direction 3: Given the high degree of concerns of individuals about the use of their data, it is required to understand, through further research, the compromise between the shift of their willingness to share their data and to enjoy better services that improve the quality of their lives.
- Policy making and sustainability implication: One of the most significant directions for the future would be to develop policies that will protect individuals from being the victims of extreme third-party insights about their habits, preferences and social behavior. In a next step, there must be a sustainability regulation, through which all big data services should comply with general requirements and this must also be included in inform consent statements of user agreements.

Research objective 4: To clarify the impact of big data in the lives of individuals and to their perceptions.

- Key finding 4: The overall understanding of this survey is, that users have a positive impression of the value of big data research in their lives, and a rather positive perception. There is, however, a critical mass of users that are not willing to use big data services due to their concerns and fear that their data will be used for unknown purposes. The social impact of big data is also perceived as a potential under specific requirements.
- Further research direction 4: The measurement of the individual and social impact of big data will require advanced heuristic methods, and a lot of support is needed from the social sciences to adopt psychological, personality, mental, cognitive and other models to incorporate metrics and key performance indicators. It is also important to adopt studies and social research about the impact of big data applications in addiction or other psychometric directions.
- Policy making and sustainability implication: We need to develop more advanced social policies and regulations for the interconnection of information and communication technology with social sciences. We must recognize that all the modern big data applications are setting a huge socio-technical system where humans are actors and their interactions are more complicated than ever.

6. Conclusions

This paper calls for a revisit into the insights related to the use of big data research from individuals. The main point of departure is that users have an intrinsic motivation to protect their privacy and ownership of their data, but from the other side they are also happy to use unique services that improve the quality of their lives. In this required compromise, the contribution of our research is that it provides evidence from an empirical survey with objective interpretation of findings.

The research, research findings and discussion presented in this paper seek to initiate a debate on the diffusion and impact of big data on our lives. The ultimate contribution of this research work is the agreement of participants that the added value of big data towards responsive social aware services to critical social problems, is just the beginning of the journey and not its end. In the next few years this line of research will have a great impact on the development of a brand new research area related to social aware big data enabled social services and analytics.

Author Contributions: M.D.L. and A.V. contributed equally to the design, implementation, research and analysis of data and the delivery of the main findings. Both coauthors contributed equally to all phases of this research.

Acknowledgments: The authors would like to thank Effat University in Jeddah, Saudi Arabia for funding the research reported in this paper through the Research and Consultancy Institute.

Conflicts of Interest: The authors declare no conflict of interest.

Appendix A. The Pilot Survey—Questionnaire

The research tool-questionnaire for our survey is available through this web link: https://www.surveymonkey.com/r/LY2R3XV.

References

1. Lytras, M.D.; Raghavan, V.; Damiani, E. Big-data and data analytics research: From metaphors to value space for collective wisdom in human decision making and smart machines. *Int. J. Semant. Web Inf. Syst.* **2017**, *13*, 1–10. [CrossRef]
2. Lytras, M.D.; Aljohani, N.R.; Hussain, A.; Luo, J.; Zhang, X.Z. Cognitive Computing Track Chairs' Welcome & Organization. In Proceedings of the Companion of the Web Conference, Lyon, France, 23–27 April 2018.
3. Lytras, M.D.; Mathkour, H.I.; Abdalla, H.; Al-Halabi, W.; Yanez-Marquez, C.; Siqueira, S.W.M. Enabling technologies and business infrastructures for next generation social media: Big-data, cloud computing, internet of things and virtual reality. *J. Univers. Comput. Sci.* **2015**, *21*, 1379–1384.
4. Lytras, M.D.; Mathkour, H.I.; Abdalla, H.; Al-Halabi, W.; Yanez-Marquez, C.; Siqueira, S.W.M. An emerging—Social and emerging computing enabled philosophical paradigm for collaborative learning systems: Toward high effective next generation learning systems for the knowledge society. *Comput. Hum. Behav.* **2015**, *5*, 557–561. [CrossRef]
5. Arnaboldi, M. The Missing Variable in Big Data for Social Sciences: The Decision-Maker. *Sustainability* **2018**, *10*, 3415. [CrossRef]
6. Wamba, S.F.; Akter, S.; Edwards, A.; Chopin, G.; Gnanzou, D. How 'big data' can make big impact: Findings from a systematic review and a longitudinal case study. *Int. J. Prod. Econ.* **2015**, *165*, 234–246. [CrossRef]
7. Calzada, I. (Smart) Citizens from Data Providers to Decision-Makers? The Case Study of Barcelona. *Sustainability* **2018**, *10*, 3252. [CrossRef]
8. Edwards, L. Privacy, Security and Data Protection in Smart Cities: A Critical EU Law Perspective. *Eur. Data Prot. Law Rev.* **2016**, *2*, 28–58. Available online: https://papers.ssrn.com/sol3/papers.cfm?abstract_id=2711290 (accessed on 17 August 2019). [CrossRef]
9. Guille, A.; Hacid, H.; Favre, C.; Zighed, D.A. Information diffusion in online social networks: A survey. *Sigmod. Rec.* **2013**, *42*, 17–28. [CrossRef]
10. Grabowicz, P.A.; Ramasco, J.J.; Moro, E.; Pujol, J.M.; Eguiluz, V.M. Social features of online networks: The strength of intermediary ties in online social media. *PLoS ONE* **2012**, *7*, e29358. [CrossRef]
11. Maynard, D.; Gossen, G.; Funk, A.; Fisichella, M. Should I care about your opinion? Detection of opinion interestingness and dynamics in social media. *Future Internet* **2014**, *6*, 457–481. [CrossRef]

12. Liu, B. *Sentiment Analysis and Opinion Mining*; Morgan & Claypool Publishers: San Rafael, CA, USA, 2012.
13. Olszak, C.M.; Mach-Król, M. A Conceptual Framework for Assessing an Organization's Readiness to Adopt Big Data. *Sustainability* **2018**, *10*, 3734. [CrossRef]
14. McAfee, A.; Brynjolfsson, E. Big Data: The Management Revolution. *Harv. Bus. Rev.* **2012**, *90*, 60–668. [PubMed]
15. Mikalef, P.; Pappas, I.O.; Krogstie, J.; Giannakos, M. Big data analytics capabilities: A systematic literature review and research agenda. *Inf. Syst. e-Bus. Manag* **2017**, 1–32. [CrossRef]
16. Kitchin, R. The Real-Time City? Big Data and Smart Urbanism. *GeoJournal* **2014**, *79*, 1–14. [CrossRef]
17. Kent, P.; Kulkarni, R.; Sglavo, U. Finding Big Value in Big Data: Unlocking the Power of High Performance Analytics. In *Big Data and Business Analytics*; Liebowitz, J., Ed.; CRC Press Taylor & Francis Group, LLC: Boca Raton, FL, USA, 2013; pp. 87–102. ISBN 9781466565784.
18. Davenport, T. Big Data at Work. In *Dispelling the Myths, Uncovering the Opportunities*; Harvard Business Review Press: Boston, MA, USA, 2014; ISBN 9781422168165.
19. Boyd, D.; Crawford, K. Critical questions for big data: Provocations for a cultural, technological, and scholarly phenomenon. *Inf. Commun. Soc.* **2012**, *15*, 662–679. [CrossRef]
20. Edelenbos, J.; Hirzalla, F.; van Zoonen, L.; van Dalen, J.; Bouma, G.; Slob, A.; Woestenburg, A. Governing the Complexity of Smart Data Cities: Setting a Research Agenda. In *Smart Technologies for Smart Governments. Public Administration and Information Technology*; Rodríguez Bolívar, M., Ed.; Springer: Cham, Switzerland, 2018; Volume 24, pp. 35–54. ISBN 978-3-319-58576-5.
21. Kharrazi, A.; Qin, H.; Zhang, Y. Urban big data and sustainable development goals: Challenges and opportunities. *Sustainability* **2016**, *8*, 1293. [CrossRef]
22. Wielki, J. The Opportunities and Challenges Connected with Implementation of the Big Data Concept. In *Advances in ICT for Business, Industry and Public Sector*; Mach-Król, M., Olszak, C.M., Pełech-Pilichowski, T., Eds.; Springer: Cham, Switzerland, 2015; pp. 171–189, ISBN 978-3-319-11327-2.
23. Spaletto, J. An Investigation of Strategies for Managing Exponential Data Growth in the Enterprise. *J. Leadersh. Org. Eff.* **2013**, *1*, 4–14.
24. Alkmanash, E.H.; Jussila, J.J.; Lytras, M.D.; Visvizi, A. Annotation of Smart Cities Twitter Microcontents for Enhanced Citizen's Engagement. *IEEE Access* **2019**, *7*, 116267–116276. [CrossRef]
25. Angelidou, M.; Psaltoglou, A.; Komninos, N.; Kakderi, C.; Tsarchopoulos, P.; Panori, A. Enhancing sustainable urban development through smart city applications. *J. Sci. Technol. Policy Manag.* **2017**. [CrossRef]
26. Bi, S.; Liu, Z.; Usman, K. The influence of online information on investing decisions of reward-based crowdfunding. *J. Bus. Res.* **2017**, *71*, 10–18. [CrossRef]
27. Yin, C.; Xiong, Z.; Chen, H.; Wang, J.; Cooper, D.; David, B. A literature survey on smart cities. *Sci. China Inf. Sci.* **2015**, *58*, 1–18. [CrossRef]
28. Visvizi, A.; Lytras, M.D. Editorial: Policy Making for Smart Cities: Innovation and Social Inclusive Economic Growth for Sustainability. *J. Sci. Technol. Policy Mak.* **2018**, *9*, 1–10.
29. Visvizi, A.; Lytras, M.D. Rescaling and refocusing smart cities research: From mega cities to smart villages. *J. Sci. Technol. Policy Mak.* **2018**. [CrossRef]
30. Van de Voorde, T.; Jacquet, W.; Canters, F. Mapping form and function in urban areas: An approach based on urban metrics and continuous impervious surface data. *Landsc. Urban Plan.* **2011**, *102*, 143–155. [CrossRef]
31. Carrasco-Sáez, J.L.; Careaga Butter, M.; Badilla-Quintana, M.G. The New Pyramid of Needs for the Digital Citizen: A Transition towards Smart Human Cities. *Sustainability* **2017**, *9*, 2258. [CrossRef]
32. Smart Cities Research: Contemporary Issues in Smart Cities Research. 2018. Available online: https://docs.google.com/forms/d/e/1FAIpQLSciGvQE78AWnUcHqt4Q_MLFC6FT52UQskZSSNlZzOBJNQTdkA/viewform (accessed on 10 June 2018).

 *sustainability*

Article

The Missing Variable in Big Data for Social Sciences: The Decision-Maker

Michela Arnaboldi

Department of Management, Economics and Industrial Engineering, Politecnico di Milano, 20156 Milano, Italy; michela.arnaboldi@polimi.it; Tel.: +39-2-23994069

Received: 23 August 2018; Accepted: 19 September 2018; Published: 25 September 2018

Abstract: The value of big data for social sciences and social impact is professed to be high. This potential value is related, however, to the capacity of using extracted information in decision-making. In all of this, one important point has been overlooked: when "humans" retain a role in the decision-making process, the value of information is no longer an objective feature but depends on the knowledge and mindset of end users. A new big data cycle has been proposed in this paper, where the decision-maker is placed at the centre of the process. The proposed cycle is tested through two cases and, as a result of the suggested approach, two operations—filtering and framing—which are routinely carried out independently by scientists and end users in an unconscious manner, become clear and transparent. The result is a new cycle where four dimensions guide the interactions for creating value.

Keywords: big data; social sciences; decision-making; data analyst; filtering; framing

---

1. Introduction

Big Data (BD) has gained wide attention since the first introduction of the definitional statement in which they were described as "high-volume, high-velocity and high-variety information assets that demand cost-effective, innovative forms of information processing for enhanced insight and decision making" [1]. This seminal definition gave BD its signature "V", originally the three Vs of Volume, to indicate the vast amount of data in play, Velocity, to indicate the rate of data generation and possible successive processing and Variety, to indicate the many data types and sources. Further "Vs" have been since be added: Veracity, to spell out the need for carefully scrutinizing the reliability of data, and Value, which can be seen as the ultimate goal when BD are used in social science [2–4]. Value is understood as the capability of BD to generate insights that can benefit decision-makers, organizations, policy-makers and other end users [2–4].

To date, an enormous number of studies have been carried out on BD [2,5], but their interest for social science is more recent [4]. The available studies offer, on the one hand, experimentations in the use of BD [6], social media, in particular, to generate Value, without involving decision-makers [7]. On the other hand, several previous works of research have pointed out that BD has the potential to generate Value for business and also to tackle societal challenges [8–10]. Further to this, an issue that has been overlooked in previous studies is that this generation of Value is linked to the way BD is used by people working inside organizations and governments. When decisions continue to be made by "humans", the value and use of the information is dependent on the characteristics and knowledge of the decision-maker [11]. This problem is not new or specifically pertinent to BD (see, for example, [12]), hence it is surprising that it has been neglected so far.

The path towards generating knowledge for decision-makers and then Value is clearly more complex in the BD arena. Many scientists from different disciplines are involved in elaborating BD in order to acquire, analyze, model and visualize data. These scientists often make arbitrary decisions [4], for example about what material to select from the entire world of social media [6] or how to model the information [13]. Without the involvement of decision-makers—who, it must be remembered, are the final users of the data—there is the high risk of creating a mismatch between the needs of the users and the information provided by the data scientists. What happens is that "obscure" areas in BD processing are created and, as a consequence, two contradictory situations can arise. In the first case, "slave" decision-makers blindly follow the indications proposed by algorithms without mastering the numbers, while in the second case, "reluctant" decision-makers [14] ignore the information extrapolated from BD entirely. Both situations are problematic and can result in sub-optimal decisions being made.

In this paper, the author argues that a new BD cycle is urgently needed to achieve social impact and, at its centre, is the true missing variable for pursuing Value: the decision-maker. When "humans" retain a role in the decision-making process, the value of information is, accordingly, no longer objective but is influenced by the end users' own knowledge and mental outlook. More specifically the research question addressed in this paper is how decision-makers can be put at the centre of the BD cycle and become the point of reference for knowledge generation? To pursue this goal, two processes that are usually carried out implicitly by data scientists—filtering and framing—become clear-cut and explicit. The new approach was tested out in two cases of wider social interest: City and Art. In the City case, the key actor (CityEx) wanted to have new insight from BD, to stimulate the public debate during the mayor election campaign. In the Art case, a cultural foundation wanted to explore its network and its role in the local scene. Both cases are located in the same major city in northern Italy.

To describe these arguments, the paper is articulated as follows: the next section sets out the methodology; section three outlines the results. The discussion is lastly presented and conclusions are drawn.

2. Methodology

This section covers the methodology employed in the study, first introducing the conceptual perspective and introducing the decision-maker centric approach and then, in the following sub-section, describing the empirical strategy.

2.1. Conceptual Perspective

Decision-makers in this paper are considered as private or public actors who make decisions and act within their social and institutional roles. This definition excludes the individual sphere of decision-making, such as purchasing items for personal needs. The relationship between the value of information and user knowledge was recognized before BD arrived on the scene (e.g., [12]). The common starting point, which remains the same as in the past, is that knowledge is subjective and the same information can be easily understood differently by two distinct decision-makers on the basis of their personal background and experience [13]. Having set out the above, attention is drawn to two processes that have acquired particular importance in the era of BD and can influence the way new knowledge is used, learnt and generated, although the decision-makers are often only involved in the initial and/or final stages. These two processes are filtering and framing, and they flow through in all phases of the BD chain (see Figure 1).

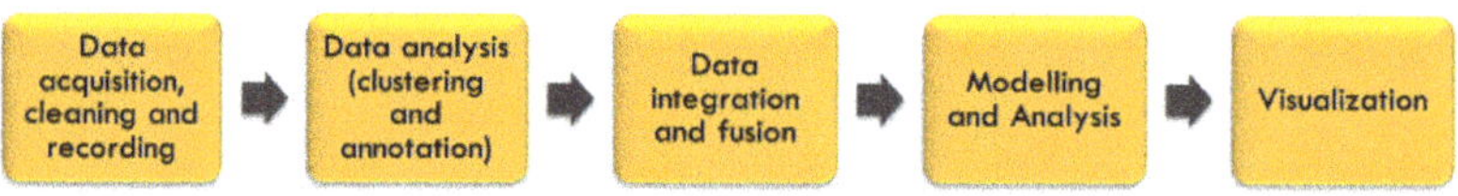

Figure 1. Phases of the BD chain.

The first process, filtering, involves the selection of data and the relative attributes and features ingrained within the mechanisms for transforming data first into information and then into knowledge that can be used by decision-makers. From the stage of data acquisition to that of data visualization, data scientists make decisions that will reduce the mass of data processed and ensure that the data selected can be easily managed and understood. One of the most obvious—and most discussed—cases of filtering concerns the collecting and analyzing of data from social media [15–17]. Whether this information is to be used for sentiment analysis or network analysis, subjective decisions come into play that impact on the type of data acquired and the kind of attributes gathered. For instance, collecting data through key words implies including or excluding a certain number of social media posts, and therefore shaping the space for decision-making in different ways. While filtering is less visible, it is nonetheless present, and not only in the data collection process. When integrating and merging data, data scientists must make decisions that affect the way temporal and spatial references are aligned [18,19]. For example, when data have a different geographical granularity, data scientists must decide either to aggregate fine-grained data or to disaggregate higher-level data. Algorithms may very well be highly sophisticated, but there will always be choices that affect the final results.

The second process, framing, is the strategy used to contextualize and communicate data so that they can be understood, trusted and enacted upon. This second process is rooted in the awareness that data do not exist "per se" and do not have an objective meaning. As highlighted by Gitelman ([11], p. 3), "data need to be imagined as data to exist and function as such, and the imagination of data entails an interpretive base". Within the broader phase of framing, visualization is the process that has been studied the most widely, starting with the seminal contribution by Frankel and Reid [20], who claimed that the role of designers is to stimulate novel thinking patterns. Since then, other studies have highlighted how different visual frames can influence the understanding of experts and businesses (e.g., [21,22]. Framing comes into action from the data collection phase onwards, to decide which data and attributes enable better decision-making and present the best annotation [23].

With the final aim of enhancing the value of BD for users, this study is proposing and evaluating a BD cycle centred on the decision-makers, where the two processes, filtering and framing, are investigated and explained clearly. Following the scheme set out in Figure 2, decision-makers are placed at the centre of the proposition and become the yardstick for the entire BD process. This, in turn, consists of five phases. The first three phases are connected to data management, and involve the operations for acquiring, storing and setting data that will be used in subsequent analyses. This modelling and analysis phase is the cornerstone of analytics, and underpins the process of presenting the data so that they can be understood and used effectively by the people making decisions.

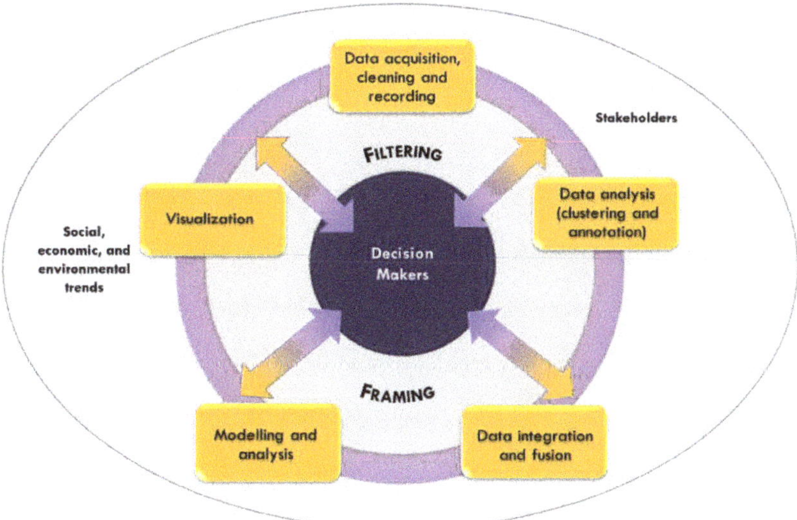

Figure 2. The new BD process: decision-makers at the centre.

According to this innovative approach, where decision-makers play a central role, the new cycle is placed within a wider context, one in which social, economic and environmental trends all need to be taken into consideration together with the relevant stakeholders.

2.2. Empirical Strategy

At the empirical level, the BD cycle introduced in the previous section was tested out in two cases through action research methodology [24–26]. The distinctive feature of this approach is that it relies on researchers and the decision-makers defining a joint work and research framework to address both practical concerns and academic problems. After conducting a preliminary screening of the possibilities and presenting an initial proposal to five decision-makers, two cases—City and Art—were chosen and, more specifically, two expert decision-makers, CityEx and ArtEx. The selection was based on a series of parameters: whether the decision-makers were available and ready to dedicate the necessary time to the interactive cycle; whether it was possible to test the new data within their sphere of decisions/actions; and their initial attitude towards BD. Both experts were willing to test (big) data, meaning that they were not "reluctant"; they did not put blind faith in the process, however, in part because of their long-time experience in making decisions on the basis of data and analytics, meaning that they had the critical mindset needed to test the cycle.

The City case deals with the construction of a new digital monitoring system with the very specific aim of studying Milan's own brand of internationalism, "In what way is Milan international?" (i.e., how does it express its inner international self); CityEx belongs to a non-profit association that works towards creating value for the city of Milan. CityEx came from a background in public management with expertise in accounting; he had just begun to approach BD at the time when the project started. Although he was a novice in the field of BD, he was incredibly curious and ready to read practitioner and academic material in order to gain a better understanding of the potential of BD and social media. CityEx wished to use data to stimulate public debate, but also to gain insights that could be applied to projects under discussion in specific areas of Milan. The City case ran between February 2014 and July 2016.

The second case, Art, is a performing arts organization and is considered one of the most important cultural institutions in Italy, boasting a superb reputation worldwide. ArtEx is the Head of Marketing and Communications at Art and he started by asking about Art's followers on social media and whether there are particular patterns or characteristics in their online behaviour. ArtEx's background is in economics and he has always worked in marketing for cultural organizations. He fully believed in digitalization and even before taking part in this project was very keen to promote it both in Art and in his previous organizations. He was particularly interested in social media, seeing them as the means to reach Art's audience when it was not actually present in the performance venue.

The overall project started in April 2016 and concluded in November 2017. The author of this paper was involved in both cases as a researcher and the coordinator of an interdisciplinary group with skills in statistics, science computing, design and management. More specifically, the City project team consisted of two science computing experts, two statisticians, two designers and two management researchers, with a total of eight people working on the project. The Art project team consisted of one science computing expert, one statistician, two management researchers and one designer, with a total of five people working on the project.

The same protocol was adopted in both cases, following action research methodology [24–26]. The cyclical process undertaken was articulated into three stages: diagnosis and planning, action taking, and evaluation and learning outcomes. The diagnosis and planning process was based upon two types of empirical methods. After initially defining the problem with the relevant expert, the formal interviews were carried out in order to learn more about the environment and context in which the expert operated. Secondary sources were then collected in order to triangulate the data collected through the formal interviews. The planning phase was carried out by the project coordinator in both cases and shared first with the data scientist team and then with the relevant expert. The action taking process was the central part of both cases and consisted of three types of interaction: (1) plenary meetings involving the decision-makers and the entire research group; (2) restricted meetings between the project coordinator and the decision-makers; (3) internal research group meetings. Overall, more than forty meetings were held for both projects to complete the BD cycle and ten meetings to revise the cycle after using the information. The final step of the action research cycle is evaluation and learning. During this step the project coordinator summarized the findings at both theoretical and managerial level. The summary took the form of presentations and reports that were first discussed with the scientist team, then revised and finally presented to the expert in each case. This cycle was reiterated several times, although the first cycle was the lengthiest in both cases.

3. Results and Analysis

This section illustrates the results of testing the new scheme in City and Art, the two cases presented in the methodology. The findings are presented according to the phases of the scheme. Please note that as the two experts involved, CityEx and ArtEx, are both men, for the sake of simplicity it was decided to refer to them using the pronouns he/him.

3.1. The City Case

The first case was set around the main theme, that is, to study in what way Milan is international. The goal of CityEx was to analyze this topic in an innovative way through digital sources. CityEx's expectations were that BD would provide more data about Milan and more quickly. After an initial meeting to share problems and objectives, the research team started on the process of acquiring data. The results were classified according to the different phases, but the decision was taken to focus selectively on the areas where filtering and framing were more visible.

3.1.1. Data Acquisition

Data were acquired from three main sources: data from social media (Twitter, Foursquare and Instagram), mobile phone data and traditional data (census data and open certified data from official sources). The decision to choose such data was aligned with CityEx's desire to pursue the three "Vs" —Volume, Velocity and Variety—with respect to the data. Filtering was evident in this first phase, especially when relating to the rules applied to acquire social media data. Focusing on Twitter initially, two types of search were carried out, in both cases using the public API (application programming interface) provided by Twitter. The first search was based on key words, searching very broadly for "Milano/Milan" in the Twitter content. The second search examined geo-tagged tweets within the boundaries of the city of Milan. The entire payload of each tweet was downloaded, including the main body of the text and a large set of metadata, consisting of geographical data, tags, user mentions, images, links, time of tweet and language.

The tweets retrieved went through an initial cleaning process to exclude inappropriate tweets, and the data were then presented to CityEx together with a basic quantitative and content analysis. This first interaction was crucial to activate a discussion about the boundaries of the search and the meaning of the question of "in what way is Milan international". It was soon evident that the question had a two-fold meaning, which had not been considered at the start. In one sense, the question is observed from a local perspective, one that is visible through the geo-tagged tweets of Milan, in the another, it is seen from a global viewpoint, where key words are used more openly to carry out searches. The data, although very raw, helped the decision-maker to realize that his organization was more interested in the digital layer embedded in the physical city, hence, in the tweets geo-tagged within the boundaries of the city of Milan. The same rule applied to Instagram and Foursquare.

The process of framing started also to be a matter of discussion at this early point, in terms of how data were initially visualized. At this stage, the visualization of data followed the two strategies for data acquisition: words and geography. CityEx did not feel that the first view, based on words, provided enough information. The geographical view was, instead, considered of value but not clear enough. In the first visualization, data (social media and mobile phone data) were located with reference to the precise longitude and latitude available for each tweet. Despite the pure amount of data, CityEx was unable to come up with any original or constructive ideas and, during a subsequent meeting, it became clear that a less detailed unit of analysis was needed to anchor the data. CityEx suggested that the various districts of Milan could provide a good reference unit. This unit was considered relevant by city managers, as it was already the reference for several traditional sets of data, and was also coherent with the need for greater detail while being aggregated enough to take decisions. From this point on, the district was framed as the main unit of analysis, guiding the other phases.

3.1.2. Data Clustering, Annotation and Merging

Starting from the main filter that had been selected (data referring to the physical city) and main framing (Milan districts), CityEx requested the research team to cluster data in such a way as to extract innovative insights and ideas. An example of this process was the analysis of Foursquare data. Foursquare is a social networking app employed by users to recommend particular venues (arts and commercial events, etc.), sharing their location with friends, via the app's "check in" function. Users select venues from a list that can be updated. The venues are all classified by group. The basic unit of analysis was represented by the downloaded "check-ins" geo-tagged within the city of Milan. The dataset downloaded for the initial cluster analysis consisted of 301,770 rows or observations featuring four variables: venue_name, the name of the venue proposed by Foursquare; category_name, the name of the venue category proposed by Foursquare; month, the month taken in the analysis as ranging from the first day of the month to the last; check-ins, the number of check-ins concerning the venue and the month; nil_name, the name chosen by the city of Milan to identify the district. The first three variables were downloaded directly from the Foursquare API, while the fourth, constructed to identify the Milan districts, was based upon the tweet's geo-tag. The dataset was pre-processed and transferred to a matrix structure, where the rows correspond to different districts and the columns correspond to the single categories. After a detailed analysis, the data scientists decided to apply a hierarchical algorithm involving Euclidean's distance and Ward's method. The output of this procedure was a dendrogram (see Figure 3), with the specific goal of reducing the 234 existing Foursquare categories.

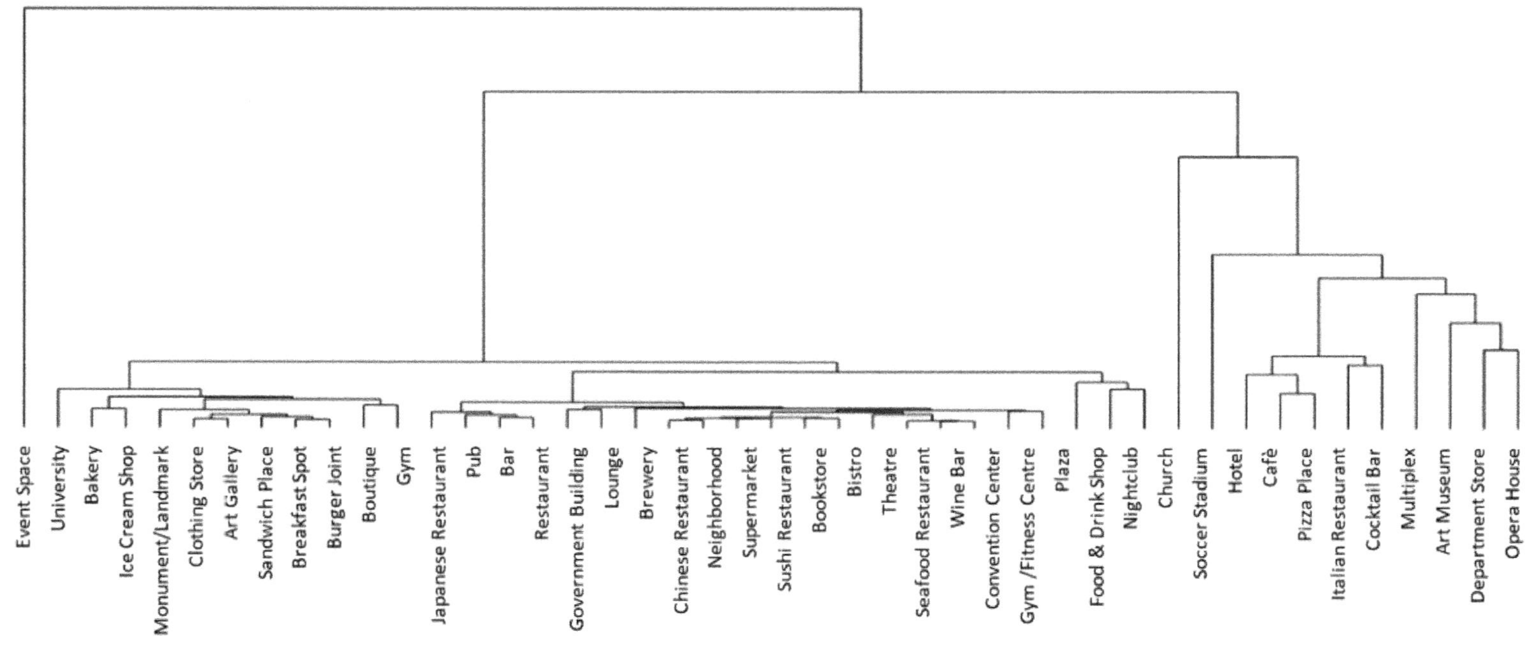

Figure 3. The dendrogram derived from the clustering.

When this first analysis was presented to CityEx, he was not convinced about the results since some of the clusters were too dense, combining categories that he wanted to see displayed separately, while other clusters contained too narrow a spectrum of categories. The final system of classification came out of discussions between analysts and experts, and CityEx proposed combining or separating clusters according to the categories typically used by city managers. The analysis of the clusters and granular data about the users gave rise to another problem, that of privacy. Individual data and messages were visible when small clusters were analyzed. Although such data are in the public domain, this situation led to some serious considerations about the ethical boundaries of the analysis, and the final decision was to exclude several categories containing too little highly-detailed data.

3.1.3. Modelling and Analysis

The modelling phases started during the clustering process, where the research group and CityEx started to rack their brains about how the data was to be used. Once placed within an appropriate visualization frame, the acquired data and clustering provide good support in terms of monitoring the phenomenon. However, the elements of Volume, Velocity and Variety of data left several of CityEx's questions about decision-making variables unanswered. A good example on this point is the density, within the various districts, of the language used in the tweets. CityEx was interested in analyzing whether there were any districts where the prevailing language was other than Italian or English. His interest derived from the assumption that the languages used on Twitter reflect the users' countries of origin, and one consequence of this is that the digital layer can provide a "weak signal" about the density of non-native Italians within a given district in a timely and evolving manner.

Starting from this initial framing, the analysts had to address an issue with implications for the entire decision-making cycle, that of the temporal framing of the data. The analysts made several considerations about the data's statistical significance. Fisher's exact test was applied to assess the stochastic independence between the districts and the language for each time unit candidate (i.e., months, two-month periods, quarters...). In this first analysis, each tweet was considered as an instance of two categorical random variables, the district and the language of the tweet. The iterative procedure clearly indicated that a unit equivalent to a quarter of a year was the optimal aggregation rate. In addition, the data had to be framed from a temporal viewpoint, meaning that a further filter had to be applied during the modelling process in order for it to attain statistical significance. In particular, several districts were excluded when the sample size was too small to draw statistically significant conclusions.

When this analysis was proposed, CityEx was surprisingly delighted by the quarterly temporal resolution, explaining that he sees very little need for data of this ilk to undertake action in real time, but that the data are useful for carrying out periodical analyses of the city's evolution. A resolution over the time span of a quarter of a year was considered sufficiently accurate to analyze this variable. CityEx was, however, more disappointed about the exclusion of some districts, as this affected the completeness of data compared to the overall frame (the geography of Milan), but no better solutions were found.

3.1.4. Visualization

In the decision-maker centric approach, the phases of the cycle overlap. This is clearly visible in the visualization of data. The author's close and frequent interaction with CityEx meant that it was possible to progressively identify his interests and his preferences about how data was visualized. For example, when analyzing the Foursquare data, discussions with CityEx brought up two different business interests. The first was to rank popular venues in Milan with the aim of observing the trends in attractiveness and visitor numbers for each category identified, these being monuments, hotels and such like (see Figure 3). The geographical framing was retained, on the right-hand side, but the central view was placed in a table format to show how the ranking of venues changes (see, for example, Duomo di Milano in Figure 4).

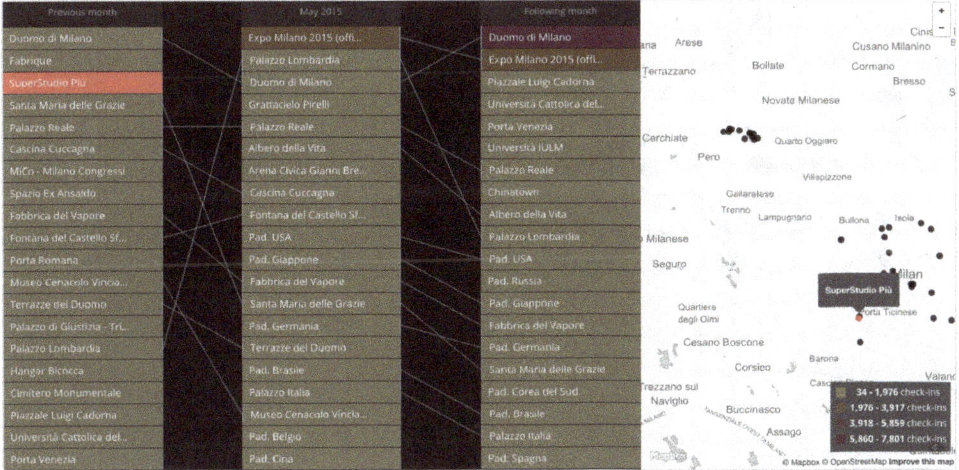

Figure 4. The visualization of leading venues from Foursquare data.

The second point of interest was the "density" and variability of interest around the city. This was constructed through a heat map linked to geographical position, and it highlighted the "hot" venues for different periods. Figure 5 shows a dark-red dot in the top-left-hand corner of the map, which came up in 2015, when Expo 2015 was running. This dot then disappeared almost totally after the Expo closed in November 2015.

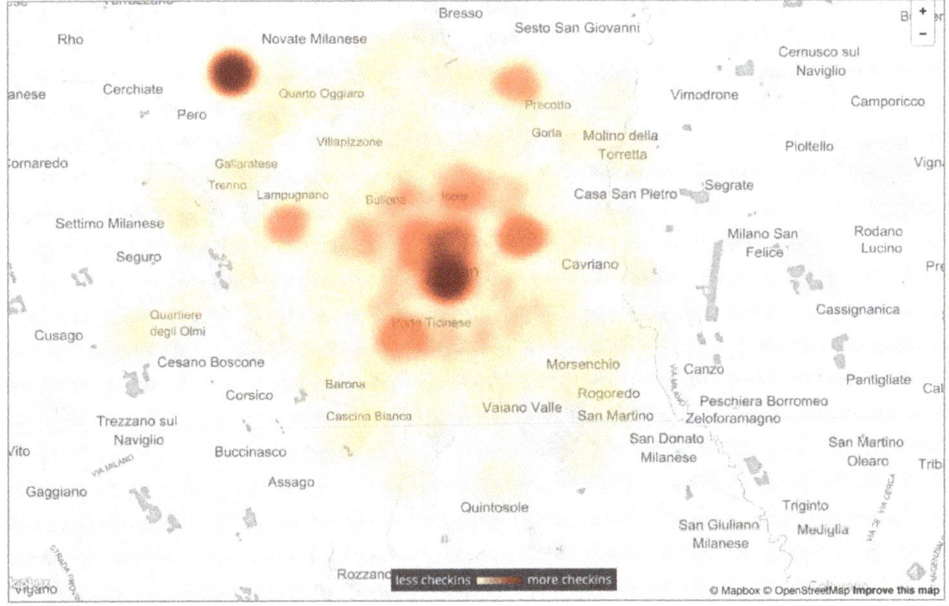

Figure 5. Heat map and Expo 2015.

The filtering and framing for these visualizations are mostly rooted in the unit of analysis and the variables previously identified. A single issue was by and large addressed here: the functions that the decision-maker wanted to apply in order to surf through the data. An example of this is the visualization of telephone calls, where CityEx asked for filters on gender, geographical area—for economic aggregation purposes—and individual nationalities (Figure 6). This was considered important for the purpose of monitoring data according to different parameters, without losing the wider picture.

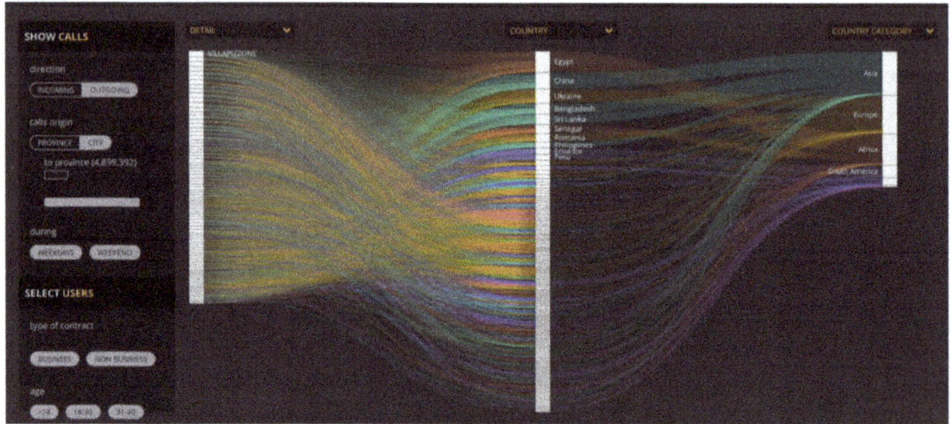

Figure 6. Filtering telephone calls: visualization.

The research group also suggested that they should develop several predictive functions, but the main interest of CityEx lay in the monitoring and communication data.

3.2. The Art Case

Our second case to investigate how decision-makers affect the BD cycle emerged, from a more specific question, in the form of "who are the followers of Art on social media and are there particular patterns in their digital traces?". Following the same procedure for City, the results for Art are set out according to the same phases, but here the decision was taken to focus selectively on issues where the filtering and framing were more visible.

3.2.1. Data Acquisition

At the beginning, ArtEx wanted to focus on their social media accounts (Twitter, Facebook and Instagram), searching for patterns and relationships with their "physical" network of ticket holders. After the first basic analysis of their social media accounts, ArtEx became aware of the potential and scope of the digital layer, going beyond the boundaries linked to their main geographical target, the north of Italy. The presence of international followers in particular, stimulated ArtEx to search for a numerical benchmark in other comparable international theatres, pointing out a recent trend that initially had been omitted: people can move physically within Europe over a weekend simply to go to the theatre and enjoy an unforgettable experience. With this new international framing, the research group suggested extending the social media analysis from proprietary (Art's own social media pages) to non-proprietary (accounts other than Art's own) social media (as before, examining Twitter, Facebook and Instagram). The operations to filter the key words required further framing because of the emergence of two, interconnected, phenomena that were considered of sufficient importance to be analyzed as if they were independent: the theatre's international reputation and the debate on specific operas. The monitoring of the theatre's reputation was based on keywords containing the name of the theatre, plus triangulating blogs and social media posts with more traditional sources (official press). The international benchmarking in this area was carried out against five major international theatres. Only one European theatre was selected for the working benchmark. Three operas from the

2016 season were selected for each of the two theatres (Art and the European benchmark). In order to monitor the six operas, the data had to be filtered according to a different methodology, developed following discussions held with ArtEx: many posts did not include the name of the opera or the theatre, but they did contain the names of singers, directors and musicians. Several iterations were needed to come up with the most appropriate boundary for the data. For example, just putting the names of important singers would have mixed the impact of the opera with the singer's reputation. Both paths were possible and meaningful at the business level, but the way they were interpreted changed. ArtEx decided, in the end, that it was always best to have a good relationship between the data and the phenomena they wanted to trace (operas), and so an association between the artists and the opera performed at Art had always to be retained.

In a final step, the data acquired was completed and integrated through the joint analysis of the network, a point that Art reached indirectly through the mentions of its social media followers. This analysis was only carried out on Twitter. More specifically, the network was built by assigning a "node" position to each social media user, as well as to other users cited in a post, while the message sent represents the interaction (i.e., the connection between two nodes). Taking the following tweet as an example: "@NewYorkTimes: outstanding Vivaldi performance @teatroArt yesterday", written by "JohnDoe", this gives rise to three nodes (@JohnDoe, @NewYorkTimes and @teatroArt) and two interactions, the one between @JohnDoe and @NewYorkTimes and the other one between @JohnDoe and @teatroArt. Following this approach for all the posts, the outcome was a weighted network where the nodes were represented by the users' posts or their mentions in other posts, and the interactions are given by the messages exchanged between users. The network structure was tested over a limited period (March–May 2016), with 3080 social media users and 13,318 interactions being counted over that period.

3.2.2. Data Clustering, Annotation and Merging

The data were collected within the revised decision-making boundaries for monitoring and managing the international (digital) network. The network was taken as the reference unit of analysis, but a number of different networks were constructed to satisfy the various business needs, that is, proprietary social media networks, the networks based on key words and the "Twitter potential network" based on mentions.

Exemplary of the interaction is the annotation of the Twitter potential network, with one European theatre chosen as the benchmark. Starting from the business need of monitoring the network and examining how information was spread, types of nodes were initially identified by the scientists, based on a social network technique called the K-core decomposition algorithm [27]. This technique was proposed as it is widely adopted to investigate the diffusion of given phenomena [28]. The idea was welcomed by ArtEx, as the research group framed this choice as offering the potential of investigating the dissemination of information emanating from Art. The assumption behind the method is that network nodes positioned in core layers can spread messages more extensively than nodes located in peripheral layers. Hence, the nodes positioned close to the centre (where the focal organization is positioned) influence the network and can easily spread messages throughout the grid. Operationally, the K-shell decomposition algorithm iteratively assigns a K-shell layer value to each node in the network: "nodes are assigned a k shell according to their remaining degree, which is obtained by successive pruning on nodes with degree smaller than the Ks value of the current layer. After removing all the nodes with K = 1, some nodes may be left with one link, so the system is continuously pruned iteratively until there is no node left with K = 1 in the network" [29]. The technique was performed iteratively for Art until K = 3, where saturation was reached. This analysis was presented to ArtEx, showing a graphical representation of Art's network and selected competitors (see Figure 7). Despite being fascinated by the outcome, ArtEx had difficulty in understanding how this information could be used in his business operations.

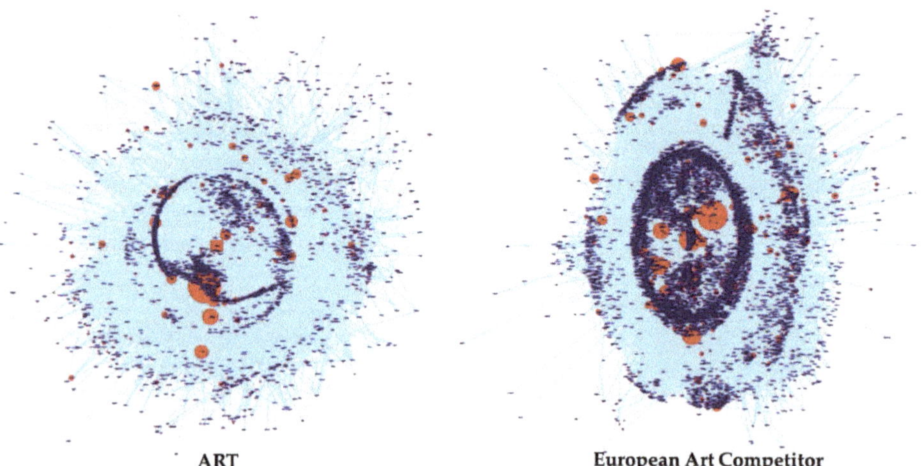

Figure 7. Networks' initial representations. Note: The red color highlighted the most popular nodes in terms of followers.

During the discussions, he started to ask questions about the methods and data, in order to reframe the data into schemes that he had mastered and used. The network always remained in the background, but "old" units of analysis emerged, linked to a higher-order characterization of the single nodes: influencers, active and passive users, and foreign users. These units of analysis relating to nodes were deemed to have a greater basis for being acted upon than the network overall, although the relationships were considered valuable. The final decision was to retain the network, but in addition, to carry out several analyses to build indicators (i.e., the specific metric of a phenomenon) that could be more easily understood by decision-makers. The next section illustrates this point.

3.2.3. Modelling and Analysis

The phase of analysis was simple in statistical terms, but complex in terms of filtering and framing. The research group had the task of filtering and simplifying the information about the network, without losing the benefits of the network itself. The solution proposed was to build network indicators that were to be inserted in Art's daily and monthly reports, enriching the toolkit of their key performance indicators (KPI). All the indicators defined were discussed with ArtEx to find the best business fit, in the form of an indicator that could highlight lags or lead actions. Higher interactions emerged when identifying indicators for the potential Twitter network. In the end, five indicators were defined by merging previous studies on the networks and ArtEx's needs: followers, active relationships, hubness, authority and centrality. The first three indicators are quite simple, but can be applied to all the networks monitored (proprietary, keyword, potential) and actionable. An active relationship identifies pairs of users that maintain an active relationship over time in terms of posting. Hubness analyzes users on the basis of the total number of messages sent. ArtEx considered this simple metric to be crucial for monitoring the vitality of each single network (proprietary, keyword, potential), their evolution in time and against other theatres. Operationally, it is computed as the eigenvector of a transformation AAT (The adjacent matrix of the network is generically called A and it is a NxN matrix where N are the number of nodes of the network. Each entry of the matrix A(i,J) accounts for the number of links between the node i and the node j. AT is the transposed matrix of A; AAT is equal to A multiplied by AT.) of the adjacent matrix of the network A. The values obtained ranged from zero to one (i.e., the lowest and highest level of hubness, respectively). Authority measures users on the basis of the number of messages they receive from within the network. Users with a high level of authority have numerous connections inside the Art network. This indicator is computed as the eigenvector of the transformation ATA (ATA is the transposed matrix of A (AT) multiplied by the matrix A.) of the adjacent matrix A associated to the network. Similarly, to hubness, authority ranged from zero to one. A high level of authority corresponded to values close to one, which detected what has been called "authority" in this context. Finally, the research group proposed the centrality indicators to translate the K-shell network in ArtEx's mental frame. More specifically, a first indicator was the distribution of types of users with reference to the layers (K = 1, 2 and 3); the central layer was labelled as "in-core", rather than K = 1, as this was easier to understand; for each layer user, the number of followers and their vitality (in and out) was then presented. As the project unfolded, a successful framing process emerged when presenting the indicators to ArtEx, which were benchmarked over time and with other theatres. Comparisons between all the indicators implied filtering the data further, but the information being presented became more immediate and meaningful for ArtEx. For example, ArtEx was surprised to see that the "in-core" network of Art's main competitor was larger (133 users against 15 for Art), although the users had similar characteristics.

3.2.4. Visualization

As in the City case, the visualization of Art data started very early on in the project, because of the high interaction between researchers and experts. However, ArtEx followed a different approach to CityEx in how the visualization structure was defined and the data reported. ArtEx was less intrigued by new ways of visualizing data, and preferred simple reports, as mentioned in the previous section. He strongly wanted to carry out particular actions on social media and also work on its current sponsors. In order to pursue this strategy, he needed plain numbers to convince people within and outside Art. For example, on analyzing the "in-core" network, it was clear that Art's main competitor was better able to exploit their relationships with institution sponsors: ArtEx wanted to propose agreements to Art's sponsors, but this required internal consensus. Another example that emerged during the process of mentoring data was that some types of communication based on backstage videos of the artists were particularly successful across the network. To push this strategy further, the CEO needed a budget and legal support. Again, they favoured short reports with few indicators over complex representations.

Regarding digital interaction and functions, while ArtEx was willing to consider real time monitoring and interaction through the web, another key issue emerged, with it becoming clear that there was the need to integrate the new methods of gathering and analyzing data into Art's existing systems. Art had already put in place three different systems to control the marketing area: an internal control system, a ticketing monitoring tool and a social media–web monitoring application. He asked the research group to use these three systems, as they were considered satisfactory despite not being integrated. This request implied limiting the research group's aspirations regarding visual creativity and, instead, study the current reporting structure. As it turned out, exploring the traditional reporting proceeding proved beneficial, as it allowed the author to reframe some information in terms of time. For example, information on operas and artists initially included in the monthly reports was actually important enough to be included in the daily monitoring proceeding, as is currently the case for the traditional press. At this stage, ArtEx involved other organizational actors in order to come up with the most suitable reporting portfolio, and this was brought into the organizational infrastructure.

4. Discussion

As explained initially, this paper is concerned with examining how decision-makers play an important role in establishing what information is to be extrapolated from large quantities of BD and how it is to be used, and to analyze the way in which the BD cycle is affected. The previous section, dealt with how the decision-maker centric approach was applied. The study started from the hypothesis that there are two processes, framing and filtering. These are carried out by scientists translating data into knowledge. However, these processes are also traditionally carried out by decision-makers using information to take action, and they are at the basis of their understanding, trust in data and, more importantly, awareness about their use, which is an essential step for creating Value [30]. The cases presented here are experimental and test an interactive approach involving decision-makers, where the filtering and framing procedures were made transparent, and the final aim was to highlight the key elements in the decision-maker-centric approach.

This application lead to the revision of the initial scheme. More specifically: (1) four spokes were introduced to the graphic scheme, as they were seen as pivotal to the filtering and framing proceedings, and, as a consequence, also to the interaction between decision-makers and scientists; these spokes were the boundary, unit of analysis, timing, and functionalities; (2) a new phase was inserted, described as the "Decision Fit" (Figure 8).

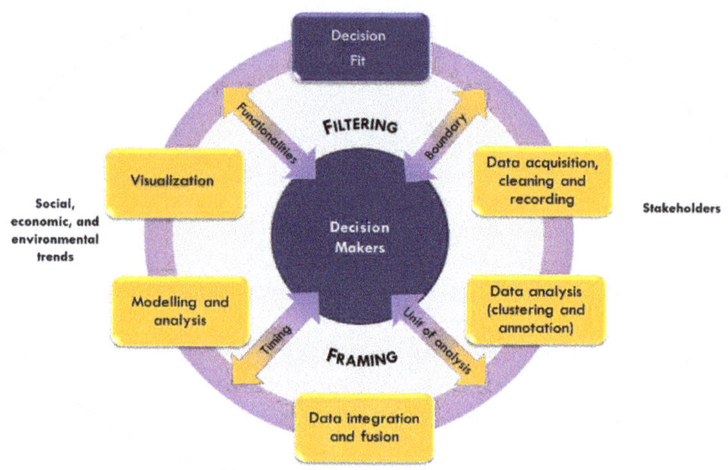

Figure 8. The revised BD cycle.

The first element is the boundary, indicating the abstract outer limit of information and decisional space. When approaching BD, despite the scope of their questions being different, both managers set no limits to the type of data to be collected to analyze their initial questions, which, to remind ourselves, were: "in what way is Milan international?", for CityEx; and "who are the followers of Art on social media and are there particular patterns in their digital traces?", for ArtEx. Since the data acquisition phase, the interactive approach led to reshaping the decision-makers' boundaries through filtering and framing. The link between filtering and boundary was more visible in City, within the key word search on social media. The data acquisition process gave rise to a two-fold perspective: global, looking at "Milan" and its reputation; and local, collecting data geo-referenced within the city itself. This first set of information, interesting though it was for its insights into Milan's reputation, was considered beyond the scope of CityEx's action boundary and so abandoned. The interaction led the author to focus instead on detailed information at district level, which had initially been overlooked.

The case of Art is interesting, as it shows how a boundary is shaped by framing. Due to the basic need of having a numerical benchmark for data on social media followers, ArtEx and the research group started by positioning the theatre among its international competition. This framing process affected the way in which data was acquired as well as the filtering process, and it also changed the type of the search being carried out, from a proprietary network to a "key word" search on social media.

The boundary is also reflected in the ethical sphere. Public data collected from APIs are subject to rules that are interpreted differently by researchers and scientists, often propounding a tradeoff between the completeness of data and ethical behaviour [31]. The problem was highlighted by CityEx and there was the general consensus about respecting rules and privacy, even if this meant reducing the volume and granularity of the data collected. This is an issue often overlooked in practice and in academic studies, but one that needs serious consideration in the decision-maker-centric approach.

The second element is the unit of analysis for decisions (and action). With traditional data, decision-makers use reference schemes where the unit of analysis is the organization or its nested sub-units (processes, organizational units, people). With BD, the starting point is a given problem and the unit of analysis is often undefined and different from common references. Filtering and framing were seen to be powerful processes where both sides (scientists and decision-makers) needed to clarify what the unit to be "controlled" actually was. This issue emerged more clearly during the process of clustering data, when trying to build the relevant categories to be operated on and managed. In the case of City, the focus on the local dimension of internationalization led to filtering the districts within the city of Milan, which became the main unit of analysis. Data then needed to be re-framed in relation to this main unit, through guided clustering. At the practical level, data were easily anchored to districts, but social media data, in particular, needed to be structured into sub-units of analysis that were meaningful in terms of analyzing one district on its own and making comparisons with other districts. In the case of Art, the main unit of analysis was the international network, which was compared over time and against other theatres of reference. After this filtering process, in this case, other variables also had to be re-framed and related to the network. Content clustering and analysis became relevant to trace the content and/or actors that activate the network; the network users were clustered in order to monitor and reshape the communication strategy, revealing "hubs" and "authorities". In both cases, the interaction on the unit of analysis uncovered a hybrid approach where qualitative choices (driven by decision-makers) are mixed with quantitative–statistical indications in the common objective of fitting data to the decision needs.

The third element is timing, indicating the appropriate temporal resolution of data in relation to the decision-makers' needs. The decision-maker centric approach implies understanding not only which data are more suited to the process, but also when and how frequently they are needed. A first choice that entered the interactive approach was the reference period of the algorithm. Starting from the decision-makers' initial desire to have data in real time, the analysts proposed techniques to frame the various frequencies of the data (ranging from yearly to real time). In practice, different algorithms were proposed to divide greater periods into sub-intervals, based on historical paths and trends. From discussions with the decision-makers, it became clear that this division of original data was considered to be "fake", even when refined division methods were proposed. The failure of this statistical approach of dividing time into periods meant that data was aggregated into longer periods, and one point that emerged was that the high frequency of some data was just an over-ambitious yearning thrown up by the potential of actual real time data. The two cases show how this approach can present a wide range of possibilities. Some data about events were retained with real time frequency: leading venues from Foursquare, for CityEx; and audience monitoring for operas, for ArtEx. Other data were aggregated on a quarterly base: language diversity among districts based on Twitter, for CityEx; and monthly reports on reputation, for ArtEx, where various sets of data are evaluated against the chosen benchmark theatre.

The final element of interaction consists of the functionalities needed by the decision-makers. This last element, although present throughout the BD cycle, finds its full expression in the visualization choices. In the case of CityEx, external communications were considered as crucial to promote awareness of the city's dynamics, leading to an intriguing and innovative interface being developed. Careful attention was placed on the type of interaction that users could come up with at every layer, in full coherence with the decisions taken on the first three spokes. In the case of Art, interactions regarding functionalities brought up another important factor of the need to integrate new reporting procedures within the existing system and, crucially, with the mindset and time-frame that the managers were used to working with.

The second proposed revision of the new BD cycle is to bring in the new phase of decision fit, where scientists and decision-makers assess the benefit and costs of BD in the context of use. This assessment emerged naturally as a need when the data started to be used. Decision fit is carried out on a mix of technical and business parameters such as completeness, precision and cost. Completeness is the value added when capturing the critical success factors within the boundaries of the decision. In both the cases, the data added new knowledge to traditional data. For example, the variety of language among the districts, for CityEx; and the drivers of international expansion, for ArtEx. Precision is instead determined by the relevance and correlation of specific data for the decision-makers' goals. For instance, the cross-over effect of City's events on various sectors (hotels, entertainment, etc.); and monitoring and using "hubs" and "authorities" as promotional vehicles in Art's network. Lastly, the term cost refers to the cost needed to process data on a routine basis.

During the concluding phase of the test, the work on the two cases confirmed the benefits of the approach, but it is also important to highlight the limitations of this study. A first such limitation is linked to the decision-making sphere. In both cases, the experiment was carried out in arenas where the decision-maker had the power to use the data with only marginal involvement of other actors. In complex decisions, it is often the case that many actors are closely involved and rational choices are mixed up by political inputs, even down to the choice of the data to be used. These contexts are typically public–private domains, such as transport and health care, where BD can heighten awareness of the impact that various policies can have. The new cycle should be tested from a shared viewpoint to observe both the interaction between scientists and decision-makers and, especially, between the decision-makers themselves.

A second limitation is linked to the actual type of decision-makers included in the tests. The two experts were chosen on purpose for their mid-way outlook towards BD: they were open to BD but remained wary. Referring to the two opposing attitudes towards BD, blind faith and reluctance, they were in a half-way house situation and ready to challenge themselves and the team. Further studies are needed to test the approach derived at when decision-makers lean more towards one side of the argument or the other, but it also opens a new question: "what manner of training do decision-makers and scientists need?". In order to interact along the spokes of the scheme, at least initially, both parties must be able to share their language and toolkit to a certain degree.

5. Conclusions

This paper has addressed an issue that had only been studied marginally in BD research: the need and manner to involve decision-makers in the data processing to avoid any misalignment between information provided by data scientists and the decision-makers' needs. This is particularly relevant in the field of social science, where BD are seen as a panacea to provide Value when addressing business and social challenges. To tackle this problem, a new BD cycle centred around the decision-maker has been proposed and applied using action research methodology in two cases.

The findings enhance previous studies in BD for social sciences at two levels. The first results show that the interaction between scientists and decision-makers when preparing BD is a reciprocal process of knowledge, which, in turn, meant that it was possible to avoid two opposing and risky behaviours: blind faith, where decision-makers overestimate the benefits of BD; and reluctance, where decision-makers treat all data they do not fully understand with suspicion. In pursuing the path of further knowledge, two operations embedded in the information processing system must be made transparent: filtering and framing. Second, the study provides evidence of the value of a quali-quantitative approach to BD for social science; the final cycle provided the dimensions of interaction, in the form of boundary, unit of analysis, timing and functionalities, which confer rigor to the whole process. These dimensions are the grey area between scientists and decision-makers. When explicitly addressed during the information preparation stage, they enable the transfer of skills necessary to make technical choices and in the business context. While the study was conducted with the involvement of only two experts, which could be seen as a major limitation, in both cases, the experts were highly engaged and positive about BD and they had sufficient backing from their organization to carry out the study and interact with the team.

To conclude, this study places the missing variable of decision-making at the centre of the process, reinforcing previous studies on information processing in the BD age and opening the way for future research in social sciences. A first area of development is to apply the cycle to more complex contexts where the decision-making power is distributed among many actors who could be reluctant to work with BD. Further areas for future research include monitoring decision-makers over a longer period of time, examining how information that originated from BD is used and, lastly, studying the impact of BD, and the knowledge that it brings about, on organizations and societal challenges.

Funding: This research received no external funding.

Acknowledgments: The author acknowledges the contribution of the two experts who took part in the study, agreeing to open collaboration, sharing their ideas and granting access to their organizations. Furthermore, the author acknowledges the contribution of the data scientists of the Urbanscope Lab involved in testing the cycle; they played a fundamental role in carrying out the various phases and were always ready to hold a critical discussion about the outcomes, in terms of learning and evaluation, provided by the author. Finally, a special acknowledgement is due to Piercersare Secchi, who encouraged me to consolidate and disseminate my work on this cycle.

Conflicts of Interest: The author declares no conflict of interest.

References

1. Gartner IT Glossary. Available online: http://www.gartner.com/it-glossary/big-data/ (accessed on 10 September 2018).
2. Wamba, S.F.; Akter, S.; Edwards, A.; Chopin, G.; Gnanzou, D. How 'big data' can make big impact: Findings from a systematic review and a longitudinal case study. *Int. J. Prod. Econ.* **2015**, *165*, 234–246. [CrossRef]
3. Sivarajah, U.; Kamal, M.M.; Irani, Z.; Weerakkody, V. Critical analysis of Big Data challenges and analytical methods. *J. Bus. Res.* **2017**, *70*, 263–286. [CrossRef]
4. Arnaboldi, M.; Busco, C.; Cuganesan, S. Accounting, accountability, social media and big data: Revolution or hype. *Account. Audit. Account. J.* **2017**, *30*, 762–776. [CrossRef]
5. Mikalef, P.; Pappas, I.O.; Krogstie, J.; Giannakos, M. Big data analytics capabilities: A systematic literature review and research agenda. *Inf. Syst. E-Bus. Manag.* **2017**, *16*, 1–32. [CrossRef]
6. Visvizi, A.; Mazzucelli, C.; Lytras, M. Irregular migratory flows: Towards an ICTs' enabled integrated framework for resilient urban systems. *J. Sci. Technol. Policy Manag.* **2017**, *8*, 227–242. [CrossRef]
7. Agostino, D.; Sidorova, Y. How social media reshapes action on distant customers: Some empirical evidence. *Account. Audit. Account. J.* **2017**, *4*, 777–794. [CrossRef]
8. De Pablos, P.O.; Lytras, M. Knowledge management, innovation and big data: Implications for sustainability, policy making and competitiveness. *Sustainability* **2018**, *10*, 2073. [CrossRef]
9. Fu, H.; Li, Z.; Liu, Z.; Wang, Z. Research on big data digging of hot topics about recycled water Use on micro-blog based on particle swarm optimization. *Sustainability* **2018**, *10*, 2488. [CrossRef]
10. Kim, K.; Lee, S. How can big data complement expert analysis? A value chain case study. *Sustainability* **2018**, *10*, 709. [CrossRef]
11. Gitelman, L. (Ed.) *Raw Data Is an Oxymoron*; MIT Press: Cambridge, MA, USA, 2013.
12. Hinton, C.M.; Kaye, G.R. The hidden investments in information technology: The role of organisational context and system dependency. *Int. J. Inf. Manag.* **1996**, *16*, 413–427. [CrossRef]
13. Bhimani, A.; Willcocks, L. Digitisation, 'Big Data' and the transformation of accounting information. *Account. Bus. Res.* **2014**, *44*, 469–490. [CrossRef]
14. Quattrone, P. Management accounting goes digital: Will the move make it wiser? *Manag. Account. Res.* **2016**, *31*, 118–122. [CrossRef]
15. Yang, M.; Kiang, M.; Shang, W. Filtering big data from social media–Building an early warning system for adverse drug reactions. *J. Biomed. Inform.* **2015**, *54*, 230–240. [CrossRef] [PubMed]
16. McCormick, T.H.; Lee, H.; Cesare, N.; Shojaie, A.; Spiro, E.S. Using Twitter for demographic and social science research: Tools for data collection and processing. *Sociol. Methods Res.* **2017**, *46*, 390–421. [CrossRef] [PubMed]
17. Zhang, Y.; Lu, H.; Luo, S.; Sun, Z.; Qu, W. Human-Scale sustainability assessment of urban intersections based upon multi-source big data. *Sustainability* **2017**, *97*, 1148. [CrossRef]

18. Chui, K.T.; Alhalabi, W.; Pang, S.S.H.; Pablos, P.O.D.; Liu, R.W.; Zhao, M. Disease diagnosis in smart healthcare: Innovation, technologies and applications. *Sustainability* **2017**, *9*, 2309. [CrossRef]

19. Dalla Valle, L.; Kenett, R. Social media big data integration: A new approach based on calibration. *Expert Syst. Appl.* **2017**, *111*, 76–90. [CrossRef]

20. Frankel, F.; Reid, R. Big data: Distilling meaning from data. *Nature* **2008**, *455*, 30. [CrossRef]

21. Killen, C.P. Managing portfolio interdependencies: The effects of visual data representations on project portfolio decision making. *Int. J. Manag. Proj. Bus.* **2017**, *10*, 856–879. [CrossRef]

22. Didimo, W.; Giamminonni, L.; Liotta, G.; Montecchiani, F.; Pagliuca, D. A visual analytics system to support tax evasion discovery. *Decis. Support. Syst.* **2018**, *110*, 71–83. [CrossRef]

23. Gandomi, A.; Haider, M. Beyond the hype: Big data concepts, methods, and analytics. *Int. J. Inf. Manag.* **2015**, *35*, 137–144. [CrossRef]

24. Argyris, C.; Putnam, R.; Smith, D. *Action Science: Concepts, Methods and Skills for Research and Intervention*; Jossey-Bass: San Francisco, CA, USA, 1985.

25. Baskerville, R.; Pries-Heje, J. Grounded action research: A method for understanding IT in practice. *Account. Manag. Inf. Technol.* **1999**, *9*, 1–23. [CrossRef]

26. Cassell, C.; Johnson, P. Action research: Explaining the diversity. *Hum. Relat.* **2016**, *59*, 783–814. [CrossRef]

27. Seidman, S.B. Network structure and minimum degree. *Soc. Netw.* **1983**, *5*, 269–287. [CrossRef]

28. Pastor-Satorras, R.; Vespignani, A. Epidemic spreading in scale-free networks. *Phys. Rev. lett.* **2001**, *86*, 3200. [CrossRef] [PubMed]

29. Kitsak, M.; Riccaboni, M.; Havlin, S.; Pammolli, F.; Stanley, H.E. Scale-free models for the structure of business firm networks. *Phys. Rev.* **2010**, *81*, 036117. [CrossRef] [PubMed]

30. Lytras, M.D.; Raghavan, V.; Damiani, E. Big data and data analytics research: From metaphors to value space for collective wisdom in human decision making and smart machines. *Int. J. Semant. Web Inf. Syst.* **2017**, *13*, 1–10. [CrossRef]

31. Williams, M.L.; Burnap, P.; Sloan, L. Towards an ethical framework for publishing Twitter data in social research: Taking into account users' views, online context and algorithmic estimation. *Sociology* **2017**, *51*, 1149–1168. [CrossRef] [PubMed]

 *sustainability*

Article

(Smart) Citizens from Data Providers to Decision-Makers? The Case Study of Barcelona

Igor Calzada [1,2]

[1] Urban Transformations ESRC & Future of Cities Programmes, COMPAS, University of Oxford, 58 Banbury Road, Oxford OX2 6QS, UK; igor.calzada@compas.ox.ac.uk; Tel.: +44-7887-661925
[2] Global Sustainable Cities, Institute for Future Cities, Faculty of Humanities and Social Sciences (HASS), Technology & Innovation Centre, University of Strathclyde, 99 George Street, Glasgow G1 1RD, Scotland, UK; igor.calzada@strath.ac.uk

Received: 30 July 2018; Accepted: 10 September 2018; Published: 12 September 2018

Abstract: Against the backdrop of the General Data Protection Regulation (GDPR) taking effect in the European Union (EU), a debate emerged about the role of citizens and their relationship with data. European city authorities claim that (smart) citizens are as important to a successful smart city program as data and technology are, and that those citizens must be convinced of the benefits and security of such initiatives. This paper examines how the city of Barcelona is marking a transition from the conventional, hegemonic smart city approach to a new paradigm—the experimental city. Through (i) a literature review, (ii) carrying out twenty in-depth interviews with key stakeholders, and (iii) actively participating in three symposiums in Barcelona from September 2017 to March 2018, this paper elucidates how (smart) citizens are increasingly considered decision-makers rather than data providers. This paper considers (i) the implications of the technopolitics of data ownership and, as a result, (ii) the ongoing implementation of the Digital Plan 2017–2020, its three experimental strategies, and the related seven strategic initiatives. This paper concludes that, from the policy perspective, smartness may not be appealing in Barcelona, although the experimental approach has yet to be entirely established as a paradigm.

Keywords: smart citizens; experimental cities; smart cities; technopolitics; big data; Barcelona; data commons; decision-makers; policy; GDPR

---

1. Introduction: Conceptual Transitions from Smart Cities to Experimental Cities

In recent years, European city authorities started claiming that (smart) citizens are as important to a successful smart city program as data and technology, and must be convinced of the benefits and security the initiatives offer to them [1–3]. As such, the main argument of this paper is that European cities like Barcelona are already restoring privacy and empowering citizens with data through a diverse set of policies and strategies [4,5]. In doing so, Barcelona seems to be leading a new digital transformation agenda called Data Commons Barcelona [6]. This works alongside the European Union (EU)'s General Data Protection Regulation (GDPR) [7–9], which mandates the ethical use of data to protect (smart) citizens from risks inherent in new, data-intensive technologies [10–17]. Hence, this paper, through a case study of Barcelona, explores the problem of how city authorities can proactively set up policies, strategies, and initiatives to locally enhance digital rights, and give citizens more control over personal data by protecting them from discrimination, exclusion, and the erosion of their privacy [18,19].

Amid the clamor for smart-city rhetoric in Europe [20–22], a deep debate about digital ethics and rights emerged in cities such as Barcelona [23,24]. Thus, the paper aims at elucidating whether Barcelona is marking a departure from the so-called smart-city paradigm [25–28] to a newly emerging paradigm called experimental cities [29–31]. The case study is examined through the systematic articulation and response to three operational research questions that concern (i) the existence of alternative data ownership regimes in the ongoing smart city model of Barcelona; (ii) the practical consequences of the grassroots innovation initiatives implemented so far in the city; and (iii) the searching for evidence of another experimental type of smart city.

Methodologically speaking, this paper deconstructs Barcelona's Digital Plan 2017–2020 through qualitative fieldwork research carried out from September 2017 to March 2018 [32]. The methodology is based on the mixed-method technique of triangulation via action research [33], which encompasses a deep literature review; the findings of twenty in-depth interviews with key stakeholders, following the Penta Helix framework (private sector, public sector, academia, civic society, and social entrepreneurs/activists) [34]; and result validation through participation in three key symposiums organized in Barcelona and related to the smart city renewed strategic digital agenda formulation [35].

In the EU, the debate on data privacy, transparency, and ultimately ownership is gaining momentum as the GDPR, based on codes such as "privacy by design" and "data portability", took effect in May 2018 [36]. The GDPR may directly influence how city authorities implement their data policies and strategies, and foster potential transitions to allow citizens a more active role in decision-making processes. The GDPR may have already shifted the conversations that city authorities have with smart-city solution providers, particularly in relation to their business models, monetization strategies, and their data-processing procedures. Thus, a consensus among developers, companies, and governments on the ethics of the underlying decisions in the application of digital technology of data seems to be at stake in the post-GDPR future.

With the backdrop of this regulation, on 24 January 2018 at the World Economic Forum in Davos, Angela Merkel announced [37] "data will be the raw material of the 21st Century" and then added "the question 'who owns that data?' will decide whether democracy, the participatory social model, and economic prosperity can be combined" (p. 1). This declaration comes at a time when data sovereignty is becoming increasingly important to colocation providers, and citizens' data rights in smart cities is attracting attention from many scholars and practitioners [38–42]. At the same international event, the billionaire investor and philanthropist George Soros argued that the large [43], multinational technological firms, Facebook and Google, are "obstacles to innovation and are a menace to society", subsequently adding "(the) days are numbered for the two geek giants" (p. 1). Moreover, the digital philosopher Eugeny Morozov routinely criticized the economics of these giants' data "extractivism" business models in recent years [44–46], claiming that the models lead to a world in which these kind of tech firms build addictive services to gather citizens' data with artificial intelligence (AI) and machine learning tools [47,48]. Most recently, Cambridge Analytica and Facebook were in the spotlight due to circumstances in which Facebook data may have been illegally obtained and used by the firm [49]. This episode is having enormous influence on the technopolitical debate and the critical perception of how data analytics could have been used for political purposes without the informed consent of users. Overall, this incident highlighted the idea that political parties, data analytics companies, and social media platforms view citizens and social media users as no more than data providers [50–52]. Hence, according to Acuto [53] "data availability does not immediately translate into better-informed urban management, nor fairer, greener, and more prosperous cities" (p. 165).

The GDPR is timely when considering the debate over controversies related to algorithmic disruption and fueled by big corporations, along with hegemonic rhetoric about the benefits of technocentric smart cities. Thus, a critical perspective on data ownership blended with a socially constructed and citizen-centric smart-city approach emerges as the context for this paper. Furthermore, over the last five years, certain cities, such as Barcelona, started implementing digital policy frameworks and programs intended to strategically overcome the side effects of the technodeterministic emphasis on smart cities [54–56]. So far, this emphasis meant understanding the city just from a dispassionate, scientific perspective—purely as "solutionist", predictable, replicable, linear, and normative urban machinery fed by citizens as pure user/data providers. Moreover, citizens' data are pervasively owned by private big data firms. As such, it becomes a mechanical entity solely governed through the market logic stemming from the public–private partnership (PPP) scheme, and occasionally and slightly including stakeholders' engagement through helix strategies such as the Triple or Quadruple Helix [34]. By contrast, and consciously in response to this technodeterministic approach [57], a critical and evolutionary transition known as experimental cities recently surfaced in cities such as Barcelona. It is characterized by initiatives encompassing (i) the awareness of the technopolitics of data for citizens [58]; (ii) potential alternative economics for city policies [39]; (iii) citizen engagement as a democratic practice [59–62]; (iv) multi-stakeholder schemes as a pervasive governance logic [34]; and (v) living lab initiatives as sites devised to design, test, and learn from social and technical innovation in real time [29,30,63–69]. This emergent amalgamation of initiatives surrounding the experimental urbanism trend regards urban transformations as inherently interdisciplinary, data-intensive, and embedded in place. However, it remains to be seen whether the top-down, hegemonic approach to the smart city is effective in transitioning to the new paradigm, which is characterized by considering (smart) citizens as decision-makers rather than data providers.

The term smart city, which became particularly hegemonic in the policy agendas of European cities, turns 26 this year. It was introduced in 1992, around the time of the historic Rio Earth Summit, and typing it into a search engine today generates more than 15 million results. Figures for the smart-city market over the last two decades are similarly impressive [70]. Persistence market research forecasts that global growth will balloon from its current $622 billion valuation to $3.48 trillion by 2026—a fivefold increase in only a single decade [71].

However, according to Evans et al. [29], while the hegemonic smart-city discourse has hitherto focused on "trialing technological solutions" in real cities and benefitting multinational corporations as urban actors (p. 3), the experimental cities approaches—embodied by grassroots movements, living labs, and co-operative platforms—consistently position local communities as the designers and proactive instigators of urban experiments [23,25,26,30]. The smart city approach is being driven by market-based urban solutions and deconstructed from many angles after recently saturating policy agendas as a concept with very little reflexivity. In response, a new smart-citizen-centric paradigm is being tested in the experimental city. This real-world urban experiment matters because it produces "a different type of city by offering novel modes of engagement, governance, and politics that both challenge and complement conventional strategies such as on-going smart city strategies" [29] (p. 9). Therefore, new techniques explored in the experimental city approach should rethink the smart city approach from the ground up [2,72]. Reframing urban development to utilize the experimental city approach seems to shift the balance of power between stakeholders, empowering some while disempowering others, and enables new forms of co-produced knowledge by raising a fundamental academic debate around two normative questions in the smart city field inspired by Kitchin [73]: "First, for whom and for what purpose are smart cities being developed? Second, are the primary goals of smart cities about, or should they be about, (i) creating new markets and profit; (ii) facilitating state control and regulation; or (iii) improving quality of life while enhancing citizens' opportunities to participate in democracy?" (p. 7).

Hence, this paper systematically presents some key concepts. Table 1 introduces literature sources and depicts an analytical framework for ten conceptual transitions from smart cities to experimental cities as follows: (i) (smart) citizens are considered decision-makers rather than users/data providers; (ii) data sovereignty replaces "extractive big" data; (iii) cities are viewed as open platforms rather than pure markets; (iv) firms owning personal data are responsive to public scrutiny; (v) stakeholder interdependencies stemming from the social innovation perspective are implemented through the Penta Helix scheme which expands on the Triple or Quadruple Helix scheme; (vi) PPP urban business models are complemented by experimental models through urban commons and urban co-operative platforms; (vii) scalability and replication are achieved by unpacking urban problems rather than "solutionism"; (viii) the sensor network is based on a citizen-sensing awareness rather than a pure internet of things (IoT); (ix) the system is based on living-lab niche experiments rather than electronic (e)-government systems; and (x) a causality logic based on emergence and complex adaptive systems is adopted rather than a linear and normative system.

Table 1. Conceptual transitions from smart cities to experimental cities. PPP—public–private partnership; IoT—internet of things.

	Conceptual Transitions	Smart Cities	Experimental Cities
1.	(Smart) citizen [62,74–80]	User/data provider	Decision-maker
2.	Technopolitics of data [12,24,58,81–93]	Big data	Data sovereignty
3.	Notion of the city [94]	As a market	As a platform
4.	Personal data ownership [41,42,49,95,96]	Owned by firms	Publicly scrutinized
5.	Stakeholder helixes [34]	Triple or Quadruple Helix	Penta Helix
6.	Business models [97–103]	PPP	Urban commons and urban co-operative platforms
7.	Scalability and replicability [22,104]	Based on urban *solutionism*	Unpacking urban problems
8.	Algorithmic coding [105–110]	IoT sensor networks	Citizen-sensing
9.	Governance [30,63,65,68,69]	E-government systems	Living labs
10.	Causality [10,11,57]	Linear and normative	Complex adaptive systems and emergence

Since the GDPR recently took effect in the EU, the transition to data sovereignty inevitably sparked a debate regarding the role of citizens and their relationship with data in their own context. As such, two general questions frame this debate in relation to implementing the transition toward (smart)-citizen-centric, data-driven urban environments: (i) Can European cities build alternatives that put citizens back in the driver's seat as decision-makers rather than relegating them to the role of data provider? (ii) Should European cities focus on building decentralized infrastructures based on blockchain to prevent extractive data practices by large technological corporations, where these practices violate citizens' digital rights [111]? In particular, the following three operational research questions about the examination of the transition of Barcelona from the smart city to the experimental city paradigm constitute the core strands of this paper [2,38,56,58]:

1. What prospects exist for an alternative data ownership regime in the current smart-city model used in Barcelona?
2. What are the practical consequences of the grassroots innovation initiatives implemented in Barcelona for businesses, local governments, academia, civic society, and social entrepreneurs/activists?
3. Is another experimental type of smart city driven by co-operative service provision models based on social innovation possible in Barcelona? That is, does a "third way" exist between the state and the market that overcomes the PPP?

To respond to these three operational research questions, the paper is organized as follows: (i) the introduction addresses the conceptual transitions from smart cities to experimental cities; (ii) a literature review on the term "smart citizens"; (iii) the results of the paper elaborate on (smart) citizens as decision-makers; (iv) Barcelona's (smart) citizens' policy framework are deconstructed through its Digital Plan 2017–2020; and (v) conclusions are presented in response to each of the three operational research questions.

2. Literature Review

2.1. Conceptualizing (Smart) Citizens: A Systematic State of the Art

Many attempts were made to conceptualize smart citizens. While a variety of researchers focused on contextually bound definitions, there seems to be a lack of agreement regarding the significance of smart citizens in practice [40,62,74–77,112,113]. Expanding on prior works, some authors fiercely countered the technodeterministic and/or neoliberal smart city rhetoric from policymakers and technology vendors by pointing out the absence of benefits from citizen-centered smart cities [114,115], whereas others more recently took a constructive approach to consider and elaborate upon the alternatives offered by experimentalism [29,64,65,116].

Back in 1969, Arnstein's ladder—a widespread conceptual frame used to examine citizen participation in place-making and city governance—offered a valuable heuristic to start integrating the role of (smart) citizens in the city [117]. Arnstein detailed three forms (citizen power, tokenism, and non-participation) and seven levels of participation (citizen control, delegated power, partnership, placation, consultation, informing, choice, therapy and manipulation). This conceptual framework should strongly resonate in the current European context, where the citizen's data debate coincides with the significant weakening of citizens' right to privacy and information self-determination. Likewise, a 1969 paper might provoke a profound debate on the real meaning of participation and democracy after more than 30 years of neoliberalism [40]. Despite the glossy citizen-centric institutional discourse, the framework of that paper analytically enhanced the evidence-based analysis by presenting a wide range of participation levels at which smart citizens' representation and involvement may find many modalities for deliberation. These included not participating, only consuming, providing feedback, suggesting ideas, negotiating between stakeholders with conflicting visions, and leading ownership regimes. These findings are particularly relevant when cities become expanded laboratories—with special efforts to equip (smart) citizens with information and skills to enable access beyond merely supplying data. In this way, different, sustainable, prosperous, and livable urban futures can be tested in real-time [118]. Coletta et al. [82] argue that "Singapore, Barcelona, Dublin, and San Francisco are (currently) but a few examples of cities undertaking experimental modes of development" (p. 8).

More recently, Ratti [38] argued that municipalities in these timely transitional examples "are starting to understand the importance that citizens, in particular, should have a key role through 'bottom-up' dynamics" (p. 142). Therefore, instead of concentrating on the installation and control of hardware, the (smart) citizens approach highlights the importance of governments getting people enthusiastic about creating apps and exploiting data themselves. Thus, an understanding of this novel paradigm of data—which involves algorithmic analysis of unrelated datasets within presently under-examined situations and social interactions—exists in the recent literature. It suggesting that the notion of (smart) citizens should be associated with the idea that governments, regardless of governance scale, should act as data facilitators rather than as providers of services. This notion connects directly with the aim of this paper insofar as (smart) citizens interact (often unwittingly) with an increasing volume of data, encounter pervasive and unethical extractive data practices, face an increasing number of privacy issues, and remain unaware of the value of their data in terms of ownership; addressing these issues opens new avenues to engage citizens with government digital policy and strategic frameworks through specific initiatives and projects [119].

Over the last few years, the idea of smart citizens profoundly influenced smart-city literature [40,75–80,113]. In summary, scholars such as Cardullo and Kitchin [40] interestingly argue that the concept of the smart citizen is often synonymous with "choice" in the market, with the following predominant roles: a "consumer" or "user"—who selects the services acquired from the marketplace of providers; "resident"—who can afford the exclusivity offered by a "smart district"; and "data provider"—who creates data through the use of smart-city technologies that companies can incorporate into products and extract value from. For instance, one study [40] recently revealed that citizens in smart cities are "consistent with neoliberal citizenship and its emphasis on personal autonomy and consumer choice" (p. 12). This remarkable interpretation may lead us to reflect upon individuals performing certain roles and taking responsibility for opportunities in their private life (entrepreneurial self), and for the marketization and privatization of amenities and infrastructures. Despite citizen participation being potentially varied, it is frequently framed in a post-political fashion that provides feedback, participation, negotiation, and creation within an instrumental frame —rather than a normative or political one. That is to say, citizens are stimulated to assist in providing solutions to everyday issues, for example by producing an app, giving an opinion on a progress plan, or performing certain functions/responsibilities that do not contest or substitute the fundamental political rationalities affecting an issue or plan. Very often, citizens are empowered by devices that consider them as consumers or testers—people to be steered, controlled, and nudged to behave in certain ways—or as raw material for algorithms that can be turned into commodities—i.e., passive data providers. In other words [40], citizens in the smart city act within the constraints of predictable and acceptable patterns "rather than transgressing or resisting social and political norms" or simply becoming active decision-makers (p. 12).

Recent literature on smart citizens reached broad consensus that technological solutions are often proposed under the sponsorship of the smart city buzzword, while neither considering citizens' needs nor socio-technical misalignment between mechanical urban solutions and the citizens' decision-making processes [56,120,121]. Moreover, the literature review of recent reports published by the European Commission, analyzing the implementation of a wide range of H2020 lighthouse projects and covering developments on smart citizens, illustrates that smart-city initiatives only foster limited forms of citizen engagement and citizen power [20–22,104]. Although some European cities are allegedly being used as experimental testbeds or living laboratories for super-connected, technologically mediated smart districts, they must continue providing common resources and benefits to all citizens. Thus, the expectations that (smart) citizens participate in these experimental laboratories should be grounded in a much more politically active discourse of rights and urban commons [102]. In other words, is it possible to reassemble the driving ethos for smart cities beyond a market rooted in data ownership?

2.2. Deciphering the Case Study of Barcelona: (Smart) Citizenship at Stake in European Cities

Hence, (smart) citizenship appears to be at stake. In response, European smart cities such as Barcelona [122] and Amsterdam [123] (Barcelona's partner city in EU project Decode) are leading an alternative movement by following experimental city policy framework practices that rely on democratic data ownership regimes establishing municipal data offices, grassroots innovation, and co-operative service provision models [29,63–65,124,125]. These cities are implementing pervasive transitions to cope with the disruption to the technopolitics of data, and address concerns that include privacy, literacy, awareness, and ownership issues while empowering local communities to avoid the side effects of the predatory sharing economy and extractive practices [2,126].

According to Cardullo and Kitchin [40], Barcelona "is presently attempting to formulate and implement a different vision of a smart city and smart citizenship" (p. 13). While under a right-wing and neoliberal government in the early 2010s, the city became an iconic smart city through its abundant initiatives and self-promotion events within the program "Smart City Barcelona" [127]. However, since the May 2015 election of Ada Colau—the new mayor, representing the left-wing, green, social movement coalition—Barcelona gradually experimented with its vision and policy framework via the Digital Plan 2017–2020, deployed as "Barcelona Ciutat Digital: A Roadmap Toward Technological Sovereignty" [32,128]. In October 2016, the new vision for Barcelona was fixed as an "open, fair, circular, and democratic city" with a special emphasis on its mission to promote citizens' data ownership and technological sovereignty. This was a conscious strategy to overcome the excesses of data extractivism [35]. Since then, the Barcelona City Council implemented a new policy framework entitled the Barcelona Data Commons program [6], which parallels an EU-funded experimental project titled Decode [129,130]. The latter project functioned as a testbed and flagship project to repoliticize the smart city and shift its creation and control toward grassroots, civic movements, and social innovation, and away from private interests and the state [131–133]. As an amalgamation of strategic initiatives, the policy framework established under the Barcelona City Council Digital Plan 2017–2020 is analyzed in the fourth section of this paper.

Based on this new shift, Barcelona is currently explicitly branding itself as an inclusive, democratic, and participative smart city by promoting accordingly [1,134,135]: "If you would rather have smart citizens than smart cities ... BITS, 'Barcelona Initiative for Technological Sovereignty', will be in your interest" [35] (p. 1). Likewise, in Amsterdam, those advocating for smart citizens suggest that (smart) citizens should be guided by the following principles [78,84,121,123]: take responsibility for the environment they live and work in, and the places they like; value access over ownership and involvement over power; ask for tolerance, not authorization; ethical algorithms rather than extractive algorithms; and offer assistance to less technologically savvy individuals. This set of ethics underlies the notion of a still-limited consideration of the interconnections between hard and soft smart infrastructures, as well as those between political, institutional, economic, and social systems on the metropolitan and regional scales. Moreover, this new paradigm advocates the significance of overcoming the often-failed smart-city-in-a-box strategy, by keeping the loop with the various stakeholders.

By examining the Digital Plan 2017–2020 of Barcelona [32], in particular, it is evident that the notion of the commons extensively influenced the three experimental strategies of Barcelona's renewed smart city plan that were put into effect since September 2016 [136–138]. Three main strategic components were identified as experimental strategies during the fieldwork research (and are explained in greater detail in the next section): (i) a data commons through three strategic initiatives; (ii) local democracy through two intertwined grassroots innovation initiatives; and (iii) political economic co-operative platforms in relation to three initiatives built around Digital Social Innovation, Barcelona Urban Commons, and a "social economy" municipal framework. Thus, this plan may open a new path for action in Barcelona based on Ostrom's prominent thinking on the commons [4,98,99,102]. For example, one of the strategic initiatives, entitled Data Commons Barcelona [1,6], is defined as "a shared resource that enables citizens to contribute, access, and use the data—for instance about air quality, mobility, or health—as a common good, without intellectual property rights restrictions" (p. 1), which clearly resonates with Ostrom's assertion that nonprofit and voluntary actions should govern the collective resources that many citizens use as data (including free software, Wikipedia, and Open Street Map). However, as Hardin warned [100], the "tragedy of the commons" could occur when data extractive practices conquer the urban metabolism. A notion of the commons, recently updated by Bollier, argues that historically embedded individualization procedures gradually shape the communal settings such that they require experimental interventions as a corrective [136].

In a bold attempt to transition from a smart city to an experimental city, conceptual explorations around the commons are likely to influence a more profound examination of the technopolitics of data. Thus, because this paper emphasizes the active role of (smart) citizenship; the conception of the commons should be encompassed in the narrative, because the rhetoric of the hegemonic smart city is uniquely based on the idea of PPPs [139]. Bollier suggests that cities are at an intersection insofar as (smart) citizens could use the ideas of the commons to retain control of the services that matter to them and safeguard the functions of these services for the people of the city, and not only for businesses or bureaucracy. However, the idea of the commons could also be argued to be rather abstract with regards to policy implementation.

One possible solution to the issues above is offered by the hypothesis of this paper: the notion of (smart) citizens as decision-makers is being explicitly included in the policy formulation of Barcelona by addressing the three main operational research questions of this paper. The fieldwork research process identified three experimental strategies that directly relate to the debate on the implications of the GDPR for citizens and on the governance structures required to critically tailor the technopolitics of these data:

- Alternative data ownership alternative regime initiatives;
- Grassroots innovation initiatives;
- Co-operative service provision models, such as urban co-operative platforms [97], living-lab initiatives [67,68,140], and proactive social economy policy frameworks.

Ultimately, this paper aims to examine whether the rhetoric of the experimental city is followed by an evidence-based impact in the case of Barcelona. The exact definition of more citizen-centered approaches is often left unstated; therefore, this paper aims to unravel the meaning from the Barcelona case to provide evidence-based insights into the city's experimental strategies. Within the scope of some approaches to experimental cities, the extent to which citizens play a more active role as co-producers rather than as mere data providers remains to be seen, although the technopolitics of data (i.e., ownership and governance) is an under-explored area of research [85,86].

After a rigorous literature review covering the case of Barcelona and its transition from the "Smart City Barcelona" strategic period in 2011–2014, led by former mayor Xavier Trias [141–143], to the current "Barcelona Data Commons" strategic period led by current mayor, Ada Colau, [58,79,80,127,134,135,144–147], this paper offers the hypothesis that the existing smart city cannot simply be reduced to the economic trade-off of the data it generates through partnerships with powerful public and private actors [139]. A key element of this shift in discourse is the need to embrace a multi-stakeholder approach by broadening the representation of social actors in decision-making processes in order to overcome the new religion of data. Coined "dataism" by Harari [87], over-reliance on data is understood as a logic oversimplifying a city's metabolism as a mere assemblage or system of data and algorithms [12,16,44,88,92,93,148,149], rather than an ecosystem of citizens [116].

3. Results: Technopolitics of Data for (Smart) Citizens as Decision-Makers Rather Than Data Providers

According to the research firm Gartner [150], 8.4 billion devices worldwide were connected to the internet by the end of 2017, and 20.4 billion are expected by 2020. However, as previous sections in this paper highlight, some hesitation persists at the center of the dataism discussion (see the article by Harari [87]). Thus, what are the real consequences of the big data for (smart) citizens? In response to this open question, Shilton [89] argued that "uncertainties about how to use increasingly large sets of personal data are at the center of social debates about the virtues of big data". She continues "Not all 'big data' are data about people, but data about people inspire much of the hope and anxiety bound up in discussions of the term" (p. 21). Thus, who controls not only data collection, analysis, storage, and usage, but ultimately ownership? Table 2, inspired by Shilton's contribution, was elaborated to enhance the key parameters of the technopolitics of data related to the degree of citizen participation in today's cities. The interpretation by Shilton is complemented in that it can be assumed that users with high participation—smart citizens—would be able to control the reuse of their own data. This assumption is a reaction to the fast development of the data extractivist practices by which (smart) citizens should not only be involved in participation processes, but also gain the knowledge to engage autonomously in the data collection processes. It may offer to bridge the gap to implementing data sovereignty, ownership, and commons policy schemes (as the case study of Barcelona depicts).

Concerning updated sources of data collection, storage, use, and ownership, the major obstacles to fostering a people-centered design of data are found in the acquisition, shareability, licensing, and knowledge boundaries of the obtained data. Thus, the requirement to consider individuals not only as citizens deliberating on their material conditions, but also as consumers agreeing and disagreeing to the specific terms of a provision should be taken further by advocating for a more human-centered perspective to the smart city—one that fosters interplay and interdependencies among multiple stakeholders.

Table 2. Citizen participation in the technopolitics of data collection, analysis, storage, reuse, and ownership. Based on Reference [89] (p. 26).

Citizen Participation in the Technopolitics of Data	Collection	Analysis	Storage	Reuse	Ownership
High participation	Subjects own or control devices; data collection can be customized	Raw data accessible; subjects can conduct their own analyses	Data stored on local devices	Individuals control reuse	Individuals own their data and customize their data policy
Low participation	Subjects aware of devices; data collection can be avoided	Subjects can see visualizations or analysis of their data	Data in cloud storage with options for deletion	Reuse is restricted to aggregated forms	Data collectors use contracts to obtain citizens' consent over their own data
Little to no participation	Subjects unaware of devices; data collection cannot be avoided	Subjects are evaluated or categorized without their knowledge	Data in cloud storage with no option for deletion	Data collectors share or sell data	Data collectors own citizens' data

Habermas opposed technocratic and democratic smartness, enabling the generalization of the smart-citizen category [151]. Traditional notions of the smart city put individual privacy at risk, and thus, citizen interaction, involvement, engagement, participation, and deliberation are the focus of the debate on the technopolitics of data. Yet, how should we address the distrust, apathy, and open indignation that turned progressively pervasive in political attitudes? Misalignments between algorithmic computation and the social desires of citizens in data generation and ownership constitute a collective challenge facing us today: Will data-driven cities and devices continue to serve citizens or vice versa [54,90]? Is computation replacing conscious thought? Are we thinking increasingly like a machine, or are we not even thinking at all? In response to these questions, different forms of engagement on the part of (smart) citizens can be discussed in relation to the technopolitics of data and the algorithmic disruption [10,11]. Morozov argued that key questions [44], such as "who implements data?" and "what kinds of data do technological solutions smuggle through the back door?" (p. 1) remain unanswered despite the plethora of technological solutions to social problems. Policy discussions highlight how seemingly simple calls for open data strategies actually challenge existing legal norms and have potentially profound implications for users down the line. For instance, liability risks might be passed to the end user of open data, but what if the end users cannot bear those risks? If the IoT generates continuous monitoring and commonly individualized data, how do we theorize, regulate, and make visible the ethical choices that emerge around the legal liability surrounding the ownership of data [18,152]? Ownership of data promises to deliver significant personal and public benefits if city authorities start reconceiving data as a new type of common good. According to the Barcelona city authorities, (smart) citizens' data can generate public value by protecting citizens' "technological sovereignty" as a whole [1].

Hence, citizens' personal data are a fundamental part of their urban experience, and their digital rights should be based on a logic of solidarity, social co-operation, and collective rights to the city [148]. Why, then, do we not naturally consider (smart) citizens as pure decision-makers rather than simply passive data providers? Despite the willingness to pursue sustainable futures that are more democratic than technocratic futures, strong inertia obstructs this alternative path. If the government offers greater data accessibility, better public services will be delivered [1,59,91]. Thus, alternative policy paths explore city government policies that could require large corporations to share data collected from their users to preserve full digital rights for citizens by developing decentralized, privacy-enhancing alternative data infrastructures based on blockchain and attribute-based cryptography [111,125] In fact, the current round of urban experimentation differs from previous incarnations [29], indicating a "specific type of governance fix for a neoliberal system that is struggling to move toward more sustainable forms of urban development" (p. 10). In Barcelona, this ongoing experimentalism addresses new business and social models that also demand previously accepted legal frames and procurement regimes be revisited. According to Francesca Bria, the Chief Technology Officer (CTO) of Barcelona [38], the technopolitics of data in Barcelona are articulated in response to an explicit strategy to "disrupt these data accumulation, making data available across vertical silos experimenting with decentralized data infrastructures through distributed ledgers such as blockchain, and proposing new frameworks and business models that reward and incentivize openness, enabling data discovery, transactions, and secure data sharing" (p. 145). This citizen-centric approach to the technopolitical policy scheme of Barcelona is called Barcelona Data Commons [6]; it is influenced by the experimental EU-funded project Decode [129], which is centered around strategic initiatives regarding technological sovereignty and data transparency.

This section underlines how data-driven and algorithmic issues open up novel challenges for implementing policy in cities all over the world. In terms of the ethical, institutional, and political dimensions of the ownership of data, urban experiments—Barcelona in particular—are gaining traction as a strategy to stimulate alternatives and steer change [124], as suggested by the Digital Plan 2017–2020. The idea of experimentation feeds on the appealing notions of innovation and creativity (both individual and collective) while also shifting the focus of sustainability from distant targets and government policies to concrete and feasible actions that can be undertaken by a wide variety of urban stakeholders in particular locations. The ability of urban experiments to be radical in ambition while limited in geographical scope underpins a challenging debate in both the policy and academic domains with respect to the potential for such experiments to initiate genuine transformation. Are these experiments simply extensions of business as usual, spatially limited, and captured by a familiar cast of dominant interests? Or, can they engender genuine alternatives and stimulate profound transformation?

Thus, the next section examines the case of Barcelona in depth by deconstructing the Digital Plan 2017–2020. I analyze the promises and hazards of experimentation as an alternative mode of urban governance that moves beyond the structural mistakes of the so-called smart city, which is the dominant mode thus far [30,57]. The section is divided into three subsections that examine the three experimental strategies of Barcelona's Digital Plan through an analysis of fieldwork material: policy documents, interviews, and symposiums (Appendix A). The first subsection examines the first operational research question presented in the introduction, situating (smart) citizens within the context of the data ownership and technological sovereignty experimental strategy; three initiatives illustrate the ways in which experimentation is occurring within Barcelona. In the second subsection, I investigate the experimental strategy of grassroots innovation in response to the paper's second operational research question; two intertwined initiatives are presented to allow dissection of the practical consequences for stakeholders (interviewed following the Penta Helix framework). In the third subsection, the paper explores the extent to which considering decision-makers could achieve a potential ecosystem of co-operative service provision models as alternatives to capital-based forms of business.

4. Methodology and Discussion: Deconstructing Barcelona's (Smart) Citizens' Digital Policy Framework Case Study

According to the Barcelona City Council's Digital Plan 2017–2020 [32], "in September 2016, Barcelona City Council embarked on an important digital transformation process, announcing that public services must be provided through digital channels from the outset, following new guidelines oriented toward citizens and the use of open standards and open-source software, and in accordance with an ethical data strategy that puts privacy, transparency and digital rights at the forefront" (p. 3). The smart-citizen concept at the core of this strategic formulation suggests that cities should implement new legal, economic, and governance schemes to nurture collaborative behaviors from citizens to contribute to the digital commons, including commons involving personal data. It is not yet clear the extent to which this move in strategic formulation is merely lip service to the critics of the term smart city versus an initiative that will actually lead to implementing real consideration of (smart) citizens. Thus, this section aims to provide empirical evidence-based insights to scrutinize the nature and intensity of this transition via the three operational research questions.

To begin, the Barcelona case study will deconstruct the Digital Plan 2017–2020 by applying three methodological techniques based on triangulation. Case study research, as defined by Yin [153], was identified as the most effective form of mixed-method research. Firstly, an analysis of its most recent policy documents related to the three experimental strategies was identified in the Digital Plan 2017–2020. Secondly, a qualitative study of twenty semi-structured interviews was conducted in Barcelona from September 2017 to March 2018. The interviewees were identified following the Penta Helix multi-stakeholders' framework and included representatives from each helix (see Figure 1) [34]. The purpose of the interview study was to validate the mixed-method action-research process with a broad representation of opinions and evidence. Finally, to reinforce the mixed-method technique of triangulation via action research [33], the author participated in three symposiums directly related to the smart-city strategic formulation in Barcelona during the fieldwork timeframe (September 2017–March 2018): (i) the celebration of the Smart City Expo World Congress 2017; (ii) the ESADE Business School Smart Cities and Data Speaker Series; and (iii) a presentation of the Barcelona Data Commons program in January 2018 (see Appendix A).

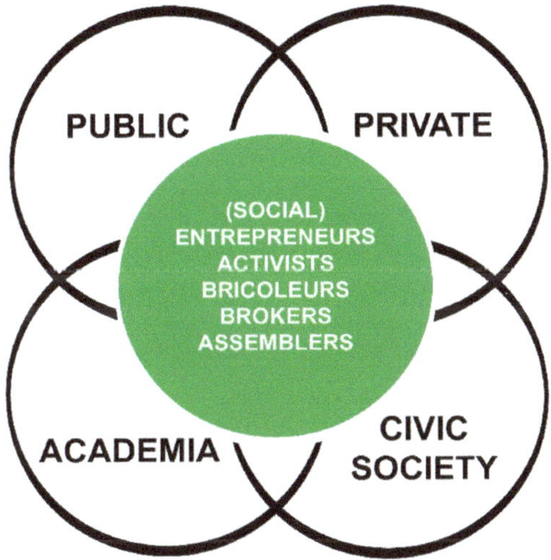

Figure 1. Penta Helix multi-stakeholders' framework [34].

The case of Barcelona was selected using a theoretical sampling approach. With slightly more than 3.2 million citizens in the metropolitan area, the city of Barcelona falls into the category of urban areas experiencing a deep transition connected to the dynamic devolution debate and vibrant bottom-up initiatives. Furthermore, its success in the smart-cities field makes the analysis of its strategic transition an ideal sample for this study; this assertion is easily demonstrated by considering the multiple smart-city awards Barcelona received, in addition to its geographical location [20].

Barcelona's smart-city strategy experienced two distinct periods. In 2011, the mayor, Xavier Trias, and his municipal administration decided to transform Barcelona into a smart city through an approach slightly biased toward top-down management and with a focus on technological capacity. However, since Ada Colau's election as mayor in May 2015, the strategy reached a clear turning point. In October 2016, the publication of "Barcelona Ciutat Digital: A Roadmap Toward Technological Sovereignty" kicked off the implementation of the new smart-citizen policy framework disseminated through the Barcelona Initiative for Technological Sovereignty (BITS) by Commissioner Francesca Bria [1,35,128].

In the extensive evaluation of the literature discussing Barcelona, considerable consensus can be found regarding the strategic shift in the smart-city pathway [38,58,79,80,122,127,134,135,141–147]. Firstly, during the 2011–2014 period and based on the corporative technocratic vendor push, the Barcelona Smart City strategy was driven more by deliberate components than by emergent and experimental components. Secondly, the current transition, called Technological Sovereignty, gained momentum. Almirall et al. [38] highlight "tensions in smart cities regarding efforts in the provision of services" during the current period (p. 152), while Bakici et al. [141] identify five components (smart districts, living-lab initiatives, infrastructures, new services for the citizen, and open data) that are clearly being updated with the recent launch of the City Data Analytics Office. Overall, this period is characterized by a critical and proactive social context formed through transversal initiatives that merge distinct municipal datasets.

Regarding the political economy resulting from the increasing influence of grassroots initiatives on the city, Eizaguirre and Parés [144], and Degen and García [142] studied several neighborhoods in Barcelona. They found great significance in the role of collective leadership practices, creative social strategies, and political dissent in the founding and the social impact of local initiatives. Likewise, Almirall and Wareham [68], Gascó-Hernandez [143], and Timeus and Gascó [147] examined the degree of efficiency in living labs in Barcelona at different periods [63,65,67,94,140]. Although the findings are primarily related to public and private organizations, they suggest that it may be worth exploring the contribution of living labs as intermediaries that mediate between multiple stakeholders in highly political environments such as Barcelona [34].

For several authors, the changing urban governance model over the two periods became the most important factor in Barcelona. At the time of the switch between the two institutional periods, Capdevilla and Zarlenga suggested a combined approach by merging top-down and bottom-up dynamics [79]. This approach is already included in the political agenda of the current smart-city strategy, despite the problematic so-called smart contradictions noted by March and Ribera-Fumaz [127]: "it is unclear now how interests of citizens are to be made compatible with the interests of private capital and of the urban political elites in Barcelona" (p. 825). Their underlying critique relies on the idea that there is little exploration of the inner contradictions in how smart interventions are inserted into a wider political economy of urban transformation. As such, in this section, I pay special attention to the central role (smart) citizens hypothetically play in the collective production and administration of the city, by identifying three experimental strategies included in the Digital Plan 2017–2020.

4.1. Establishing the Data Commons Barcelona Program, Launching the City Data Analytics Office, and Experimenting with Decode—the EU's Scientific Flagship

My fieldwork revealed that the data ownership and technological sovereignty experimental strategies comprise three intertwined initiatives [32].

The first initiative, Data Commons Barcelona, focuses on ethical data ownership and technological sovereignty principles achieved by opening the public procurement process, and ensuring that 70% of investments in new software development go toward free and open-source software in municipal systems. Barcelona is attempting to include small and medium-sized enterprises (SMEs) in the public procurement process, generating a virtuous cycle where any SME, co-operative, or self-employed citizen can work under equal conditions while avoiding vendor lock-in and path dependency. Interviews revealed that the procurement process is opening promising pathways, based on the upcoming creation of an open digital marketplace for small-scale providers designed to meet the needs of the 13,000 existing technological companies, and to offer employment increases of 26.3% in the digital sector compared to 2016 (#I20) (See Appendix A). This initiative is implementing services such as a help for new citizens and an information portal, mobile phone services for citizens, a new calendar of city events and facilities, an open city dashboard for citizens, a technology-provider portal, and a new mobile digital identity service. Likewise, a main principle of the technopolitics of data involves explicitly improving access to the authority's data, respecting privacy, and evaluating the ethical risks of smart cities and large databases by establishing a code of ethical technological practices, including legal compliance with data protection regulations and defining a data strategy.

Thus, Barcelona is experimenting with socializing previously collected data via sensors operated by citizens, with the city taking the lead in aggregating and acting upon such data to promote new co-operative approaches that solve common urban issues, such as tackling noise levels and improving air quality. The plan is to keep this common data infrastructure open to local companies, co-operative platforms, and social organizations, allowing them to build data-driven services and create long-term public value [1] (#I4). The strategy is leading the transition to the GDPR by motivating initiatives creating a data commons that incorporates self-determination and the digital right of citizens to use municipal data for the public good of all citizens and stakeholders [6] (#I7,#I8,#I9,#I11,#I14).

Secondly, the city council opened the City Data Analytics Office on 13 February 2018. The office houses a Chief Data Officer and 40 staff members from different departments. They offer new privacy-related data improvement and analysis capabilities and can make predictions covering all areas related to managing the city. My research revealed that civil servants seem to understand the challenge of breaking down data "silos" (energy, noise, housing, garbage, meteorological, parking, air quality, water, bicycle flow, people flow, vehicle flow, and gentrification, among others) to manage three million data records per day. Despite that, it remains to be seen whether the external agile entrepreneurial culture and the bureaucratic public culture will find common ground [154] (#I5,#I6,#I18). Furthermore, the new office created an Open Data Challenge program [119] to include SMEs in solving the city's challenges and to successfully position the Data Protection Delegate in line with the GDPR. The Data Protection Delegate is in charge of coordinating data protection policy within and across departments.

Thirdly, the Decode project is an EU-funded experimental project led by Barcelona (in partnership with Amsterdam) to develop a blockchain-based architecture for data sovereignty in parallel with the operational activity of the city council. It is attempting to build new foundations for data sharing on the internet, drawing inspiration from decentralized technologies such as Bitcoin, Blockstack, and Sovrin. The project enables a system whereby all interactions involving any citizen's data are fully auditable on a public ledger (though raw data itself remains hidden). Decode is testing how people could allow their data to be used for specific social purposes, such as informing local policymaking, and could overcome the municipalities' failure to foster local alternatives to certain services (such as Uber and Airbnb) due to the lack of raw data [130]. However, interviews revealed that, despite this remarkable, cutting-edge initiative, citizens and even some stakeholders currently believe that the initiative is too abstract and experimental, and question the feasibility of business models based on this scheme (#I1,#I2,#I3,#I15).

To summarize this subsection, the fieldwork research revealed that Barcelona is motivated by compliance with EU legislation, particularly the GDPR. Business models that rely on the collection and commercial exploitation of personal data based on neoliberal data extractivist practices will face clear boundaries. GDPR seems to offer to Barcelona a way to build alternative digital infrastructures that enhance privacy and can guarantee data sovereignty for citizens, giving them back control over their data. Data Commons Barcelona could be seen as a custodian of the digital rights of citizens and a milestone in the strategic formulation of Barcelona stemming from GDPR, since it is based on remarkable principles such as privacy by design, data portability, and the right to be forgotten or "unplugged" [56]. The city authorities proactively advocate application of the GDPR. For them, "it represents an opportunity to create a city data infrastructure that puts citizens' right to data at the center of the game" (#I4). Another interviewee described the situation as follows: "We have created a new City Data Analytics Office, appointed a Chief Data Officer, and introduced a new data directive that is implemented across city hall that includes transparency, ethics, security, and privacy as fundamental principles" (#I7,#I8). Moreover, other interviewees representing the private sector, academia and (social) entrepreneurship/activism agree upon the following visionary statement (#I1,#I3,#I13,#I18,#I20): "We need to make sure that GDPR does not hinder social and technological innovation, but foster data-driven innovation while making sure citizens' rights are respected. The competitive advantage of Europe in the future world of AI and data should be this rights-based democratic framework, a new deal on data that can shape a people-centric digital future and make GDPR an international standard" (See Table 3 and Appendix A).

4.2. Experimentation within Grassroots Innovation Initiatives: "Decidim Barcelona"; and "Metadecidim Barcelona"

According to Aragon et al. [101], Decidim Barcelona is an "online participatory-democracy platform launched by the City Council of Barcelona" (p. 277) on 1 February 2016 as the main grassroots innovation experimental strategic initiative [155]. In parallel fashion, Metadecidim Barcelona [156] was launched in November 2017 as a watchdog, technopolitical research community positioned to monitor Decidim Barcelona and reflect on its work. During its short tenure, Decidim Barcelona achieved the active participation of 40,000 citizens. Overall, 12 participatory processes were initiated and 11,873 proposals were analyzed, with 70% of proposals approved as public policy for a wide range of urban issues. Interviews revealed the difficulties of integrating representative participation from the private sector, particularly from large companies with a stake in city-making (#I1,#I2,#I3,#I13,#I19,#I20). Including the multi-stakeholders' framework suggests further challenges regarding representation and the constitution of socioeconomic co-operative firms and platforms beyond the civilian volunteer activist [157].

4.3. Fostering Urban Co-Operative Platform Initiatives: "Digital Social Innovation", "Barcelona Urban Commons", and "Social Economy" Policy Framework

The third identified experimental strategy centers around the emergence of three urban co-operative platform initiatives. According to Scholz [97], these initiatives "encompass new ownership models for the internet in the urban realm" (p. 1).

The first initiative, Digital Social Innovation [158], is an EU-funded project to map and support cases of digital social innovation in Europe [159]. In Barcelona, this initiative was established as the "Barcelona Maker District" around the FabLabs in the Poblenou neighbourhood. FabLabs aims to develop and expand new educational models and skills by democratizing access to fabrication through work with targeted segments, such as women, elderly people, and children.

Similarly, the second initiative, the Barcelona Urban Commons [160], was established in the broader metropolitan area of Barcelona in 2010 [161,162]. A wide range of economic activities affecting citizens and communities was identified, including "Guifi.net", "Som Energia", and "Coop57", among many others. Thus, over the last decade, these example activities offered digital connectivity, energy, and financial services, respectively, in a socioeconomic model that comprises 10% of the city economy. However, interviews revealed that some participants do not consider these initiatives as having successfully scaled up due to the shortcomings in transitioning from a grassroots communitarian movement driven by voluntarism to entrepreneurial co-operatives driven by professionalism (#I1,#I13).

The third initiative corresponds to the social economy policy framework which consists of 4718 socioeconomic projects and 861 co-operative firms [163]. This policy framework aims to support transformation of these projects from serving local economies to inclusion in the socioeconomic model of the entire city. My interviews point out the gap between the modus operandi of the corporate private sector and the initiatives conducted at the neighborhood level by the city council. One example is the recent launch of the new digital local currency called "rec", promoted by the EU-funded project B-Mincome [164], that addresses guaranteed minimum income in deprived urban areas of the city.

Despite the strong efforts of the council, the EU funds received to foster urban co-operative platforms, and the rapid pollination of projects, my findings suggest that the metropolitan ecosystem still needs to mature, particularly because scaling up an alternative socioeconomic model seems to be a process that takes longer than one institutional period. Although the strategic formulation attempts to line up a large number of initiatives in a transversal fashion, in the short term, it is likely to require further corrective measures.

Table 3. Deconstructing Barcelona's (smart) citizen digital policy framework. GDPR—General Data Protection Regulation.

Three Operational Research Questions	Policy Analysis		Scientific Analysis	Qualitative Analysis
				Based on twenty interviews of key stakeholders in Barcelona from September 2017 to March 2018
	Experimental Strategies	Strategic Initiatives	Literature Review	Summary of key responses structured through the Penta Helix framework [34] (Appendix A): private sector (2); public sector (9); academia (2); civic society (3); (social) entrepreneurs/activists (3).
1. What prospects exist for an alternative data ownership regime in the ongoing smart city model of Barcelona? (#E3)	(Section 4.1.) Data Ownership and Technological Sovereignty	1. Barcelona Data Commons 2. City Data Analytics Office 3. Decode	[1,2,6,32,35,111,128–130,154]	1. The majority of interlocutors clearly addressed the positive and irreversible transition as Barcelona is already adapting to the GDPR (#I7,#I8,#I9,#I11,#I14). 2. The feedback was very positive in relation to the public procurement reform (#I20). 3. The transition to open-source software presents remarkable obstacles (#I5,#I6,#I18). 4. Despite the positive initiatives and irreversible transition toward GDPR, Decode may have created overly high expectations, as the initiative is perceived as "too experimental" (#I1,#I2,#I3,#I15).
2. What are the practical consequences of the grassroots innovation initiatives implemented in Barcelona for businesses, local governments, academia, civic society, and social entrepreneurs/activists? (#E2)	(Section 4.2.) Grassroots Innovations	1. Decidim Barcelona 2. Metadecidim Barcelona	[34,101,155,156]	1. Interlocutors argued that these initiatives lack an overall degree of representation because they omit "stakeholders' power relation interdependencies" (#I19,#I20). 2. Others asked "How is the private sector represented in these initiatives?" (#I1,#I2,#I3,#I13). 3. The established channel could be seen as a seed to scale up a broader deliberative governance model based on "multi-stakeholder" representation (#I12,#I15,#I16,#I17).
3. Is another experimental type of smart city driven by co-operative service provision models based on social innovation possible in Barcelona? That is, does a "third way" exist between the state and the market that overcomes the PPP? (#E1)	(Section 4.3.) Urban Co-Operative Platforms	1. Digital Social Innovation Barcelona 2. Urban Commons 3. Social Economy	[2,97,116,132,159,161–164]	1. Interlocutors revealed the lack of a critical mass for co-operative platforms (#I4,#I7,#I8,#I10,#I20). 2. Although Barcelona is shifting its smart-city model, co-operative niche experiments are having difficulties scaling up (#I10,#I16). 3. Setting up a "third way" requires moving on from volunteer grassroots initiatives to professional entrepreneurial co-operatives (#I1,#I13).

To summarize this policy analysis and integrate it with the fieldwork research findings, Table 3 illustrates the responses to each research question proposed in this paper. The identified experimental strategies and the strategic initiatives respond directly to the initiatives that connect the three operational research questions. Despite the individual identification of initiatives, I must acknowledge the vast amalgamation of intertwined initiatives, confirmed through interviews, which made the fieldwork complex. Thus, Table 3 does not provide a deep analysis of each initiative but instead serves to deconstruct Barcelona's digital policy framework to examine strategic evidence-based insights by providing the main findings grouped through the coding system of interviews/interlocutors accordingly.

5. Conclusions

This paper's analysis of the current smart-city strategy and digital policy in Barcelona is exploratory and prospective because existing initiatives in Barcelona are new, with more awaiting implementation or existing only as concepts rather than material projects. Nonetheless, in light of the Barcelona study results, this paper argues that deconstructing the policy's framework provides further research directions to expand understanding of the ongoing urban challenges arising from the direct technopolitical implications of these initiatives for (smart) citizens. As a result, this paper concludes by highlighting these future research directions potentially shaped as follows: (i) post-GDPR privacy, technological sovereignty, and data commons alternative regimes in the establishment of more tailored, human-centered, and context-aware algorithmic governance platforms for new co-operative socio-economic models within and among European cities and regions; (ii) publicly scrutinized new bottom-up democratic mechanisms through blockchain infrastructure to conduct decentralized and networked decision-making and deliberation processes on integrated sectorial urban policies (energy, mobility, health, entrepreneurship, voting, taxation, residency, migration, etc.); (iii) practical urban experiments through living labs at the neighborhood and district level using blockchain technology to foster open social innovation ecosystems among a broad representation of agents through the Penta Helix multi-stakeholders' framework.

The research findings from the fieldwork (Appendix A) articulated (through interviews coded as #I) and validated (through events coded as #E) responses to the three operational research questions presented in the introduction, as summarized below.

In response to the first question (validated via #E3), in the current experimental phase in Barcelona, the degree to which strategic initiatives modify technopolitical awareness of who owns data could offer interesting pathways for establishing an alternative data ownership regime. In addition, based on the fieldwork research findings, Barcelona City Council's new procurement process is designed to incentivize responsible innovation and respect for privacy, and is currently undergoing complete internal migration to open-source software by spring 2019. Furthermore, by the end of 2019, the city government plans to launch new online tools allowing people to selectively disclose the information they desire to share when using the council's official e-participation platform, Decidim Barcelona, while preserving citizens' anonymity. Nonetheless, the fieldwork revealed that, despite the positive transition reforming public procurement service provisions, technological sovereignty—citizens' ability to have a voice in surrounding technological infrastructure operation and end results—still has growth opportunities regarding integrating open-source software and emerging blockchain data architecture in the new EU context, characterized by the GDPR. As such, employing infrastructures such as blockchain is likely to require a new interpretation from the urban and data science perspective, which will be key to future experiments and predicting alternative data ownership regimes in cities. Likewise, technologies such as blockchain could improve access to data held by authorities while respecting privacy. Ultimately, this can also lead to proper evaluation of the ethical and political risks of smart cities with large databases by establishing a code of ethics for technological practices, including legal compliance with data protection regulations such as the GDPR and a defined data strategy benefitting citizens.

In response to the second question (validated via #E2), in light of the grassroots initiatives examined, interdependencies among stakeholders have a thorough policy focus while fostering ecosystems of entrepreneurial starts-ups and SMEs. However, despite the effort to bring together a wide range of stakeholders, interviews revealed that initiatives based on a collaborative, polycentric and deliberative governance model may need to evolve toward a more integrated, multi-stakeholder framework through further collaboration with the private sector (potentially by living labs bringing in entrepreneurial academics and enthusiastic civil servants). These initiatives might also need to foster the active role of social entrepreneurs, activists, "bricoleurs", brokers, and assemblers (see the fifth helix embedded in the Penta Helix multi-stakeholders' framework in Figure 1) to further explore the opportunities of an emerging and open social innovation ecosystem in Barcelona.

Ultimately, in response to the third question (validated via #E1), thus far, PPPs were the leading smart-city model underpinning the alliance between the state and market in many cities. Yet, these deals may have been driven by neoliberal and extractivist data policy implementation. Questions such as "who controls the data?" are still rarely subject to any public oversight and scrutiny. Thus, to revert this situation, Barcelona could implement two steps: (i) build a critical mass of locally rooted co-operative platforms based on the data commons to scale-up initiatives, while taking back control of citizens' data; (ii) explore alternative, professional, entrepreneurial, co-operative business cultures to complement the volunteer grassroots seeds already flourishing [157,165].

Regarding the lessons learnt from the Barcelona case study and policy implications of the field of interdisciplinary and global smart cities research [166], this paper, drawing from detailed policy analysis and fieldwork study, suggests that Barcelona, in the post-GDPR context, appears to be leading a cutting-edge digital transformation agenda that may soon influence other European or international cities and regions on data technopolitical implications for (smart) citizens.

Whereas, generally, the pervasive algorithmic disruption consciously or unwittingly reveals the lack of critical reflective smart-city policy-making, or indicates the hegemonic mainstreamed position of the PPPs, cities like Barcelona are challenging, and thus, embarked in experimenting an alternative policy route. Hence, it is likely the interdisciplinary and global smart cities research focus may gradually evolve toward the increasing study of a typology of (smart) citizens, defined as decision-makers and concerned users not only about data protection and security issues [167], but also data ownership and technological sovereignty policy implications directly related to digital rights to the city.

In summary, whether Barcelona's current strategy is little more than a declaration of intentions of a progressivist smart-city policy agenda remains to be seen. This paper concludes that, in the ongoing algorithmic disruptive context characterized by extractive data practices, smartness may not be appealing in cities such as Barcelona, although the experimental approach has yet to be firmly fixed as an alternative paradigm. Recognizing that smartness may not be appealing and having been viewed as technocratic, the experimental approach has yet to be entirely established. Thus, an open question remains: how will (smart) citizens decide, control, govern, manage, and ultimately, own their data by being both conscious of digital rights to the city and aware of duties in the technopolitical processes of city making?

Funding: The author's work was supported by the ESRC under the grant "Urban Transformations"; the RSA under grant "Smart City-Regional Governance for Sustainability" Research Network; the Brussels Centre for Urban Studies under the grant "Visiting Research Fellowship Scheme 2016"; and the European Commission under the Grant "H2020-SCC-01-2015-Smart Cities and Communities solutions integrating energy, transport, ICT sectors through lighthouse project-691735-REPLICATE".

Acknowledgments: The author is deeply indebted to the interlocutors/interviewees in Barcelona who kindly co-operated for interviews and data provision. I extend my gratitude also to Esteve Almirall and Ramón Sangüesa for help during the fieldwork research process in Catalonia.

Conflicts of Interest: The author declares no conflicts of interest.

Appendix A. Lists of Interviews and Symposiums. The List of Interviewed Twenty Interlocutors Structured through Penta Helix Multi-Stakeholders' Framework [34] and the List of Three Symposiums Related to This Study during the Fieldwork Research from September 2017 to March 2018

Table A1. Details of direct sources: fieldwork research through interviews and symposiums from September 2017 to March 2018.

Interviews			
Interview/Interlocutor Coding System [168]	**Description of the Stakeholder**	**Date**	**Penta Helix Multi-Stakeholders' Framework**
#I1	City Protocol Society	4 September 2017	Private sector
#I2	CISCO	5 September 2017	
#I3	Telefónica Alpha	6 September 2017	
#I4	Technology & Digital Innovation Commissioner, Barcelona City Council	7 September 2017	Public sector
#I5	Social Economy, Local Development and Consumption Commissioner, Barcelona City Council	8 September 2017	
#I6	Institut Municipal d'Informàtica (IMI), Barcelona City Council	17 January 2018	
#I7	Data Commons Program, Barcelona City Council	19 March 2018	
#I8	Municipal Data Office, Barcelona City Council	20 March 2018	
#I9	General Secretary of Telecommunications, Cybersecurity, and Digital Society of the Regional Government of Catalonia	21 March 2018	
#I10	Democratic Innovation department, Barcelona City Council	22 March 2018	
#I11	Smart Catalonia, Regional Government of Catalonia	17 January 2018	
#I12	Digital Social Innovation department, Barcelona City Council	17 January 2018	
#I13	ESADE Business School	23 November 2017	Academia
#I14	Universitat Pompeu Fabra (UPF)	5 March 2018	
#I15	Urban Commons facilitator	5 February 2018	Civic society
#I16	Urban Co-operative representative	6 February 2018	
#I17	Social Economy representative	7 February 2018	
#I18	Technological activist	8 February 2018	(Social) entrepreneurs/activists
#I19	Algorithmic activist/entrepreneur	9 February 2018	
#I20	Data scientist/entrepreneur	12 February 2018	
Symposiums			
Event Coding System [168]	**Details of Each Symposium: Direct Participation with the Detailed Scientific Contribution**	**Date**	**Organizer**
#E1	*Smart City Expo World Congress 2017 (#SCEWC2017)* Title: "*Unplugging, Technopolitics of Data and Smart City Devolution: Comparing Barcelona, Bilbao, Glasgow, and Bristol*" (https://www.igorcalzada.com/barcelona/)	16 November 2017	SCEWC
#E2	*ESADE Smart Cities and Data Speaker Series (#MIBASpeakerSeries)* Title: "*Data Commons and Devolution in Europe: GDPR*" (https://www.igorcalzada.com/esade/ & https://www.youtube.com/watch?time_continue=6&v=iP8LVQWrdJ0)	23 November 2017	ESADE Business School
#E3	*Barcelona Data Commons program with the Board of Directors of the Barcelona City Council (#DataCommons)* Title: "*Cities & Data: How the Digital, Big Data & Data Science are Transforming the Government/Com el Digital, Big Data & Data Science està Transformant els Governs*" (https://www.igorcalzada.com/speaker-on-the-board-of-directors-of-the-barcelona-city-council-on-data-driven-cities-17-01-2018-catalonia-spain/)	17 January 2018	Barcelona City Council

References

1. Bria, F. Our Data Is Valuable. Here's How We Can Take That Value Back. 2018. Available online: https://www.theguardian.com/commentisfree/2018/apr/05/data-valuable-citizens-silicon-valley-barcelona (accessed on 5 April 2018).
2. Morozov, E.; Bria, F. *Rethinking the Smart City: Democratizing Urban Technology*; Rosa Luxemburg Stiftung: New York, NY, USA, 2018.
3. Xnet, Xnet installs a Whistleblowing Platform against Corruption for the City Hall of Barcelona—Powered by GlobaLeaks and TOR Friendly. 2017. Available online: https://blog.p2pfoundation.net/xnet-installs-whistleblowing-platform-corruption-city-hall-barcelona-powered-globaleaks-tor-friendly/2017/01/19 (accessed on 27 July 2018).
4. Bass, T.; Sutherland, E.; Symons, T. *Reclaiming the Smart City: Personal Data, Trust and the New Commons*; NESTA: London, UK, 2018.
5. Andrews, J. How Will Europe's New Data Laws Affect Cities? 2018. Available online: https://cities-today.com/how-will-europes-new-data-laws-affect-cities/ (accessed on 17 August 2018).
6. Barcelona City Council, Plan Digital de Ayuntamiento de Barcelona: Medida de Gobierno de Gestión Ética y Responsable de Datos: Barcelona Data Commons. 2018. Available online: http://ajuntament.barcelona.cat/digital/en/blog/ethical-and-responsible-data-management-barcelona-data-commons (accessed on 31 March 2018).
7. General Data Protection Regulation (GDPR). 2018. Available online: https://www.eugdpr.org/ (accessed on 1 April 2018).
8. Edwards, L. Privacy, Security and Data Protection in Smart Cities: A Critical EU Law Perspective. *Eur. Data Prot. Law Rev.* **2016**, *2*, 28–58. Available online: https://papers.ssrn.com/sol3/papers.cfm?abstract_id=2711290 (accessed on 17 August 2018). [CrossRef]
9. Schiek, D.; Gideon, A. Outsmarting the gig-economy through collective bargaining: EU competition law as a barrier to smart cities? *Int. Rev. Law Comput. Technol.* **2018**, *32*, 275–294. [CrossRef]
10. Bridle, J. *New Dark Age*; Verso: London, UK, 2018.
11. Finn, E. *What Algorithms Want: Imagination in the Age of Computing*; The MIT Press: Cambridge, MA, USA, 2017.
12. Van Dijck, J. Datafication, dataism and dataveillance: Big Data between scientific paradigm and ideology. *Surveill. Soc.* **2014**, *12*. [CrossRef]
13. Buttarelli, G. Big Tech Is Still Violating Your Privacy. 2018. Available online: https://www.washingtonpost.com/news/theworldpost/wp/2018/08/14/gdpr/?utm_term=.9e9ddd6cf937 (accessed on 17 August 2018).
14. Evenstad, L. Cities Key to Making Data a New form of Infrastructure, Says Nesta. 2018. Available online: https://www.computerweekly.com/news/252445624/Cities-key-to-making-data-a-new-form-of-infrastructure-says-Nesta (accessed on 17 August 2018).
15. Hintz, A.; Dencik, L.; Wahl-Jorgensen, K. Digital citizenship and surveillance society. *Int. J. Commun.* **2017**, *11*, 731–739.
16. O'Neill, C. *Weapons of Math Destruction: How Big Data Increases Inequality and Threatens Democracy*; Penguin Random House: London, UK, 2008.
17. Taplin, J. *Move Fast and Break Things: How Facebook, Google, and Amazon Have Cornered Culture and What It Means for All of Us*; Little Brown: New York, NY, USA, 2017.
18. Eubanks, V. *Automating Inequality: How High-Tech Tools Profile, Police, and Punish the Poor*; St. Martin's Press: New York, NY, USA, 2017.
19. Noble, S.U. *Algorithms of Oppression: How Search Engines Reinforce Racism*; NYU Press: New York, NY, USA, 2018.
20. European Commission. *The Making of a Smart City: Policy Recommendations*; European Commission: Brussels, Belgium, 2017.
21. European Commission. *The Making of a Smart City: Best Practices across Europe*; European Commission: Brussels, Belgium, 2017.
22. European Commission. *The Making of a Smart City: Replication and Scale-Up of Innovation in Europe*; European Commission: Brussels, Belgium, 2017.
23. Trencher, G. Towards the Smart City 2.0: Empirical Evidence of Using Smartness as a Tool for Tackling Social Challenges. *Technol. Forecast. Soc. Chang.* **2018**. [CrossRef]

24. Kitchin, R. *The Data Revolution: Big Data, Open Data, Data Infrastructures & Their Consequences*; Sage: London, UK, 2014.

25. Karvonen, A.; Cugurullo, F.; Caprotti, F. *Inside Smart Cities: Place, Politics and Urban Innovation*; Routledge: London, UK, 2018.

26. Marvin, S.; Luque-Ayala, A.; McFarlane, C. *Smart Urbanism: Utopian Vision or False Dawn?* Routledge: New York, NY, USA, 2015.

27. Willis, K.S.; Aurigi, A. *Digital and Smart Cities*; Routledge: Oxon, UK, 2018.

28. Sadowski, J.; Pasquale, F. The Spectrum of Control: A social theory of the smart city. *First Monday* **2015**, *20*. [CrossRef]

29. Evans, J.; Karvonen, A.; Raven, R. *The Experimental City*; Routledge: London, UK, 2016.

30. Marvin, S.; Bulkeley, H.; Mai, L.; McCormick, K.; Palgan, Y.V. *Urban Living Labs: Experimenting with City Futures*; Routledge: London, UK, 2018.

31. Karvonen, A. The city of permanent experiments? In *Innovating Climate Governance: Moving Beyond Experiments*; Turnheim, B., Kivimaa, P., Berkhout, F., Eds.; Cambridge University Press: Cambridge, UK, 2018; pp. 201–215.

32. Barcelona City Council, Barcelona City Council Digital Plan 2017–2020: A Government Measure for Open Digitization: Free Software and Agile Development of Public Administration Services. 2017. Available online: http://ajuntament.barcelona.cat/digital/en/blog/barcelona-digital-government-open-agile-and-participatory (accessed on 31 March 2018).

33. Fielding, N.G. Triangulation and Mixed Methods Designs: Data Integration with New Research Technologies. *J. Mixed Methods Res.* **2012**, *6*, 124–126. [CrossRef]

34. Calzada, I.; Cowie, P. Beyond Data-Driven Smart City-Regions? Rethinking Stakeholder-Helixes Strategies. *Regions* **2017**, *308*, 25–28. [CrossRef]

35. Barcelona City Council, BITS, Barcelona Initiative for Technological Sovereignty. 2016. Available online: https://bits.city/ (accessed on 10 December 2016).

36. Wallace, N.; Castro, D. *The Impact of the EU's New Data Protection Regulation on AI*; Centre for Data Innovation: Washington, DC, USA, 2018.

37. Chu, B. Angela Merkel at Davos. 2018. Available online: https://www.independent.co.uk/news/business/analysis-and-features/angela-merkel-davos-2018-speech-trump-what-did-she-say-germany-a8176021.html (accessed on 10 March 2018).

38. Almirall, E.; Wareham, J.; Ratti, C.; Conesa, P.; Bria, F.; Gaviria, A.; Edmondson, A. Smart cities at the crossroads: New tensions in the city transformation. *Calif. Manag. Rev.* **2017**, *59*, 141–152. [CrossRef]

39. Caragliu, A.; Del Bo, C. The Economics of Smart City Policies. *Sci. Reg.* **2018**, *17*, 81–104.

40. Cardullo, P.; Kitchin, R. Being a 'Citizen' in the Smart City: Up and Down the Scaffold of Smart Citizen Participation in Dublin, Ireland. *GeoJournal* **2018**, *78*, 1–13. [CrossRef]

41. Christl, W.; Spiekermann, S. Networks of Control: Corporate Surveillance, Digital Tracking, Big Data and Privacy. 2016. Available online: http://crackedlabs.org/dl/Christl_Spiekermann_Networks_Of_Control.pdf (accessed on 17 August 2018).

42. Christl, W. Corporate Surveillance in Everyday Life. How Companies Collect, Combine, Analyze, Trade, and Use Personal Data on Billions. A Report by Cracked Labs. 2017. Available online: http://crackedlabs.org/en/corporate-surveillance/info (accessed on 17 August 2018).

43. Solon, O. Georges Soros: Facebook and Google a Menace to Society. 2018. Available online: https://www.theguardian.com/business/2018/jan/25/george-soros-facebook-and-google-are-a-menace-to-society (accessed on 26 February 2018).

44. Morozov, E. The Rise of Data and the Death of Politics. 2014. Available online: https://www.theguardian.com/technology/2014/jul/20/rise-of-data-death-of-politics-evgeny-morozov-algorithmic-regulation (accessed on 10 November 2016).

45. Morozov, E. So You Want to Switch Off Digitally? I'm Afraid That Will Cost You. 2017. Available online: https://www.theguardian.com/commentisfree/2017/feb/19/right-to-disconnect-digital-gig-economy-evgeny-morozov (accessed on 19 March 2017).

46. Morozov, E. *Capitalismo Big Tech ¿Welfare o Neofeudalismo Digital?* Enclave de Libros Ediciones: Madrid, Spain, 2018.

47. Stucke, E.; Grunes, A.P. Data-Opolies. 2017. Available online: https://ssrn.com/abstract=927018orhttp://dx.doi.org/10.2139/ssrn.2927018 (accessed on 3 March 2017).

48. Stucke, E. Should We Be Concerned About Data-Opolies? 2018. Available online: https://ssrn.com/abstract=3144045 (accessed on 8 April 2018).

49. Bucher, T. Want to Be on Top? Algorithmic Power and the Threat of Invisibility on Facebook. *New Media Soc.* **2012**, *14*, 1164–1180. [CrossRef]

50. Cadwalladr, C.; Graham-Harrison, E. Revealed: 50 million Facebook Profiles Harvested for Cambridge Analytica in Major Data Breach. 2018. Available online: https://www.theguardian.com/news/2018/mar/17/cambridge-analytica-facebook-influence-us-election (accessed on 19 March 2018).

51. Lupton, D. You Are Your Data: Self-Tracking Practices and Concepts of Data. In *Lifelogging: Digital Self-Tracking and Lifelogging—Between Disruptive Technology and Cultural Transformation*; Seike, S., Ed.; Springer: Wiesbaden, Germany, 2016; pp. 61–79.

52. Lupton, D.; Michael, M. Depends on Who's Got the Data: Public Understandings of Personal Digital Dataveillance. *Surveill. Soc.* **2017**, *15*, 254–268. [CrossRef]

53. Acuto, M. Global science for city policy. *Science* **2018**, *359*, 165–166. [CrossRef] [PubMed]

54. Coletta, C.; Kitchin, R. Algorithmic governance: Regulating the 'heartbeat' of a city using the Internet of Things. *Big Data Soc.* **2017**, *4*, 1–16. [CrossRef]

55. Danaher, J.; Hogan, M.J.; Noone, C.; Kennedy, R.; Behan, A.; De Paor, A.; Felzmann, H.; Haklay, M.; Khoo, S.M.; Morison, J.; et al. Algorithmic governance: Developing a research agenda through the power of collective intelligence. *Big Data Soc.* **2017**, *4*, 1–21. [CrossRef]

56. Calzada, I.; Cobo, C. Unplugging: Deconstructing the Smart City. *J. Urban Technol.* **2015**, *22*, 23–43. [CrossRef]

57. Nilssen, M. To the smart city and beyond? Developing a typology of smart urban innovation. *Technological Forecast. Soc. Chang.* **2018**. [CrossRef]

58. Calzada, I. The Technopolitics of Data and Smart Devolution in City-Regions: Comparing Glasgow, Bristol, Barcelona, and Bilbao. *Systems* **2017**, *5*, 18. [CrossRef]

59. Joss, S. Future cities: Asserting public governance. *Palgrave Commun.* **2018**, *4*, 1–4. [CrossRef]

60. Shin, Y.; Shin, D. Modelling Community Resources and Communications Mapping for Strategic Inter-Organizational Problem Solving and Civic Engagement. *J. Urban Technol.* **2016**, *23*, 47–66. [CrossRef]

61. Shin, D. Ubiquitous city: Urban technologies, urban infrastructure and urban informatics. *J. Inf. Sci.* **2009**, *35*, 515–526. [CrossRef]

62. Thomas, V.; Wang, D.; Mullagh, L.; Dunn, N. Where's Wally? In Search of Citizen Perspectives on the Smart City. *Sustainability* **2016**, *8*, 207. [CrossRef]

63. Bulkeley, H.; Breitfuss, M.; Coenen, L.; Frantzeskaki, N.; Fuenfschilling, L.; Grillitsch, M.; Hartmann, C.; Kronsell, A.; McCormick, K.; Marvin, S.; et al. Working Paper on Urban living Labs and Urban Sustainability Transitions. GUST Governance of Urban Sustainability Transitions. 2015. Available online: http://www.urbanlivinglabs.net/p/publications.html (accessed on 9 March 2018).

64. Evans, J. Trials and Tribulations: Problematizing the City through/as Urban Experimentation. *Geogr. Compass* **2016**, *10*, 429–443. [CrossRef]

65. Kronsell, A.; Mukhtar-Landgren, D. Experimental Governance: The Role of Municipalities in Urban Living Labs. *Eur. Plan. Stud.* **2018**, 1–20. [CrossRef]

66. Scholl, C.; Kemp, R. City Labs as Vehicles for Innovation in Urban Planning Processes. *Urban Plan.* **2016**, *1*, 89–102. [CrossRef]

67. Schuurman, D.; De Marez, L.; Ballon, P. The Impact of Living Lab Methodology on Open Innovation Contributions and Outcomes. *Technol. Innov. Manag. Rev.* **2016**, *1*, 7–16. [CrossRef]

68. Almirall, E.; Wareham, J. Living labs: Arbiters of mid-and ground-level innovation. *Technol. Anal. Strat. Manag.* **2011**, *23*, 87–102. [CrossRef]

69. Von Wirth, T.; Fuenfschilling, L.; Frantzeskaki, N.; Coenen, L. Impacts of urban living labs on sustainability transitions: Mechanisms and strategies for systemic change through experimentation. *Eur. Plan. Stud.* **2018**, 1–29. [CrossRef]

70. Mora, L.; Bolici, R.; Deakin, M. The First Two Decades of Smart-City Research: A Bibliometric Analysis. *J. Urban Technol.* **2017**, *24*, 3–27. [CrossRef]

71. De Jong, M.; Joss, S.; Schraven, D.; Zhan, C.; Weijnen, M. Sustainable-smart-resilient-low carbon-eco-knowledge cities; making sense of a multitude of concepts promoting sustainable urbanization. *J. Clean. Prod.* **2015**, *109*, 25–38. [CrossRef]
72. Saunders, T.; Baeck, P. *Rethinking Smart Cities from the Ground Up*; NESTA: London, UK, 2015.
73. Kitchin, R. Reframing, Reimagining and Remaking Smart Cities. The Programmable City Working Paper 20. 2016. Available online: https://osf.io/cyjhg/ (accessed on 7 February 2017).
74. Bull, R.; Azennoud, M. Smart citizens for smart cities: Participating in the future. *Proc. Inst. Civ. Eng. Energy.* **2016**, *169*, 93–101. [CrossRef]
75. Hemment, D.; Townsend, A. *Smart Citizens*; Future Everything Publications: Manchester, UK, 2013.
76. Noveck, B.S. *Smart Citizens, Smarter State: The Technologies of Expertise and the Future of Governing*; Harvard University Press: Cambridge, MA, USA, 2015.
77. Niederer, S.; Priester, R. Smart Citizens: Exploring the Tools of the Urban Bottom-Up Movement. *Comput. Support. Cooperative Work (CSCW)* **2016**, *25*, 137–152. [CrossRef]
78. Waag. A Manifesto for Smart Citizens. 2016. Available online: http://waag.org/en/blog/manifesto-smart-citizens (accessed on 20 December 2016).
79. Capdevilla, I.; Zarlenga, M. Smart City or Smart Citizens? The Barcelona Case. 2015. Available online: http://dx.doi.org/10.2139/ssrn.2585682 (accessed on 7 July 2018).
80. Gutiérrez-Rubí, A. *Smart Citizens: Ciudades a Escala Humana*; Barcelona: Catalonia, Spain, 2017.
81. The Royal Society. *Data Management and Use: Governance in the 21st Century*; The Royal Society: London, UK, 2017.
82. Coletta, C.; Heaphy, L.; Perng, S.-Y.; Waller, L. Data-driven cities? Digital urbanism and its proxies: Introduction. *Tecnoscienza Ital. J. Sci. Technol. Stud.* **2017**, *8*, 5–18.
83. Burns, R.; Daltan, C.M.; Thatcher, J.E. Critical Data, Critical Technology in Theory and Practice. *Prof. Geographer.* **2018**, *70*, 126–128. [CrossRef]
84. Ojo, A.; Curry, E.; Zeleti, F.A. A Tale of Open Data Innovations in Five Smart Cities. In Proceedings of the 48th Hawaii International Conference on Systems Sciences, Kauai, HI, USA, 5–8 January 2015; pp. 2326–2335. [CrossRef]
85. Baeck, P. *Data for Good: How Big and Open Data Can Be Used for the Common Good*; NESTA: London, UK, 2015.
86. Gray, J.; Lämmerhirt, D. *Data and The City: How Can Public Data Infrastructures Change Lives in Urban Regions*; Open Knowledge: Cambridge, UK, 2017.
87. Harari, Y.N. Yuval Noah Harari on Big Data, Google and the End of Free Will. 2016. Available online: https://www.ft.com/content/50bb4830-6a4c-11e6-ae5b-a7cc5dd5a28c (accessed on 10 November 2016).
88. Morozov, E.; Bria, F. Roundtable Session—A New Deal on Data: What Role for Cities? Smart City Expo World Congress. 2017. Available online: https://www.youtube.com/watch?v=1cakaaip2Vw (accessed on 1 February 2017).
89. Shilton, K. When They Are Your Big Data: Participatory Data Practices as a Lens on Big Data. In *Big Data is Not a Monolith*; Sugimoto, C., Ekbia, H., Mattioli, M., Eds.; MIT Press: Boston, MA, USA, 2016.
90. PWC. *From Concept to Applied Solutions: Data-Driven Cities*; PWC: London, UK. 2016.
91. Ashton, P.; Weber, R.; Zook, M. The cloud, the crowd, and the city: How new data practices reconfigure urban governance. *Big Data Soc.* **2017**, *4*, 1–5. [CrossRef]
92. Kelleher, J.D.; Tierney, B. *Data Science*; The MIT Press: Cambridge, MA, USA, 2018.
93. Lohr, S. *Data-ism: Inside the Big Data Revolution*; Oneworld Publications: London, UK, 2015.
94. Acuto, M.; Steenmans, K.; Iwaszuk, E.; Ortega-Garza, L. Informing urban governance? Boundary-spanning organisations and the ecosystem of urban data. *Area* **2018**, 1–10. [CrossRef]
95. Kitchin, R. Thinking critically about and researching algorithms. *Inf. Commun. Soc.* **2017**, *20*, 14–29. [CrossRef]
96. Gerlitz, C.; Helmond, A. The like economy: Social buttons and the data-intensive web. *New Media Soc.* **2013**, *15*, 1348–1365. [CrossRef]
97. Scholz, T. *Platform Cooperativism: Challenging the Corporate Sharing Economy*; Rosa Luxemburg Stiftung: New York, NY, USA, 2016.
98. Ostrom, E. Beyond Markets and States: Polycentric Governance of Complex Economic Systems. *Am. Econ. Rev.* **2010**, *100*, 1–33. [CrossRef]
99. Subirats, J.; Rendueles, C. *Los Bienes Comunes ¿Oportunidad o Espejismo?* Icaria: Barcelona, Spain, 2016.

100. Hardin, G. The tragedy of the commons. *Science* **1968**, *162*, 1243–1248. [CrossRef] [PubMed]

101. Aragón, P.; Kaltenbrunner, A.; Calleja-López, A.; Pereira, A.; Monterde, A.; Barandiaran, X.; Gómez, V. Deliberative Platform Design: The Case Study of the Online Discussions in Decidim Barcelona. In *Social Informatics*; Ciampaglia, G.L., Mashhadi, A., Yasseri, T., Eds.; Springer International Publishing: Cham, UK, 2018.

102. Bianchi, I. The post-political meaning of the concept of commons: The regulation of the urban commons in Bologna. *Space Polity* **2018**, 1–20. [CrossRef]

103. Alosi, A. Commoditized Workers: Case Study Research on Labor Law Issues Arising from a Set of on-Demand/Gig Economy Platforms. *Comp. Labor Law Policy J.* **2016**, *37*, 653.

104. Replicate EU Project: City-to-City-Learning Programme. 2018. Available online: www.replicate-project.eu (accessed on 18 August 2018).

105. Atkin, R. Stop Replacing London's Phone Boxes with Corporate Surveillance. 2018. Available online: https://www.wired.co.uk/article/linkuk-bt-google-free-wifi-and-calls-london (accessed on 17 August 2018).

106. Cherry, M. Beyond misclassification: The digital transformation of work. *Comp. Labour Law Policy Rev.* **2016**, *37*, 1–27.

107. Levy, K.; Barocas, S. Refractive surveillance: Monitoring customers to manage workers. *Int. J. Commun.* **2018**, *12*, 1166–1188.

108. Andrews, L. *I Know Who You Are and I Saw What You Did: Social Networks and the Death of Privacy*; Free Press: New York, NY, USA, 2013.

109. Lightfoot, G.; Wisnieski, T.P. Information asymmetry and power in a surveillance society. *Inf. Organ.* **2014**, *24*, 214–235. [CrossRef]

110. Manzerolle, V.; Smeltzer, S. Consumer databases and the commercial mediation of identity: A medium theory analysis. *Surveill. Soc.* **2011**, *8*, 323–337. [CrossRef]

111. Jun, M. Blockchain government—A next form of infrastructure for the twenty-first century. *J. Open Innov. Technol. Market Complex.* **2018**, *4*, 1–7. [CrossRef]

112. Vanolo, A. Is there anybody out there? The place and role of citizens in tomorrow's smart cities. *Futures* **2016**, *82*, 26–36. [CrossRef]

113. De Waal, M.; Dignum, M. The citizen in the smart city. How the smart city could transform citizenship. *Inf. Technol.* **2017**, *59*, 263–273. [CrossRef]

114. Greenfield, A. *Radical Technologies: The Design of Everyday Life*; Verso: London, UK, 2017.

115. Hollands, R.G. Critical interventions into the corporate smart city. *Camb. J. Reg. Econ. Soc.* **2014**, *8*, 61–77. [CrossRef]

116. Calzada, I. From Smart Cities to Experimental Cities? In *Co-Designing Economies in Transition: Radical Approaches in Dialogue with Contemplative Social Sciences*; Giorgino, V.M.B., Walsh, Z.D., Eds.; Palgrave Macmillan: Cham, UK, 2018; pp. 191–217.

117. Arnstein, S.R. A ladder of citizen participation. *J. Am. Inst. Plan.* **1969**, *35*, 216–224. [CrossRef]

118. Kitchin, R. The Realtimeness of Smart Cities. *Tecnoscienza Ital. J. Sci. Technol. Stud.* **2017**, *8*, 19–41.

119. Datacity Numa. 2018. Available online: http://datacity.numa.co/ (accessed on 19 March 2018).

120. Campbell, T. *Beyond Smart Cities: How Cities Network, Learn and Innovate*; Earthscan: Oxon, UK, 2012.

121. Hajer, M.A.; Dassen, T. *Smart about Cities: Visualising the Challenge for 21st Century Urbanism*; Nai010 Publishers: Amsterdam, The Netherlands, 2014.

122. Mora, L.; Bolici, R. The Development Process of Smart City Strategies: The Case of Barcelona. In Proceedings of the 1st International City Regeneration Congress, Tampere, Finland, 3–4 September 2015.

123. Mora, L.; Bolici, R. How to Become a Smart City: Learning from Amsterdam. In *Smart and Sustainable Planning for Cities and Regions: Results of SSPCR 2015*; Bisello, A., Vettorato, D., Stephens, R., Elisei, P., Eds.; Springer International Publishing: Cham, UK, 2017; pp. 251–266.

124. Caprotti, F.; Cowley, R. Interrogating Urban Experiments. *Urban Geogr.* **2017**, *38*, 1441–1450. [CrossRef]

125. Decode (DEcentralised Citizens Owned Data Ecosystem). *D.5.3. Data Analysis Methods and First Results from Pilots*; Decode: Barcelona, Spain, 2018.

126. Sassen, S. Predatory Formations Dressed in Wall Street Suits and Algorithmic Math. *Sci. Technol. Soc.* **2017**, *22*, 6–20. [CrossRef]

127. March, H.; Ribera-Fumaz, R. Smart contradictions: The politics of making Barcelona a self-sufficient city. *Eur. Urban Reg. Stud.* **2016**, *23*, 816–830. [CrossRef]

128. Barcelona City Council, Barcelona Ciutat Digital: A Roadmap Toward Technological Sovereignty. 2016. Available online: http://ajuntament.barcelona.cat/estaregiadigital/upload_Digital.pdf (accessed on 21 March 2018).

129. DEcentralised Citizens Owned Data Ecosystem (Decode). 2018. Available online: www.decodeproject.eu (accessed on 19 March 2018).

130. Roio, D. *Algorithmic Sovereignty*; University of Plymouth: Plymouth, UK, 2018.

131. Islar, M.; Irgil, E. Grassroots practices of citizenship and politicization in the urban: The case of right to the city initiatives in Barcelona. *Citizsh. Stud.* **2018**, *22*, 491–506. [CrossRef]

132. Mazzucato, M. *The Entrepreneurial State: Debunking Public vs. Private Sector Myths*; Public Affairs: New York, NY, USA, 2015.

133. Calzada, I. Problematizing and Politicizing Smart City-Regions: Is Devolution Smart? *Territorio* **2017**, *83*, 37–47. [CrossRef]

134. Eizaguirre, S. Comparing Social Innovation Initiatives in Barcelona and Bilbao. Looking at Associative Participation in the Governance of Citizens' Rights. *Rev. Catal. Sociol.* **2016**, *31*, 19–33. [CrossRef]

135. Blanco, I.; Salazar, Y.; Bianchi, I. Transforming Barcelona's Urban Model? Limits and Potentials for Radical Change under a Radical Left Government. 2017. Available online: http://www.urbantransformations.ox.ac.uk/blog/2017/transforming-barcelonas-urban-model-limits-and-potentials-for-radical-change-under-a-radical-left-government/ (accessed on 19 March 2017).

136. Bollier, D. The City as a Commons. 2016. Available online: https://www.youtube.com/wtch?v=z3itmhDuem8 (accessed on 1 February 2017).

137. Borch, C.; Kornberger, M. *Urban Commons: Rethinking the City*; Routledge: London, UK, 2015.

138. Foster, S.; Iaione, C. The City as a Commons. *Yale Law Policy Rev.* **2016**, *34*, 2. [CrossRef]

139. Rossi, U. The Variegated Economics and the Potential Politics of the Smart City. *Territ. Politics Gov.* **2016**, *4*, 337–353. [CrossRef]

140. Tõnurist, P.; Kattel, R.; Lember, V. Innovation Labs in the Public Sector: What They Are and What They Do? *Public Manag. Rev.* **2017**, *19*, 1455–1479. [CrossRef]

141. Bakıcı, T.; Almirall, E.; Wareham, J. A Smart City Initiative: The Case of Barcelona. *J. Knowl. Econ.* **2013**, *4*, 135–148. [CrossRef]

142. Degen, M.; García, M. The Transformation of the 'Barcelona Model': An Analysis of Culture, Urban Regeneration and Governance. *Int. J. Urban Reg. Res.* **2012**, *36*, 1022–1038. [CrossRef]

143. Gascó-Hernandez, M. Building a Smart City: Lessons from Barcelona. *Commun. ACM* **2018**, *61*, 50–57. [CrossRef]

144. Eizaguirre, S.; Parés, M. Communities making social change from below. Social innovation and democratic leadership in two disenfranchised neighbourhoods in Barcelona. *Urban Res. Pract.* **2018**, 1–19. [CrossRef]

145. Tieman, R. Barcelona: Smart City Revolution in Progress. 2017. Available online: https://amp.ft.com/content/6d2fe2a8-722c-11e7-93ff-99f383b09ff9 (accessed on 10 January 2018).

146. Jones, E. Smart City Barcelona. A City of the Future? 2015. Available online: http://www.barcelona-metropolitan.com/features/smart-city-Barcelona/ (accessed on 21 March 2018).

147. Timeus, K.; Gascó, M. Increasing innovation capacity in city governments: Do innovation labs make a difference? *J. Urban Affairs* **2018**. [CrossRef]

148. Morozov, E.; Harvey, D. Conversation between Evgeny Morozov and David Harvey. 2016. Available online: http://davidharvey.org/2016/11/video-conversation-between-david-harvey-evgeny-morozov-on-post-neoliberalism-trump-infrastructure-sharing-economy-smart-city/ (accessed on 10 November 2016).

149. Morozov, E.; Eno, B. Debat. Brian Eno i Evgeny Morozov Una Conversa Sobre Tecnologia i Democràcia. CCCB. 2017. Available online: https://vimeo.com/206060710 (accessed on 28 February 2017).

150. Gartner. Gartner Says 8.4 Billion Connected "Things" Will Be in Use in 2017, up 31 Percept from 2016. 2017. Available online: https://www.gartner.com/en/newsroom/press-releases/2017-02-07-gartner-says-8-billion-connected-things-will-be-in-use-in-2017-up-31-percent-from-2016 (accessed on 20 August 2018).

151. Habermas, J. *The Lure of Technocracy*; Polity Press: New York, NY, USA, 2015.

152. Hughes, B. *The Bleeding Edge: Why Technology Turns Toxic in an Unequal World*; New Internationalist Publications Ltd.: Markham, ON, Canada, 2016.

153. Yin, R.K. *Case study Research: Design and Methods*; SAGE Publications: Thousand Oaks, CA, USA, 1984.

154. Wiseman, J. *Lessons from Leading CDOs: A Framework for Better Civic Analytics*; Civic Analytics Network, Ash Centre for Democratic Governance and Innovation: Cambridge, MA, USA, 2017.
155. Decidim Barcelona. 2018. Available online: www.decidim.barcelona (accessed on 19 March 2018).
156. Metadecidim Barcelona. 2018. Available online: http://www.metadecidim.barcelona (accessed on 21 March 2018).
157. Mondragon Co-Operative Corporation. *Mondragón del Futuro*; MCC: Mondragon, Spain, 2018.
158. Digital Social Innovation. 2018. Available online: http://ateneusdefabricacio.barcelona.cat/ (accessed on 19 March 2018).
159. Stokes, M.; Baeck, P.; Baker, T. *What Next for Digital Social Innovation*; NESTA: London, UK, 2017.
160. Barcelona Urban Commons. 2018. Available online: www.bcncomuns.net (accessed on 31 March 2018).
161. Observatorio Metropolitano de Barcelona, Comunes Urbanos de Barcelona. 2018. Available online: http://www.bcncomuns.net (accessed on 21 March 2018).
162. Cámara, C. *Urban Commons: Lessons from Barcelona at the Beginning of 21st Century*; UOC: Barcelona, Spain, 2017.
163. Barcelona City Council, Plan de Impulso Economía Social y Solidaria. 2016. Available online: http://ajuntament.barcelona.cat/economia-social-solidaria/es/objectivos-generales (accessed on 3 May 2018).
164. B-Mincome (Combining Guaranteed Minimum Income and Active Social Policies in Deprived Urban Areas). 2018. Available online: http://www.uia-initiative.eu/en/uia-cities/barcelona (accessed on 19 March 2018).
165. Calzada, I. Knowledge Building & Organizational Behaviour: Mondragon Case. In *International Handbook of Social Innovation. Social Innovation: Collective action, Social Learning and Transdisciplinary Research*; Moulaert, F., Maccallum, D., Mehmood, A., Hamdouch, A., Eds.; Edward Elgar Publishing: Cheltenham, UK, 2013; pp. 219–229.
166. Lytras, M.; Visvizi, A. Who Uses Smart City Services and What to Make of It: Toward Interdisciplinary Smart Cities Research. *Sustainability* **2018**, *10*, 1998. [CrossRef]
167. Visvizi, A.; Lytras, M. It's Not a Fad: Smart Cities and Smart Villages Research in European and Global Contexts. *Sustainability* **2018**, *10*, 2727. [CrossRef]
168. Campbell, J.L.; Quincy, C.; Osserman, J.; Pedersen, O.K. Coding in-depth semistructured interviews: Problems of unitization and intercoder reliability and agreement. *Sociol. Methods Res.* **2013**, *42*, 294–320. [CrossRef]

 *sustainability* 

Article

Towards Sustainable Development of Online Communities in the Big Data Era: A Study of the Causes and Possible Consequence of Voting on User Reviews

Jie Zhao [1], Jianfei Wang [1], Suping Fang [1] and Peiquan Jin [2,*]

[1]   School of Business, Anhui University, Hefei 230601, China; zhaojie@ahu.edu.cn (J.Z.); m16201026@stu.ahu.edu.cn (J.W.); m16201039@stu.ahu.edu.cn (S.F.)
[2]   School of Computer Science and Technology, University of Science and Technology of China, Hefei 230026, China
*   Correspondence: jpq@ustc.edu.cn

Received: 14 July 2018; Accepted: 31 August 2018; Published: 4 September 2018

Abstract: This paper focuses on the review voting in online communities, which allows users to express their own opinions in terms of User-generated Content (UGC). However, the sustainable development of online communities is likely to be affected by the social influence of UGC. In this paper, we study the so-called crowd intelligence paradox of review voting in online communities. The crowd intelligence paradox means that the quality of reviews is not highly connected with the increasing of review votes. This implies that a review with many votes is likely to be of low quality, and a review with few votes is likely to be of high quality. The crowd intelligence paradox existing in online communities inhibits users' wishes of participating in social networks and may impact the sustainable development of online communities. Aiming to demonstrate the existence of the crowd intelligence paradox in online communities, we first analyzed a large set of reviews crawled from Net Ease Cloud Music, which is one of the most popular online communities in China. The maximum likelihood (ML) and the hierarchical regression approaches are used in this step. Then, we construct a new research model called the Voting Adoption Model (VAM) to study how different factors impact the crowd intelligence paradox in online communities. Particularly, we propose six hypotheses based on the VAM model and conduct experiments based on the measurement model and the structural model to evaluate the hypotheses. The results show that the quality of reviews is not influential to review votes, and the hot-site attribute is a dominant factor influencing review voting. In addition, the variables of the VAM model, including information credibility, perceived ease of use, and social influence have significant impacts on review voting. Finally, based on the empirical study, we present some research implications and suggestions for online communities to realize healthy and sustainable development in the future.

Keywords: Social network; sustainable development; review voting; online community; paradox

---

1. Introduction

With the rapid development of social networks, more and more people tend to spend a lot of time in online communities, such as online video sharing platforms or online music platforms. For example, Net Ease Cloud Music (http://music.163.com) is one of the most popular online entertainment communities in China and has attracted millions of users. The large number of users in online communities can bring huge marketing values and business values to enterprises, e.g., through online advertising. So far, online communities are mostly based on the application of User-Generated Content (UGC), which allows users to create personalized information and share them to online communities [1].

Many UGC systems adopt a voluntary voting mechanism to identify valuable reviews that can help users browse content efficiently [2]. However, we note that there is a paradox of review voting in online communities, i.e., the number of votes is not consistent with the quality of the corresponding content. Especially, some content with low quality surprisingly receives more votes than others with high quality. Such a paradox has a close relationship with the crowd intelligence of online communities, and might have an impact on the sustainable development of online communities. Previous studies have pointed out that the bias of ratings existed in online reviews [3]. However, so far there are no studies focusing on the paradox of review voting in online communities.

In this paper, we regard the paradox of review voting in online communities as the crowd intelligence paradox. The crowd intelligence paradox has negative effects on resource allocation and the sustainable development of online communities, because it may lower the quality of UGC, which violates the initial objective of the review voting mechanism, i.e., to recommend high quality content. Furthermore, online communities have more users than traditional communities. Therefore, the crowd intelligence paradox in online communities will lead to an unfair distribution of community resources such as content ranking, exposure time, and bonus points. As a result, the crowd intelligence paradox of review voting might inhibit users who wish to participate in UGC and hinders the development of online communities.

This paper aims to address the crowd intelligence paradox and to study the inconsistency between review voting and review quality in online communities. Particularly, we aim at answering two research questions:

(1) What is the paradox of review voting?

(2) Why does the crowd intelligence paradox exist?

In order to answer what the paradox of review voting is, we designed two aspects of work. Firstly, we analyzed the distribution of review votes in a real online community (Net Ease Cloud Music) using the power law distribution theory, in which a large data set including 351, 578 reviews are used. This aims at exploring the distribution of review votes in online communities and matches the 80/20 rule—only a few reviews can receive votes. Secondly, we also propose a hierarchical-regression-based method to verify the existence of the crowd intelligence paradox. This aims at indicating that review voting is not closely connected with the quality of reviews. The above successive issues will quantitatively measure the existence of the crowd intelligence paradox in online communities.

Furthermore, aiming at answering factors influencing the crowd intelligence paradox, we present a new research model named Voting Adoption Model (VAM), which is an integration of the information adoption model (IAM), the self-determination theory (SDT), and the signal theory. Based on the VAM model, we propose six hypotheses and analyze the imbalance of voting motives and the signals of the crowd intelligence paradox based on tests with respect to measurement model evaluation and structural model evaluation. In summary, we make the following contributions in this paper:

(1) We first study the crowd intelligence paradox in online communities, and experimentally reveal the distribution features of the review votes in online communities.

(2) We conduct experiments over a reviews dataset and discover that the number of review votes is not consistent with the quality of corresponding UGC, which demonstrates the existence of the crowd intelligence paradox of review voting in online communities.

(3) We propose a new research model called Voting Adoption Model (VAM) and six hypotheses to study why the crowd intelligence paradox exists in online communities. Compared with information credibility, perceived ease of use, and social influence, the influence of commentary quality is not significant. Hence, high-quality reviews do not always receive high votes.

(4) We present discussions on the experimental results and give some suggestions for the sustainable development of online communities.

To the best of our knowledge, this work provides the first empirical study on the crowd intelligence paradox in online communities. The crowd intelligence paradox of review voting reflects an unreasonable allocation of community resources. High-quality resources are not necessarily recognized, while low-quality ones get a lot of attention. This is a potential crisis for the UGC ecological environment and may restrict the traffic monetizing. Understanding the mechanism of the existence of the crowd intelligence paradox in review voting can help people solve the voting problem by increasing the cost of the voting signals and enhancing the demand of autonomy and competence.

The remainder of the paper is organized as follows. In Section 2, we survey the related work. In Section 3, we combine the distribution of review votes in online communities with a hierarchical regression to discuss the measurement of the crowd intelligence paradox. In Section 4, we construct a research model to study why the crowd intelligence paradox exists. Section 5 presents discussions of the data analysis as well as some suggestions for the sustainable development of online communities. Finally, Section 6 concludes the paper.

2. Related Work

2.1. User-Generated Content (UGC)

The network economy is increasingly dependent on creativity. With the technological advances provided by Web 2.0, users can creatively share UGC in communities. The most commonly cited definition for UGC is given by the Organization for Economic Co-operation and Development (OECD): Publicly available content, creative effort, and non-professional staff [4]. UGC can be divided into rational users (e.g., knowledge sharing), perceptual users (e.g., social activities, entertainment), group collaboration (e.g., Wikipedia), and individual creation (e.g., blog, reviews) by user type [5]. Online communities can be divided into trading communities, interest communities, fantasy communities, and relationship communities [6]. Traditional UGC research was more focused on online reviews and eWOM of trading communities. This paper focuses on review voting in online communities.

User-generated content has many research areas and can be divided into four categories: Users' participant motivation [7,8], the influence of content [9–11], content mining [12,13], and content management [14–16]. In these studies, online reviews [17] and eWOM [18] are the most abundant themes, which focus on three areas: Consumers' acts of releasing reviews [19], consumers' perceived value of reviews [20], and consumers' shopping decisions under the effect of UGC [21]. Among them, the helpfulness of online reviews is one of the important works of UGC research in business areas. The helpfulness of online reviews refers to identify helpful reviews to support consumer purchase decisions. [17]. For example, Amazon asks users a question about whether this review is helpful to you.

Many scholars have discussed the helpfulness of the comment from three perspectives: Information source (e.g., platform reputation [22], publisher characteristic [23], and product type [20]), information content (e.g., quality [24], quantity [25], valence [26], attribute [27], and expression [28]), and information receiver (e.g., consumer knowledge [29] and involvement [30]). They found the above factors have a significant impact on the helpfulness of reviews.

With the rapid increasing of the volume and scale of Web data, massive UGC content increased the information overload, so the industry began to use the "votes" of this social voting mechanism to identify helpful content (e.g., Facebook's thumb up). It has been studied through software usage data and Amazon commentary data that the information characteristics (e.g., basic characteristics, style features and semantic features) and commentator characteristics (e.g., commentator confidence) have a significant influence on the number of votes [2,31].

While many studies have looked at what makes an online review helpful (the helpfulness of content), little is known on what makes an online review receive votes (review voting). In addition, some scholars measure helpful content by a percentage of useful votes in the total number of votes [17]. We want to know whether the vote is as deviant as the rating data in trading reviews [3]. While it has been argued that more helpful comments are more likely to be voted for [2], the scope of the study is limited to the online reviews of electronic business platforms. Furthermore, it ignored the characteristics of review recipients. In addition, previous studies lacked a study of non-commercial information content in online communities. The number of votes has significantly affected the value perception of UGC content publishers [22] and studying review voting of online communities helps to improve the UGC ecosystem, which will help improve users' stickiness and traffic monetizing.

2.2. IAM, SDT and Signal Theory

Review voting reflects recipients' acceptance of content information. The theory of information communication points out that in the course of information communication, information content, information sources, and receiver characteristics all influence recipients' acceptance and the persuasion of information. Among many persuasion process models, the Heuristic-Systematic Model (HSM) [32] and the Elaboration Likelihood Model (ELM) [33] are the most representative. The Information Adoption Model (IAM) [34] is an extension of ELM on network environments. The process of influencing people's decision-making is seen as a process of information adoption. The model regards the quality of information as a central path and the information source as an edge path. While the model is generally used [35], it only focuses on characteristics of information, such as quality, credibility, and usefulness. The influence of information, however, should not be limited to characteristics of information [23].

Thus, we add information receiver characteristics to explore influences about what is the feature of review voting and why. Those receiver characteristics mainly relate to voting motives, perceived ease of use and so on. Theoretical explanation comes from a set of theories described below, including self-determination theory (SDT), technology acceptance model (TAM), and signal theory.

Scholars apply the Theory of Motivation to explain individual behavioral intentions, including UGC behavior. For example, the self-determination theory (SDT) was used to successfully explain how UGC publishers perceive and respond to others' feedback [22]. Self-determination theory is composed of basic psychological need theory, cognitive evaluation theory, organismic integration theory, and causality orientation theory [36]. The organismic integration theory points out the internalization of external motivation, and the basic psychological need theory points out the important influence of autonomy, competence, and relatedness. Social relations are the main part of the demand for users in online communities, and social relations have a significant driving effect on UGC generation [10]. We also believe that online users will exchange their voting to get some kind of reward, including community honor, altruism satisfaction, and social entertainment.

The technology acceptance model (TAM) focuses on behavioral characteristics of users and influencing factors. Perceived usefulness and perceived ease of use affect the uses' motivation and action by affecting the users' attitude of information technology [37]. Review voting needs to rely on network platforms. The voting function is easy to understand and the perceived usefulness affects the users' acceptance.

Review voting is based on the original and old votes, so the number of votes becomes a signal, which affects users' judgment and analysis. Signal theory points out that there are two indispensably key features: High cost (e.g., higher education) and being readily observable (e.g., company financial statement) [38]. The latest research discussed the degree of consumer acceptance of expert blogs that form the perspective of the cost of signals [9]. The expert blog is to consumers what the UGC is to users. In addition, users want to know others' views on the review and the information between users is asymmetric. So the number of votes becomes a clear signal, a representation of how many people recommend the comment. Therefore, this paper will use signal theory to explain reasons for review voting.

In conclusion, this paper focuses on the review voting in online communities. We not only uncovered the paradox of review voting from the perspective of information characteristics, but also explored the influencing factors of why the paradox exists from the perspective of receiver characteristics, on the basis of SDT and signal theory.

3. Validation of the Crowd Intelligence Paradox

In this section, we validate the following two propositions: (1) The distribution of review votes in online communities matches the 80/20 rule and only a few comments can receive votes; and (2) the inconsistency between review voting and review quality exists in the reviews receiving many votes.

Practically, we study the distribution of review votes in online communities by conducting data analysis on a real online community, Net Ease Cloud Music, which is one of the most popular online entertainment communities in China. In Section 3.1, we give the details about the vote dataset and subsequent survey data, and in Section 3.2, we present the results of distribution tests on the reviews dataset. Finally, we measure the crowd intelligence paradox in Section 3.3.

3.1. Dataset

Net Ease Cloud Music was officially released in 23 April 2013. Compared with other music software, Net Ease Cloud Music pays more attention to music socialization and creates music community with UGC, especially its music comment function attracting so many people. As of April 2017, Net Ease Cloud Music users exceeded 300 million people, singles exceeded 400 million. We choose Net Ease Cloud Music community as a research platform to study the paradox of review voting. We need to prepare two kinds of data, one is the vote data from Net Ease Cloud Music, and the other is the community survey data. The respondents to the survey are randomly selected from reviewers of above 17 songs, and all the reviews delivered by the respondents are among the total 351, 578 reviews.

3.1.1. Vote Data

The vote dataset is prepared by two steps: Data crawling and text mining. We crawled review votes from Net Ease Cloud Music and finally got a dataset containing 351, 578 reviews on 17 songs. The format of crawled data is shown in Table 1.

Table 1. Data crawling format.

ID	Time	Votes	Hot Site	Content
335940343	2017/8/27 21:04	13,051	1	The music is g
. . . . . .	. . . . . .	. . . . . .	. . . . . .	. . . .

ID	Activity	Like	Fans	Level
335940343	46	17	24,871	7
. . . . . .	. . . . . .	. . . . . .	. . . . . .	. . . . . .

In order to effectively distinguish high quality reviews from low quality ones, we use text classification to perform review classification on the crawled data. There are many classification models proposed in the area of data mining, such as SVM (Support Vector Machine), KNN (K nearest neighbors), and Naive Bayes Model (NB) [39]. The process is shown in Figure 1.

In the manual annotation step, we extract a random sample of 20,000 reviews from our set (351,578 reviews). Five postgraduates do the manual annotation about quality in five days from 25 August 2017 to 30 August 2017. The quality score is labeled by a five-point Likert-scale, ranging from strongly disagree (1) to strongly agree (5), according to personal perception. Because they belong to users of Net Ease cloud music, personal labeling is a simple but effective way. We roughly split the annotated reviews according to the average score [9]. Basically, the reviews with above-average scores are marked as high-quality reviews, and those with below-average scores are regarded as low-quality reviews. We did not use the full scoring scale because the annotated data was not large enough to conduct a full-scale analysis. Consequently, we get 8500 high quality reviews (above-average scores) and 8500 low quality reviews (below-average scores). The data is tested by the significance testing of an internal consistency reliability (Cronbach's α = 0.809).

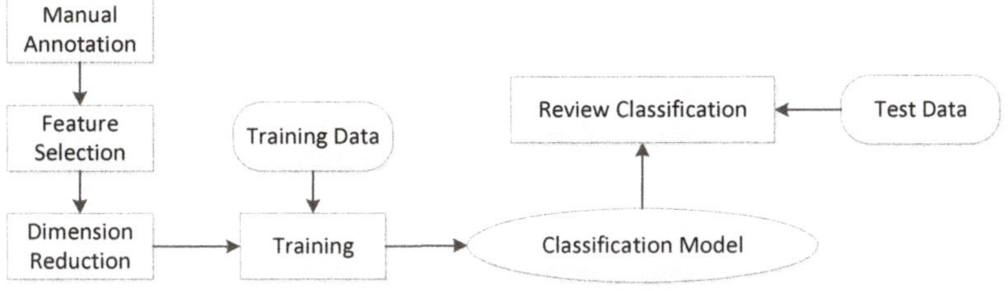

Figure 1. The process of reviews classification.

In the feature selection step, we select single and double words as features. In the step of dimension reduction, we use a Chi-square method based on the features. We select 60% data as the training data and the rest are used as test data. Then, we construct classifiers with different classification algorithms over the training data. To ensure the generality of the experiments, we randomly change the sequence of records in the training dataset and repeat the training process five times to get the average results.

Table 2 shows the classification results on reviews, where four classification models are compared, including Bernoulli Native Bayes (denoted as *BernoulliNB* in Table 2), Multinomial Native Bayes (denoted as *MultinomialNB* in Table 2), Logistic Regression, and SVM. We use the metrics of precision, recall, and the F-measure to indicate the performance of classification. These metrics are commonly used in classification and text mining.

Table 2. Classification of reviews.

	Precision	Recall	F1-Score	Support		Precision	Recall	F1-Score	Support
	BernoulliNB's accuracy is 0.96					MultinomiaNB's accuracy is 0.91			
low	1.00	0.93	0.96	3661	low	0.89	0.92	0.91	3291
high	0.92	1.00	0.96	3139	high	0.92	0.89	0.91	3509
ratio	0.96	0.96	0.96	6800	ratio	0.91	0.91	0.91	6800
	Logistic Regression's accuracy is 0.96					SVM's accuracy is 0.71			
low	0.98	0.93	0.96	3579	low	1.00	0.63	0.77	5359
high	0.93	0.98	0.96	3221	high	0.42	0.99	0.59	1441
ratio	0.96	0.96	0.96	6800	ratio	0.87	0.71	0.73	6800

3.1.2. Survey Data

The survey dataset is designed using a multi-item approach. All variables were carried out by a seven-point Likert-scale, ranging from strongly disagree (1) to strongly agree (7). Items were borrowed from previous literature and modified for the context of this study.

The questionnaire consists of three parts:

(1) Sample selection: Ask a question of whether user vote on the hot review in a song.
(2) Sample characteristic: This part mainly measures the sex, age, and community age of user.
(3) Variable questionnaire: As shown in Table 3, this study includes six latent variables, which are information quality (IQ), information credibility (IC), perceived ease of use (EOU), social influence (SI), information usefulness (IU), and vote adoption (A). Items were borrowed from previous literature and modified for the context of online reviews. Specifically, 'EOU' is adapted from the study of Gefen et al. (2003). 'IQ' is assessed by adapting three items used by Park et al. (2007). 'IC' is based on Prendergast et al. (2010). Finally, to examine 'IU' and 'A', six statements were adopted from Ismail Erkan et al. (2016).

Table 3. Variables used in the VAM model.

Variable	Item
Perceived ease of use [40]	EOU1: The vote function is easy to use. EOU2: It is easy to become skillful at using the vote function. EOU3: It is easy to interact with the vote function.
Social influence	SI1: Voting on the hot review will help the announcer. SI2: I like to vote the content for connecting other people SI3: There are many people visiting what I vote.
Information quality [41]	IQ1: I think they are understandable. IQ2: I think they are clear. IQ3: In general, I think the quality of them is high.
Information credibility [42]	IC1: I think they are convincing. IC2: I think they are credible. IC3: I think they are familiar.
Information usefulness [23]	IU1: I think they are generally useful. IU2: I think they are generally informative. IU3: I think they are generally appreciative.
Vote adoption [23]	A1: It is very likely that I will vote the review. A2: I will definitely vote the review. A3: I will recommend the review to my friends.

The 500 questionnaires are distributed by artificial network through the short message of Net Ease Cloud Music. Respondents are randomly equably selected form reviewers of above 17 songs. A total of 273 questionnaires are recovered and finally obtains 244 after the cleaning work. The effective recovery rate was 48.8%. We then examined if the common method bias is a concern in this study. An exploratory factor analysis of all items extracted six factors which explain 75.08% of all the variance, with no single factor accounting for significant loadings ($p < 0.10$) for all items. We conclude that the common method variance (CMV) is probably not a concern in this data set. The sample characteristics are shown in Table 4.

Table 4. Sample characteristics.

Measure	Gender		Age				Years of Registration			
	Male	Female	18–23	24–28	29–33	34–39	1	2	3	4
Frequency	132	112	113	83	22	26	29	31	85	99
Percentage (%)	54.1	45.9	46.3	34.03	9.02	10.66	11.89	12.7	34.84	40.57

3.2. Distribution of Review Votes

Table 5 shows the descriptive statistic of the vote data, which contains 351,578 reviews. We can see that the distribution of the number of votes is extremely skewed. Among the whole data set, the maximum value is 125,381 while the minimum value is zero. Additionally, 99.5% of the votes are less than 21. Figure 2 shows the distribution of votes ranging from 0 to 20.

Table 5. Descriptive statistics of the vote data.

Vote	Mean	Median	Max	Min	Sum	Standard Deviation	Reviews#
[0, 20]	1.76	1	20	0	617,283	2.18	349,581
[21, 125,381]	1330.19	64	125,381	21	2,656,404	6485.78	1997
All	9.31	1	125,381	0	3,273,687	476.66	351,578

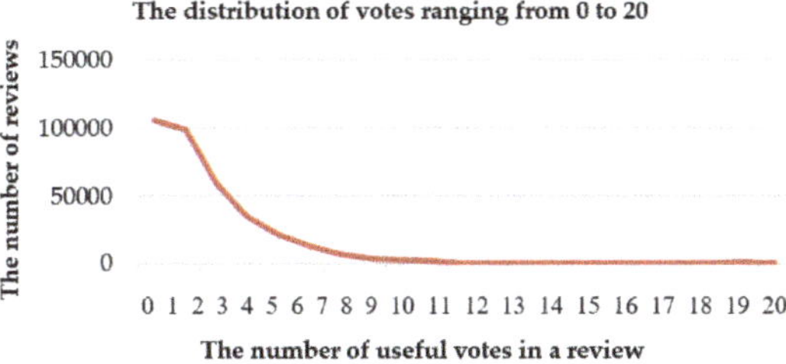

Figure 2. Distribution of votes (0, 20).

By observing review voting in Net Ease Cloud Music, we suppose that review voting may match the power law distribution. Power law distribution implies that the probability of occurrence of an event is extremely skewed. The event of small observation occurs in large numbers. However, the event of big observation occurs in small numbers. As UGC connects massive users, users' commentary reply behaviors over time intervals have been found to match the power law distribution [43].

As a long tail of the power law distribution is complex and has a greater volatility, there may be a large error when using the least squares method. In addition, it cannot be effectively compared with other forms of distributions. Thus, according to the method of fitting the power law distribution data [44], we apply the Maximum Likelihood (ML) method on the dataset and evaluate the results with the Kolmogorov-Smirnov testing method. Based on the algorithm developed by Python 3.6 [45], we examine the votes k and the corresponding occurrence number $P(k)$ for the 351, 578 reviews. The results are shown in Table 6 and Figure 3. We discovered that the data distribution has a multi-segment truncation feature and does not conform to a strict power law distribution.

Table 6. The power law index table over the dataset.

Power Law Index		
Vote range	[1, 303]	[16,068, 125,381]
power law index	−1.31	−2.68
Sig.	0.16	0.057
Logarithmic Likelihood Ratio Compared with Other Distributions *		
	R, p	R, p
Exponential	1173.38, 0.0	3.74, 0.00
Log-normal	−40.78, 0.0	−0.29, 0.44
Truncated power law	−22.50, 1.96	−0.57, 0.28
Stretched exponential	−45.94, 0.0	−0.35, 0.39

* R is logarithmic likelihood ratio of two distributions. When R is positive, then the first distribution is preferred; otherwise, we should consider the second distribution. P is the significance.

As shown in Table 6, in the high range from 16,068 to 125,381, the power law is satisfied ($\alpha = -2.68$, $P = 0.057$). However, compared to the entire data set, it discards a lot of data. In the lower range from 1 to 303, the power law is not satisfied ($\alpha = -1.31$, $P = 0.16$).

Therefore, we conclude that the distribution of review voting is not always a power law distribution. There are two possible reasons. First, the power law distribution of online behavior changes with time, meaning that different voting behavior may occur in different time. Second, the dataset is relatively small compared to the entire community data; thus it cannot cover all review characteristics. However, we can see that review voting matches the 80/20 rule in our study. An event of small votes occurs in large numbers while an event of big vote occurs in small numbers.

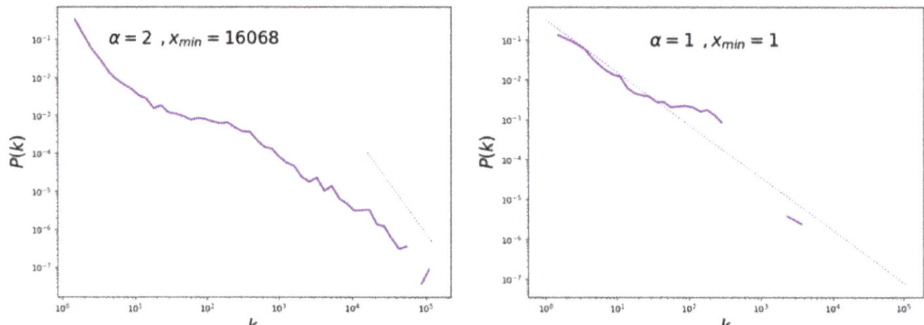

Figure 3. The power law distribution of the dataset.

3.3. Measuring the Crowd Intelligence Paradox

In this, Section 3.3, we aim to measure the existence of the crowd intelligence paradox in online communities. From Section 3.2, we know the distribution of review votes in online communities and matches the 80/20 rule and only a few reviews can receive votes. As to reviews with many votes, if they failed to reflect the quality of review content, this voting mechanism cannot reflect crowd intelligence, and it means that there is a crowd intelligence paradox in online communities.

3.3.1. Measures

Based on previous literature about the characteristics of information and information publisher, we choose content length, release time, content quality, hot-site (a mark of whether the content is/was entering the hot list), and the characteristics of information publisher, including activity, fans, likes, and levels. We construct a hierarchical regression model to analyze the effect of variables by the ordinary least squares method (OLS).

The multiple regression model is set as follows:

$$\ln(Voting_i) = \beta_0 + \beta_1 \ln(Level_i) + \beta_2 \ln(Activity_i) + \beta_3 \ln(Fans_i) + \beta_4 \ln(Like_i) + \beta_5 \ln(Time_i) + \\ \beta_6 \ln(Words_i) + \beta_7 \ln(Words_i)^2 + \beta_8 \ln(Quality_i) + \beta_9 Hotsite_i + u_i \tag{1}$$

There is a dependent variable in our model: Voting. We measure voting by the increased number of votes [2] to a music review between two periods—20 August 2017 and 21 September 2017 in our experimental dataset. For instance, if a review had 100 votes on 20 August 2017, and had 300 votes on 21 September 2017, the increased number of the votes is 200. There are eight independent variables, including quality, time, words, activity, fans, hot-site, level, and like, as shown in Table 7. The variable quality is measured by the score of content quality, which comes from our text mining of content. High quality review valence is the value of the probability of quality classification. We measure time by a time difference between the release time of some review in a song and the earliest review in the same song. The variable words are measured by the number of words in the review. The variable hot-site is measured by a binary value indicating whether a review was included in the hot list on 20 August. "1" denotes that the review was in the hot list and "0" means that the review did not appear in the hot list. In addition, and through the collection of community data, we measured four commentator variables: Level, by the level of community account; activity, by the mark that often logs in and browses the community content; fans, by the fans of the user; and likes, by the number of users who take an initiative to pay attention to someone (e.g., Facebook's Likes). Table 7 presents the descriptive statistics and correlations of the variables.

Table 7. Correlations and descriptive statistics over the vote data.

Variables	Type	1	2	3	4	5	6	7	8	9
1. Voting	Dependent	1								
2. Quality	Independent	0.10	1.00							
3. Time	Independent	0.07	0.09	1.00						
4. Words	Independent	0.13	0.58	0.03	1.00					
5. Activity	Independent	−0.01	−0.02	−0.10	−0.02	1.00				
6. Fans	Independent	0.02	−0.02	−0.08	−0.04	0.57	1.00			
7. Hot-site	Independent	0.44	0.03	−0.02	0.04	0.02	0.06	1.00		
8. Level	Independent	0.00	−0.01	−0.01	−0.03	0.38	0.48	0.01	1.00	
9. Like	Independent	−0.02	−0.04	−0.14	−0.06	0.55	0.56	0.01	0.37	1.00
Mean		2.49	0.27	54.81	23.03	13.07	37.43	0	5.99	21.47
Max		11,392	1	948.81	2.5	5248	136,089	1	10	2333
Min		0	0	0	1	0	0	0	0	0
Std. Dev.		88.74	0.39	170.57	26.18	69.53	795.7	0.04	1.9	65.81

3.3.2. Results

We analyzed the sub-dataset with the hierarchical regression method using Stata 14.0 and the ordinary least squares method (OLS). Table 8 shows the results of the regression model between the independent variables and the dependent variable, as shown in the left-most column of Table 7.

Table 8. Results of the regression model over the vote data.

Variable	(1)	(2)	(3)
Voting	−0.085 (−11.59)	−0.075 (−10.30)	−0.067 (−10.44)
Level	−0.004 (−0.88)	−0.004 (−0.97)	0.004 (1.13)
Activity	−0.008 (−4.94)	−0.008 (−4.94)	−0.006 (−4.52)
Fans	0.023 (8.76)	0.023 (8.79)	0.008 (5.00)
Like	−0.012 (−5.67)	−0.019 (−5.85)	−0.007 (−4.33)
Time	0.020 (25.93)	0.020 (25.40)	0.022 (34.49)
Words	0.595 (18.8994)	0.557 (17.61)	0.508 (18.01)
Words2	−0.239 (−17.4796)	−0.225 (−16.37)	−0.205 (−16.71)
Quality		0.036 (4.99)	0.022 (3.47)
Hot site			2.498 (41.28)
$\overline{R}^2$	0.0264	0.0267	0.2149
$\Delta\overline{R}^2$	0.0264	0.0003	0.1882
Prob $(F > 0)$	0.0000	0.0000	0.0000

Considering the problem of heteroscedasticity, we applied the logarithmic transform on all variables (except the *Hot-site*). We used heterosexuality-robust standard error to fix the model. Based on the test of Variance Inflation Factor (VIF ranges from 1.3 to 4.9) and the above correlations of variables, we argue that there is no severe multiple collinearity.

We first estimated a basic model that contains time, words and the other four commentator variables. We report the results in column (1) of Table 8. Those results are similar to the previous studies [2,17,30]. Specifically, the effect of content length is an inverted U curve (Words: $\beta = 0.595$, $P < 0.001$; Words2: $\beta = -0.239$, $P < 0.001$) [17]. The more words the review has, the more information it can present. However, if a sentence had too much information, it would bring an information burden to users and affects the acceptance of information [2]. Four commentator variables have different influences to review voting. Fans has a significant influence ($\beta = 0.023$, $t = 8.76$, $P < 0.001$), which is corresponding with the fact, because the commentator will have more personal channels to get votes from fans. Activity ($\beta = -0.008$, $t = -4.52$, $P < 0.001$) and Like ($\beta = -0.012$, $t = -5.67$, $P < 0.001$) have a significant negative influence, which may indicate that paying attention to external activities does not attract others to focus on themselves. Additionally, the action reduces to a time to think and write. Furthermore, because of the weak ties of network community, people do not pay more attention to others who concern about themselves. Additionally, the level ($\beta = -0.004$, $t = -0.88$, $P > 0.05$) has no significant influence. As sharing and listening to music is the main activity in Net Ease Cloud Music, the review part is relatively un-associated with experience level.

Second, we add quality to the estimation and the results are presented in column (2) of Table 8. We can see that quality has a positive influence on voting ($\beta = 0.036$, $t = 4.99$, $P < 0.001$). However, the coefficient value is low and the model fitting degree changes very little ($\Delta \overline{R}^2 = 0.0003$, $P < 0.001$).

Third, we add hot-site to the estimation and the results are shown in column (3) of Table 8. The model fitting degree changes a lot ($\Delta \overline{R}^2 = 0.1882$, $P < 0.001$) and hot-site has a significant positive influence on voting ($\beta = 2.498$, $t = 41.28$, $P < 0.001$). This indicates that the acquisition of hot-site will make votes increase by 249.8%. But 1% promotion of quantity only lets votes increase by 0.022%, which indicates that hot-site are more important than quality. In addition, the words ($\beta = 0.508$, $t = 18.01$, $P < 0.001$), time ($\beta = 0.022$, $t = 34.49$, $P < 0.001$) and fans ($\beta = 0.008$, $t = 5.00$, $P < 0.001$) have much smaller coefficients than hot-site. Therefore, hot-site is a dominant factor of review voting. This result is easy to understand. The hot-site value of a review is either one (i.e., the review is listed in the hot list) or zero (the review is not listed in the hot list). In our study, if a review is listed in the hot list of the online community "Net Ease Cloud Music" on 20 August 2017, its hot-site value is set to one, otherwise is set to zero. Generally, the reviews in the hot list are more likely to be viewed and commented on by users, yielding the increasing of the review votes to the hot-site reviews. This indicates that the hot-site value of reviews has a highly positive influence on review votes.

What is more, time ($\beta = 0.022$, $t = 34.49$, $P < 0.001$) indicates that early reviews are easier to get votes, because users are more likely to be attracted by new songs.

Combined with the distribution of review votes, reviews with many votes fail to reflect the quality of review content, and this voting mechanism cannot reflect crowd intelligence. Hence, the crowd intelligence paradox exists in review voting in online communities. Additionally, the crowd intelligence paradox is different from previous studies [2,30,31], which argued that the quality of reviews was the main factor influencing review voting in trading communities.

In addition, we found out that hot-site have a dominant influence on review voting. The distribution of review votes matches an extreme 80/20 rule and the content is inconsistent with quality. The voluntary voting mechanism is a way to recommend high quality content, but this mechanism sometimes results in content with low quality. Thus, the crowd intelligence paradox affects the UGC ecological environment.

4. Factors Influencing the Crowd Intelligence Paradox

In this section, we analyze why the crowd intelligence paradox exists. From Section 3, we know that the crowd intelligence paradox exists. However, it is much more important for enterprises to know the reasons behind the paradox. Thus, we construct a research model that includes a set of variables to study the influences of different factors on the crowd intelligence paradox. Compared with previous research, this fresh model explores the influencing factors of why the paradox exists from the perspective of receiver characteristics, on the basis of SDT and signal theory.

4.1. Research Model and Hypotheses

First, based on the organismic integration theory of SDT [36], the internalization of external motivation has four steps, including external regulation, introjection regulation, identification regulation, and integration regulation. Autonomy, competence and relatedness are three important factors to promote the internalization of external motives and influence actions. Usually, these three needs make up an organic entirety. However, relatedness is a decisive factor for review voting. Autonomy and competence almost do not work because of the special environment.

Specifically, social relations are the main part of the demand for user relations in online communities [18]. First, because of the community's emphasis on the users' experience, the vote does not require operational capabilities. Second, comment on some songs is easy to understand, so it does not need much cognitive effort. Third, people are completely autonomous in voting. As previous studies pointed out, self-efficacy is not related with the comment [19]. Therefore, relatedness is relatively more important. Meanwhile, community points externally influenced review voting. What is more, altruism and herd mentality strengthen this voting behavior. Taking into account the tendency of collectivism in Chinese culture, this strengthening is more significant. That means only relatedness can affect users' voting behaviors to some extent, at least to some degree. "Readily praise (unconditionally vote)" becomes a custom under the imbalance of voting motives. People use voting rights to exchange social satisfaction.

On the other hand, except for the number of votes, there is no more signal in Net Ease Cloud Music. People tend to judge the quality of the content based on the number of votes. But the signal is useless because it does not have a higher time cost, operating cost, reputation cost, and cognitive cost [38]. Thus, the fake votes influence user behaviors.

Specifically, while watching a comment, users can vote without waiting. Moreover, operating costs can be ignored due to the strengthening of perceived ease of use. The mechanism of content voting is extremely convenient, for increasing community traffic and interactions. In addition, voting has anonymity. Users' reputation will not change after voting. Furthermore, people do not want to think about the true value of the content for the herd mentality.

Therefore, the low cost signals cannot be an effective tool for reducing or eliminating information asymmetry [9]. That means that the votes cannot represent the quality of the content. But the community usually tends to design rankings according to the number of votes. The paradox comes with the useless signal.

In summation, we propose the following variables, including information quality, information credibility, perceived ease of use, and social influence to study the factor influences on the crowd intelligence paradox. Based on this, we present the voting adoption model (VAM) in Figure 4.

Next, we propose the following assumptions.

Hypothesis 1 (H1). *Information quality of review significantly increases users' decision on review usefulness.*

Hypothesis 2 (H2). *Information credibility of review significantly increases users' decision on review usefulness.*

Hypothesis 3 (H3). *Perceived ease of use significantly increases users' decision on review usefulness.*

Hypothesis 4 (H4). *Social influence significantly increases users' decision on review usefulness.*

Hypothesis 5 (H5). *Social influence significantly increases users' decision to provide a vote on the review.*

Hypothesis 6 (H6). *Information usefulness significantly increases users' decision to provide a vote on the review.*

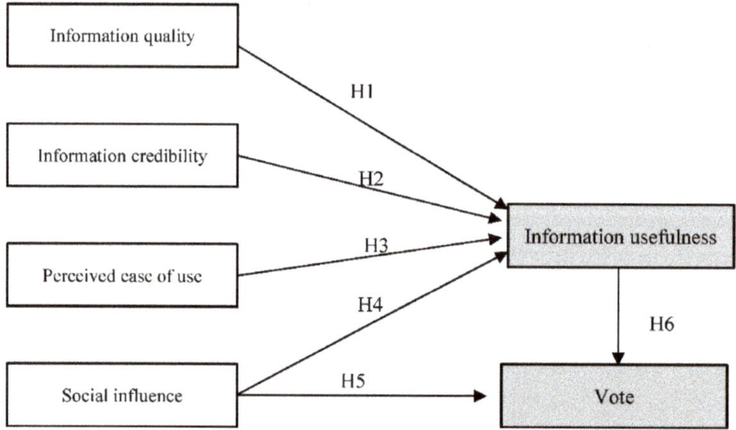

Figure 4. The voting adoption model (VAM).

4.2. Measurement Model Evaluation

The measurement model evaluation mainly includes three pieces of content: Reliability analysis, convergent validity, and discriminant validity.

Reliability is usually examined by using internal consistency reliability and composite reliability. As shown in Table 9, the internal consistency reliability is over 0.70. The composite reliability (CR) of all constructs is also over 0.70. Therefore, the reliability validity is achieved, which indicates that the variables within the VAM model are consistent.

Convergent validity is usually examined by using a composite reliability (CR) and an average variance is extracted (AVE). As shown in Table 9, all items load significantly on their respective constructs and none of the loadings are below the cutoff value of 0.60. The AVE of each variable is over 0.50 (except one value which is 0.485). Thus, the convergent validity is achieved.

Table 9. Factor loadings, CR and AVE values.

Variable	Item	Cronbach's α	Item Loading	CR	AVE
Information quality	IQ1 IQ2 IQ3	0.844	0.77 0.83 0.81	0.845	0.646
Information credibility	IC1 IC2 IC3	0.900	0.86 0.93 0.81	0.901	0.754
Perceived ease of use	EOU1 EOU2 EOU3	0.807	0.76 0.83 0.71	0.811	0.590
Social influence	SI1 SI2 SI3	0.770	0.78 0.71 0.66	0.761	0.516
Information usefulness	IU1 IU2 IU3	0.710	0.68 0.72 0.69	0.739	0.485
Vote adoption	A1 A2 A3	0.854	0.75 0.93 0.79	0.865	0.684

Discriminant validity is proposed to examine whether a measurement is a reflection of any other measurement. As shown in Table 10, the square root of AVE for each variable is greater than other correlation coefficients. Thus, the discriminant validity is achieved.

Table 10. Correlation matrix of the key variables.

Variable	SI	IC	EOU	IQ	IU	A
Social influence (SI)	0.718					
Information credibility (IC)	0.625	0.868				
Perceived ease of use (EOU)	0.438	0.405	0.768			
Information quality (IQ)	0.396	0.23	0.282	0.804		
Information usefulness (IU)	0.667	0.598	0.562	0.329	0.696	
Vote adoption (A)	0.693	0.473	0.363	0.289	0.583	0.827

Note: Diagonal elements are the square root of AVE for each variable.

4.3. Structural Model Evaluation

Next, we perform structural model evaluation using AMOS 23.0. The results of hypotheses verification are presented in Table 11, and Table 12 shows the goodness of fitting.

Five hypotheses between variables are found statistically significant while one hypothesis is not significant.

H1 states that the information quality significantly increases users' decision on review usefulness, which is not supported ($b = 0.04$, $p > 0.5$).

H2 states that information credibility significantly increases users' decision on review usefulness, which is supported ($b = 0.24$, $p < 0.001$).

H3 states that perceived ease of use significantly increases users' decision on review usefulness, which is supported ($b = 0.29$, $p < 0.001$).

H4 states that social influence significantly increases users' decision on review usefulness, which is supported ($b = 0.37$, $p < 0.001$).

H5 states that social influence significantly increases users' decision to provide a vote on the review, which is supported ($b = 0.55$, $p < 0.001$).

H6 states that information usefulness significantly increases users' decision to provide a vote on the review, which is supported ($b = 0.22$, $p < 0.001$).

Table 11. Results of hypotheses verification.

Hypotheses	Relationship	Standard Regression Coefficient	CR	p
H1	Information quality→ Information usefulness	0.04	0.613	0.54
H2	Information credibility→ Information usefulness	0.24	2.598	<0.001
H3	Perceived ease of use→ Information usefulness	0.29	3.545	<0.001
H4	Social influence→ Information usefulness	0.37	3.377	<0.001
H5	Social influence→ Vote adoption	0.55	2.159	<0.001
H6	Information usefulness→ Vote adoption	0.22	5.058	<0.01

Table 12. Goodness of fitting.

Item	Value
$\chi^2 / d.f.$	1.434
Goodness-of-fit index (GFI)	0.928
Adjusted GFI (AGFI)	0.900
Comparative fit index (CFI)	0.975
RMSEA	0.042

Additionally, in order to test the mediating effect of the information usefulness in the model, we choose the bootstrapping method (using 2000 samples) [46]. At a significant level of 95%, the mediation of information usefulness exists between information quality and vote adoption (0.057, 0.218), and between perceived ease of use and vote adoption (0.160, 0.520). The mediating effect of the information usefulness is achieved, which means that voting the hot review is usually as a result of information usefulness.

5. Discussions and Suggestions

5.1. Discussions

According to the data analysis in Section 4, the survey data confirms that reliability, perceived ease of use, and social influence are the main factors influencing review voting. However, the correlation between behavior and text quality has not been validated. The imbalance of voting motives and the failure of voting signals may be the causes of the uselessness of information quality. Therefore, the analysis supports our assumption on the causes of the crowd intelligence paradox.

Specifically, strong social needs repress the work of autonomy and competence. An economic person is embedded in a social structure [47]. In online communities, everyone is deeply embedded in a community environment. They constantly collect information and make decisions in the interaction process between individuals and then consciously or unconsciously make a choice under the influence of others. When people provide a vote to a song, they are implicitly connected with other users who follow the song. As the voters and the followers of a song are supposed to have similar interests, they are much likely to interact with each other in the online music community. So the quality of a review is insignificant. The satisfaction of social need is much greater than the satisfaction of the cognitive value of a review [19]. At the same time, group praises form a huge pressure of identity. Voting for a hot review is a way to escape the pressure of judgment. Therefore, "Readily praise (unconditionally vote)" becomes a custom under an imbalance of voting motives.

Information reliability has a significant impact, which is similar to previous studies. The reliability of the information source (as an edge path) has an impact on information receivers [30]. However, there is a lack of heterogeneous signals in online communities, compared to the trading community. Users assess the reliability and often rely on the number of previous points. Because the low-cost signals cannot be an effective tool for reducing or eliminating information asymmetry, the number of votes cannot represent the quality of content. A comment that is relatively low or even of no value will receive the majority of votes under the fake signal. It may mislead users, which is similar to the bias of the ratings [3]. This principle can also be observed by a counter example. For example, Stack Overflow (https://stackoverflow.com/) is the largest, most trusted online community for developers to learn, share their programming knowledge, and build their careers. A voting mechanism is relatively good in Stack Overflow. Because the knowledge of communication is more professional and standard, users need to spend a certain time to do a mental work. Based on a professional and technical literacy, users pay more attention to their reputations. Therefore, the cost of voting is high and the signal is relatively good.

5.2. Suggestions for the Sustainable Development of Online Communities

Based on the empirical research, we provide some research implications and advice for the sustainable development of online communities.

We suggest increasing the cost of the voting signal. Based on the cost point, we can set a time interval to avoid voting for three reviews in a second. This may increase a selection cost and a time cost, while potentially not affecting normal voting. We can also set a negative vote to amend the integral of votes by forcing users to actively and subjectively evaluate. In order to avoid a negative feedback from negative votes to publishers, we can only open to the recipient and cannot display information publicly.

We suggest setting up multiple signals for review voting in online communities. Due to a lack of signals, users usually judge the value of content based on the number of votes. We can set multiple signals to weaken the strength of the voting signal. For example, give a preset score for content (through arithmetic) and then encourage people to independently revise the score. Because changing the score requires some certain ability, it may enhance the satisfaction of autonomy and competence. Therefore it may reduce the impact of social voting motives.

It is better to provide behavioral data, especially the record of voting data, in the review voting environment. For example, we can list the historical content of the votes, as well as the feedback of voting, specifying how many people vote for the same content. This may make users reflect their voting behavior and enhance the demand of autonomy and competence.

Enterprises need to avoid traffic traps. A key indicator of community development is user traffic. Community administrators are accustomed to seeing a large number of frequent voting interactions as a help of improving community activity—however, it ignores the harm of the "Readily praise (unconditionally vote)". Content publishers often face many praises, but only a small amount of feedback content. This long-term, frequent, and ineffective interaction affects the publishers' enthusiasm. Community managers should carefully avoid traffic traps and maintain the UGC ecological environment. Additionally, paying more attention to traffic and less attention to quality is an extensive development pattern. At early stages, massive users provide natural resources for online communities. Additionally, network externality has strengthened the power of a crowd. However, the Internet is becoming more and more mature and the demographic dividend is relatively disappearing. Refined management will become a feasible way to get competitive advantages.

6. Conclusions and Future Work

In this paper, we discovered what the crowd intelligence paradox of review voting in online communities is and why it exists. We first testify that the distribution of review votes in online communities matches the 80/20 rule and only a few comments can receive votes. Then, as to those reviews with many votes, not all reviews are related to high-quality comments. We found that they fail to reflect the quality of review content and this voting mechanism cannot reflect crowd intelligence, and it means that there is a crowd intelligence paradox in online communities. The above successive aspects quantitatively measure the existence of the crowd intelligence paradox in online communities.

Furthermore, we constructed a new research model called VAM (voting adoption model) to uncover the imbalance of voting motives and the failure of voting signals. The survey data confirms that reliability, perceived ease of use, and social influence are main factors influencing review voting. Based on SDT, strong social needs repress the work of autonomy and competence. The satisfaction of social need is much greater than the satisfaction of the cognitive value of a review. Therefore, "Readily praise (unconditionally vote)" becomes a custom under an imbalance of voting motives. On the other hand, because the low-cost signals cannot be an effective tool for reducing or eliminating information asymmetry, the number of votes cannot represent a quality of content. A review that is relatively low or even of no value will receive the majority of votes under the fake signal.

One limitation of this study is that the manual annotation might not reflect users' real intention, because the involved volunteers for annotation are different from users in online communities. This is always a crucial issue in classification and other text mining tasks, and in the future we will explore other feasible approaches that can get results close to natural voting. Another future work is to investigate more online communities to verify the crowd intelligence paradox. In addition, we will also explore the application of the research results on a real online community to improve the efficiency and effectiveness of online businesses.

Author Contributions: J.Z.: Conceptualization, Project administration, Supervision, and Writing—original draft. J.W.: Data curation and Formal analysis. S.F.: Investigation and Software. P.J.: Funding acquisition, Supervision, and Writing—review & editing.

Funding: This paper is partially supported by the National Science Foundation of China (Nos. 71273010 & 61672479) and the Doctor Start-up Fund of Anhui University.

Acknowledgments: We would like to thank the editors and anonymous reviewers for their suggestions and comments to improve the quality of the paper.

Conflicts of Interest: The authors declare no conflict of interest.

References

1. Choi, B.; Lee, I. Trust in open versus closed social media: The relative influence of user-and marketer-generated content in social network services on customer trust. *Telemat. Inform.* **2017**, *34*, 550–559. [CrossRef]
2. Kuan, K.; Hui, K.L.; Prasarnphanich, P.; Lai, H.Y. What makes a review voted? An empirical investigation of review voting in online review systems. *J. Assoc. Inf. Syst.* **2015**, *16*, 48–71. [CrossRef]
3. Hu, N.; Pavlou, P.; Zhang, J. On self-selection biases in online product reviews. *MIS Q.* **2017**, *41*, 449–471. [CrossRef]
4. Vickery, G.; Wunsch-Vincent, S. *Participative Web and User-Created Content: Web 2.0 Wikis and Social Networking*; Organization for Economic Cooperation and Development (OECD): Paris, France, 2007.
5. Krishnamurthy, S.; Dou, W. Note from special issue editors: Advertising with user-generated content: A framework and research agenda. *J. Interact. Advert.* **2008**, *8*, 1–4. [CrossRef]
6. Armstrong, A.; Hagel, J. The real value of online entertainment communities. *Knowl. Communities* **2000**, *74*, 85–95.
7. Wang, W.; Li, Y. How trust and need satisfaction motivate producing user-generated content. *J. Comput. Inf. Syst.* **2017**, *57*, 49–57. [CrossRef]
8. Li, G.; Yang, X. Effects of social capital and community support on online community members' intention to create user-generated content. *J. Electron. Commer. Res.* **2014**, *15*, 190.
9. Luo, X.; Gu, B.; Zhang, J.; Phang, C. Expert Blogs and Consumer Perceptions of Competing Brands. *MIS Q.* **2017**, *41*, 371–395. [CrossRef]
10. Shriver, S.; Nair, H.; Hofstetter, R. Social ties and user-generated content: Evidence from an online social network. *Manag. Sci.* **2013**, *59*, 1425–1443. [CrossRef]
11. Christodoulides, G.; Jevons, C.; Bonhomme, J. Memo to marketers: Quantitative evidence for change. *J. Advert. Res.* **2012**, *52*, 53–64. [CrossRef]
12. Zhao, J.; Wang, X.; Jin, P. Feature selection for event discovery in social media: A comparative study. *Comput. Hum. Behav.* **2015**, *51*, 903–909. [CrossRef]
13. Jin, P.; Mu, L.; Zheng, L.; Zhao, J.; Yue, L. News feature extraction for events on social network platforms. In Proceedings of the Companion Proceedings of the 26th International Conference on World Wide Web Companion (WWW'17), Perth, Australia, 3–7 April 2017; pp. 69–78.
14. Yang, Y.; Wang, X.; Guan, T.; Shen, J.; Yu, L. A multi-dimensional image quality prediction model for user-generated images in social networks. *Inf. Sci.* **2014**, *281*, 601–610. [CrossRef]
15. Ahn, D.; Duan, J.; Mela, C. Managing user-generated content: A dynamic rational expectations equilibrium approach. *Mark. Sci.* **2015**, *35*, 284–303. [CrossRef]
16. Chen, J.; Xu, H.; Whinston, A.-B. Moderated online entertainment communities and quality of user-generated content. *J. Manag. Inf. Syst.* **2011**, *28*, 237–268. [CrossRef]
17. Mudambi, S.; Schuff, D. What makes a helpful online review? A study of customer reviews on amazon.com. *MIS Q.* **2010**, *34*, 185–200. [CrossRef]
18. Hennig-Thurau, T.; Gwinner, K.P.; Walsh, G.; Gremler, D.D. Electronic word-of-mouth via consumer-opinion platforms: What motivates consumers to articulate themselves on the internet? *J. Interact. Mark.* **2004**, *18*, 38–52. [CrossRef]
19. Cheung, C.; Lee, M. What drives consumers to spread electronic word of mouth in online consumer-opinion platforms. *Decis. Support Syst.* **2012**, *53*, 218–225. [CrossRef]
20. Luan, J.; Yao, Z.; Zhao, F.T.; Liu, H. Search product and experience product online reviews: An eye-tracking study on consumers' review search behavior. *Comput. Hum. Behav.* **2016**, *65*, 420–430. [CrossRef]
21. Jabr, W.; Zheng, Z. Know yourself and know your enemy: An analysis of firm recommendations and consumer reviews in a competitive environment. *MIS Q.* **2014**, *38*, 635–654. [CrossRef]
22. Wang, X.; Li, Y. Trust, psychological need, and motivation to produce user-generated content: A self-determination perspective. *J. Electron. Commer. Res.* **2014**, *15*, 241–253.

23. Erkan, I.; Evans, C. The influence of eWOM in social media on consumers' purchase intentions: An extended approach to information adoption. *Comput. Hum. Behav.* **2016**, *61*, 47–55. [CrossRef]
24. Xu, X.; Yao, Z. Understanding the role of argument quality in the adoption of online reviews: An empirical study integrating value-based decision and needs theory. *Online Inf. Rev.* **2015**, *39*, 885–902. [CrossRef]
25. Fan, Y.; Miao, Y.; Fang, Y.; Lin, R. Establishing the adoption of electronic word-of-mouth through consumers' perceived credibility. *Int. Bus. Res.* **2013**, *6*, 58–65. [CrossRef]
26. Wang, F.; Liu, X.F.; Fang, E. User reviews variance, critic reviews variance, and product sales: An exploration of customer breadth and depth effects. *J. Retail.* **2015**, *91*, 372–389. [CrossRef]
27. Korfiatis, N.; GarcíA-Bariocanal, E.; Sánchez-Alonso, S. Evaluating content quality and helpfulness of online product reviews: The interplay of review helpfulness vs. Review content. *Electron. Commer. Res. Appl.* **2012**, *11*, 205–217. [CrossRef]
28. Moore, S. Attitude predictability and helpfulness in online reviews: The role of explained actions and reactions. *J. Consum. Res.* **2015**, *42*, 30–44. [CrossRef]
29. Zhu, F.; Zhang, X. Impact of online consumer reviews on sales: The moderating role of product and consumer characteristics. *J. Mark.* **2010**, *74*, 133–148. [CrossRef]
30. Cheung, M.; Sia, C.; Kuan, K. Is this review believable? A study of factors affecting the credibility of online consumer reviews from an ELM perspective. *J. Assoc. Inf. Syst.* **2012**, *13*, 618. [CrossRef]
31. Cao, Q.; Duan, W.; Gan, Q. Exploring determinants of voting for the "helpfulness" of online user reviews: A text mining approach. *Decis. Support Syst.* **2011**, *50*, 511–521. [CrossRef]
32. Chaiken, S. The heuristic model of persuasion. *Ont. Symp. Soc. Influ.* **1987**, *5*, 3–39.
33. Cacioppo, J.; Petty, R.; Kao, C.; Rodriguez, R. Central and peripheral routes to persuasion: An individual difference perspective. *J. Pers. Soc. Psychol.* **1986**, *51*, 1032–1043. [CrossRef]
34. Sussman, S.; Siegal, W. Informational influence in organizations: An integrated approach to knowledge adoption. *Inf. Syst. Res.* **2003**, *14*, 47–65. [CrossRef]
35. Knoll, J. Advertising in social media: A review of empirical evidence. *Int. J. Advert.* **2016**, *35*, 266–300. [CrossRef]
36. Deci, E.; Eghrari, H.; Patrick, B.; Leone, D. Facilitating internalization: The self-determination theory perspective. *J. Pers.* **1994**, *62*, 119–142. [CrossRef] [PubMed]
37. Zhao, J.; Fang, S.; Jin, P. Modeling and quantifying user acceptance of personalized business modes based on TAM, trust and attitude. *Sustainability* **2018**, *10*, 356. [CrossRef]
38. Connelly, B.; Certo, S.; Ireland, R.; Reutzel, C. Signaling theory: A review and assessment. *J. Manag.* **2011**, *37*, 39–67. [CrossRef]
39. Reyes, A.; Rosso, P. Making objective decisions from subjective data: Detecting irony in customer reviews. *Decis. Support Syst.* **2012**, *53*, 754–760. [CrossRef]
40. Gefen, D.; Karahanna, E.; Straub, D. Trust and TAM in online shopping: An integrated model. *MIS Q.* **2003**, *27*, 51–90. [CrossRef]
41. Park, D.; Lee, J.; Han, I. The effect of on-line consumer reviews on consumer purchasing intention: The moderating role of involvement. *Int. J. Electron. Commer.* **2007**, *11*, 125–148. [CrossRef]
42. Prendergast, G.; Ko, D.; Yin, V. Online word of mouth and consumer purchase intentions. *Int. J. Advert.* **2010**, *29*, 687–708. [CrossRef]
43. Wu, Y.; Zhou, C.; Chen, M. Human comment dynamics in on-line social systems. *Phys. A Stat. Mech. Appl.* **2010**, *389*, 5832–5837. [CrossRef]
44. Clauset, A.; Shalizi, C.; Newman, M. Power-law distributions in empirical data. *SIAM Rev.* **2009**, *51*, 661–703. [CrossRef]
45. Alstott, J.; Bullmore, E.; Plenz, D. Powerlaw: A Python package for analysis of heavy-tailed distributions. *PLoS ONE* **2014**, *9*, e85777. [CrossRef] [PubMed]
46. Preacher, K.; Rucker, D.; Hayes, A. Addressing moderated mediation hypotheses: Theory, methods, and prescriptions. *Multivar. Behav. Res.* **2007**, *42*, 185–227. [CrossRef] [PubMed]
47. Granovetter, M. Economic action and social structure: The problem of embeddedness. *Am. J. Sociol.* **1985**, *91*, 481–510. [CrossRef]

 *sustainability*

Article

SentiFlow: An Information Diffusion Process Discovery Based on Topic and Sentiment from Online Social Networks

Berny Carrera and Jae-Yoon Jung *

Department of Industrial and Management Systems Engineering, Kyung Hee University, Yongin-si, Gyeonggi-do 446-701, Korea; berny@khu.ac.kr
* Correspondence: jyjung@khu.ac.kr; Tel.: +82-31-201-2537

Received: 3 July 2018; Accepted: 31 July 2018; Published: 2 August 2018

Abstract: In this digital era, people can become more interconnected as information spreads easily and quickly through online social media. The rapid growth of the social network services (SNS) increases the need for better methodologies for comprehending the semantics among the SNS users. This need motivated the proposal of a novel framework for understanding information diffusion process and the semantics of user comments, called SentiFlow. In this paper, we present a probabilistic approach to discover an information diffusion process based on an extended hidden Markov model (HMM) by analyzing the users and comments from posts on social media. A probabilistic dissemination of information among user communities is reflected after discovering topics and sentiments from the user comments. Specifically, the proposed method makes the groups of users based on their interaction on social networks using Louvain modularity from SNS logs. User comments are then analyzed to find different sentiments toward a subject such as news in social networks. Moreover, the proposed method is based on the latent Dirichlet allocation for topic discovery and the naïve Bayes classifier for sentiment analysis. Finally, an example using Facebook data demonstrates the practical value of SentiFlow in real world applications.

Keywords: information diffusion; community detection; topic analysis; sentiment analysis; social networks

---

1. Introduction

Today, social network services (SNS) are an effective medium through which new information, such as opinions, news, and advertisements, is easily and quickly disseminated. The spread of these information starts when users create new posts. Subsequently, all subscribers and users who comment are notified of the new posts. To better understand the spread of these ideas, it is important to analyze how people propagate their thoughts based on their opinions and topics of interest, which are the underlying context and information flow.

Some applications for the proposed technique are viral marketing, where marketers quantify the impact of released products by applying sentiment analysis to understand unsatisfied consumers; influence analysis, determining how groups of users influence other groups; and trend detection, in which with the application of topic models, discussions, and opinions can be uncovered. In particular, in terms of social science, it is possible to understand the relationship among the users' behaviors, the distinctions of communities and the information diffusion in social networks. Understanding users' reactions are valuable since opinions can influence the news trend or purchase decisions. Therefore, the users' opinions are vital to understanding the way information spreads and how communities interact among them.

There have been several studies on information flow modeling based on the structure of social network or the discovery of information diffusion processes without analyzing the structure of communication [1–6]. However, their studies have not shown the relation between context and information flow. A few studies have considered how to model information diffusion with a process structure. Kim et al. [7] presented an information diffusion model using data of a blog to analyze the reposting behaviors of people. Although other studies have been carried out on information flows and SNS data [8–11], they can be used only to infer an information diffusion flow without considering the probability that communities will communicate or showing the contextual information. Kim et al. [12] analyzed the behavioral patterns in SNS, News and Blog sites. However, the disadvantage in their study is the absence of opinions in the users' comments. Opinions made by users in SNS can have a huge impact in society, therefore the analysis of these emotions are important to monitor the response of users to a specific news or product [13]. There exist some studies in information diffusion based on sentiment [14–16], but they do not consider the opinion flow between communities. Other researchers analyzed how to visualize topics and opinions in SNS [17–21], but they lack in the information process flow. In this research, a new semantic hidden Markov model (HMM) for discovering information diffusion, named SentiFlow, is introduced to discover probabilistic information flow in consideration of topics and sentiment. It is an extension of HMM [22] using text mining and process mining [23]. The probabilities in the SentiFlow are computed based on maximum likelihood (ML) [22]. In our previous studies [24,25], a method for probabilistic information flow of the communication between users and communities is presented. A method to underline the semantics and opinions in the interactions among user groups is suggested in this paper by applying community clustering algorithms to find user communities and by undertaking two different analyses, topic modeling and sentiment analysis, for the user comments. Finally, the traces of these communications are analyzed, and different information flow process models are generated. The goal is to answer the following important questions: (1) "What topics promote communication among user communities?" (2) "How are the positive, neutral, and negative opinions shared in the information diffusion process from a probabilistic point of view?"

The rest of the paper is organized as follows. Section 2 describes the proposed methodology used in this work. Section 3 describes the general algorithm used. Section 4 provides the experimental results. Finally, Section 5 concludes this work.

2. Framework

In this research, log data collected from SNS are used to discover information diffusion process. Generally, users in SNS wrote posts and their friends comment on the posts. Therefore, it can be assumed that, in a SNS log, each post of a specific user is characterized by many comments of his/her friends and the comments are ordered chronologically. From the SNS log, we first find the user communities based on their interaction and draw the information diffusion process. The topics of interest are found and annotated on the discovered process. Finally, the sentiment for the topics is analyzed for topics and users. The overall framework is depicted in Figure 1.

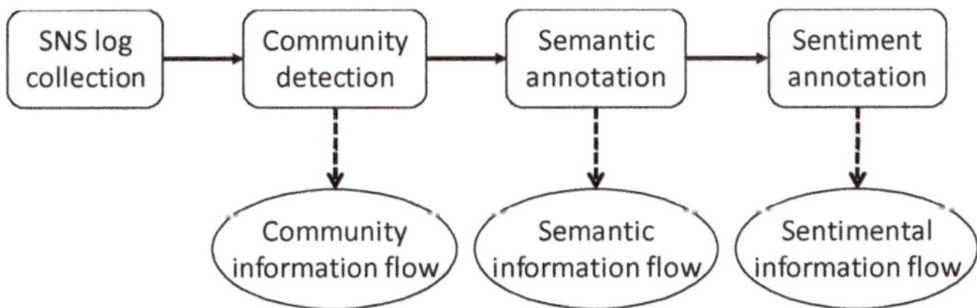

Figure 1. A framework of the information diffusion process discovery with topic and sentiment.

2.1. SNS Log Collection

In SNS such as Facebook or Twitter, all posts are obtained along with the user's name, user's comments, and the timestamp indicating when it was posted to create an SNS log. In this research, it is assumed that users write comments to reply to other users and thus create or continue a discussion about a related post. Each post published by the fan page owner is characterized with a sequence of SNS events. An SNS event contains a user, the user comment, and the time when the comment was published. This sequence is ordered using the timestamp of the comment publication. An SNS log is defined below.

Definition 1. *(SNS log) Let $P = \{p_1, \ldots, p_K\}$ and $U = \{u_1, \ldots, u_V\}$ be the finite sets of all possible post identifiers and users, respectively. K and V are the numbers of posts and users, respectively. Posts are characterized by SNS events e, which in turn are characterized by various attributes att. For any SNS event $e \in E$, #att(e) is the value of attribute att in event e to have SNS event $e = (\#u(e), \#user\_comment(e), \#time(e))$. Additionally, each post has an attribute action trace for a specific post p, denoted by σ_k, and is defined as the sequence of SNS events in p, i.e., $\sigma_k = \langle e_1, \ldots, e_H \rangle$ for $1 \leq k \leq K$, where H is the number of events for p. An SNS log, denoted by $L = [\sigma_k]$, is a multi-set of action traces over U and P in the SNS.*

To illustrate the operation of the proposed framework, Table 1 presents an example SNS log with synthetic data. The example SNS log contains the post identification and the comment traces. The comment traces show the structure User $_{comment}$, where the user's name is written and followed by the comment in subscript. Each user and comment are ordered by the timestamp for when the comment was published. The example SNS log contains six posts and 23 comments written by five users: Angela, George, John, Paul, and Ringo.

Table 1. An example SNS log L_1. A SNS log contains many action traces, which are sequential comments replies to specific posts.

Post ID	Action Trace
p_1	John $_{I\ like\ it}$, Angela $_{This\ is\ amazing!}$, George $_{I\ think\ this\ is\ absurd}$
p_2	John $_{We\ need\ to\ be\ persistent}$, Ringo $_{I\ think\ this\ is\ very\ aggressive}$, Paul $_{I\ am\ ashamed}$, George $_{We\ need\ to\ demand\ our\ rights!}$
p_3	John $_{It's\ better\ if\ we\ reform\ the\ laws}$, Paul $_{I\ am\ relaxed}$, Ringo $_{This\ is\ a\ revolution}$, George $_{pitiful}$
p_4	John $_{Wow\ this\ is\ perfect}$, Paul $_{That\ is\ bad}$, Ringo $_{Nice}$, George $_{Excellent}$
p_5	John $_{Too\ much\ stress}$, Paul $_{I\ am\ afraid}$, Ringo $_{Superficial}$, George $_{It\ is\ ok\ everything\ will\ be\ fine}$
p_6	John $_{Terrorism}$, Ringo $_{I\ am\ so\ tired}$, Paul $_{I\ am\ so\ happy}$, George $_{Cool}$

2.2. Information Flow among Communities

In this step, the communication of users and the interaction between them are analyzed. For this, users with similar behavior can be clustered into communities. The community detection analysis performs the next activities: identify the network structure inside the SNS log by applying community-detection algorithms, determine how the people across the comments are related, and help minimize the complexity of the discovered process model. The discovered communities represent the community states in the process model.

In the proposed framework, the Louvain modularity (LM) algorithm is used, which is often applied as a community detection method in social network analysis [26]. LM detects and extracts communities in a network by providing the optimal number of communities and optimizing the value of modularity [26], the results of which is used for the best grouping of users in this research. The user communities for this research also represent the information diffusion states for the process model. Moreover, the objective in this step is achieved by creating the information diffusion matrix. The information diffusion matrix represents the frequency of communication inside, outside, and among the communities. Thereby, the action traces in the posts can represent the sharing of information between user communities, which creates an information diffusion matrix to represent the information flow frequencies from one community to another. A process model is then obtained as output. The user community and the diffusion community matrix are defined as follows.

Definition 2. *(User community) Let U be a finite set of users in SNS log L and C = {c_1, …, c_N} for $1 \leq i \leq N$ be a finite set of communities of users in L. A user community $c_i \subseteq U$ is a subset of users grouped by the results of a community detection algorithm community(u).*

Definition 3. *(Information diffusion matrix) Let C be user communities of SNS log L. The information diffusion matrix contains the information flow frequencies between two communities in C, which is denoted by A = (a_{ij}), where $a_{ij} = \sum_{\forall \sigma \in L} |c_i \rightarrow c_j|$ represents the sum of the frequencies of information diffusion from c_i to c_j and $\pi = \{c_1, \ldots, c_N\}$, where π_i is the probability of being in c_i at time 1 in every action trace $\sigma \in L$.*

$$a_{ij}' = \frac{a_{ij}}{\sum_{n=1}^{N} a_{in}} \tag{1}$$

The information diffusion process model indicates the beginning of the information diffusion and how the information spreads among the communities. The model mines the initial probability π that describes the probability of which user community starts the information diffusion in each action trace. The calculation of the parameter values using A for the process model is shown in Equation (1). $a_{i,j}'$ is obtained from the information flow frequency of the community c_j, which follows community c_i, divided by the total information flow frequency of all communities that follow c_i.

From the example introduced in Table 1, it is seen that the communities with LM, for this example the resolution parameter value = 0.02, is used to show a better structure of the information diffusion from smaller clusters. The results of the LM algorithm are c_1 = {John, Angela}, c_2 = {George, Ringo}, and c_3 = {Paul}. Table 2 presents the first findings for the process model obtaining the matrix diffusion community A and the probability of state transition A'. In addition, the transition probability distribution is $\pi = (1, 0, 0)$ where c_1 always initiates the information diffusion.

Table 2. Matrices extracted from L_1: (a) information diffusion matrix A; and (b) state transition probability matrix A'.

	(a)			(b)		
	c_1	c_2	c_3	c_1	c_2	c_3
c_1	0	3	3	0.0	0.5	0.5
c_2	0	0	2	0.0	0.0	1.0
c_3	0	5	0	0.0	1.0	0.0

The information diffusion process model can be drawn in Figure 2. It is clear that the thickness of c_2 is greater because it has more incoming information diffusion than do the other communities. Moreover, the information diffusion between communities c_2 and c_3 is notable.

The information flow in this research is based on the detected communities as shown in Figure 2. However, the communities may be changing over time. To obtain more reliable structure of the information flow, data in enough long period is needed for community detection.

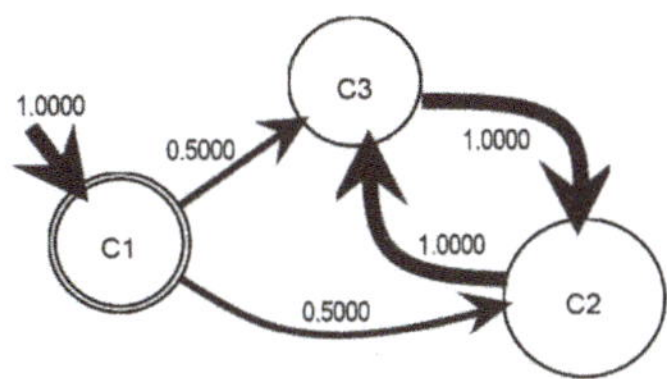

Figure 2. The information diffusion process model generated from log L_1.

2.3. Semantic Information Flow

The third step is the discovery of underlying topics where the meaning and sense of data are analyzed by extracting the principal keywords from comments to understand what people's interests are and how the topics relate to them. As input, the user comments are collected, and tokenization of the words is conducted by breaking the comments into sentences and then into tokens to remove the English non-words, punctuation, and stop words. In this study, the probabilistic topic model technique called the latent Dirichlet allocation (LDA) is used to properly assign and discover the hidden context from the data. LDA can represent documents, signified by the comments of users as mixtures of topics, and then assigns the words with certain probabilities [27]. Furthermore, in the semantic information diffusion process model, the frequent topics inside a community and the probability of those topics are analyzed.

To construct this semantic process model, analysis of the comments needs to be completed as described above. The first goal of this step is to find a small number of topics from the observations of user comments. Hence, the LDA algorithm is used to discover these topics. Thus, each topic found is assigned to each comment to finally obtain the topic matrix as a second goal. The topic matrix represents the frequency with which each user from a community publishes a comment for a specific topic. Therefore, the topic matrix can be defined.

Definition 4. *(Topic matrix) Let C be information diffusion states of the information diffusion process of SNS log L and $T = \{t_1, \ldots, t_M\}$, with $1 \leq m \leq M$ being a finite set of topics discovered from the LDA algorithm $lda(\#user\_comment(L))$. The topic matrix is denoted by $B = (b_{im})$, where each element $b_{im} = \sum_{\forall \sigma \in L} f(c_i, t_m)$ contains the sum of the frequencies in which a topic t_m exists in the user comments of a community c_i in every action trace $\sigma \in L$.*

$$b_{im}' = \frac{b_{im}}{\sum_{q=1}^{M} b_{iq}} \tag{2}$$

The calculation of the parameter values using B for the information diffusion process model is shown in Equation (2). b_{im}' is obtained from the frequency of community c_i, which is paired with a topic t_m, divided by the total frequency of all topics paired with c_i.

This step uses the community states, topic observations, and SNS log as inputs and generates a semantic process model as output. Moreover, a topic matrix is constructed and shows the frequency between the observed context and each community. The topics are determined from the user's comments. To discover the topics, the LDA algorithm is used, and two topics are obtained from the comments of L_1. The top eight keywords of the two topics are t_1 = {need, nice, demand, happy, need demand, need persistent, ok, everything fine} and t_2 = {think, wow perfect, pitiful, better, better reform, cool, everything, excellent}.

Table 3. Matrices extracted from log L_1 for constructing a semantic information diffusion process model: (a) topic matrix B; and (b) observation symbol probability matrix B'.

	(a)		(b)	
	t_1	t_2	t_1	t_2
c_1	4	3	0.57	0.43
c_2	3	8	0.27	0.73
c_3	4	1	0.80	0.20

Table 3 presents the first findings for the semantic process model obtaining the topic matrix B and the observation symbol probability matrix B'. Figure 3 shows the semantic information flow drawn from B'. In the graphical representation, the dashed arcs created from the community state of the topic represent the interest of the community in the specific topic, and the thickness of the arc represents the probability between them.

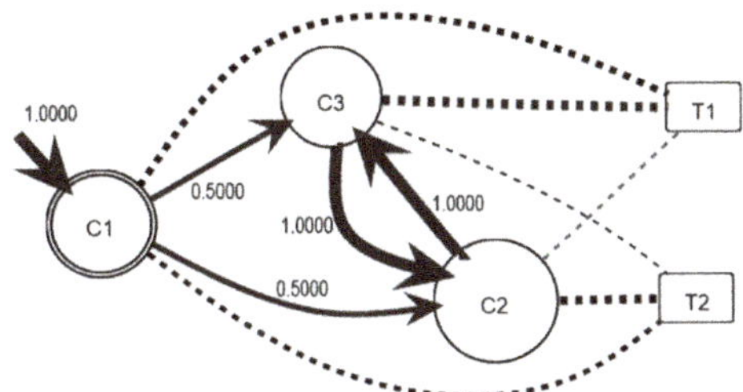

Figure 3. The semantic information diffusion process model generated from log L_1. t_1 and t_2 are the topics that were discovered from the texts exchanged among users.

2.4. Sentimental Information Flow

In this step, sentiment analysis, which is classification of the user comments based on sentiment, is performed to gain a better understanding of the user. The analysis is performed using the naïve Bayes sentiment classifier described in [28] because this method shows good performance in many applications [29]. In this research, the polarity of the comment is classified as a positive, neutral, or negative impression of the topics on which the users have commented.

The last output is the sentimental information flow. The model uses the previous matrices and creates a sentiment matrix that represents the probabilities of the sentiments of each community for a specific topic. The sentiment matrix is described below.

Definition 5. *(Sentiment matrix) For SNS log L, let C be a set of user communities in L, T be a finite set of topics in L, and $S = \langle s_1, s_2, s_3 \rangle$ be a tuple of positive, neutral, and negative sentiments discovered from naïve Bayes sentiment classifier nbsc(#user_comment(e)). The sentiment matrix is a three-dimensional matrix $D = (d_{imr})$, where $d_{i,m,r} = \sum_{\forall \sigma \in L} f(c_i, t_m, s_r)$ is the sum of the frequencies in which a sentiment s_r exists in the user comments of a community c_i for a topic t_m in every action trace $\sigma \in L$.*

$$d_{imr}{}' = \frac{d_{imr}}{\sum_{q=1}^{3} d_{imq}} \tag{3}$$

The calculation of the parameter values using D for the sentimental information flow model is shown in Equation (3). $d_{imr}{}'$ is obtained from the frequency of community c_i and topic t_m, that is paired with a sentiment s_r, divided by the total frequency of all sentiments paired with c_i and t_m. After the construction of the diffusion community matrix, topic matrix, and sentiment matrix, the last step is the modeling of the sentimental information flow, called SentiFlow. A SentiFlow model can be defined as follows.

Definition 6. *(SentiFlow) A SentiFlow model of SNS log L is an extension of HMM for representing semantic and sentimental information diffusion. A SentiFlow model is denoted by $\Lambda(L) = (\pi, C, T, A', B', D')$, where π is the transition probability distribution of initial states, C is a set of user communities, T is a set of discovered topics, A' is the matrix of state transition probability distribution from information diffusion matrix A, B' is the matrix of observation symbol probability distribution from topic matrix B, and D' is the three-dimensional matrix of sentiment probability distribution from sentiment matrix D. Note that $\sum_{j} a'_{ij} = 1$ for $\forall i$, $\sum_{m} b'_{im} = 1$ for $\forall i$, and $\sum_{r} d'_{imr} = 1$ for $\forall (i, m)$.*

$\pi = (\pi_i)$ for $1 \leq i \leq N$, where π_i is the probability of being in c_i at time 1.

$C = \{c_1, \ldots, c_N\} \in L$ for $1 \leq i \leq N$, where $\{c_1, \ldots, c_N\}$ are the information diffusion states of L.

$T = \{t_1, \ldots, t_M\} \in L$ for $1 \leq m \leq M$, where $\{t_1, \ldots, t_M\}$ are the observed topics of L.

$A' = (a'_{ij})$ for $1 \leq i, j \leq N$, where a'_{ij} is the probability of state transition from c_i to c_j.

$B' = (b'_{im})$ for $1 \leq i \leq N, 1 \leq m \leq M$, where b'_{im} is the probability of observing t_m in state c_i.

$D' = (d'_{imr})$ for $1 \leq i \leq N, 1 \leq m \leq M, 1 \leq r \leq 3$, where d'_{imr} is the probability of observing a sentiment s_r from a topic t_m in a state c_n.

Figure 4 provides the representational model with the notation used in this research. It should be noted that the probabilities of B' are shown, but simply indicate the thickness of the arc from a community to the respective topic.

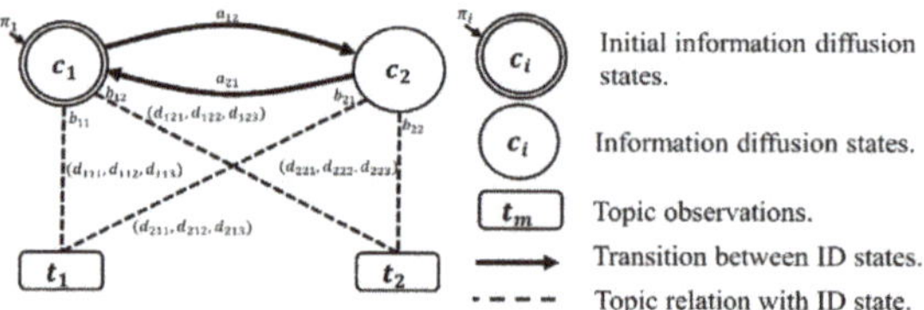

Figure 4. Representation of a SentiFlow model.

The sentimental analysis step in our framework is for understanding the user opinions written in the comments and obtaining a sentiment matrix that has the frequencies from three types of sentiments (positive, negative, and neutral) from the discovered topics in the semantic annotation step. For the sentiment analysis, the naïve Bayes sentiment classifier is used to classify 12 positive, 10 negative, and 1 neutral commentaries. Table 4 presents the findings for the sentimental annotation by obtaining a sentiment matrix D and a sentiment probability matrix D'.

Table 4. Matrices extracted from log L_1 for sentiment annotation: (a) sentiment matrix D; and (b) sentiment probability matrix D'.

		(a)			(b)		
		s_1	s_2	s_3	s_1	s_2	s_3
c_1	t_1	3	0	1	0.75	0.00	0.25
	t_2	2	0	1	0.67	0.00	0.33
c_2	t_1	1	1	1	0.33	0.33	0.33
	t_2	4	0	4	0.50	0.00	0.50
c_3	t_1	1	0	3	0.25	0.00	0.75

Figure 5 presents the graphical representation of the SentiFlow model constructed from log L_1. The difference of color between the communities and topics is shown. For example, c_1 to t_1 shows a bluish color representing predominant positive commentaries (0.75) compared to negative commentaries (0.25). However, c_3 to t_1 presents predominantly negative commentaries for t_1 (0.75), with 0.25 positive commentaries. Additionally, community c_2 to topic t_2 has a mixture of sentiments in the comments, with 0.5 for both. The mapping color of the arc from a community to a topic represents the type of sentiment; the arc is red if the sentiment is negative, lime if neutral, and blue if positive.

A SentiFlow model provides the required information to answer the two questions presented at the end of Section 1. The first question is about the topics that promote communication between communities. In this example, the communication between communities c_2 and c_3 was about topics t_1 and t_2, although c_2 mainly focused on t_2 and c_3 mainly focused on t_1. The second question is about how the sentiment is shared in the information diffusion process from a probabilistic perspective. As an example, the information diffusion from communities c_1 to c_2 for topic t_2 is used. Considering the sequence, <positive, positive>, the result can be analyzed using the forward algorithm [22], and the probability of the sequence is P (<positive, positive> | Λ) = 1.0 × 0.67 × 0.5 × 0.5 = 0.1675. In the case of the sequence <neutral, neutral>, the probability is 0, and the sequence <negative, negative> is

$P(<\text{negative, negative}> | \Lambda) = 1.0 \times 0.33 \times 0.5 \times 0.5 = 0.0825$. Therefore, the probability that community c_2 responds positively to a positive comment of community c_1 is higher because community c_1 has a higher probability of posting a positive comment.

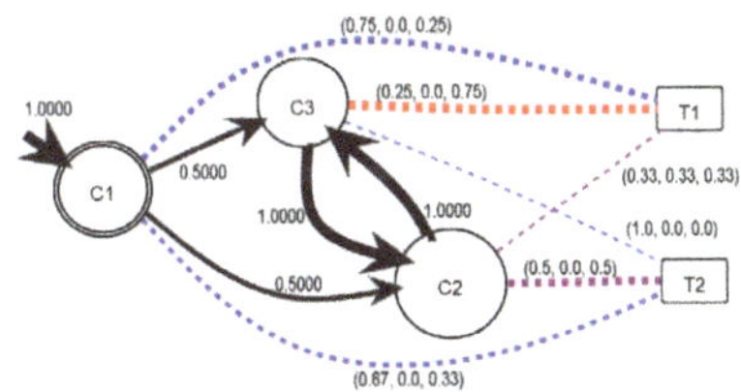

Figure 5. The SentiFlow model generated from log L_1.

3. Algorithm

In this section, the overall procedure that was introduced in the proposed framework is described as an algorithm. The SentiFlow algorithm creates the structure $\Lambda(L) = (\pi, C, T, A', B', D')$ from an SNS log $L = [\sigma]$ similar to an HMM structure $\lambda(L) = (\pi, States, Observations, A', B')$, as shown in Algorithm 1. In the algorithm, LM is adopted for community algorithm detection, and it starts with the clustering of users U in the log (Lines 2–3). Next, for all traces in the log, the algorithm discovers from the user comments, first, the topics T as a result of the LDA algorithm $lda\left(\#user_{comment(e)}\right)$ and, second, the classification of the sentiments S from the naïve Bayes sentiment classifier $nbsc(\#user\_comment(e))$ for each user comment (Lines 4–6). Then, for each trace in the log, the algorithm finds initial communities π, diffusion community matrix a_{ij}, topic matrix b_{im}, and sentiment matrix d_{imr} (Lines 8–25). In particular, if two adjacent users belong to the same community, the algorithm skips the count in the diffusion community matrix, and the last SNS event is counted for its topic and sentiment. Afterwards, the state transition probability matrix A', the observation symbol probability B', and the opinion probability matrix D' are calculated from A, B, and D using ML. Finally, the algorithm returns a SentiFlow model, $\Lambda(L) = (\pi, C, T, A', B', D')$.

Algorithm 1. SentiFlow

1:    **Input:** SNS log $L = [\sigma]$, which is a multi-set of action traces σ in the SNS.
2:    **Output:** A SentiFlow model, $\Lambda(L) = (\pi, C, T, A', B', D')$
3:    Insert all users in L into a user set U.
4:    Detect communities C from users U, and prepare function $c = community(\#u(e))$.
5:    **For** each trace $\sigma = \langle e_1, \ldots, e_H \rangle$ in L **Do**
      Discover topics T from user comment, and prepare a function
6:        $t = lda(\#user\_comment(e))$.
7:        Discover sentiments S, and prepare a function $s = nbsc(\#user\_comment(e))$.
8:    **End For**
9:    **For** each trace $\sigma = \langle e_1, \ldots, e_H \rangle$ in L **Do**
10:       **If** e_1 **Then**
11:        Increase π_i in $community(\#u(e_1))$.
12:       **End If**
13:       **For** each adjacent SNS event (e_h, e_{h+1}) in σ for $1 \leq h \leq H - 1$ **Do**
14:        $c_i = community(\#u(e_h))$ and $c_j = community(\#u(e_{h+1}))$.
15:        $t_m = lda(\#user\_comment(e_h))$ and $s_r = nbsc(\#user\_comment(e_h))$.

16:	**If** $c_i \neq c_j$ **Then**
17:	Increase a_{ij} in A by 1.
18:	**End If**
19:	Increase b_{im} in B by 1.
20:	Increase d_{imr} in D by 1.
21:	**If** $e_{h+2} = null$ **Then**
22:	Increase b_{jm} in B by m and $t_m = lda(\#user\_comment(e_{h+1}))$.
23:	Increase d_{jmr} in D by m,r and $s_r = nbsc(\#user\_comment(e_{h+1}))$.
24:	**End If**
25:	**End For**
26:	**End For**
27:	Calculate the state transition probability matrix $A' = (a'_{ij})$ based on $A = (a_{ij})$.
28:	Calculate the observation symbol probability matrix $B' = (b'_{im})$ based on topic matrix $B = (b_{im})$.
29:	Calculate the sentiment probability matrix $D' = (d'_{imr})$ based on the sentiment matrix $D = (d_{imr})$.
30:	**Return** a SentiFlow model, $\Lambda(L) = (\pi, C, T, A', B', D')$

4. Experiments

In this research, the SentiFlow algorithm was implemented as a plug-in of the ProM platform to verify the proposed framework. ProM is the open source platform that provides practical applications for process mining and supports many kinds of process discovery algorithms [23].

To illustrate the proposed algorithm, the posts of the CNN Facebook page from 1–5 April 2017 were used. The data contain 208 posts with a total of 67,831 users participating with 143,876 comments from 1 April to 6 June 2017.

To obtain information flow among communities, the community detection was analyzed by applying the LM algorithm. Then, the data were filtered to reduce the noise generated by the infrequent users; as a result, six communities were detected using a resolution parameter of 0.8 [26]. The six detected communities, c_1 to c_6, contain 203, 1048, 13, 9, 25, and 121 users, respectively, among a total of 1419 users.

The result of the information diffusion process discovery based on detected communities is shown in Figure 6; the number of comments in a community is represented by the size of the corresponding node in the figure, and the thickness of an arrow denotes the probability of information diffusion from one community to another. Community c_2 concentrates most of the information flows from c_1, c_3, c_4, c_5, and c_6, revealing a larger size from the higher incoming information flow from smaller communities and the number of user comments. Moreover, the information flow received from c_3 to c_2 shows the highest information diffusion probability among all communities. The threshold of information diffusion probability used for the process model visualization in Figure 6 is 0.04. The threshold is used to present a readable process model removing the arcs with lower probability.

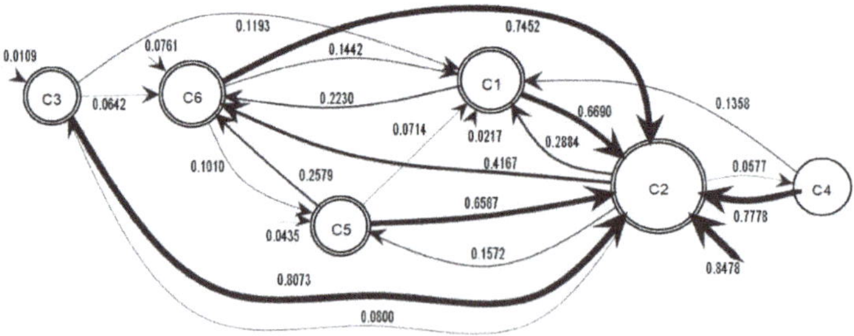

Figure 6. The information flow among communities generated from the CNN Facebook page.

The topic annotation step started with analysis of the comments. First, the stop words, English non-words, and punctuation were removed. Second, duplicate and empty comments were removed. As a result, 13,706 comments and 92 posts were evaluated. To find the different topics of the comments, each word was tokenized as an input for the LDA algorithm. Figure 7 presents a cloud word visualization of the token results for user comments.

Table 5 presents the five topics discovered from the LDA algorithm with their top eight keywords from the discovered comment topics. As shown, topics t_2 and t_3 share two keywords. The word "Trump" is repeated in t_1, t_2, t_3, and t_4 with notable importance in a mixture of topics, but relays in categorize individually each topic.

Figure 7. Word cloud of comment keywords of the CNN Facebook page.

Table 5. Top 8 keywords discovered by the LDA algorithm.

Topic	Top 8 Keywords
t_1	money, Trump, troll, pay, make, blah, need, wall
t_2	people, like, get, would, Trump, one, go, women
t_3	Trump, Obama, president, war, Syria, world, people, us
t_4	rice, Susan, Trump, Susan Rice, CNN, Obama, Russia, story
t_5	CNN, news, fake, fake news, Fox, Clinton, lol, lemon

Figure 8 describes illustrates the semantic information flow between user communities. Here, the discovered topics from the LDA algorithm are shown as rectangles along with their identification name. The dashed arcs indicate the use of the topics from the communities. The topic t_2 has greater importance because it has many thicker arcs connecting communities than do other topics. Communities c_1, c_3, and c_4 present frequent use of keywords for topics t_2 and t_3. As in the previous step, the threshold used to present the information flow is 0.04.

The last step is the generation of the sentimental information flow shown in Figure 9. The model describes the probability of opinions by drawing the arc to a positive community in blue, neutral in green, and negative in red. In more detail, a label with three probabilities of positive, neutral, and negative comments is added on the corresponding arc in order. An example of a negative opinion is shown by a reddish dashed arc representing c_3 over t_2. Conversely, a positive probability opinion can be observed from c_3 toward t_3 with a bluish color. The figure shows that c_5 toward t_2 shows a mixture of opinions and has relative balance between positive and negative opinions. In general, neutral opinions show a lower probability than positive and negative opinions.

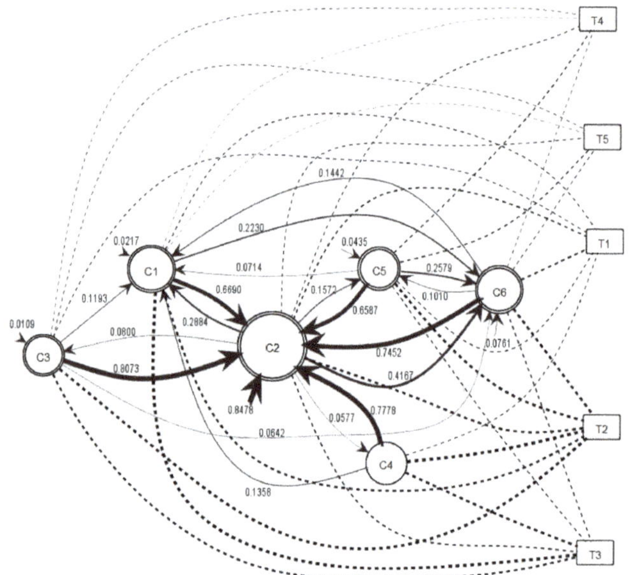

Figure 8. The semantic information flow generated from the CNN Facebook page.

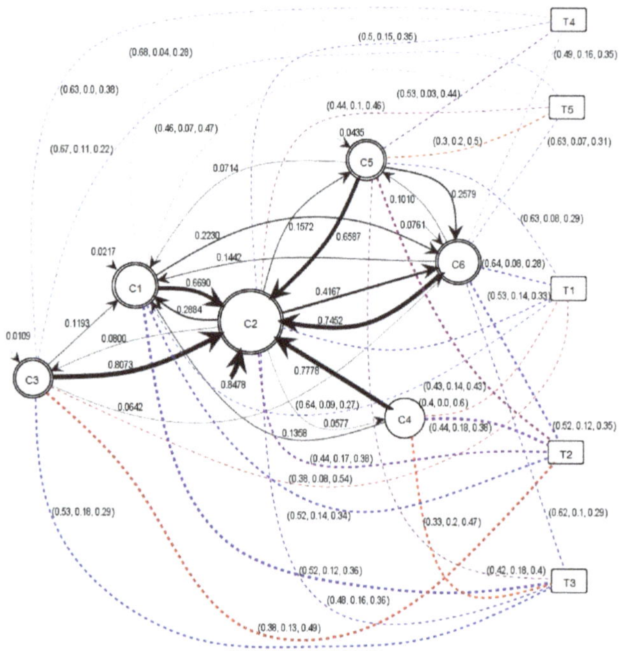

Figure 9. The sentimental information flow based on polarity for topics generated from the CNN Facebook page.

Figure 10 illustrates the relationship between the five discovered topics and the opinions with a total of 6820 positive, 4957 negative, and 1910 neutral comments, separated into the six communities. There are two trend topics, t_2 and t_3, followed by t_1 and t_4 and finally t_5 as the least discussed. In addition, community c_5 has a similar amount of interest between topics t_1, t_3, t_4, and t_5. Furthermore, topic t_2 is the most commented upon among all the communities with the exception of community c_1, which focuses on topic t_3.

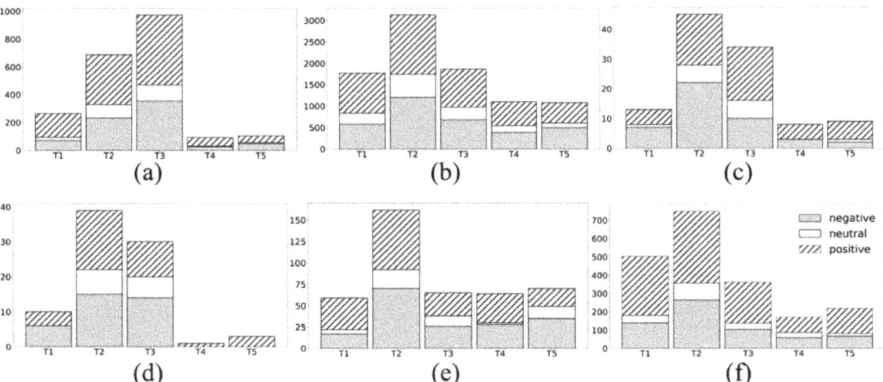

Figure 10. Sentiment analysis across user opinions by each topic among six communities: (**a**) community c_1; (**b**) community c_2; (**c**) community c_3; (**d**) community c_4; (**e**) community c_5; and (**f**) community c_6.

The individual information flows for each topic are shown in Figure 11 using the threshold of 0.04. In the different information flows, community c_2 is continuously the largest community and concentrates most of the information flows from c_1, c_3, c_4, c_5, and c_6 from the different topics. Figure 11a shows a SentiFlow model from topic t_1 with a different flow from Figure 9, where community c_3 does not provide an initial probability and indicates an information diffusion to c_4 with probability 0.0769. Additionally, c_3 and c_4 have a predominant negative opinion in contrast to c_1, c_2, c_5, and c_6 with a positive opinion. In Figure 11b, the initial information diffusion π changed for community c_4 not presented in other SentiFlow models with probability of 0.0110. In addition, c_4, c_5, and c_2 have a purple color to note they have an opinion divided between negative and positive. However, c_6 shows a greater positive opinion, whereas c_3 presents a greater negative opinion. In Figure 10a, community c_1 has the most comments for topic t_3, but, in Figure 11c, c_1 is smaller than c_2 because there are more user comments than in c_1. Positive opinions are expressed in c_1, c_2, c_3, and c_6, whereas while negative opinions are expressed in c_4, and mixed opinions are expressed by users from c_5. In Figure 11d,e, good information flow is observed between all communities with the exception of c_4. In this case, the community does not show an incoming information flow because the probabilities are below 0.04. For Figure 11d, a general predominant positive opinion can be seen for almost all communities, even though c_5 has a combination of positive and negative opinions. In the sentiment information diffusion for topic t_5, c_3, c_4, and c_6, have a positive opinion, in contrast to c_5 with negative comments and c_1 and c_2 with a balanced opinion between positive and negative, as shown in Figure 11e.

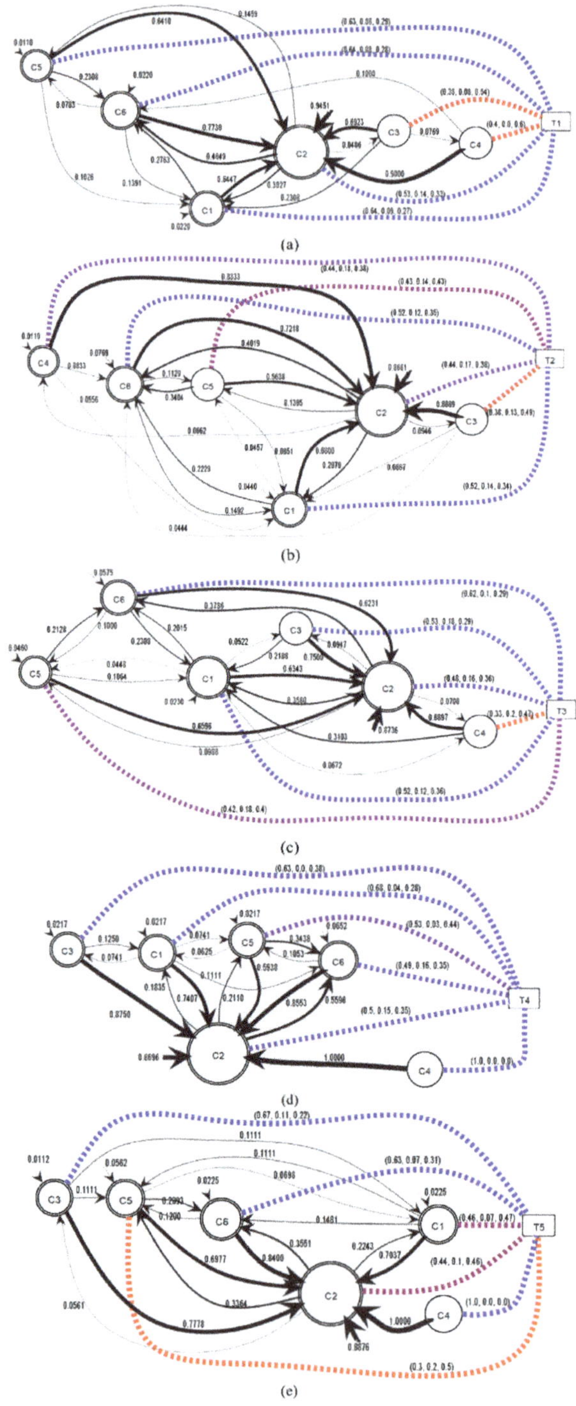

Figure 11. A SentiFlow model for each topic: (**a**) topic t_1; (**b**) topic t_2; (**c**) topic t_3; (**d**) topic t_4; and (**e**) topic t_5.

Figure 11 shows how the topics promote communication between the communities. This answers the first question in this study. For example, Figure 11c presents a description of the information diffusion for topic t_3, with the communication between community c_4 and community c_1 with a probability of 0.3103 and with a response communication probability of 0.0672, which is not observed in the other information diffusion flows. As a response for the second question about how the sentiments are shared from a probabilistic view, the example of information diffusion from community c_2 to community c_4 for topic t_1 shown in Figure 11a is analyzed. Taking the sequence of communities $<c_2, c_3, c_4>$ and the sequence of sentiments <positive, positive, positive>, the probability of the sequence is P (<positive, positive, positive> | Λ) = 0.9451 × 0.53 × 0.0486 × 0.38 × 0.0769 × 0.4 = 2.8455 × 10^{-4}. Moreover, if the communication from community c_2 to community c_6 and the sequence of sentiments <positive, positive> are analyzed, the probability is P (<positive, positive> | Λ) = 0.9451 × 0.53 × 0.4649 × 0.64 = 0.1490 because the only way that c_2 can communicate with c_4 is through c_3, decreasing the probability of the positive sentiment, instead of from c_2 to c_6, where the information diffusion does not need an intermediary community.

5. Conclusions

In this work, an information diffusion process discovery method for SNS was proposed to understand information flow among users better. A SentiFlow model is developed by extending the HMM technique to include process mining by adapting the information from SNS. To understand how the context of user groups is connected with the information flow, different techniques such as an LM for community detection, LDA for natural language preprocessing, and the naïve Bayes classifier for sentiment analysis were used. The proposed method suggested the use of these algorithms, but, in the future, new algorithms can easily be adapted for more accurate and helpful analysis.

The proposed framework has the advantage of allowing users to understand the information flows by displaying the different paths and possible sequences of information delivery obtained from the different users' comments with corresponding probabilities. Analysis of the community of users who plays significant roles in the discovered process shows their sentiment for a related topic.

Three types of information flow diagrams provide the following information. The community information flow describes how the user communities spread their ideas among each other. Moreover, the semantic information flow shown demonstrates how the topics are related with the communities, distinguishing the importance of the topics in each community. Finally, the sentimental information flow presents the potential information to find the focus groups with positive, neutral, or negative opinions and how they influence other user groups according to topic. Additionally, different information diffusion models can be separated and analyzed for each topic.

However, this research still has some limitations. This research focused on understanding the information flow inside a single SNS page, although it can be extended to analyze multiple sites or the whole SNS service. The user profiles of gender, age, and region were not considered in this research, although they may be useful to understand the interactions among users in more detail. In addition, a broader range of human emotions such as anger, joy, and sadness could be used to study the effects of emotions on public opinion. Another limitation is that this research is based on community detection, but the communities may not be stable over time. The study of the reliable community detection can be conducted. In addition, this study focused on the architecture of information diffusion with topic and sentiment, while the analysis methods such as information diffusion process discovery and topic and sentiment analysis were not evaluated. To show the reliability of the analysis result, the detailed methods may be able to be evaluated with evaluation measures such as precision, recall, and F-score.

In future work, a hierarchical model of information flow can be induced to provide different views according to level of abstraction. An integrated approach to capture major interactions among user can be developed without separating the community detection stage and the information flow mining step since the two steps are closely dependent with each other. The dynamics of information flow can also be analyzed to detect the changes of information diffusion in SNS over time.

Sustainability **2018**, *10*, 2731

Author Contributions: B.C. and J.J. conceived and designed the research purposes; B.C. wrote the Proposed Framework and Algorithm Sections; B.C. and J.J. analyzed the data; and B.C. and J.J. wrote the Results and Conclusions Sections.

Funding: This work was supported by the National Research Foundation of Korea (NRF) grant funded by the Korea government (MSIP) (No. 2013R1A2A2A03014718).

Conflicts of Interest: The authors declare no conflict of interest.

References

1. Zafarani, R.; Abbasi, M.A.; Liu, H. *Social Media Mining: An Introduction*, 1st ed.; Cambridge University Press: Cambridge, UK, 2014; ISBN 1107018854.
2. Guille, A.; Hacid, H.; Favre, C.; Zighed, D.A. Information diffusion in online social networks: A survey. *Sigmod. Rec.* **2013**, *42*, 17–28. [CrossRef]
3. Grabowicz, P.A.; Ramasco, J.J.; Moro, E.; Pujol, J.M.; Eguiluz, V.M. Social features of online networks: The strength of intermediary ties in online social media. *PLoS ONE* **2012**, *7*. [CrossRef] [PubMed]
4. Tafti, A.; Zotti, R.; Jank, W. Real-time diffusion of information on Twitter and the financial markets. *PLoS ONE* **2016**, *11*, e0159226. [CrossRef] [PubMed]
5. Zhang, X.; Han, D-D.; Yang, R.; Zhang, Z. Users' participation and social influence during information spreading on Twitter. *PLoS ONE* **2017**, *12*. [CrossRef] [PubMed]
6. Jafari, S.; Navidi, H. A Game-Theoretic Approach for Modeling Competitive Diffusion over Social Networks. *Games* **2018**, *9*, 8. [CrossRef]
7. Kim, K.; Jung, J.-Y.; Park, J. Discovery of information diffusion process in social networks. *IEICE Trans. Inf. Syst.* **2012**, *95*, 1539–1542. [CrossRef]
8. Kim, K.; Obregon, J.; Jung, J.-Y. Analyzing information flow and context for Facebook fan pages. *IEICE Trans. Inf. Syst.* **2014**, *97*, 811–814. [CrossRef]
9. Ullah, F.; Lee, S. Social Content Recommendation Based on Spatial-Temporal Aware Diffusion Modeling in Social Networks. *Symmetry* **2016**, *8*, 89. [CrossRef]
10. Kimura, M.; Saito, K.; Nakano, R.; Motoda, H. Extracting influential nodes on a social network for information diffusion. *Data Min. Knowl. Discov.* **2010**, *20*, 70. [CrossRef]
11. Li, D.; Zhang, S.; Sun, X.; Zhou, H.; Li, S.; Li, X. Modeling information diffusion over social networks for temporal dynamic prediction. *IEEE Trans. Knowl. Data Eng.* **2017**, *29*, 1985–1997. [CrossRef]
12. Kim, M.; Newth, D.; Christen, P. Modeling dynamics of diffusion across heterogeneous social networks: News diffusion in social media. *Entropy* **2013**, *15*, 4215–4242. [CrossRef]
13. Li, M.; Wang, X.; Gao, K.; Zhang, S. A Survey on information diffusion in online social networks: Models and methods. *Information* **2017**, *8*, 118. [CrossRef]
14. Zhao, J.; Dong, L.; Wu, J.; Xu, K. Moodlens: An emoticon-based sentiment analysis system for Chinese tweets. In Proceedings of the 18th ACM SIGKDD, Beijing, China, 12–16 August 2012; pp. 1528–1531.
15. Fan, R.; Zhao, J.; Chen, Y.; Xu, K. Anger is more influential than joy: Sentiment correlation in Weibo. *PLoS ONE* **2014**, *9*, e110184. [CrossRef] [PubMed]
16. Kramer, A.D.; Guillory, J.E.; Hancock, J.T. Experimental evidence of massive-scale emotional contagion through social networks. *Proc. Natl. Acad. Sci. USA* **2014**, *111*, 8788–8790. [CrossRef] [PubMed]
17. Vitale, P.; Guarasci, R.; Iannotta, I.S. Visualizing research topics in Facebook conversations. *Proceedings* **2017**, *1*, 895. [CrossRef]
18. Maynard, D.; Gossen, G.; Funk, A.; Fisichella, M. Should I care about your opinion? Detection of opinion interestingness and dynamics in social media. *Future Internet* **2014**, *6*, 457–481. [CrossRef]
19. Zeng, F.; Zhao, N.; Li, W. Effective social relationship measurement and cluster based routing in mobile opportunistic networks. *Sensors* **2017**, *17*, 1109. [CrossRef] [PubMed]
20. Hernandez-Suarez, A.; Sanchez-Perez, G.; Toscano-Medina, K.; Martinez-Hernandez, V.; Perez-Meana, H.; Olivares-Mercado, J.; Sanchez, V. Social sentiment sensor in Twitter for predicting cyber-attacks using $\ell 1$ regularization. *Sensors* **2018**, *18*, 1380. [CrossRef] [PubMed]
21. Ren, G.; Hong, T. Investigating Online destination images using a topic-based sentiment analysis approach. *Sustainability* **2017**, *9*, 1765. [CrossRef]

22. Bishop, C.M. *Pattern Recognition and Machine Learning*, 1st ed.; Springer: New York, NY, USA, 2006; ISBN 9780387310732.

23. Van der Aalst, W.M.P. *Process Mining: Data Science in Action*, 2nd ed.; Springer: Berlin, Germany, 2016; ISBN 9783662498507.

24. Carrera, B.; Lee, J.; Jung, J.-Y. Discovering information diffusion processes based on hidden Markov models for social network services. In Proceedings of the Asia-Pacific Conference BPM, Busan, Korea, 24–26 June 2015; pp. 170–182.

25. Carrera, B.; Lee, J.; Jung, J.-Y. Discovery of gatekeepers on information diffusion flows using process mining. *Int. J. Ind. Eng.* **2016**, *23*, 253–269.

26. Newman, M. *Networks: An Introduction*, 1st ed.; Oxford University Press: Oxford, UK, 2010; ISBN 9780199206650.

27. Blei, D.M. Probabilistic topic models. *Commun. ACM* **2012**, *55*, 77–84. [CrossRef]

28. Jurka, T. Sentiment: Tools for Sentiment Analysis. R Package Version 02. 2012. Available online: https://github.com/timjurka/sentiment (accessed on 6 March 2018).

29. Liu, B. *Sentiment Analysis and Opinion Mining*; Morgan & Claypool Publishers: San Rafael, CA, USA, 2012.

 *sustainability*

Article

Big Data Approach as an Institutional Innovation to Tackle Hong Kong's Illegal Subdivided Unit Problem

Yung Yau [1],* and Wai Kin Lau [2]

[1] Department of Public Policy, City University of Hong Kong, Hong Kong, China
[2] Department of Construction Technology and Engineering, Technological and Higher Education Institute of Hong Kong, Hong Kong, China; wkelau@vtc.edu.hk
* Correspondence: y.yau@cityu.edu.hk; Tel.: +852-3442-8958

Received: 8 June 2018; Accepted: 31 July 2018; Published: 1 August 2018

Abstract: While applications of big data have been extensively studied, discussion is mostly made from the perspectives of computer science, Internet services, and informatics. Alternatively, this article takes the big data approach as an institutional innovation and uses the problem of illegal subdivided units (ISUs) in Hong Kong as a case study. High transaction costs incurred in identification of suspected ISUs and associated enforcement actions lead to a proliferation of ISUs in the city. We posit that the deployment of big data analytics can lower these transaction costs, enabling the government to tackle the problem of illegal accommodations. We propose a framework for big data collection, analysis, and feedback. As the findings of a structured questionnaire survey reveal, building professionals believed that the proposed framework could reduce transaction costs of ISU identification. Yet, concerns associated with the big data approach like privacy and predictive policing were also raised by the professionals.

Keywords: big data; illegal accommodation; institutional innovation; transaction costs; housing problem; building stock management; Hong Kong

---

1. Introduction

"Big data" is a poor term, lacking a universally agreed definition [1,2]. It is "an all-encompassing term for any collection of data that is very large or complex, and therefore difficult to analyze using conventional data-processing applications" [3]. Some others defined big data as "data that can no longer be captured, stored, managed and analyzed using conventional methods" [4]. Big data has been commonly characterized by its huge volume, high velocity, diverse variety, high complexity, and fine-grained resolution [2,5]. Nonetheless, the definition of what constitutes big data should be relative to our abilities to process it [6]. As a matter of fact, big data should not be limited to data that is so big and exceeds our capacity to handle, search and aggregate it [1]. Instead, it represents a new landscape of the data ecosystem. Big data should comprise a wide spectrum of data sets with varying characteristics [5].

Big data is everywhere, though we may not realize its existence immediately. The big data approach is a novel way to combine digital data sets from different sectors, including governments and businesses, and apply analytic techniques to extract or mine hidden information in the data sets [7]. For example, big data analytics can facilitate advanced decision making which is essential for better-informed policy making [8]. In the arena of urban management, the use of big data analytics can transform mega-cities into resilient smart cities [9,10]. While applications of big data in urban management have been widely explored and discussed in the literature, most of the research has concentrated on technical and legal issues. Perspectives of informatics, computer science and Internet services dominate the big data literature. Big data has been often regarded as a kind of information technology. No previous attempt has been made to frame big data application as an institutional change or innovation. In fact, big data can be a social technology for better urban management. Using the case of enforcement against illegal subdivided units (ISUs) in Hong Kong, we illustrate in this article that the high transaction costs incurred in various stages of public enforcement lead to the enforcement failure. For more effective control of ISUs in the city, we propose a big data approach, which can significantly reduce the costs of identifying ISUs in the existing building stock of Hong Kong. The results of a structured questionnaire survey conducted in Hong Kong generally support the feasibility of the proposed approach. Yet, the respondents have some concerns about the application of big data in fighting ISUs.

This article is organized as follows. First, the ISU problem in Hong Kong is overviewed and relevant literature on the applications of big data approach in urban management is reviewed. What comes next is the outline of the research design. Then, the failure of the public authority in halting ISUs is analyzed from the perspective of transaction cost economics. Afterwards, an alternative approach, which is built upon big data analytics to facilitate ISU enforcement, is detailed. The results of the empirical study are presented and discussed before the article is concluded. An agenda for further research is suggested at the end of the article.

2. Issue and Literature Review

In this section, the subject issue of the research is briefly discussed. It is then followed by a review of literature relevant to the research.

2.1. The Rise of Subdivided Units

In Hong Kong, some people have no choice but to live in transient accommodation in subdivided units as a result of multiple forces like protracted waiting time for public rental housing and highly unaffordable rentals of private housing. Here, the term 'subdivided unit' (*tongfang*) refers to accommodation that was originally designed for single-family occupation and has "been subdivided into two or more smaller units for rental" [11]. It was estimated that, as at 30 April 2013, about 2.4% of the total population in Hong Kong or 171,300 persons lived in 66,900 micro units produced by flat subdivision [11]. The number of quarters that are subdivided amounted 18,800 [11]. However, these estimations did not cover those subdivided units in industrial buildings, residential buildings constructed before 1988, and village houses. Therefore, the number of micro units in Hong Kong is expected to far exceed the official figure. The number of micro units created through flat subdivision was estimated again in 2015; the new estimate jumped to 88,800 [12]. As shown in Table 1, the estimated population living in sub-divided rose from 171,300 in 2013 to 199,900 in 2015.

Table 1. Estimates of the number of subdivided units in Hong Kong [11–13]

Estimated Figure	2013 Estimate	2014 Estimate	2015 Estimate
Number of quarters with micro units	18,800	24,600	25,200
Number of micro units	66,900	86,400	88,800
Number of households living in micro units	66,900	85,500	87,600
Number of persons living in micro units	171,300	195,000	199,900
Average area of micro unit per capital (m^2)	Unavailable	5.7	5.8

The typical sizes of these micro units range from 70 ft^2 to 120 ft^2. Relatively smaller micro units are commonly known as 'coffin homes'. Most of the micro units have an independent toilet and a few come with a kitchen as well. The general living conditions of the subdivided units are rough. Residents in subdivided units complain of different problems associated with their living environment, such as water seepage and concrete spalling [11]. Flat subdivision can create potential death traps and the cramped living environment in subdivided units triggers many social conflicts among residents [14]. A review of the news reports in local newspapers reveals that there were at least 40 incidents of fire within subdivided units in Hong Kong during 1 January 2010 to 30 June 2017. These incidents resulted in 9 deaths and 72 injuries. In the Policy Address 2015, it was emphasized that the safety problems associated with subdivided units should be addressed without further delay. Moreover, some works point out that living in a subdivided unit could impose long-term impacts on the residents' physical and mental health [15,16].

2.2. Illegality of Subdivided Units

Many subdivided units are essentially unauthorized building works (UBWs). In Hong Kong, the *Buildings Ordinance* (Chapter 123 of the *Laws of Hong Kong*) and its subsidiary legislations such as *Building (Planning) Regulations* and *Building (Construction) Regulations* form the statutory framework of building control. The building control regime covers all building works, ranging from demolition, new building construction to alterations carried out in existing buildings. According to the *Buildings Ordinance*, the Director of Building—i.e., the head executive of the Buildings Department—serves as the Building Authority and the Buildings Department executes and enforces the provisions in the *Buildings Ordinance*. To make sure that the minimum acceptable standards are met in the design and execution of a building work, prior approval and consent granted by the Building Authority are necessary before the work can be carried, unless the work is exempted from this requirement by the ordinance [17–19]. Building works carried out in contravention with this requirement are generally referred to "UBWs" [20]. The creation of micro units in existing buildings commonly involves the subdivision of a dwelling unit or flat into two or more smaller, individual units. As Figure 1 illustrates, the related building works typically comprise setting up of new bathrooms/toilets, modification of existing drainage and plumbing systems for the new bathrooms/toilets, thickening of floor screeding to house the diverted or new drains, demolition of original nonstructural partitions and erection of new nonstructural partitions.

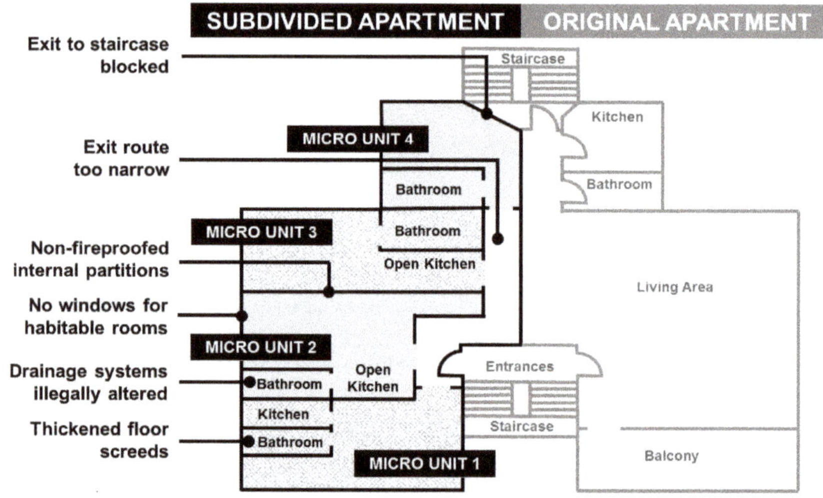

Figure 1. Common illegal alterations made in flat subdivision.

In fact, many of these building works are 'minor works' in nature. They can be performed in the absence of the Building Authority's prior approval and consent. However, the property owners and qualified contractors have to follow a set of simplified statutory procedures (e.g., notification of commencement of works and certification of completion of works) under the Minor Works Control System with regard to the undertaking of minor works. The Minor Works Control System is a relatively new part of the building control system in Hong Kong. Its idea was first incepted in the late 1990s.

After decade-long discussions and legislation, the system finally came into effect in December 2010. Under the current Minor Works Control System, there are three classes of minor works which are classified according to their scale, nature, complexity and potential safety risk. Table 2 summarizes the streamlined submission requirements of different classes of minor works. Yet, minor works undertaken without following the statutory procedures are still taken as UBWs.

Table 2. Classification of minor works [21]

	Class I	Class II	Class III
Degree of complexity and risk	High	Medium	Low
Number of items	44	40	42
Example	Making an opening to staircase enclosure	Repair of a column or beam	Erection of drying rack
Persons required to prepare and sign prescribed plans	Prescribed building professional and prescribed registered contractor	Prescribed registered contractor	Prescribed registered contractor
Document submission before work commencement	Minimum 7 days before work commencement	Minimum 7 days before work commencement	Not required
Document submission after work completion	Within 14 days after work completion	Within 14 days after work completion	Within 14 days after work completion

In addition, many flat subdivision works are illegal in nature because the products do not comply with the building regulations. For example, those new rooms produced as a result of flat subdivision very often do not have windows or have inadequate window areas so the statutory requirements regarding the provision of ventilation and lighting cannot be fulfilled. The partitions separating different occupancies in a subdivided flat are supposed to be fire-resistant according to the requirements laid down in the *Building (Construction) Regulations*. However, most of these partitions in real-life cases are not fire-resistant. As for subdivided units in industrial buildings, they are all illegal for their nonconformance with the *Buildings Ordinance* and statutory land-use zoning.

2.3. Government Efforts to Fight against ISUs

To a certain extent, the contemporary illegal micro units evolve from other types of illegal accommodations like caged homes, which have existed in Hong Kong for over 50 years [22,23]. Since 1994, caged homes have been controlled through a licensing regime under the *Bedspace Apartments Ordinance* (Chapter 447 of the *Laws of Hong Kong*) and the number of caged homes in the territory has been dropping. However, ISUs and cage homes are different in some ways. First, building works are usually involved in the former but not the latter. Second, the leasing subjects in the ISU case are micro units while bedspaces are rented in the case of cage homes. Third, the *Bedspace Apartments Ordinance* does not apply to ISUs. Therefore, other measures to deal with the ISU problem are needed. In response to the ISU problem in the territory, the Hong Kong Special Administrative Region (HKSAR) Government has taken a multipronged approach.

On the community education side, the Buildings Department has produced television advertisements and publications to promote the Minor Works Control System as a proper pathway to do alterations in a flat. The Buildings Department has also distributed numerous pamphlets to educate the public on the deathtraps in subdivided flats and their prevention [24]. Furthermore, the Buildings Department launched various public education programs. For example, the website www.careyourbuilding.bd.gov.hk came into use in 2013 to disseminate useful information and promote building safety to the general public. Building Safety Certificate Courses were also organized for property owners and homeowner associations.

Apart from community education, the HKSAR Government has also attempted to deal with the ISU problem through the existing building control system. Under the *Buildings Ordinance*, knowingly failing to appoint a prescribed building professional and/or a prescribed registered contractor to carry out a minor work is an offence. The person who arranges for the minor work without making prescribed appointments is liable to a maximum fine of HK$100,000 upon conviction. If the illegal flat subdivision is not a minor work in nature, the person commits an offence if he or she carries out the work without going through the submission process for the Building Authority's prior approval and consent properly. Imprisonment for two years and a fine of HK$400,000 are imposed on the offender who is found guilty [25]. In case of a continuing offence, a further fine of HK$20,000 is imposed for each day during which the law violation persists [25].

The Building Authority may serve a statutory order, which is usually known as a reinstatement order, mandating the offending property owner to rectify the violations by a specified deadline. The order may be registered against the title of the property in the Land Registry. As per the *Buildings Ordinance*, it is a serious offence if one person fails to comply with a statutory reinstatement order without any reasonable excuse. Upon conviction, the offender is subject to a fine of HK$200,000 and one year's imprisonment as the maximum penalty [25]. There is also a further fine of HK$20,000 per day for any continuing violation [25]. In case the property owner concerned refuses to carry out the required rectification works, the Building Authority may also engage a government contractor to rectify the contraventions directly. The costs incurred together with supervision charge will then be recovered from the property owner [25].

The Buildings Department investigates the cases involving UBWs associated with subdivided units after receiving reports or complaints about subdivided units. Suitable enforcement actions will then be taken in accordance with the prevailing UBW enforcement policy [26]. In addition, the Buildings Department has launched large-scale operations since April 2011 to identify, inspect and enforce against ISUs [27]. Heavy emphasis has been placed on the means of evacuation from the buildings in these operations. Once irregularities are identified, the Buildings Department will take necessary enforcement actions. Table 3 summarizes the number of target buildings and buildings actually inspected during the period between 2011 and 2016. Up to 31 December 2016, around 9000 subdivided flats have been inspected during these large-scale operations since April 2011. At least 1900 statutory orders were issued and 439 prosecutions were instigated. As mentioned above, in spite of the government actions, the number of ISUs did not go down in the past years [11–13]. The decreasing trends of the number of target buildings and number of buildings actually inspected from 2014 to 2016 actually reflected the difficulties of ISU enforcement and limited resources of the Buildings Department for coping with the ISU problem. In order not to give the general public any false expectation, the HKSAR Government consecutively lowered the enforcement targets.

Table 3. Statistics of large-scale operations targeting ISUs

Year	Number of Target Buildings to be Inspected [1]	Number of Buildings Actually Inspected [1]
2011	150	116
2012	200	369
2013	200	300
2014	330	308
2015	330	210
2016	100	100

[1] The figures were compiled based on the information contained in the Controlling Officer's Reports (Head 82–Buildings Department) in the government budgets of various years.

2.4. Applications of Big Data in Urban Management

In view of the unresolved problem of ISU proliferation in Hong Kong, we propose a big data approach to facilitate the public enforcement. In point of fact, 'big data' is now a buzzword in many different disciplines, ranging from business management and public administration to public health though there is still a lot of noises and misunderstandings about the use of big data [28]. Many governments have committed to make use of big data to develop their smart cities [29,30]. Big data can improve organizational efficiency, operational effectiveness, and decision-making [31]. It can also enhance productivity [32,33]. The application of big data and associated analytics also significantly improves government services in different arenas.

Big data has been regarded as an important source of insights into urban management. It allows better understanding of urban problems and provides actionable and sustainable solutions [34]. Governments have employed big data in many different areas in urban management and planning—e.g., land-use planning [3], land administration [35,36], and traffic operations [37,38]. For example, a big data approach was proposed to ease the problem of illegal parking in Goyang City, South Korea [16]. In American cities such as Chicago, Los Angeles, and Manchester, big data was employed to predict where crime would take place before it occurred [39,40]. This helped policing agencies to decide smartly the places for their officers to patrol. Evidence shows that crime rates could be lowered with the same manpower input upon the application of big data analytics. From above, it is clear that data-driven urban management is becoming a global trend [35,41]. Big data has a high potential to facilitate urban management efforts aimed at tackling city problems [42].

3. Research Design

The current research has five stages and its design is graphically illustrated in Figure 2. First of all, under the broad umbrella of big data applications in urban management, a specific issue was picked for in-depth investigation. ISU enforcement in Hong Kong was chosen eventually because we expected that the big data approach could have promising implications in this aspect though its application was still at the conception stage. Background information about the issue of ISU proliferation in Hong Kong and what the HKSAR Government had done to cope with the issue was overviewed. In the second stage, relevant literature about the applications of big data analytics in urban management was reviewed. Besides, the literature review covered previous works that explained failures of public enforcements. What comes next is the theorization of research issue. In this research, the transaction cost theory in the discipline of institutional economics was adopted to explain why HKSAR Government's efforts in combating ISUs were in vain. An alternative approach drawn upon the transaction cost theory was then proposed.

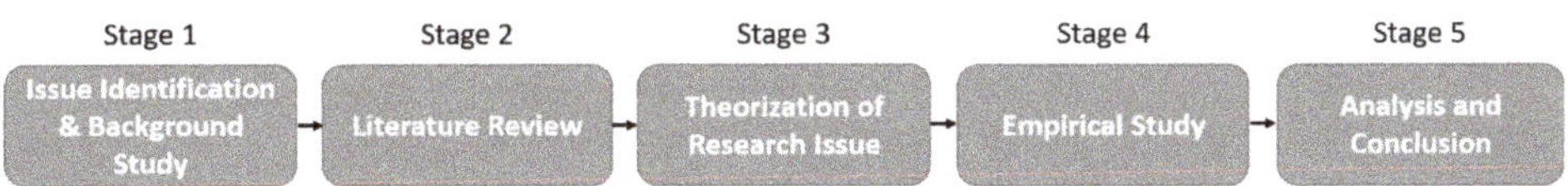

Figure 2. Research design.

In Stage 4, the feasibility of the proposed approach was evaluated. Nonetheless, it was impossible to obtain any real-life data to evaluate if the proposed approach could really trim down the number of ISUs in Hong Kong because the proposed approach was yet implemented at the time when the research was conducted. Therefore, views of local building professionals (e.g., architects, building surveyors, builders, fire engineers, structural engineers, and property and facility managers) towards the proposed approach were collected through a semi-structured questionnaire survey which was conducted in the period between January 2018 and March 2018. To achieve a balanced sample with building professionals from different disciplines, purposive sampling was adopted for selecting the invitees. A total of 120 building professionals with a profound understanding of Hong Kong's ISU issue were chosen. An invitation letter was sent to each of these 120 professionals to complete an online questionnaire. A follow-up invitation was sent to the selected professionals if we had not received any replies from them within one month after the first invitation. Eventually, 88 invitees (73.3%)

completed the questionnaire fully. The profile of the invitees and respondents is shown in Table 4. The questionnaire was predesigned to comprise three parts. The first part asked the respondent to rate the feasibility of the proposed big data approach with reference to technical practicality and political acceptability (i.e., whether the proposal would be received by relevant policy stakeholders). The respondent was required to give a rating to these two aspects using a four-point scale, with 1 = very low; 2 = low, 3 = high, and 4 = very high. The second part concerns whether the proposed approach could reduce transaction costs incurred in different stages of ISU enforcement. A four-point Likert scale (with 1 = strongly disagree; 2 = disagree; 3 = agree and 4 = strongly agree) was employed to indicate respondent's degree of agreement (or disagreement) with the argument that the proposed big data approach would help reduce the transaction costs of ISU enforcement. The last part contained an open-ended question, asking respondents to raise their concerns about the proposed big data approach. Before the official survey started, the questionnaire had been pretested and modified by taking the testers' comments and suggestions into account. In the last stage of the research, the findings of the survey were analyzed and interpreted. Implications were drawn based on the analysis results.

Table 4. Profiles of the survey invitees and respondents

Characteristic		Invitees		Respondents	
		Number	%	Number	%
Gender	Male	87	72.5%	67	76.1%
	Female	33	27.5%	21	23.9%
Profession	Architect	21	17.5%	16	18.2%
	Building surveyor	24	20.0%	19	21.6%
	Builder	17	14.2%	12	13.6%
	Fire engineer	16	13.3%	9	10.2%
	Structural engineer	22	18.3%	13	14.8%
	Property & facility manager	20	16.7%	19	21.6%
Professional experience	10 years or less	33	27.5%	24	27.3%
	11–20 years	34	28.3%	25	28.4%
	21–30 years	30	25.0%	22	25.0%
	More than 30 years	23	19.2%	17	19.3%
Working sector	Public sector	72	60.0%	46	52.3%
	Private sector	48	40.0%	42	47.7%

4. Transaction Cost Model of ISU Enforcement

In this section, the failure of government interventions in curbing ISU problem in Hong Kong was explained from the perspective of institutional economics. In view of such failure, we put forward an institutional innovation.

4.1. High Transaction Costs of Government's Enforcement

In spite of the efforts of the HKSAR Government in fighting against ISUs, the ISU problem in the city still remains very serious. The failure of government's enforcement actions can be explained from the angle of institutional economics. Institutional economics supplements classical economics with the concepts of institutions and transaction costs [43–45]. Institutions are the game rules in a society, shaping the contexts for economic behavior [44]. Institutions come in different forms, including formal rules (e.g., laws and constitutions) and informal rules (e.g., societal norms and customs) [46]. On the other hand, transaction costs are the costs of making and enforcing agreements, which also include the rules such as laws and regulations. The costs of searching and information collection are also important transaction costs in many different institutional settings [47–49]. In the building control system, transaction costs are incurred in the law making and law enforcement processes. This research focusses on the law enforcement process only.

As Figure 3 shows, there are several stages in a law enforcement process against ISUs and different enforcement costs are involved in these stages. For example, at the start of the enforcement exercise, the public officials in the Buildings Department need to figure out which properties in the existing building stock have been subdivided unlawfully. Then, the public officials have to collect sufficient evidence for subsequent enforcement actions, such as the issuance of statutory reinstatement orders or direct prosecutions. These two initial stages of the enforcement exercise involve prohibitively high transaction costs, leading to law enforcement incapacity.

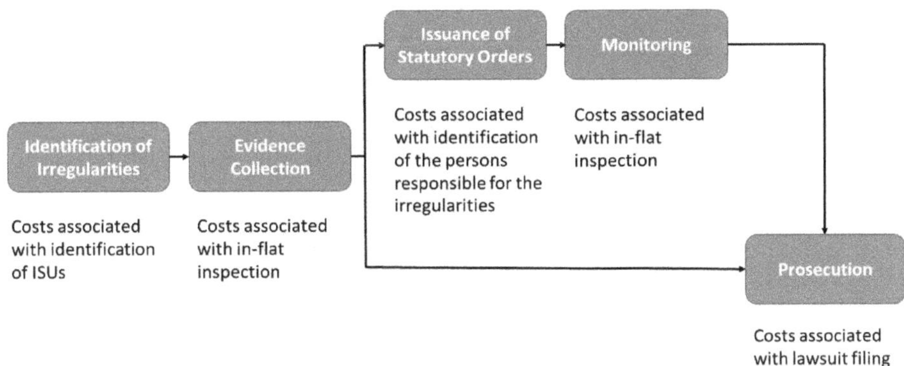

Figure 3. Transaction costs incurred in different stages of ISU enforcement.

Unlike other types of UBWs, such as flower racks and metal cages erected on the external walls of buildings, ISU works are undertaken inside a flat so their presence is not so readily observable from the outside. Several years ago, public officials relied on some noticeable signs to identify ISUs in a building. These signs included multiple doorbells, mailboxes, and water meters installed for a single flat or dwelling unit, as illustrated in Figure 4. Through learning from previous enforcement experience, landlords and renters of ISUs are getting smarter. Households of different micro units within an ISU now share the same doorbell, mailbox, and water meter, hiding the existence of the ISU from outsiders. Therefore, it becomes more and more difficult for the public officials to identify ISUs without entering a premise.

In practice, the officials of the Buildings Department inspect a property or properties in a building for suspected ISUs either because the building is targeted in a large-scale operation or the department receives complaints from the public. In the first scenario, the Buildings Department picks a sample of target buildings based on a number of criteria, such as building age and building management regime (e.g., formation of incorporated owners and appointment of third-party management agent). However, these criteria may not be good predictors for the level of ISU proliferation in a building. Older, unmanaged buildings do not necessarily have more ISUs. Erroneous shortlisting may result in inefficient use of public resources. More importantly, there is a need for the public officials to enter the properties in the targeted building for inspection. Otherwise, the public officials cannot ascertain whether the properties have been illegally subdivided for subsequent actions.

In the second scenario, although the suspected ISU has been spotted by a member of the public, the government officials still need to collect sufficient hard facts or evidence on the existence of ISUs (e.g., number and dimensions of the illegal micro units and types of UBWs carried out for flat subdivision) for further enforcement actions. In other words, for both scenarios, getting access to the property interior is crucial. Nonetheless, the ISU residents, in most cases, deny the access of public officials to the subdivided flats because they do not want to risk losing their current relatively affordable accommodation.

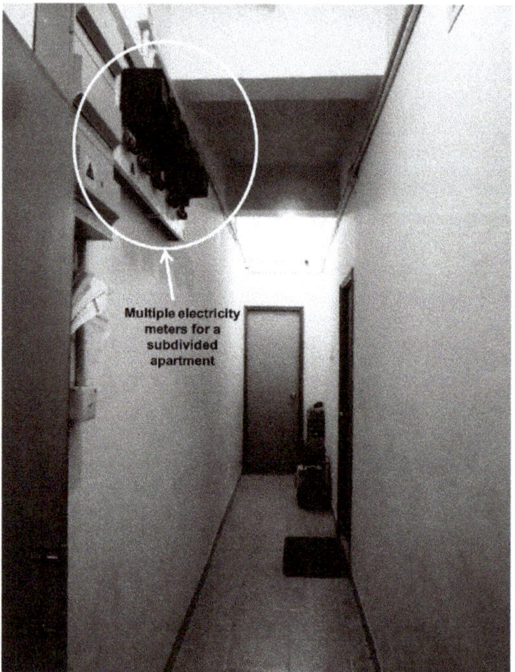

Figure 4. Multiple electricity meters as a signal of flat subdivision.

On the other hand, Section 2.2 of the *Buildings Ordinance* stipulates that the power of entry or breaking into the premises or upon land by the Building Authority in the presence of a police officer is restricted to emergency situations only. To further facilitate in-flat inspection of suspected ISUs, the *Buildings (Amendment) Ordinance 2011* introduced a new measure to empower the Building Authority to apply to the court for a warrant for entry into the interior of individual premises for inspection or other enforcement actions. Before the issuance of the warrant, a magistrate must be satisfied by information on oath that:

1. There are reasonable grounds for suspecting any of the following matters:

   - building works have been or are being carried out to the premises or land in contravention of any provision of the Buildings Ordinance;
   - the use of the premises or land has contravened any provision of the Buildings Ordinance;
   - the premises have been, or the land has been, rendered dangerous, or the premises are, or the land is, liable to become dangerous;
   - the drains or sewers of the premises or land are in a defective or insanitary condition; or
   - a notice or order served under the Buildings Ordinance has not been complied with;

2. The Building Authority has made attempted entry on at least two different days; and
3. A notice of intention to apply for a warrant has been served on the owner or occupier of the premises.

In spite of the facilitation of in-flat inspection through law amendment, transaction costs for in-flat inspections for ISUs are still very high. The Buildings Department officials need to visit the property suspected of having illegal subdivisions twice before applying to the court for an entry warrant.

Moreover, to substantiate their warrant application, they have to provide sufficient evidence to the court that there are UBWs in the subject property. As discussed above, the task of evidence collection in ISU enforcement is a really thorny issue. From the above, it is clear that the existing enforcement system in Hong Kong fails to stop the ISU problem due to the prohibitively high transaction costs of searching and information collection.

4.2. Institutional Innovation for ISU Enforcement with the Use of Big Data Analytics

In order to make ISU enforcement in Hong Kong more effective, we propose that the ISU search can be facilitated with the use of big data analytics. Big data has been used in many cities for combating different urban problems, such as crime, illegal parking, and traffic congestion [5,37,38,50,51]. In some cities, the local governments also employed big data to fight against substandard housing. For example, big data is used to register and track each housing-related complaint in New York City [52]. The data collected are then analyzed to point out those properties with the highest chances of code violation. In the past, building inspectors used their personal experience or gust feelings to prioritize cases for follow-up. Nevertheless, this mode resulted in a low enforcement efficiency as the inspectors could not find property conditions adverse enough to warrant a vacate order in 87% of the cases. Later, the employment of big data analysis brought about a fivefold improvement in the building inspectors' efficiency. Prioritizing inspections based upon the results of big data analysis, building inspectors served vacate orders on over 70% of the properties they inspected. In New South Wales of Australia, the state government harnesses big data to blitz Sydney's illegal boarding houses [53]. Data from the utility bills, electoral rolls, and other sources are collected and used to find where there is an unusual increase in the number of residents.

The same idea can be applied to the identification of the ISUs in the existing building stock in Hong Kong. The big data approach proposed in this article has five key stages, as shown in Figure 5. In Stage 1, information required to address the ISU problem and data necessary to derive the required information are identified. In the case of an ISU search, the information required is the level of risk of a property or building with ISU proliferation. To derive such information, useful data such as monthly utility consumption of each property or building are needed. If a property is subdivided to create more micro units, the total number of occupants in the property is expected to rise. In normal situations, utility consumption increases with the number of occupants. Consequently, it can be the first alert of a subdivided unit if a dwelling unit has a much higher monthly consumption of water, gas, and electricity than a similar standard unit. Abnormal rises in the utility bills can indicate the illegal subdivision of a dwelling unit. To facilitate the subsequent comparative analysis, both historical and current data should be obtained.

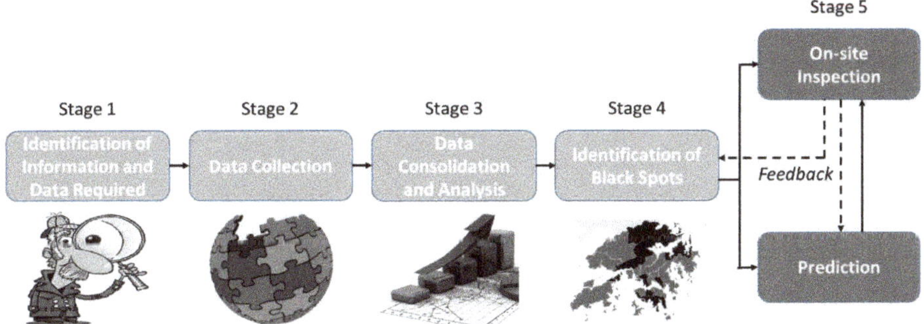

Figure 5. Five stages of the proposed big data approach.

Apart from the utility consumption, other information, such as the number of complaints or reports about water seepage in a building, may also give public officials some hints about the existence of ISUs in the building. In most cases, the conversion of a lawful flat into illegal micro units involves the installation of new toilets or bathrooms and alterations to existing pipework. All these works may increase the chances of water leakage from the pipes and water seepage through walls and slabs. Accordingly, it is reasonable to expect that a dramatic rise in the number of water seepage reports in a building is a good indication of ISU proliferation in that building.

Stage 2 is the collection of the required data. Utility consumption and usage is usually reflected in the utility bills. In this regard, the data required can be collected from various utility providers (e.g., Water Supplies Department, CLP Power Hong Kong Limited (Hong Kong, China), HK Electric Investments Limited (Hong Kong, China) and Hong Kong and China Gas Company Limited (Hong Kong, China)). As for the number of reports about water seepage, the data can be collected from the Joint Offices for Investigation of Water Seepage Complaints. We assume that monthly records of water, electricity, and town gas bills and water-seepage reports of all dwelling units in Hong Kong in a three-year moving window are necessary for a meaningful analysis for identifying problematic properties. Given that there are around 1,174,628 dwelling units in private housing in the city at as the end of 2017 [54], the dataset for analyses in the subsequent stage will have at least 211,433,040 entries (1,174,628 buildings × 12 months × 3 years × 5 columns) in the spreadsheet.

In Stage 3, the data collected in Stage 2 are consolidated and analyzed. In the data analysis, both cross-sectional and longitudinal comparisons can be made. In the cross-sectional analysis, utility bills of similar properties (e.g., properties of similar ages and sizes) are compared. Those properties with monthly or quarterly utility bills far exceeding the average figures are screened out. Similarly, properties with abnormally high number of water-seepage reports are identified. In the longitudinal analysis, the quarterly utility bills of the same property in different time periods are compared after adjusting for the seasonal factors. An abrupt, significant increase in utility bills may indicate potential illegal flat subdivision in the property. Apart from the unit-based analysis, data from the same building can be aggregated such that the big data analysis can be conducted on a building basis. In addition to utility bills, numbers of water-seepage reports can be deployed in the building-based analysis.

In Stage 4, properties or buildings at higher risk of illegal subdivision are identified based on the analysis results in Stage 3. These high-risk premises will become the black spots that warrant the government's priority enforcement actions. In Stage 5, the inspectors of the Buildings Department investigate those black spots to see if ISUs exist or not. There is a feedback loop such that the inspection results can be taken as an input, which helps the government improve black spot identification. For instance, the government can learn by trial and error what the optimal differential threshold for the utility bills should be. At the same time, the big data analytics can be further extended to enable predictive analysis. Predictive analysis can be a powerful tool for law enforcement or crime prevention. The public authorities can apprehend law-breakers based on foreknowledge of their future misdeeds. For instance, by looking into the characteristics of buildings with higher risks of ISU proliferation, the public authorities can identify a set of determinants. These determinants can be used to predict which buildings will be riskier and will warrant more attention in the future. Again, the accuracy of the prediction can be improved through the feedback mechanism.

The proposed big data approach is a kind of institutional innovation that aims to reduce the transaction costs incurred in the early stages of ISU enforcement by the public authorities. By analyzing the big data, the HKSAR Government can identify the ISU black spots on a property basis and a building basis with significantly reduced information costs. The property-based identification helps the Buildings Department to locate the properties that have very probably been subdivided illegally. This trims down the search costs for ISU enforcement. Moreover, based on the findings of building-based identification, the Buildings Department can have a more informed selection of target buildings for large-scale operations, directing the limited public resources to the neediest buildings.

In addition, the valuable information of abnormal utility consumption is a piece of evidence that the Building Authority can use for substantiating its applications to the court for entry warrants.

5. Research Findings and Discussion

5.1. Feasibility of the Proposed Big Data Approach

As Table 5 demonstrates, over 70% of the respondents in the survey rated the technical practicality of the proposed big data approach high or very high. Around 80% opined that the political acceptability of the proposed approach was high or very high. The mean scores for technical practicality and political acceptability were 2.97 and 3.11, respectively. They were both significantly greater than 2.5 which was the mid-value of the scale ($p < 0.01$). Therefore, the survey results reveal a strong collective view that the proposed big data approach is technically and politically feasible. The perceived high level of technical practicality might be ascribed to the fact that data about utility consumption in dwelling units and water-seepage reports was already easily available or obtainable in Hong Kong. There should be no need for the government to establish any new protocol for data collection. On the other hand, the proposed approach was considered politically acceptable probably because no introduction of new law or legislative amendment was envisaged. Moreover, it was believed that with the application of big data analytics, identification of ISUs would not cause any gratuitous nuisances to the building occupants.

Table 5. Respondents' responses regarding feasibility of the proposed big data approach

Response	Technical Practicality		Political Acceptability	
	Number	%	Number	%
Very high (4)	28	31.8%	32	36.4%
High (3)	35	39.8%	38	43.2%
Low (2)	19	21.6%	14	15.9%
Very low (1)	6	6.8%	4	4.5%
Total	88	100.0%	88	100.0%

Chi-squared tests were conducted to probe if the responses were dependent on the respondents' backgrounds. Gender was found to have significant impacts on both technical practicality and political acceptability ratings ($p < 0.10$). Technical practicality rating was found to be contingent on profession ($p < 0.10$) while there was no significant relationship between political acceptability rating and profession. As far as professional experience is concerned, it had no significant impact on technical practicality and political acceptability ratings. On the other hand, the sector in which a respondent was working was found to bear a significant association with the political acceptability rating ($p < 0.05$) but no impact on the technical practicality rating.

5.2. Transaction Cost Implications

Respondents' views towards transaction cost implications of the proposed big data approach varied across different stages of ISU enforcement. Over 80% of the respondents agreed or strongly agreed that the proposed approach could reduce the transaction costs incurred in the identification of ISUs. The proportion dropped to around 60% for in-flat inspection and further to around 30% for identification of persons responsible for the irregularities and lawsuit filing. Table 6 summarizes the mean scores of the respondents' responses. The mean scores for identification of ISUs and in-flat inspection were 3.14 and 2.74, respectively. They were both significantly greater than the mid-value of 2.5 ($p < 0.10$ at least), implying that respondents generally believed in the high potential of the use of big data in trimming down the transaction costs incurred in these two aspects of ISU enforcement. Conversely, the mean scores for identification of responsible persons and lawsuit filing were significantly less than the mid-value of 2.5 ($p < 0.01$). That means the respondents negated the argument that the proposed big data approach could reduce the transaction costs incurred in the identification of persons responsible for the irregularities and lawsuit filing. These findings generally confirm our expectations. With the use of big data analytics, dwelling units with high potentials of being illegally subdivided can be identified more easily. With reference to the case in New York City discussed in Section 4.2, this protocol helps the government officials prioritize their enforcement actions, and thus enhance the overall enforcement efficiency. Although in-flat inspection is still indispensable in the process of ISU enforcement, the transaction costs incurred can be reduced with the aid of big data analytics. Signs spotted in big data analysis like abrupt upsurges in utility bills can be used as evidence for applying to the court for an entry warrant, lowering the costs of evidence collection.

Table 6. Perceived transaction cost implications of the proposed big data approach

Transaction Cost	Mean Score
Identification of ISUs	3.14
In-flat inspection	2.73
Identification of responsible persons	2.07
Lawsuit filing	1.99

Based on the survey results presented above, we need to admit that the proposed big data approach might not be able to reduce the transaction costs incurred in all stages of ISU enforcement. Yet, the survey findings still give a strong indication that the use of big data have can moderate the problems associated with high costs of identification of ISUs and in-flat inspection in the early stages of ISU enforcement. Overall, the opinions of the surveyed professionals support that the proposed big data approach is an institutional innovation which helps the government lower transaction costs of ISU enforcement.

5.3. Concerns about the Application of Big Data Approach

Uses of big data are always accompanied by challenges. The respondents raised some concerns with the proposed application of big data in fighting ISUs in Hong Kong; 36 respondents (40.9%) opined that big data could pose high privacy risks. In fact, the same concern has been named by many other studies [1,55,56]. Using big data may clash with the principle of data minimization for privacy protection, blurring the boundary between personal and non-personal data [7]. Privacy issues can constrain the use of big data, even for government administration and services. Collecting and manipulating sensitive data is a contentious topic of interest to many communities outside and inside the government [31]. It is probable that government departments are not willing to share data that they regard proprietary with other public agencies. The same applies to the private organizations. These organizations in competitive markets tend to use their big data internally and are reluctant to give access to outsiders for different reasons, such as loss of competitive advantage and potential public-relation disasters [3]. Even if the utility providers agree to share their data with the Buildings Department, whether it is lawful under the current privacy law in Hong Kong to use property-based utility consumption data for risk assessment is questionable. Utility usage is closely related to one's lifestyle, which is a private matter. To a large extent, the property-based utility bills may reflect one's lifestyle. Therefore, there can be privacy concerns associated with the employment of property-based data. We anticipate that the use of aggregate figures like building-based utility consumption is less problematic and attracts fewer criticisms.

Given that data for crime prevention or detection is generally exempted from the protection of privacy law, one possible way to get around the potential privacy issue is to treat the investigation of illegal flat subdivision as criminal investigation. A special taskforce or authority can be setup within the government and empowered to collect the private data from the utility companies or related government departments for ISU investigation. Analogous to the police force, the taskforce or authority uses the 'potentially sensitive' data for the public interest, i.e., ensuring public safety. That is why the access and use of the utility usage data is not subject to the protection of the privacy law. In spite its legitimate right to collect and use the sensitive data, the authority can make it clear that it will observe important data protection principles concerning data collection, data use, data security, and openness in order to lessen the worries of the general public.

Moreover, although the proposed big data approach enables prediction, predictive analysis results should be used in a very careful manner. Thirteen respondents (14.8%) alerted that ISU proliferation prediction might lead to problems. On one hand, predictive analysis may trigger or aggravate redlining [56]. On the other hand, even with non-sensitive data, predictive analysis may have stifling effects on certain types of buildings in the long run [55]. For example, older buildings will have the deck stacked against them even more so than before. This will encourage premature redevelopment of old but well-maintained buildings. It is perhaps debatable whether or not we should apply the principle 'innocent until proven guilty' to buildings in the ISU war. This concerns how the balance between public interest (say, public safety) and private interest is maintained.

Apart from the issues of data privacy and data discrimination discussed above, there are some relatively minor concerns regarding the proposed big data approach. For instance, five respondents (5.7%) worried about the inadequate analytical know-how in the Buildings Department to handle big data. A few respondents also alerted that credibility of the big data analytics might be undermined by dirty data. This issue could be caused by incorrect data linking, duplicate data, or input errors.

6. Conclusions and Agenda for Further Research

The era of big data is still underway. While more and more literature has been dedicated to the discussion of the employability of big data analytics in the field of urban management, big data is often regarded as an information technology rather than a social technology. Besides, no attempt has been made to analyze big data application from a transaction cost perspective. Using the case of ISU enforcement in Hong Kong, this article aimed to expound why the public authorities fail in addressing the ISU problem and to recommend an institutional change in the enforcement system. With an eye to lowering the costs of information search and collection, the public authorities, particularly the Buildings Department, can collect information regarding, for example, consumption of water, gas, and electricity. Sharp increases in utility usage and the number of water seepage cases can be clear indications of ISU proliferation in a building. Use of big data in this way helps the HKSAR Government predict the likelihood of ISU proliferation in different buildings. The public authorities can improve the efficiency of their actions by targeting those buildings with greater risks in large-scale operations based on the analysis results. Results of a survey on local building professionals confirmed that the use of big data in curbing the ISU problem in Hong Kong is practical and promising. Nonetheless, while there are enormous benefits from big data analytics, the use of big data is not free of problems. For instance, the associated privacy risks should never be overlooked. Therefore, it is necessary for the policymakers to strike a balance between public interest and privacy. Lastly, one should bear in mind that the value of big data does not lie in its size. Bigger data is not always better data. It is the methodological design that actually determines the quality of big data analytics.

This article makes three major contributions. First, it offers deeper theoretical insights into the failure of public enforcement actions in the arena of building control. Transaction cost economics, which has been accorded little attention by academics in the field, is employed for the explanation. Yiu and Yau's work in 2005 [19] was the first attempt to expound why UBWs proliferate in Hong Kong from the transaction cost perspective. They argued that poorly written laws resulted in ambiguities and increased the enforcement costs. Studying the information search and collection processes of the whole enforcement exercise, this article is an extension of their work. Second, the article discusses the application of big data analytics in building control. Again, this is an innovative idea, which may stimulate more research in the field in the future. Third, the big data approach advocated can inform policy making with regard to long-term management of building stock in Hong Kong and other high-rise cities. Vigilant building stock management can ensure safety and health of the community and is thus critical for urban sustainability of a city [57].

On the other hand, the opinion survey reported in the current article indicates the views of the local building professionals only. It is necessary to collate the views of data scientists towards the practicality of the proposed big data approach. Moreover, a true validation of the proposed approach is needed. The effectiveness of the proposed big data approach cannot be evaluated at this moment because the approach has yet been put in place and there are some privacy concerns which hinder data sharing. Once the proposed approach is executed by the government, further research is warranted to investigate if the approach can really reduce the amount of man-hours incurred in the identification and enforcement against ISUs. Besides, it will be interesting to investigate how different stakeholders, particularly landlords and tenants of ISUs, respond to this new tactic to crack down on ISU proliferation when it is executed. It is expected that the number of ISUs should drop as a result of increased efficiency of irregularity identification. The empirical inquiry into the deterrent effects can further confirm the effectiveness of the proposed approach. Last but not least, the applicability of the big data approach in controlling other types of building illegality or informal settlements—such as urban villages, small property right housing, and illegal accommodations in industrial buildings—can be explored in the future.

Author Contributions: The authors' individual contributions are outlined as follows. Conception and design: Y.Y. Literature review: Y.Y. Data collection and analysis: W.K.L. Manuscript writing: all authors. Final approval of the manuscript: all authors.

Funding: The work described in this article was fully supported by a grant under the Public Policy Research Funding Scheme administered by the Central Policy Unit of the Hong Kong Special Administrative Region, China (project no. 2014.A1.019.15B).

Conflicts of Interest: The author declares no conflict of interest.

References

1. Boyd, D.; Crawford, K. Critical questions for big data: Provocations for a cultural, technological, and scholarly phenomenon. *Inf. Commun. Soc.* **2012**, *15*, 662–679. [CrossRef]
2. Kitchin, R. The real-time city? Big data and smart urbanism. *GeoJournal* **2014**, *79*, 1–14. [CrossRef]
3. Samarajiva, R.; Lokanathan, S.; Madhawa, K.; Kreindler, G.; Maldeniya, D. Big data to improve urban planning. *Econ. Political Wkly.* **2015**, *50*, 42–48.
4. Dixon, T.; van de Wetering, J.; Sexton, M.; Lu, S.-L.; Williams, D.; Duman, U.; Chen, X. *Smart Cities, Big Data and the Built Environment: What's Required?* Royal Institution of Chartered Surveyors: London, UK, 2017.
5. Desouza, K.C.; Jacob, B. Big data in the public sector: Lessons for practitioners and scholars. *Adm. Soc.* **2017**, *49*, 1043–1064. [CrossRef]
6. Batty, M. Does big data lead to smarter cities? Problems, pitfalls and opportunities. *ISJLP* **2015**, *11*, 127–151.
7. Rubinstein, I.S. Big data: The end of privacy or a new beginning? *Int. Data Priv. Law* **2013**, *3*, 74–87. [CrossRef]
8. Lytras, M.; Visvizi, A. Who uses smart city services and what to make of it: Toward interdisciplinary smart cities research. *Sustainability* **2018**, *10*, 1998. [CrossRef]
9. Visvizi, A.; Lytras, M. Rescaling and refocusing smart cities research: From mega cities to smart villages. *J. Sci. Technol. Policy Manag.* **2018**, *9*, 134–145. [CrossRef]
10. Visvizi, A.; Mazzucelli, C.; Lytras, M. Irregular migratory flows: Towards an ICTs' enabled integrated framework for resilient urban systems. *J. Sci. Technol. Policy Manag.* **2017**, *8*, 227–242. [CrossRef]
11. Policy 21 Limited. *Report on Survey on Subdivided Units in Hong Kong*; Policy 21 Limited: Hong Kong, China, 2013.
12. Census and Statistics Department. *Thematic Household Survey No. 60: Housing Conditions of Subdivided Units*; Census and Statistics Department: Hong Kong, China, 2016.
13. Census and Statistics Department. *Thematic Household Survey No. 57: Housing Conditions of Sub-Divided Units*; Census and Statistics Department: Hong Kong, China, 2015.

14. Leung, K.K.; Chow, C.L. A Brief Discussion on Fire Safety Issues of Subdivided Housing Units in Hong Kong. In Proceedings of the 3rd Residential Building Design and Construction Conference, State College, PA, USA, 2–3 March 2016.

15. Ho, K.Y.; Li, W.H.C.; Chung, J.O.K.; Lam, K.K.W.; Chan, S.S.C.; Xia, W. Factors contributing to the psychological well-being for Hong Kong Chinese children from low-income families: A qualitative study. *Int. J. Ment. Health Syst.* **2016**, *10*, 1–7. [CrossRef] [PubMed]

16. Lai, K.M.; Lee, K.M.; Yu, W. Air and hygiene quality in crowded housing environments: A case study of subdivided units in Hong Kong. *Indoor Built Environ.* **2017**, *26*, 32–43. [CrossRef]

17. Ho, D.W.C.; Chau, K.W.; Yau, Y. Evaluating unauthorized appendages in private apartment buildings. *Build. Res. Inf.* **2008**, *36*, 568–579. [CrossRef]

18. Chan, J.C.C.; Chan, W.T. Building safety and timely maintenance. In Proceedings of the HKIE Building Division 2nd Annual Seminar, Hong Kong, China, 28 March 2003; pp. 1–7.

19. Yiu, C.Y.; Yau, Y. Exemption and illegality—The dividing line for building works in Hong Kong. *CIOB (HK) Q. J.* **2005**, *3*, 16–19.

20. Yiu, C.Y.; Kitipornchai, S.; Sing, C.P. Review of the status of unauthorized building works in Hong Kong. *J. Build. Surv.* **2004**, *4*, 28–34.

21. Minor Works Control System. Available online: https://www.bd.gov.hk/english/services/index_mwcs.html (accessed on 14 May 2018).

22. Kennett, P.; Mizuuchi, T. Homelessness, housing insecurity and social exclusion in China, Hong Kong, and Japan. *City Cult. Soc.* **2010**, *1*, 111–118. [CrossRef]

23. La Grange, A.; Yung, B. Aging in a tiger welfare regime: The single elderly in Hong Kong. *J. Cross-Cult. Gerontol.* **2001**, *16*, 257–281. [CrossRef] [PubMed]

24. Audit Commission. *Director of Audit's Report No. 61*; Audit Commission: Hong Kong, China, 2013.

25. Buildings Department. *What Are Unauthorised Building Works?* Buildings Department: Hong Kong, China, 2014.

26. Measures to Deal with Subdivided Flats. Available online: https://www.legco.gov.hk/yr12-13/english/panels/hg/papers/hg0107cb1-367-3-e.pdf (accessed on 18 June 2018).

27. Buildings Department Determined to Combat Unauthorised Subdivision of Flat Units. Available online: http://www.bd.gov.hk/english/documents/news/20110802ae.htm (accessed on 18 October 2017).

28. Lytras, M.; Raghavan, V.; Damiani, E. Big data and data analytics research: From metaphors to value space for collective wisdom in human decision making and smart machines. *Int. J. Semant. Web Inf. Syst.* **2017**, *13*, 1–10. [CrossRef]

29. Morabito, V. *Big Data and Analytics: Strategic and Organizational Impacts*; Springer: Cham, Switzerland, 2015.

30. Al Nuaimi, E.; Al Neyadi, H.; Mohamed, N.; Al-Jaroodi, J. Applications of big data to smart cities. *J. Internet Serv. Appl.* **2015**, *6*, 1–15. [CrossRef]

31. Joseph, R.C.; Johnson, N.A. Big data and transformational government. *IT Prof.* **2013**, *15*, 43–48. [CrossRef]

32. Chen, C.L.P.; Zhang, C.-Y. Data-intensive applications, challenges, techniques and technologies: A survey on big data. *Inf. Sci.* **2014**, *275*, 314–347. [CrossRef]

33. Manyika, J.; Chui, M.; Brown, B.; Bugin, J.; Dobbs, R.; Roxburgh, C.; Byers, A.H. *Big Data: The Next Frontier for Innovation, Competition, and Productivity*; McKinsey & Company: New York, NY, USA, 2012.

34. Kim, G.-H.; Trimi, S.; Chung, J.-H. Big-data applications in the government sector. *Commun. ACM* **2014**, *57*, 78–85. [CrossRef]

35. Chakraborty, A.; Wilson, B.; Sarraf, S.; Jana, A. Open data for informal settlements: Toward a user's guide for urban managers and planners. *J. Urban Manag.* **2015**, *4*, 79–91. [CrossRef]

36. Franks, B. *Taming the Big Data Tidal Wave*; John Wiley: Hoboken, NJ, USA, 2012.

37. Kim, K.W.; Park, W.J.; Park, S.T. A study on plan to improve illegal parking using big data. *Indian J. Sci. Technol.* **2015**, *8*, 1–5. [CrossRef]

38. Flowers, M. Beyond open data: The data-driven city. In *Beyond Transparency: Open Data and the Future of Civic Innovation*; Brett, G., Dyson, L., Eds.; Code for America Press: San Francisco, CA, SUA, 2013; pp. 185–198.

39. Can Big Data Help Us Predict Where Crime Will Strike? Available online: http://www.bbc.co.uk/guides/zqsg9qt (accessed on 14 July 2018).

40. Facing the Threat: Big Data and Crime Prevention. Available online: https://www.ibm.com/blogs/internet-of-things/big-data-crime-prevention/ (accessed on 14 July 2018).

41. Shi, Q.; Abdel-Aty, M. Big data applications in real-time traffic operation and safety monitoring and improvement on urban expressways. *Transp. Res. Part C* **2015**, *58*, 380–394. [CrossRef]

42. Long, Y. Big/open data for urban management. *J. Urban Manag.* **2015**, *4*, 73. [CrossRef]

43. Alexander, E.R. A transaction-cost theory of land use planning and development control. *Town Plan. Rev.* **2001**, *72*, 45–75. [CrossRef]

44. North, D.C. The new institutional economics. *J. Inst. Theor. Econ.* **1986**, *142*, 230–237.

45. Van der Burg, T. Neo-classical economics, institutional economics and improved fisheries management. *Mar. Policy* **2000**, *24*, 45–51. [CrossRef]

46. Williamson, O.E. The new institutional economics: Taking stock, looking ahead. *J. Econ. Lit.* **2000**, *38*, 595–613. [CrossRef]

47. Cordella, A. Transaction costs and information systems: Does IT add up? *J. Inf. Technol.* **2006**, *21*, 195–202. [CrossRef]

48. Krier, J.E.; Montgomery, W.D. Resource allocation, information cost and the form of government intervention. *Nat. Resour. J.* **1973**, *13*, 89–105.

49. Liang, T.-P.; Huang, J.-S. An empirical study on consumer acceptance of products in electronic markets: A transaction cost model. *Decis. Support Syst.* **1998**, *24*, 29–43. [CrossRef]

50. Lv, Y.; Duan, Y.; Kang, W.; Li, Z.; Wang, F.-Y. Traffic flow prediction with big data: A deep learning approach. *IEEE Trans. Intell. Transp. Syst.* **2015**, *16*, 865–873. [CrossRef]

51. Zhou, K.; Yang, S. Understanding household energy consumption behavior: The contribution of energy big data analytics. *Renew. Sustain. Energy Rev.* **2016**, *56*, 810–819. [CrossRef]

52. Mayer-Schönberger, V.; Cukier, K. *Big Data: A Revolution That Will Transform How We Live, Work, and Think*; Houghton Mifflin Harcourt: Boston, MA, USA, 2013.

53. Sydney's Slumlord Solution: Big Data to Crack down on Illegal Boarding Houses. Available online: https://www.domain.com.au/news/sydneys-slumlord-solution-big-data-to-help-crack-down-on-illegal-boarding-houses-20160506-goo2ja/ (accessed on 18 October 2017).

54. Rating and Valuation Department. *Hong Kong Property Review 2018*; Rating and Valuation Department: Hong Kong, China, 2018.

55. Crawford, K.; Schultz, J. Big data and due process: Toward a framework to redress predictive privacy harms. *Boston Coll. Law Rev.* **2014**, *55*, 93–128.

56. Tene, O.; Polonetsky, J. Big data for all: Privacy and user control in the age of analytics. *Northwest. J. Technol. Intellect. Prop.* **2013**, *11*, 239–275.

57. Kohler, N.; Hassler, U. The building stock as a research object. *Build. Res. Inf.* **2002**, *30*, 226–236. [CrossRef]

 *sustainability*

Article

Opinion Mining on Social Media Data: Sentiment Analysis of User Preferences

Vasile-Daniel Păvăloaia [1,*], Elena-Mădălina Teodor [2], Doina Fotache [1] and Magdalena Danileț [3]

[1] Department of Accounting, Business Information Systems and Statistics,
Faculty of Economics and Business Administration, Alexandru Ioan Cuza University of Iasi,
700506 Iaşi, Romania
[2] Web Department, Falcon Trading Company, 700521 Iaşi, Romania
[3] Department of Management, Marketing and Business Administration,
Faculty of Economics and Business Administration, Alexandru Ioan Cuza University of Iasi,
700506 Iaşi, Romania
* Correspondence: danpav@uaic.ro

Received: 23 July 2019; Accepted: 14 August 2019; Published: 17 August 2019

Abstract: Any brand's presence on social networks has a significant impact on emotional reactions of its users to different types of posts on social media (SM). If a company understands the preferred types of posts (photo or video) of its customers, based on their reactions, it could make use of these preferences in designing its future communication strategy. The study examines how the use of SM technology and customer-centric management systems could contribute to sustainable business development of companies by means of social customer relationship management (sCRM). The two companies included in the study provide a general consumer good in the beverage industry. As such, it may be said that users interacting with the posts these companies make on their official channels are in fact customers or potential customers. The study aims to analyze customer reaction to two types of posts (photos or videos) on six social networks: Facebook, Twitter, Instagram, Pinterest, Google+ and Youtube. It brings evidence on the differences and similarities between the SM customer behaviors of two highly competitive brands in the beverage industry. Drawing on current literature on SM, sCRM and marketing, the output of this study is the conceptualization and measurement of a brand's SM ability to understand customer preferences for different types of posts by using various statistical tools and the sentiment analysis (SA) technique applied to big sets of data.

Keywords: opinion mining; social media; social networks; sentiment analysis; sentiment polarity classification

---

1. Introduction

The current research deals with an essential topic of interest for any company that intends to have a sustainable development and also to thrive in the current economic landscape which is ever more competitive. Though it belongs to the artificial intelligence (AI) domain, namely is considered a machine learning (ML) technique, sentiment analysis (SA) is a social media (SM) analytics tool that involves checking how many negative and positive keywords are included in a text message associated with a SM post. This in-depth analysis also involves finding opinions in SM content and extracting the sentiment they contain. Furthermore, those interested in this type of information can process it in real-time and take actions in the benefit of their company.

In light of the above information, the study looks into why organizations should change the way they develop business in order to be in line with the requirements of the new digital age. As such, according to Forrester experts, it is estimated that approximately 1 million B2B sales agents will lose their jobs by 2020 in favor of e-commerce. Adopting a selling strategy leveraging SM channels can lead to an increase in sales (social selling) [1] and therefore to a more sustainable development. Future digital experiences are impacted by fast developments in mobile technology and by developments of SM. Therefore, the organizations should strive towards providing a consistent experience across communication channels and integrated business platforms. This way, companies are able to reach a new level of competitive customer management as customers are no longer passive recipients but mostly proactive.

The exponential growth of individual SM users and SM active companies (only Facebook reported over 2.27 billion active accounts in early 2019) has turned SM into a company/brand and into a customer interaction space [2,3]. SM facilitates vivid communication with customers through text, sound, photo and video [4], in which the SM user emotional response influences brand-page engagement [5,6], brand advocacy and loyalty [5] and, indirectly, purchase behaviors [3,7,8]. Although there is an abundant source of materials [9,10] focusing on strong relationship between purchase behaviors and customer preferences for the content displayed by companies on SM, this study only covers the preference of users (as potential customers) for photo compared to video posts.

As social networks play an important role for the sustainable development in business, this study provides a deep analysis of customer reactions to SM posts using a set of artificial intelligence (AI) techniques, such as sentiment analysis, sentiment polarity classification (SPC) and mosaic plots.

Social networks represent an abundant source of big data that could be difficult to handle without automation. ML algorithms may be an effective method for managing big data and interpreting the results in such a way that it could be beneficial for companies.

To conduct the study, the following set of premises were formulated: ML algorithms for big data can be successfully applied to social network posts (massive source of big data) as to identify the emotional reactions of customers depending on the type of posts (1); the ML approach is effective and reliable for opinion mining and sentiment classification of user posts on any company's official channel (2); SM users tend to express their sentiments differently depending on the type of post. Some are reluctant to view videos and embrace photos, while others react positively (3); SM users consider that photos, compared to videos, express clearer and more concise messages so they perceive the message faster (4); a company's ability to capture and analyze user preferences as regards to the two different types of posts on official SM channels leads to sustainable business development (5).

Recently, rich literature has been produced on the use of SM for customer interactions with companies, in general, and also on SA. Further, this study used SA for studying customer preferences (video versus photo SM posts), which is in line with internationally accepted research practices. For the first months of 2019, there is evidence on the existence of a high number of literature reviews discussing the use of SA combined with text mining (TM), as well as the use of natural language processing (NLP) techniques for analyzing customers from several perspectives. Table 1 shows a few studies published in early 2019 by top journals on topics similar to this study. The studies have been extracted from ISI Web of Science using the keywords, sentiment analysis and customer.

The authors believe that this study fills the identified gap in knowledge and contributes to the literature in the field by focusing on the use of various SM platforms by two competitive brands in the beverage industry in order to analyze the preferences of their customers towards photo versus video posts.

The article is divided into five sections and comprises a study analyzing the emotions and sentiments expressed by customers on SM official channels of Coca-Cola and Pepsi by focusing on customer reactions (number of likes, commentary and distributions, retweets, repins) and on posts (photos or videos) on six SM platforms.

Table 1. Similar articles published in 2019 using the key-words, sentiment analysis (SA) and customer.

Research Objectives of the Study	CRD * Analyzed	R **
Identifies key factors in User Generated Content on Twitter for the creation of successful startups.	SM User Content	[11]
Examines consumer reviews of three different competitive automobile brands and analyzes the advantages and disadvantages of each vehicle using TM *** and association rule methods.	Customer reviews	[12]
A novel method that uses SA procedures in order to automatically create fuzzy ontologies from free texts provided by users on SM.	SM User Content	[13]
Addresses the issue of SM domain by identifying the potential customer base for advertisement activities.	SM User Content	[14]
Prediction of customer satisfaction has been proposed using fusion of EEG and sentiments.	Customer reviews	[15]
Develop a fine-grained SA supervised by semantic knowledge, context sensitive sentiments are extracted from online customer reviews.	Customer reviews	[16]
A case of TM on Airbnb user reviews to analyze and understand various aspects that drive customer satisfaction.	Customer reviews	[17]
Explore and decode the sentiment dynamics of Twitter users regarding online retailing brands.	Customer reviews	[18]
Leverage the relationship between user-generated reviews and the ratings of the reviews to associate the reviewer sentiment with certain entities.	Customer reviews	[19]
Classify all reviews and comments of customers to extract most popular category, theme and feature.	Customer reviews	[20]
Identify hotel attributes that contribute to customer satisfaction or dissatisfaction using online reviews for hotels in India.	Customer reviews	[21]
A guide for implementing the TM approach highlighting 6 key insights practitioners need in order to manage their customers' journey.	Customer interview	[22]
TM and SA techniques used to analyze the SM data set and to visualize relevant insights and patterns in order to identify customer knowledge.	SM User Content	[23]
The discovery of valuable user experience data, and their relations to product design and business strategic planning by analyzing a large volume of customer online data.	Customer reviews	[24]

* CRD—Customer related data; ** R—Reference; *** TM—Text mining.

2. Theoretical Framework

Can and Alatas [25] argued that a high source of big data provided by user interactions on social network platforms created a new concept (big social network data) which, through its effective use of specific technologies, leads to a sustainable development of a business.

Some authors [26] stress that continued development of social instruments has led to a multiplication of human interactions, while others [27] argue that this is the reason why new business models allow a customization of transactions by consumer preference. With the dawn of SM [28], the power seems to have shifted from marketing managers to individuals and communities. SM is a group of applications built on the ideological and technological foundations of Web 2.0 which aims to create and share content generated by a community of users. Therefore, Kaplan and Haenlein [29] believe that this new technology has improved the way companies deal with their customers. According to Kornum and Mulbacher [30], the changing role of marketing from the perspective of online communities should be studied, where the participants with different interests and resources wish to increase their influence on company decisions. Other authors investigated the role of SM, while Casteleyn et al. [31] put forward the idea that among the main reasons why people use Facebook are social interaction, professional advancement and entertainment. These are all tools that can be leveraged by companies to gain a competitive advantage. Another interesting study [32] brought evidence on the fact that many social tools are used by companies to communicate directly with customers, increase brand loyalty, find new sales opportunities and develop new marketing research paths.

In terms of the classification of social networks, there are two main viewpoints. The first is based on theories of media research and theories on social processes developed by Kaplan and Haenlein [29]. By combining the two theoretical perspectives, according to the contact level and image-building, SM tool opportunities can be classified into four categories: Social communities, content, virtual reality, and 3D corporate virtual games. The second classification is based on common features of social tools which identify the type of used media [33]. This classification includes ten categories of SM tool types: Social communities (Facebook, LinkedIn, Google+, Yammer, Twitter); Blogs (Quora, WordPress, Blogger); microblogs (Twitter, Tumblr); photo publishing tools (Pinterest, Instagram, Flickr, Picasa); audio Publishing Tools (Spotify, iTunes, Podcast.com), video publishing tools (Youtube, Vimeo, Vine); on-line gaming (World of Warcraft), RSS (Google, FeedBurner); second life (Kaneva) and document storage tools and forms (Google Docs, SurveyMonkey, Doodle). Specialists sustained that a company does not have to use all of these tools [33], but rather it should focus on the most important or beneficial ones for the business model that is being employed.

As has been observed earlier, a substantial amount of literature has been published on SM. A recent systematic literature review concluded that the authors, by drawing on the concept of SM [34], were able to show that a new term emerged, namely, strategic social media marketing (SSMM). In the same study [34], the authors devised a framework for SSMM comprising four central dimensions: Social media marketing (SMM) scope, SMM culture, SMM structure and social SMM governance. They concluded that cross-functional collaborations along the four dimensions of SMM were needed to successfully navigate in this dynamic arena, where stakeholders play an essential role. Another author also presents a new trend [34] in SM which points out that the term SM should no longer be used, but social business and social enterprise. The main feature of social enterprise is its wide openness towards customers. Even more, a social enterprise thinks and lives in SM and integrates it into all of its process. In order to do so, many AI techniques are used on big sets of data for pursuing different types of complex analyses.

Acker and Gröne [35] underline that some of the main benefits of social CRM are building trust and gaining knowledge on customers, establishing customer loyalty, developing customer retention and their involvement into new products or services development process, improving the organization's reputation and lowering service costs. The aspects related to the identification of business needs and most suitable technology were studied by Kietzmann et al. [36]. Other researchers [37] are further arguing that in order to be able to implement a business strategy, companies have to integrate algorithms into CRM platforms, such as SA and predictive modeling, if they want to increase the efficiency and effectiveness of customer relationship management activities. However, Belch and Belch [38] believe that these terms are used inconsistently as the costs and results are extremely different, being dependent on which SM tools are used. Cui, Lui and Guo et al. [39–43] consider that there are three categories of values that can be used to measure earned involvement in SM: Volume, valence and dispersion. Volume metrics is quantity-related and it measures the number of consumer reviews. Valence values refer to the positive or negative opinions, and dispersion measures the speed at which community impressions and opinions are spread.

Related to the use of opinion mining techniques on SM posts, Pozzi et al. [44] state that social networks represent an emerging challenging sector, where natural language expressions of users can be easily reported through short but meaningful text messages. Further, they argue that the key information that can be grasped from social environments relates to the polarity of text messages (i.e., positive, negative or neutral). In this respect, there are many approaches in the literature, some of them referring to the use of lexicons for sentiment polarity classification.

In fact, many authors applied different ML algorithms or other hybrid techniques to data collected from SM for various reasons. A further study [45] researched a multilingual sentiment detection framework used to compute the European gross national happiness (GNH) of Twitter users. Their framework consists of a novel data collection, filtering and sampling method, and a newly constructed multilingual sentiment detection algorithm for SM big data, tested in some EU countries over a six-year period. Carrera and Jung [46] applied on Facebook users the SentiFlow algorithm as a plug-in of the ProM platform to verify their proposed framework. ProM is an open source platform providing practical applications for process mining and supporting many kinds of process discovery algorithms. Gamalet et al. [47] used ten different ML algorithms with two feature extraction algorithms that were implemented on four SA datasets (IMDB, Cornell movies, Amazon and Twitter) in a comparative analysis of their methodology. Sobhani et al. [48] investigate the problem of jointly predicting the stance expressed toward multiple targets using Twitter posts. Stance detection is the task of automatically determining from the text whether the author of the text is for, against, or has a neutral view towards a proposition or target.

Concerning lexically-based approaches, in [49], the seed-word selection for semi-supervised sentiment classification is addressed through a joint lexicon corpus learning approach. Some authors [40,50,51] pursue in their research an approach that combines lexicons, labeled and un-labeled data for sentiment transfer across different domains. They first extract automatically labeled samples by using emotion keywords. Then, both the automatically-labeled samples from the target domain and the real labeled samples from the source domain were combined to create a new labeled data set. The updated methods rely on the automatic construction of lexicons [44]. Lu et al. [52] tackled the problem of deriving a sentiment lexicon that was not only domain-specific but also aspect-dependent. To achieve this aim, an optimization framework was suggested to combine different sources of information for learning context-dependent sentiment lexicons.

To sum up, the opinion mining and SA techniques in the past decade have become immensely popular and have been viewed as the most active areas of research due to the following reasons [20]: These two techniques have a wide array of applications in very different domains (1); it is considered to be a highly challenging research problem that has scarcely been studied in the past (2); due to the advent of the big data technologies, a massive volume of opinionated data is easily accessible in digital formats on the Web (3).

Whilst some research has been carried out on the use of SA on SM [39,40] and [44–46], no studies have been identified that attempt to analyze the interactions with a company posts of users on six social networks. Further, very few studies extract user preferences on the two types of posts analyzed in this study (photos and videos), and which are used by companies to get in touch with its customers on their SM official channels.

As the literature review highlights, there is a lack of studies that analyzed whether SM participants reacted differently to posts containing photos versus posts containing videos. The analysis of those two features are considered as they are the most frequently-used elements of SM posts. Due to the advent of technology, especially in the smartphone industry, people manifest a tendency to posting photos and/or videos when interacting in the online environment. The questions that the study intends to answer are which type of posts are the most preferred and what is the intensity of their preference for SM users across a wide variety of SM platforms.

3. Research Methodology

The study aims to analyze CocaCola's and PepsiCo's users or potential customer preferences for different types of posts (photos or videos) on six SM official pages posts, namely Facebook, Twitter, Instagram, Pinterest, Google+ and Youtube. Consequently, the paper intends to emphasize that user interactions with companies through SM posts bring a relevant contribution to both society and the business sector by achieving a social good.

3.1. Research Problem

To reach its aims, a two-step methodology was designed: A detailed literature review (used to make an overview of the domain and to identify the gap in knowledge) and a study [53]. For each step of the study, several inductive methods based on cause-effect have been used.

This way, the qualitative traffic on Coca-Cola (coca-cola.com) and PepsiCo (pepsico.com) web sites have been analyzed considering the global rank (in the hierarchy of searching tools), the traffic differences regarding mobile tools, the number of unique visitors, the search words used by users to get to company pages, and the types of channels used by companies to reach their customers. Some of the results are shown in Figure 1.

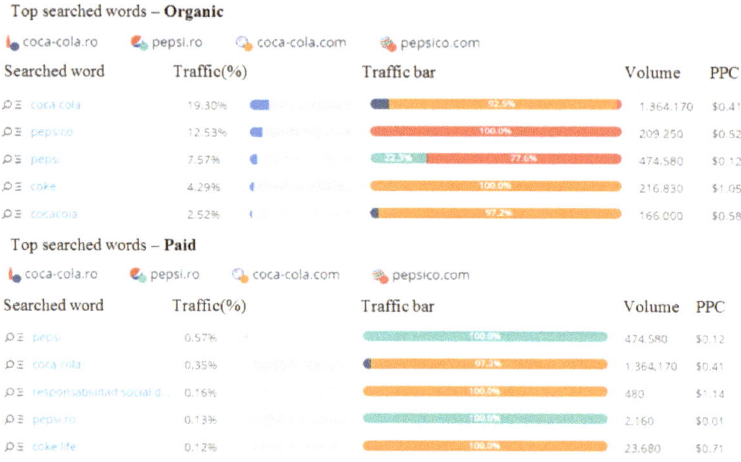

Figure 1. Top searched words organic versus paid searches; Source: SimilarWeb analysis.

The highest global rank was recorded for Coca-Cola, worth 742,309, followed by PepsiCo, with less than half the score, and worth 317,978. Considering the relation between the searched words (organic versus paid), the highest traffic percentages were found in organic searches (ranging between 2.52% and 19.30%), while in paid searches, the percentages started from 0.12%, and 0.57%. The overall results are shown in Figure 1. Based on the above results, the two analyzed websites do not require significant assistance from paid traffic as these companies get good results from unpaid traffic. Considering our results, it may be concluded that social tools have not been used to their highest capacity by PepsiCo and Coca-Cola companies.

The potential number of users that can be reached on main social networks can be visualized at [54]. In order to understand the customer experience on social networks (especially the analysis of reactions to posting photos or video), this study took into account the analysis of specific key performance indicators (KPI) based on different categories.

The aim is to investigate emotional responses to photos or video posts distributed via main SM channels of two well-known international brands in the beverages industry. The research question was associated with the following primary (main) hypothesis (PH): On social networks, the expressed emotions of users have a considerable impact on their activities by type of posts (photos or videos). Consequently, the main hypothesis had been tested for each SM investigated in this study, so the following secondary research hypotheses (one for each social network) was formulated: There are significant differences in the user emotional reactions in its activities on Facebook-SH1, Instagram (number of likes, commentary and distributions)-SH2, Twitter (number of likes and retweets)-SH3, Pinterest (the number of repins and comments)-SH4, Youtube-SH5 and Google+(SH6) for each type of post (photos or videos).

Initially, the data was analyzed for a normal distribution, and since the result was positive, the Student test was further applied to validate the secondary and, implicitly, the main hypothesis. Moreover, for a more in-depth analysis, the Wilcoxon test was applied to assess the intensity (by using ranks) (user preference for)/preference for photo versus video posts.

3.2. Data Analysis and Research Methodology

The data analysis was conducted using two main methods: Qualitative method—an analysis of existing theoretical concepts and methods; and quantitative method—an analysis of data collection and processing. The mathematical and statistical methods were the basis for the processing and interpretation of data, especially the statistical estimation for the hypotheses demonstration, using statistical tests for statistical comparison. The source of data for the above-mentioned methods was obtained from an application developed in C#. The computer application accommodates on a single platform all the charts and, more importantly, the laborious calculations required for determining the rankings of each post which have been made using KPI's for each company and social network.

The research methodology adopted for this study is a mixed one. Figure 2 shows the conceptual model used to achieve the aims of the study.

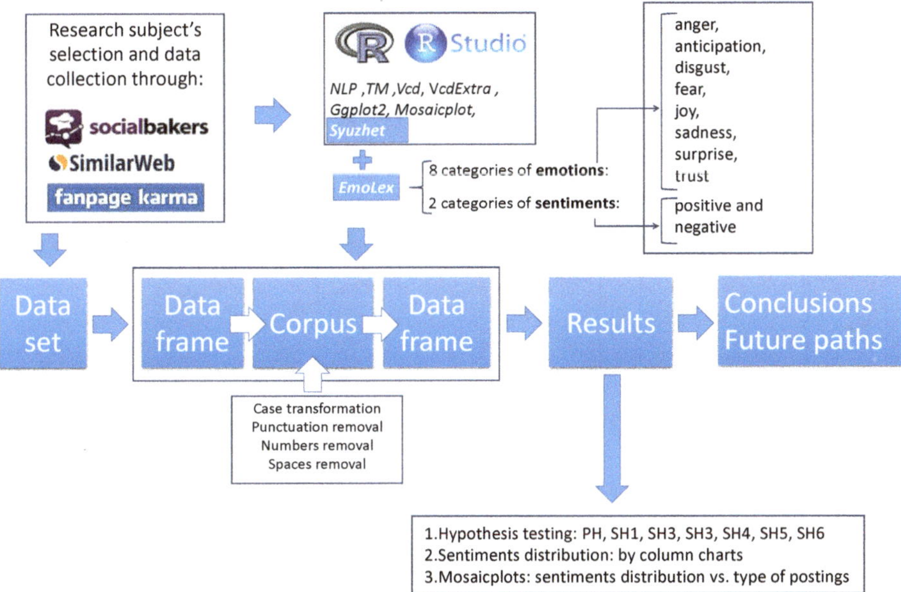

Figure 2. Research methodology.

The case-study was used as the main method of investigation for collecting necessary social network data. The first phase of the study was focused on the activities carried out on the official websites of Coca-Cola and PepsiCo (global market with the domains.com) in the first part of 2018, namely, between January and May. The research subjects were nominated according to the rankings made by Social Barker's experts obtained by targeting Facebook pages with the most followers. In this ranking, Coca Cola was ranked the 4th (with approximately 105 million followers), and its competitor PepsiCo being ranked the 31st (47.8 million followers). Although Facebook is the main SM tool used in the interaction with actual fans, the two analyzed companies have active pages on other social network platforms, such as: Twitter, Instagram, Pinterest, Google+ and Youtube. Moreover, according to top social network demographics, the two brands had the highest number of active users in the beverage industry in 2017. It should be noted that the above ranking was conducted globally in 2017, although it was published in early 2018.

The analyzed data was collected through the Similar Web site (https://www.similarweb.com/), a well-known marketing intelligence tool. The second phase of this study was the assessment of the activities for the two companies on official SM channels. The collection of data on social networks required the use of an online tool needed to analyze and monitor SM data, namely, FanPageKarma [55] which, according to reviews from g2Crowd, provides comprehensive services, and a complete picture of fan interactions on company pages. The data collected with FanPageKarma were exported and stored in spreadsheet format. The collected data was extracted from the top 50 posts from January to May 2018 of the official social network pages of Coca-Cola and Pepsi (@CocaColaCo, @PepsiCo) at a global level, and it is based on the key performance indicators divided by categories. The model based on the KPI matrix has a set of indicators and to each of them, points were awarded. Based on this, a top of the 50 posts was made for each brand and for each SM page. Depending on the carried-out activities, KPI's can be classified, as presented in Figure 3.

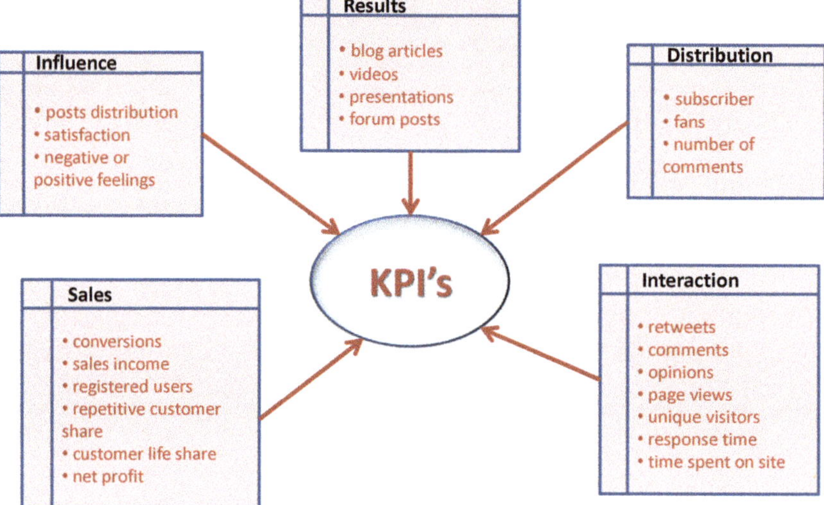

Figure 3. The classification of key performance indicators (KPIs).

The data have been the source for interpreting the information in SA conducted in R language and R studio. The SA aims to establish whether the words (within posts, in this case) have a positive, negative or neutral significance. For this purpose, the Syuzhet package available in R was used. From this package, the NRC Word-Emotion Association Lexicon (aka EmoLex) [41] was applied, which contains a list of words associated with eight categories of emotions: Anger, fear, anticipation, trust, surprise, sadness, joy, disgust, as well as two categories of sentiments, positive and negative.

Considering the above, the focus was to highlight the following features in the analyzed text as presented in Figure 4.

The data for analysis was converted into a data frame, and then into a corpus (six csv files, one for each social network), which required certain minor changes, such as, turning small letters into capitals, elimination of punctuation signs, numbers, blanks, etc.

In terms of data processing and interpretation, especially for the KPI's, several mathematical and statistical methods were used, especially statistical estimation for defining statistical hypotheses that had to be confirmed or rejected on the basis of statistical tests applied for statistical comparisons. The development environment used for hypothesis testing was R Studio, together with the R language that comprises a pack of tools for SA purposes (NLP—helps with data processing; TM—provides text mining features; Syuzhet, sentiment—performs the analysis of sentiments on the text; vcd, vcdExtra—provides features for statistical tests, ggplot2, mosaic plot—libraries for graphing).

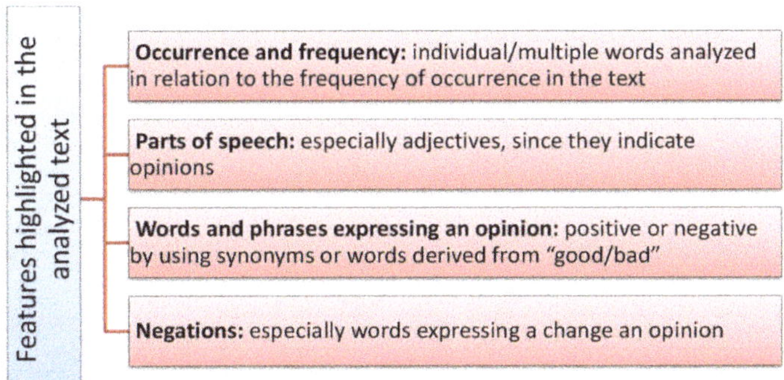

Figure 4. The highlighted features in the analyzed text.

4. Results

The results of analysis presented below are divided into three categories: Hypothesis testing (1), sentiment distribution by histogram charts (2); and polarity categories distribution by Mosaic plots (3).

4.1. Hypothesis Testing

The secondary hypothesis (SH1, SH2, SH3, SH4, SH5 and SH6) were tested. There are significant differences in the user emotional reactions in their activities on Facebook-SH1, Instagram (number of likes, commentary and distributions)-SH2, Twitter (number of likes and retweets)-SH3, Pinterest (the number of repins and comments)-SH4, Youtube-SH5, and Google+ SH6 for each type of post (photos or videos). Consequently, each secondary hypothesis was tested for every social network platform included in the study.

First, the *t*-test was used to analyze the relationship between the user's reactions and the two types of posts, and the Wilcoxon test was applied to test the intensity of preference for photo vs. video posts. Second, the Chi-square test was applied for the analysis of the variable (sentiment categories) and the type of post. Third, the significance levels were set at the 5% using both the student *t*-test, the Wilcoxon test and the Chi-square test.

The first analyzed social platform was Facebook. As can be seen from Table 2, the results of the independence tests (*t*-test results) show the following: (1) The variable number of likes is noticed in all cases, the secondary hypothesis has been validated; (2) in the case of number of comments, it is noted in almost all cases, the second hypothesis has been validated, except the *t* test for @PepsiCo; and (3) for the variable number of distributions, in almost all cases, the secondary hypothesis has been validated, except the *t*-test for @PepsiCo. Therefore, based on the results shown in Table 2, it could be assumed that, with an assumed risk of 5%, there is a correlation between the three types of analyzed reactions and the type of post (photo or video).

Table 2 also presents the analysis of the results for the independence tests conducted on the Instagram platform. The *t*-test results show the following: (1) There is a correlation between the number of likes and the type of posting, as in almost all cases, the secondary hypothesis has been confirmed, except for the *t*-test for @PepsiCo; (2) it shows a strong correlation between the number of comments and the type of posts, as in all cases, the secondary hypothesis has been validated (*P*-value < 0.05).

The analysis of the results regarding the association test between the sentiment category and the type of posting (for both Facebook and Instagram pages) is shown in the last line of Table 2 (Chisq test). From the data, it can inferred that in all cases, for each category of sentiments, the secondary hypotheses (SH1 and SH2) have been confirmed. Therefore, it can be concluded, with a 5% assumed risk, that there is strong correlation between sentiment categories and the type of post (photo or video).

Table 2. Testing the type of post (photos or videos) influence on the Facebook and Instagram user's reactions.

| | Facebook Page for the Company: | | | | Instagram Page for the Company: | | | |
| | @CocaColaCo | | @PepsiCo | | @CocaColaCo | | @PepsiCo | |
Reactions — Stat. test	Student test (P value)	Wilcoxon test (P value)	Student test (P value)	Wilcoxon test (P value)	Student test (P value)	Wilcoxon test (P value)	Student test (P value)	Wilcoxon test (P value)
Likes	2.9998 (0.0037 ***)	108 (0.001 ***)	2.1537 (0.0357 **)	150 (0.044 **)	2.4305 (0.0028 **)	128 (0.019 **)	0.17804 (0.8754)	150 (0.044 **)
Comments	3.883 (0.0303 **)	142 (0.044 **)	0.3458 (0.730)	146 (0.030 **)	3.2365 (0.0245 **)	135 (0.001 ***)	0.27735 (0.0363 **)	151 (0.01 **)
Distributions	2.045 (0.0172 **)	127 (0.011 **)	0.1234 (0.902)	115 (0.001 ***)	-	-	-	-
Variable — Stat. test	Chisq test	P value	Chisq test	P value	Chisq test	P value	Chisq test	P value
Sentiment categories	3.6813	0.0458 **	0.4219	0.0191 **	3.7310	0.0474 **	2.000	0.036 **

* *P*-value < 0.1, ** *P*-value < 0.05, *** *P*-value < 0.001.

Furthermore, what is striking about the intensity (Wilcoxon test results in Table 2) of the user's preferences for the type of posts is that on both SM platforms (Facebook and Instagram), for both companies (@CocaColaCo, @PepsiCo), the reactions are very different. Consequently, the reactions for photo versus video are very well-differentiated so that the intensity of the user's preference for photo posts is significantly different compared to users who prefer video posts.

Table 3 shows the summary statistics of the analysis conducted for the variables on Pinterest and Twitter pages (*t*-test results). Notably, what stands out in the table for the Pinterest pages is: (1) The results of testing the variable, the number of Repins, show that in all cases the secondary hypothesis (SH3) has been validated; (2) the results of testing the variable, the number of comments, shows that in almost all cases, the secondary hypothesis has been validated (except for the *t* test for @PepsiCo). As a result, these findings suggest that, with an assumed risk of 5%, the two analyzed variables (Reactions: Repins and Comments) are correlated with the type of post (photos or video). Clearly, what is apparent for the Twitter pages is that: (1) for the first variable, the number of likes, it can be stated that, in all cases, the secondary hypothesis has been validated; (2) for the second variable, the number of retweets, it has been observed that, in almost all cases, the secondary hypothesis (SH4) has been validated (except for the *t*-test for @PepsiCo). Therefore, the analysis proves that there is a strong correlation between the number of likes and retweets, and the type of posting (photo or video).

Table 3. Testing the type of post (photos or videos) influence on the Pinterest and Twitter user's reactions.

| | Pinterest Page for the Company: | | | | Twitter Page for the Company: | | | |
| | @CocaColaCo | | @PepsiCo | | @CocaColaCo | | @PepsiCo | |
Reactions — Stat. test	Student test (P value)	Wilcoxon test (P value)	Student test (P value)	Wilcoxon test (P value)	Student test (P value)	Wilcoxon test (P value)	Student test (P value)	Wilcoxon test (P value)
RePins	1.039 (0.0315 **)	144 (0.0142 **)	1.980 (0.0127 **)	42.5 (0.8341)	-	-	-	-
Comments	3.883 (0.0303 **)	128 (0.032 **)	0.3457 (0.7303)	183 (0.262)	-	-	-	-
Likes	-	-	-	-	1.1248 (0.0232 **)	201 (0.049 **)	1.893 (0.0163 **)	195 (0.963)
Retweets	-	-	-	-	3.883 (0.0303 **)	128 (0.032 **)	0.3457 (0.7303)	183 (0.262)
Variable — Stat. test	Chisq test	P value	Chisq test	P value	Chisq test	P value	Chisq test	P value
Sentiment categories	2.2154	0.0251 **	1.8196	0.0163 **	1.262	0.0268 **	1.261	0.0213 **

* *P*-value < 0.1, ** *P*-value < 0.05, *** *P*-value < 0.001.

The association analysis (also included in Table 3—Chisq test) between the category of expressed sentiments by customers and the type of postings (for both Pinterest and Twitter pages) shows that, in all cases, and for each category of sentiments, the secondary hypothesis has been validated. Therefore, it can be asserted, with a 5% assumed risk, that for both Pinterest and Twitter accounts, there is a strong correlation between the categories of sentiments that the user displays for the two types of posts.

The most interesting aspect related to the intensity of the user's preference for photo versus video posts on Pinterest and Twitter pages (Wilcoxon test results in Table 3) is that: (1) In the case of @CocaColaCo, the users have significantly different reactions (RePins, Comments, Likes, Retweets) for the two type of posts while (2) in case of @PepsiCo, the users do not manifest significant differences in their preferences for photo versus video posts.

Regarding the analysis conducted for the variables, the number of comments and the number of views on the Youtube page, the results are summarized in Table 4 (*t*-test results). The results of the independence test for both variables, the number of comments and the number of views, show that in all cases the secondary hypothesis (SH5) has been validated. Further, it can be asserted, with an assumed risk of 5%, that the two analyzed variables are correlated with the type of post (photos or video).

Table 4. Testing the type of post (photos or videos) influence on the Youtube and Google+ user reactions.

	Youtube Page for the Company:				Google+ Page for the Company:			
	@CocaColaCo		@PepsiCo		@CocaColaCo		@PepsiCo	
Reactions — *Stat. test*	*Student test (P value)*	*Wilcoxon test (P value)*	*Student test (P value)*	*Wilcoxon test (P value)*	*Student test (P value)*	*Wilcoxon test (P value)*	*Student test (P value)*	*Wilcoxon test (P value)*
Distributions	-	-	-	-	3.373 (0.0470 **)	117 (0.0145)	2.172 (0.042 **)	159 (0.0143)
Comments	1.0658 (0.0161 **)	163 (0.048)	2.887 (0.0203 **)	123 (0.0362)	2.692 (0.04 **)	170 (0.0187)	3.883 (0.0252 **)	113 (0.028)
Views	2.127 (0.0162 **)	141 (0.0034)	3.211 (0.0072 **)	236 (0.0011)	-	-	-	-
Variable — *Stat. test*	*Chisq test*	*P value*	*Chisq test*	*P value*	*Chisq test*	*P value*	*Chisq test*	*P value*
Sentiment categories	3.1203	0.0103 **	2.9865	0.0278 **	4.4545	0.0348 **	3.909	0.0105 **

* *P*-value < 0.1, ** *P*-value < 0.05, *** *P*-value < 0.001.

Similarly, Table 4 includes the *t*-test results of testing the variables, the number of distributions and the number of comments on the type of posts (photos and videos) on Google+ official channels owned by @CocaColaCo and @PepsiCo. In all cases, it is noticed that the results of the *t*-test show a strong statistical significance. Therefore, it can be asserted that there is a correlation between the analyzed variables and the types of posts on the Google+ platform validating the secondary hypothesis (SH6).

The last line of Table 4 clearly shows that the result of the independence test between the sentiment categories and the type of posts indicates a strong correlation between the variables, as in all cases, the significance value of Chi-square test is < 0.05.

As Table 4 shows (based on the Wilcoxon test results), there is no significant difference between the two SM users, in terms of intensity of the preference for photo versus video posts. Therefore, it can be concluded that, on Youtube and Google+ platforms, the users of @CocaColaCo and @PepsiCo SM channels, react similarly in terms of intensity of their preference for photo versus video posts.

Together, these results (summarized in Tables 2–4) and the above interpretation provide an insight into the analysis aimed to validate (or invalidate) the secondary hypotheses (SH1-SH6), and ultimately to validate (or not) the main hypothesis (PH). Therefore, the analyzed variables (the number of likes, comments, distributions, retweets, repins, views) on the six SM platforms (Facebook, Instagram, Pinterest, Twitter, Youtube and Google+) are relatively dependent on the variables, the type of posts and expressed emotions, which lead to the validation of the secondary hypothesis for each analyzed social tool. Moreover, the above results implicitly lead to the validation of the main research hypothesis (PH).

It can be seen in Tables 2–4 that SM users have different and random reactions to the posts of the two analyzed companies containing photos and/or video on all studies SM platforms.

Additionally, the intensity of the user preference for photo versus video posts has also been tested. The results show that, besides Facebook, Instagram and Pinterest (only for @CocaColaCo), users do not react significantly differently in terms of their preference for one of the two types of posts that have been studied.

Finally, a more refined distribution of sentiments valences (positive, negative and neutral) of the users due to the type of post (photos or videos) should be made. Therefore, a Mosaic Plot analysis was performed and the results of the sentiments valence distribution are described in Section 4.3.

4.2. Sentiment Distribution by Histogram Charts

This section contains a sentiment distribution analysis shown via histogram charts. Consequently, the use of the function get_nrc_sentiment returns another data frame where, for each analyzed term from the original (spreadsheet) file containing the posts, a new column for each type of emotion (eight in total) and the sentiment polarity (positive and negative) is created.

For each social network page of the two companies, a histogram was developed displaying the sentiments distribution for the analyzed posts by using ggplot2 package. Due to space constraints, this paper presents the results in two cumulative histogram charts, instead of individual ones. The histogram is shown in Figure 5. For Facebook, @CocaColaCo has predominantly positive posts with a score of 2420 points. The highest scores for expressed emotions are: Trust with 1020 points; anticipation with 830 points and joy with 810 points. The lowest score was recorded by the emotions of disgust (70 points), which is a positive fact for the company. For @PepsiCo, a similar trend was noticed compared to @CocaColaCo, as also the positive (720 points) sentiments predominated and the negative ones had only 140 points. The emotions of anticipation achieved the highest score of 320, followed by the emotions of trust (290 points) and joy (280 points).

On Twitter, the situation of the two analyzed companies is slightly different, but only in terms of the value of points. The trend line is similar, namely, @CocaColaCo channel has mostly positive sentiments, with a total of 240 points, while the negative ones score a total of 100 points. In terms of emotions, the emotions of anticipation are the highest (200 points), then trust (140 points), joy (130 points) and surprise (100 points). In the case of @PepsiCo, positive sentiments are also ranked first (60 points), while negative ones have a score of 30 points. The emotions of joy are first ranked, with 50 points, followed by trust (40 points) and anticipation (40 points), surprise (30 points), anger (30 points), fear (30 points), sadness (10 points) and disgust (10 points).

As for the sentiment distribution analysis developed on Instagram posts, it can be said that for the @CocaColaCo channel, the postings predominantly show positive sentiments as they score 1090 points, while the negative ones score only 210 points. The emotions of joy have 400 points, trustworthiness 370 points, and anticipation 350 points. In the case of @PepsiCo, there is an increased score for positive sentiments (1240 points), which is supported by the emotions of joy (510 points), anticipation (470 points), trust (450 points) and surprise (250 points). Negative sentiments scored 280 points, and the emotions of fear 140 points, sadness 150, anger 90 points and disgust 20 points.

The histograms display for the channels of the two companies on Pinterest platform that in the case of @CocaColaCo, positive sentiments are ranked the first (with 230 points). In terms of expressed emotions, both emotions of joy (150 points) and anticipation (120 points) predominated, while trust was given 90 points. In the case of @PepsiCo, it was found that positive sentiments are prevalent (170 points), while negative ones are at a distance of 100 points. The emotions of joy received 100 points, anticipation 80 points, trust 60 points, and there is equality in points for anger, fear and surprise (40 points each).

Regarding Youtube, the results show that both @CocaColaCo and @PepsiCo pages are similar, in the sense that there is a uniform distribution of points in the range 10–50 points, and a similarity of point distributions for both categories (sentiments and emotions). Positive sentiments predominate (50 points) over the emotions of anticipation (40 points for @CocaColaCo and 30 for @PepsiCo), joy (30 points for @CocaColaCo and 20 for @PepsiCo), trust (30 points @CocaColaCo and 40 for @PepsiCo). In Figure 5, the Youtube histogram for @PepsiCo showed that zero words were identified for the emotions of disgust and fear.

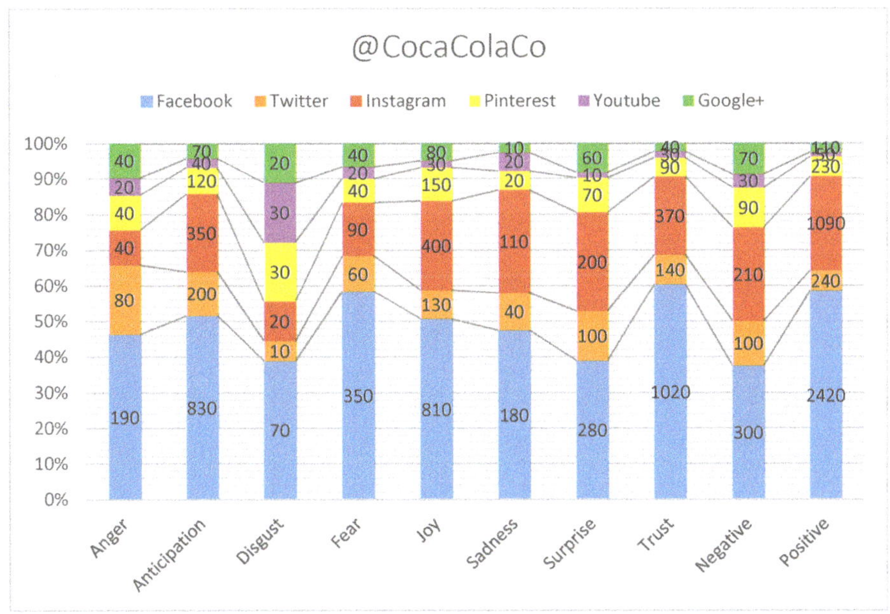

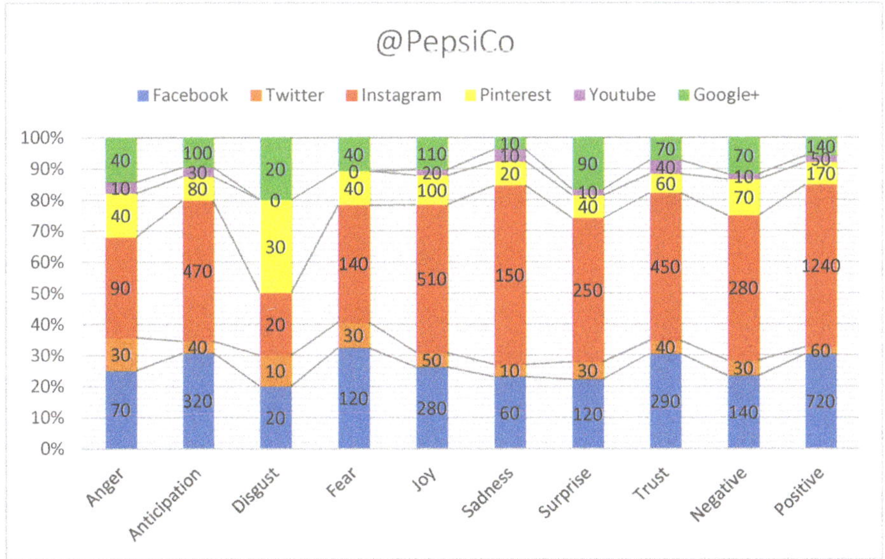

Figure 5. Cumulative histogram charts displaying the sentiment distribution on each of the six social network pages for Coca-Cola and PepsiCo official channels.

For Google+, the results in Figure 5 were obtained after the analysis of the activity for the two companies on global channels. In the case of @CocaColaCo, there is a distribution oriented towards positive sentiments, with 110 points, while the emotions of joy have 80 points, anticipation 70 points and surprise 60 points. Negative sentiments recorded 70 points, while the emotions of sadness had 10 points, anger and fear 40 points, and disgust 20 points. @PepsiCo first shows positive sentiments (140 points), while negative sentiments have half the score (70 points). The emotions of joy were given 110 points, while anticipation (100 points), surprise (90 points), and at equality, were found the emotions of anger and fear (40 points).

Overall, based on the sentiment histograms analysis, a specific pattern of positive and negative sentiments expressed by customers for the two analyzed companies was observed. The results are shown below in Table 5.

Table 5. Shares of sentiment analysis on social media (SM) platforms.

SM platform / Sentiments	@CocaColaCo Positive	@PepsiCo Positive	@CocaColaCo Negative	@PepsiCo Negative
Facebook	89%	84%	11%	16%
Twitter	71%	67%	29%	33%
Instagram	84%	82%	16%	18%
Pinterest	72%	71%	28%	29%
Youtube	63%	83%	38%	17%
Google+	61%	67%	39%	33%

The general conclusion drawn based on the analysis shown in Table 5 is that for most (four out of six) SM platforms, the highest share of positively expressed sentiments appear in favor of @CocaColaCo channels compared to @PepsiCo. Google+ and Youtube channels are the only ones with an advance of 20% and 6%, respectively.

4.3. Polarity Categories Distribution by Type of Posts: The Mosaic Plot

In order to analyze the distribution of the expressed sentiments by the type of postings (photos and videos), mosaic plots have been developed. In the original corpus, this study used the naive Bayes algorithm for assessing the polarity of information conveyed by words. Therefore, the words included into the corpus were classified as positive, negative and neutral. From the package sentiment (from R Studio), classify emotion and classify polarity methods have been applied. The result was stored into a data frame that generated the scores obtained for each post and each emotion, and ultimately for each valence of sentiment.

Figure 6 shows an example of post classifying by its valence, positive, negative or neutral for CocaCola's Facebook official page.

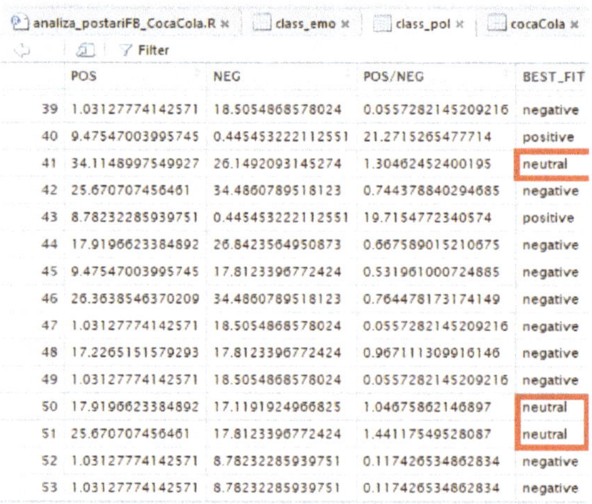

	POS	NEG	POS/NEG	BEST_FIT
39	1.03127774142571	18.5054868578024	0.0557282145209216	negative
40	9.47547003995745	0.445453222112551	21.2715265477714	positive
41	34.1148997549927	26.1492093145274	1.30462452400195	neutral
42	25.670707456461	34.4860789518123	0.744378840294685	negative
43	8.78232285939751	0.445453222112551	19.7154772340574	positive
44	17.9196623384892	26.8423564950873	0.667589015210675	negative
45	9.47547003995745	17.8123396772424	0.531961000724885	negative
46	26.3638546370209	34.4860789518123	0.764478173174149	negative
47	1.03127774142571	18.5054868578024	0.0557282145209216	negative
48	17.2265151579293	17.8123396772424	0.967111309916146	negative
49	1.03127774142571	18.5054868578024	0.0557282145209216	negative
50	17.9196623384892	17.1191924966825	1.04675862146897	neutral
51	25.670707456461	17.8123396772424	1.44117549528087	neutral
52	1.03127774142571	8.78232285939751	0.117426534862834	negative
53	1.03127774142571	8.78232285939751	0.117426534862834	negative

Figure 6. An example of post classification by its valence.

The mosaic plots provide an overview of the data and make it possible to emphasize the relationships between the analyzed variables. For this purpose, in the R language, the packages grid and vcd were used to generate the plots.

For the Facebook platform, the mosaic plots obtained for the posts of the two companies can be interpreted, as follows (in the first line of Figure 7): @CocaColaCo—the share of positive sentiments for the posts with photos is predominant, the share of neutral sentiments is equal for both types of posts (photo and video), and negative sentiments are mostly developed for video posts; @PepsiCo—there is no difference in the distribution of negative and neutral sentiments for both types of posts (photo and video), and for positive sentiments, it is noticeable that a small difference is favorable to video postings.

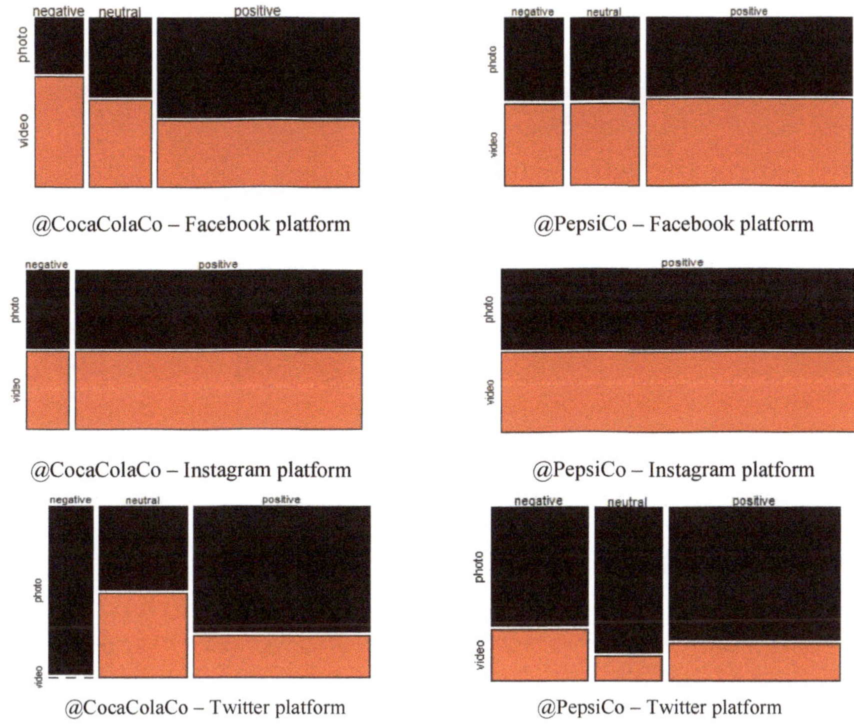

@CocaColaCo – Facebook platform   @PepsiCo – Facebook platform

@CocaColaCo – Instagram platform   @PepsiCo – Instagram platform

@CocaColaCo – Twitter platform   @PepsiCo – Twitter platform

Figure 7. Mosaic plot representations for sentiment analysis (SA) for photo and video posts of Facebook, Instagram and Twitter platforms.

On Instagram accounts, the share of sentiments displayed in the second line of Figure 7: @CocaColaCo—it can be said that there are no neutral sentiments, and positive sentiments are equal to the negative ones for both the photos and the video posts; @PepsiCo—negative and neutral sentiments are missing, and both the types of posts are equal in weight for positive sentiments.

Regarding the share of expressed sentiments on Twitter accounts, the results show the following (the third line Figure 7): @CocaColaCo—the share of positive sentiments for the posts with photos is predominant, the share of words with neutral valence is equal for both types of postings, and the negative sentiments appear only for posts with photos; @PepsiCo—positive sentiments associated with the posts with photos prevail, negative ones are more for photos and the same trend is noted for neutral sentiments (negative valence is higher for photo).

For Pinterest, the mosaic charts express the following (in the first line of Figure 8): @CocaColaCo—there are no neutral sentiments, while the negative and positive sentiments are equal for both photo and video posts; @PepsiCo—neutral sentiments are missing for the video posts, positive sentiments are higher than negative sentiments for the photo posts.

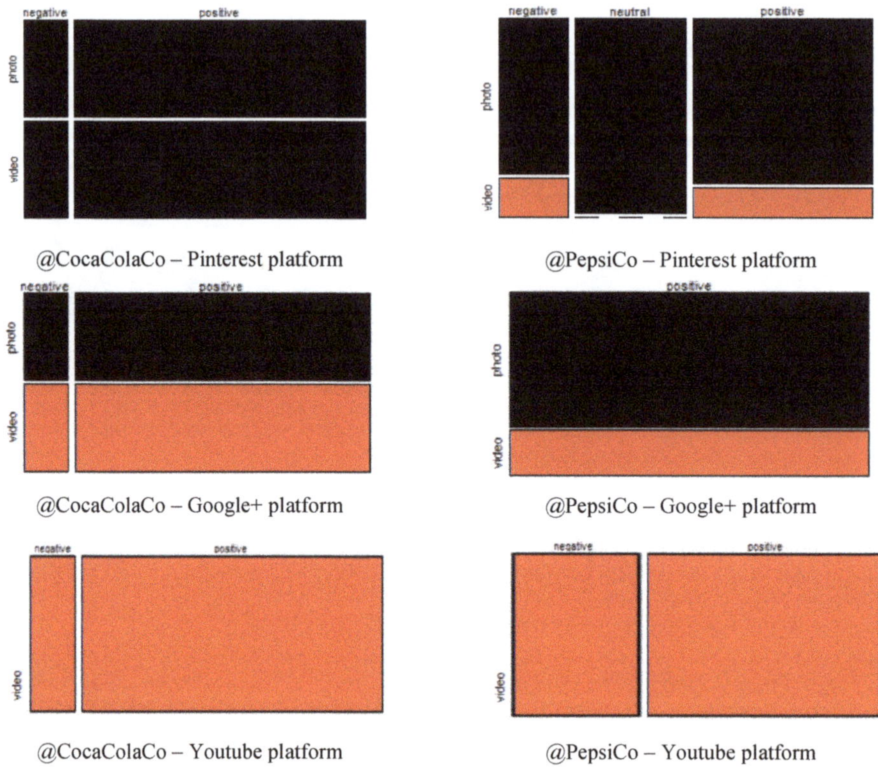

Figure 8. Mosaic plot for SA of photo and video posts on Pinterest, Google+ and Youtube platforms.

As for the share of sentiments by the type of posts on Google+, the results are only global for the two companies and can be interpreted, as follows (in the second line of Figure 8): @CocaColaCo—both types of posts, photos and video, are predominantly associated with positive sentiments; @PepsiCo—there are only positive sentiments, a higher quantity of posts with photos than videos.

As only videos can be loaded on the Youtube channel, the polarity of sentiments could not be distributed by the two types of posts (photos and videos). Therefore, for video posts, there is a slight distinction by the type of posting for the two brands. As the last line shows in Figure 8, @PepsiCo has a higher share of negatively expressed sentiments compared to @CocaColaCo. Further, neutral sentiments are not present in any of the two companies on their Youtube channels.

5. Discussion and Conclusions

Social tools, along with appropriate performance metrics (KPI), can be used to capture and then analyze big data, using ML techniques, generated by SM posts as to ensure sustainable development of a company's relationship with its customers. Metrics, such as the number of likes per post, or the number of fans on the Facebook page, are not enough to confirm a company's success on the social market. Social CRM is a tool linking public sentiments to engagement, subsequent purchase intent and ultimately to product purchasing.

The two analyzed companies (Coca-Cola and PepsiCo) have different approaches in terms of SM activity. Coca-Cola focuses on paid traffic, while PepsiCo uses a more organic one. In terms of used platforms, it was observed that PepsiCo prefers promotion on Facebook, Twitter and Instagram, while its direct competitor prefers posts on Facebook, Pinterest, and other social networks.

This study found that there are significant differences in user emotions and sentiments expressed on different SM networks in the two types of posts (photos or videos). Related to the level of KPI' dependence on the types of posts (photos or videos), as shown by the independence tests (*t*-test and Chi-square), in almost all cases, the secondary hypothesis was confirmed (the variables analyzed in turn, two by two, are relatively dependent on each analyzed social instrument).

This study found that the users tend to express their sentiments differently depending on the type of post. Some are reluctant to use videos, while others embrace photos. A possible explanation for this might be the fact that photos express clearer and more concise messages, and users perceive them faster. However, in terms of intensity of the user's preferences for photo versus video posts, the results display a significant difference for the users of Facebook, Instagram and Pinterest (only for @CocaColaCo), and not a significant difference for other SM platforms included in the study.

The study confirms that the activity of the two brands in the online environment has an emotional impact on current or potential customers and always generates new data sources. This data could be analyzed with a CRM platform and could assist companies in correct segmentation of customers. Consistent with the results, a CRM platform is required to integrate all accounts of a company on SM and to automate its interaction with followers or subscribers in order to learn in a structured way what the market thinks (react, feel) about the provided products and services. This way, important information can be targeted at the right people to be analyzed and, eventually to be used in the strategic decision-making.

Finally, emotional reactions on SM of users, in general, and especially of customers can influence purchasing decisions, taking into account that the number of people using mobile devices to exchange opinions in online communities has been increasing almost exponentially.

Marketing research has already pointed out the driving force of customer emotions and sentiments in the buying decisions. As such, the emotional reaction of customers to company posts on various social networks becomes an important input in the decision-making process at a strategic management level.

The methodology (presented in Figure 2) used in this research can be extended to the analysis of customer reactions to posts of any company's official social network channel. As long as SM posts are captured through FanPage Karma or any similar platform, a new corpus can be developed and analyzed based on the proposed methodology. Furthermore, the proposed methodology can be applied for the purpose of identifying other company's potential customer preferences for this type of posts (videos or photos) on the SM official channel. This way, companies can better understand and address the requests of their customers. As such, the massive amount of data that is daily being sent through the official company's SM channels can be captured and analyzed through ML techniques leading to sustainable development of a business.

In the future, the SM platforms are likely to be reinventing themselves and with the advances of new communication gadgets, SM media will probably shift into photo and video communication (ShapChat, WeCHat, Whatsup, etc.). The AI advances are changing our online experiences into good or bad ones. Bots and fake likes may change the real perception of companies towards true reactions of their customers to posts. Laghate [56] warns that if companies rely too much on metrics, they can be easily gamed. Unfortunately, as a study [57] highlights, there are companies that have employed armies of low-paid workers in developing countries to create fake SM profiles and boost a company or product metrics. They can boost any company's followers and reactions to specific posts on SM, or post favorable reviews and boost ratings on third party aggregators and review platforms. Undoubtedly, SM platforms are fighting back with advanced algorithms based on AI and mass purges, with mixed success. As a further study [55] declares, it is a game of cat and mouse because whenever these platforms fight against fake followers, the click farms devise a new way to game the system and the menace grows.

Similar unwanted situations can be prevented by law enforcement and ML algorithms for detecting and isolating bots and fake SM reactions. As bots and other fake accounts have been running rampant on SM platforms, law enforcement has started to pay off. As such, Keith [58] stated in 2019, the New York General Attorney underlined for the first time that a law enforcement agency has found that selling fake social media engagement is illegal. Therefore, as the authors in [59] declare, some industry sources hope this marks a turning point in a long-running battle against bot and sock puppet accounts that do not reflect genuine opinions of real people.

The cases illustrated above are beyond the scope of this study and represent this study's first limitation. Consequently, the second limitation of our study is that the research does not engage with the examination of the negative impact of bots and fake likes (or other similar aspects) on altering the analysis of customers' reactions to different types of posts on a brand SM channel.

This study has brought up many questions that need further investigation. Thus, in the context of a broader research, the analysis could be extended by continuing to investigate the interactions of customers of Coca-Cola and Pepsi-Cola on other social networks compared to those investigated in this study. The explanation could be that the more social networks, client profiles, tracking comments, posts or hash tags are included in the study, the more comprehensive is the generated behavioral analysis.

Another interesting line of a future research could be a predictive analysis. The models for customer predisposition towards choosing and buying a specific product have already started to raise interest for a wide range of market segments.

Author Contributions: Conceptualization, D.F. and E.-M.T.; methodology, V.-D.P. and E.-M.T.; software, E.-M.T.; validation, V.-D.P. and E.-M.T.; writing-original draft preparation, D.F. and V.-D.P.; writing-review and editing, M.D.; visualization, M.D.; supervision, D.F.

Funding: This research received no external funding.

Conflicts of Interest: The authors declare no conflicts of interest.

References

1. Minsky, L.; Quesenberry, K.A. How B2Bsales can benefit from social selling. *Harv. Bus. Rev.* **2016**. [CrossRef]
2. Ladhari, R.; Rioux, M.C.; Souiden, N.; Chiadmi, N.-E. Consumers' motives for visiting a food retailer's Facebook page. *J. Retail. Consum. Serv.* **2018**. [CrossRef]
3. Chang, S.-H.; Chih, W.-H.; Liou, D.-K.; Hwang, L.-R. The influence of web aesthetics on customers' PAD. *Comput. Hum. Behav.* **2014**, *36*, 168–178. [CrossRef]
4. Gangadharbatla, H.; Hopp, T.; Sheehan, K. Changing user motivations for social networking site usage: Implications for internet advertisers. *Int. J. Internet Mark. Advert.* **2012**, *7*, 120. [CrossRef]
5. Gutiérrez-Cillán, J.; Camarero-Izquierdo, C.; San José-Cabezudo, R. How brand post content contributes to user's Facebook brand-page engagement. The experiential route of active participation. *BRQ Bus. Res. Q.* **2017**, *20*, 258–274. [CrossRef]
6. Brodie, R.J.; Ilic, A.; Juric, B.; Hollebeek, L. Consumer engagement in a virtual brand community: An exploratory analysis. *J. Bus. Res.* **2013**, *66*, 105–114. [CrossRef]
7. Poecze, F.; Ebster, C.; Strauss, C. Social media metrics and sentiment analysis to evaluate the effectiveness of social media posts effectiveness of social media posts. *Procedia Comput. Sci.* **2018**, *130*, 660–666. [CrossRef]
8. He, J.; Shao, B. Examining the dynamic effects of social network advertising: A semiotic perspective. *Telemat. Inform.* **2018**, *35*, 504–516. [CrossRef]
9. Anojan, V.; Subaskaran, T. Consumers Preference and Consumers Buying Behavior on Soft Drinks: A Case Study in Northern Province of Sri Lanka. *Glob. J. Manag. Bus. Res.* **2015**. Available online: https://journalofbusiness.org/index.php/GJMBR/article/view/1687 (accessed on 18 March 2019).
10. Puschmann, C.; Powell, A. Turning Words Into Consumer Preferences: How Sentiment Analysis Is Framed in Research and the News Media. *Soc. Media Soc.* **2018**, *4*, 2056305118797724. [CrossRef]
11. Saura, J.R.; Palos-Sanchez, P.; Grilo, A. Detecting Indicators for Startup Business Success: Sentiment Analysis Using Text Data Mining. *Sustainability* **2019**, *11*, 917. [CrossRef]

12. Kim, E.-G.; Chun, S.-H. Analyzing Online Car Reviews Using Text Mining. *Sustainability* **2019**, *11*, 1611. [CrossRef]
13. Morente-Molinera, J.; Kou, G.; Pang, C.; Cabrerizo, F.; Herrera-Viedma, E. An automatic procedure to create fuzzy ontologies from users' opinions using sentiment analysis procedures and multi-granular fuzzy linguistic modelling methods. *Inf. Sci.* **2019**, *476*, 222–238. [CrossRef]
14. Goswami, S.; Nandi, S.; Chatterjee, S. Sentiment Analysis Based Potential Customer Base Identification in Social Media. In *Contemporary Advances in Innovative and Applicable Information Technology*; Advances in Intelligent Systems and Computing; Mandal, J., Sinha, D., Bandopadhyay, J., Eds.; Springer: Singapore, 2019; Volume 812. [CrossRef]
15. Kumar, S.; Yadava, M.; Roy, P.P. Fusion of EEG response and sentiment analysis of products review to predict customer satisfaction. *Inf. Fusion* **2019**, *52*, 41–52. [CrossRef]
16. Sun, Q.; Niu, J.; Yao, Z.; Yan, H. Exploring eWOM in online customer reviews: Sentiment analysis at a fine-grained level. *Eng. Appl. Artif. Intell.* **2019**, *81*, 68–78. [CrossRef]
17. Joseph, G.; Varghese, V. Analyzing Airbnb Customer Experience Feedback Using Text Mining. In *Big Data and Innovation in Tourism, Travel, and Hospitality*; Sigala, M., Rahimi, R., Thelwall, M., Eds.; Springer: Singapore, 2019.
18. Ibrahim, N.F.; Wang, X. Decoding the sentiment dynamics of online retailing customers: Time series analysis of social media. *Comput. Hum. Behav.* **2019**, *96*, 32–45. [CrossRef]
19. Wang, C.H.; Fan, K.C.; Wang, C.J.; Tsai, M.F. UGSD: User Generated Sentiment Dictionaries from Online Customer Reviews. 2019. Available online: https://www.aaai.org/Papers/AAAI/2019/AAAI-WangC.3191.pdf (accessed on 14 April 2019).
20. Madan, D.; Jobanputra, M.; Shah, H.; Rathod, S. COMM-AN Opinion Mining of Customer Feedback. In Proceedings of the 2nd International Conference on Advances in Science & Technology (ICAST-2019), Maharashtra, India, 9 April 2019. Available online: https://ssrn.com/abstract=3368898 (accessed on 10 May 2019).
21. Gunasekar, S.; Sudhakar, S. Does hotel attributes impact customer satisfaction: A sentiment analysis of online reviews. *J. Glob. Sch. Mark. Sci.* **2019**, *29*, 180–195. [CrossRef]
22. McColl-Kennedy, J.R.; Zaki, M.; Lemon, K.N.; Urmetzer, F.; Neely, A. Gaining customer experience insights that matter. *J. Serv. Res.* **2019**, *22*, 8–26. [CrossRef]
23. He, W.; Zhang, W.; Tian, X.; Tao, R.; Akula, V. Identifying customer knowledge on social media through data analytics. *J. Enterp. Inf. Manag.* **2019**, *32*, 152–169. [CrossRef]
24. Yang, B.; Liu, Y.; Liang, Y.; Tang, M. Exploiting user experience from online customer reviews for product design. *Int. J. Inf. Manag.* **2019**, *46*, 173–186. [CrossRef]
25. Can, U.; Alatas, B. Big Social Network Data and Sustainable Economic Development. *Sustainability* **2017**, *9*, 2027. [CrossRef]
26. Colliander, J.; Dahlén, M. Following the fashionable friend: The power of social media. *J. Advert. Res.* **2011**, *51*, 313–320. [CrossRef]
27. Curras-Perez, R.; Ruiz-Mafe, C.; Sanz-Blas, S. Determinants of user behavior and recommendation in social networks: An integrative approach from the uses and gratifications perspective. *Ind. Manag. Data Syst.* **2014**, *114*, 1477–1498. [CrossRef]
28. Georgescu, M.; Popescul, D. Students in Social Media: Behavior, Expectations and Views. In Proceedings of the International Conference on Informatics in Economy, Cluj-Napoca, Romania, 2–3 June 2016; pp. 84–98. [CrossRef]
29. Kaplan, A.M.; Haenlein, M. Users of the world, unite! The challenges and opportunities of Social Media. *Bus. Horiz.* **2010**, *53*, 59–68. [CrossRef]
30. Kornum, N.; Mühlbacher, H. Multi-stakeholder virtual dialogue: Introduction to the special issue. *J. Bus. Res.* **2013**, *66*, 1460–1464. [CrossRef]
31. Casteleyn, J.; Mottart, A.; Rutten, K. Forum-How to Use Facebook in your Market Research. *Int. J. Mark. Res.* **2009**, *51*, 439–447.
32. Hyllegard, K.H.; Ogle, J.P.; Yan, R.; Reitz, A.R. An exploratory study of college students' fanning behavior on Facebook. *Coll. Stud.* **2011**, *45*, 601–616.
33. Safko, L.; Brake, D. *The Social Media Bible: Tactics, Tools, and Strategies for Business Success*; John Wiley & Sons: Hoboken, NJ, USA, 2009.
34. Felix, R.; Rauschnabel, P.; Hinsch, C. Elements of strategic social media marketing: A holistic framework. *J. Bus. Res.* **2016**. [CrossRef]

35. Acker, O.; Gröne, F.; Akkad, F.; Pötscher, F.; Yazbek, R. Social CRM: How companies can link into the social web of consumers. *J. Direct Data Digit. Mark. Pract.* **2011**, *13*, 3–10. [CrossRef]

36. Kietzmann, J.H.; Hermkens, K.; McCarthy, I.P.; Silvestre, B.S. Social media? Get serious! Understanding the functional building blocks of social media. *Bus. Horiz.* **2011**, *54*, 241–251. [CrossRef]

37. Rodriguez, M.; Peterson, R.M.; Krishnan, V. Social Media's influence on business-to-business sales Performance. *J. Pers. Sell. Sales Manag.* **2012**, *32*, 365–378. [CrossRef]

38. Belch, G.E.; Belch, M.A. *Advertising and Promotion: An Integrated Marketing Communications Perspective*, 11th ed.; McGraw-Hill Education: New York, NY, USA, 2017; pp. 567–589.

39. Cui, G.; Lui, H.-K.; Guo, X. The Effect of Online Consumer Reviews on New Product Sales. *Int. J. Electron. Commer.* **2012**, *17*, 39–58. [CrossRef]

40. Genc-Nayebi, N.; Abran, A. A systematic literature review: Opinion mining studies from mobile app store user reviews. *J. Syst. Softw.* **2017**, *125*, 207–219. [CrossRef]

41. Mohammad, S.M.; Turney, P.D. Emotions evoked by common words and phrases: Using Mechanical Turk to create an emotion lexicon. In Proceedings of the NAACL HLT 2010 Workshop on Computational Approaches to Analysis and Generation of Emotion in Text, Los Angeles, CA, USA, 5 June 2010; pp. 26–34.

42. Trainor, K.J. Relating Social Media Technologies to Performance: A Capabilities-Based Perspective. *J. Pers. Sell. Sales Manag.* **2012**, *32*, 317–331. [CrossRef]

43. Carp, M.; Pǎvǎloaia, L.; Afrǎsinei, M.-B.; Georgescu, I.E. Is Sustainability Reporting a Business Strategy for Firm's Growth? Empirical Study on the Romanian Capital Market. *Sustainability* **2019**, *11*, 658. [CrossRef]

44. Pozzi, F.A.; Fersini, E.; Messina, E.; Liu, B. Beyond Sentiment: How Social Network Analytics Can Enhance Opinion Mining and Sentiment Analysis. In *Sentiment Analysis in Social Networks*, 1st ed.; Morgan Kaufmann Publishers Inc.: Los Angeles, CA, USA, 2016.

45. Coskun, M.; Ozturan, M. #europehappinessmap: A Framework for Multi-Lingual Sentiment Analysis via Social Media Big Data (A Twitter Case Study). *Information* **2018**, *9*, 102.

46. Carrera, B.; Jung, J.-Y. SentiFlow: An Information Diffusion Process Discovery Based on Topic and Sentiment from Online Social Networks. *Sustainability* **2018**, *10*, 2731. [CrossRef]

47. Gamal, D.; Alfonse, M.; M El-Horbaty, E.S.; M Salem, A.B. Analysis of Machine Learning Algorithms for Opinion Mining in Different Domains. *Mach. Learn. Knowl. Extr.* **2019**, *1*, 224–234. [CrossRef]

48. Sobhani, P.; Inkpen, D.; Zhu, X. Exploring deep neural networks for multitarget stance detection. *Comput. Intell.* **2019**, *35*, 82–97. [CrossRef]

49. Ju, S.; Li, S.; Su, Y.; Zhou, G.; Hong, Y.; Li, X. Dual word and document seed selection for semi-supervised sentiment classification. In Proceedings of the 21st ACM International Conference on Information and Knowledge Management, Maui, HI, USA, 29 October–2 November 2012; pp. 2295–2298. [CrossRef]

50. Zhu, Z.; Dai, D.; Ding, Y.; Qian, J.; Li, S. Employing emotion keywords to improve cross-domain sentiment classification. In *Workshop on Chinese Lexical Semantics*; Springer: Berlin/Heidelberg, Germany, 2012; pp. 64–71.

51. Medhat, W.; Hassan, A.; Korashy, H. Sentiment analysis algorithms and applications: A survey. *Ain Shams Eng. J.* **2014**, *5*, 1093–1113. [CrossRef]

52. Lu, Y.; Castellanos, M.; Dayal, U.; Zhai, C. Automatic construction of a context-aware sentiment lexicon: An optimization approach. In Proceedings of the 20th International Conference on World Wide Web, Hyderabad, India, 28 March–1 April 2011; pp. 347–356. [CrossRef]

53. Fotache, M.; Strimbei, C. SQL and Data Analysis. Some Implications for Data Analysits and Higher Education. *Procedia Econ. Financ.* **2015**, *20*, 243–251. [CrossRef]

54. Statista. Most Popular Social Networks Worldwide as of July 2019, Ranked by Number of Active Users. Available online: https://www.statista.com/statistics/272014/global-social-networks-ranked-by-number-of-users/ (accessed on 22 July 2019).

55. Fanpage Karma. The Allround—Tool for Strong Social-Media Management. Available online: http://www.fanpagekarma.com/ (accessed on 18 June 2018).

56. Laghate, G. Shadow of bot followers and fake likes mars social media influencers. *The Economic Times*, 21 June 2018. Available online: https://economictimes.indiatimes.com/articleshow/64674668.cms?utm_source=contentofinterest&utm_medium=text&utm_campaign=cppst (accessed on 17 August 2019).

57. Edwards, J. A Flaw in Facebook Lets Anyone Create as Many Fake 'Likes' as They Want without Using a Bot Army. *Business Insider*, 25 March 2015. Available online: https://www.businessinsider.com/how-fake-facebook-likes-are-created-2015-3 (accessed on 11 June 2019).

58. Keith, K. AG Letitia James: Selling Fake Social Media Engagement is Illegal. *New York Post*, 31 January 2019. Available online: https://nypost.com/2019/01/31/ag-letitia-james-selling-fake-social-media-engagement-is-illegal/ (accessed on 22 July 2019).
59. Stempel, J. New York Settles with Sellers of 'Fake' Online Followers, 'Likes'. *Reuters*, 31 January 2019. Available online: https://www.reuters.com/article/us-new-york-socialmedia-settlement/new-york-settles-with-sellers-of-fake-online-followers-likes-idUSKCN1PP01O (accessed on 21 July 2019).

 *sustainability*

Article

Towards Sustainable Urban Communities: A Composite Spatial Accessibility Assessment for Residential Suitability Based on Network Big Data

Yu Zhao [1,2], Guoqin Zhang [1], Tao Lin [1,*], Xiaofang Liu [1,2,3], Jiakun Liu [1,2], Meixia Lin [1,2], Hong Ye [1,*] and Lingjie Kong [3]

[1]   Key Laboratory of Urban Environment and Health, the Institute of the Urban Environment, Chinese Academy of Sciences, Xiamen 361021, China; yuzhao@iue.ac.cn (Y.Z.); gqzhang@iue.ac.cn (G.Z.); xfliu@iue.ac.cn (X.L.); jkliu@iue.ac.cn (J.L.); mxlin@iue.ac.cn (M.L.)

[2]   University of Chinese Academy of Sciences, Beijing 100049, China

[3]   School of Architecture, Huaqiao University, Xiamen 361021, China; jaykongud@gmail.com

*   Correspondence: tlin@iue.ac.cn (T.L.); hye@iue.ac.cn (H.Y.); Tel.: +86-592-619-0651 (T.L.)

Received: 22 October 2018; Accepted: 10 December 2018; Published: 13 December 2018

Abstract: Suitable allocation of residential public services is vital to realizing sustainable communities and cities. By combining network big data and spatial analysis, we developed a composite spatial accessibility assessment method for residential suitability of urban public services covering healthcare, leisure, commerce, transportation, and education services. Xiamen City, China is the test site. We found that although most facilities were concentrated on Xiamen Island, there were shortages in the per capita transportation and education service supplements compared with the average performance of Xiamen City because of the high local population. Meanwhile, Tong'an had advantages in the amount of public facilities due to its long history of regional development. However, high-quality facilities were deficient there as well as in other off-island districts. The residential communities surrounding transportation, commerce, and healthcare facilities had a similar allocation pattern in Xiamen City, whereas the residential accessibility of education and leisure services showed regional differences. Due to unbalanced regional development, evident inequality could be witnessed by comparing the composite assessment results of residential suitability between the communities on Xiamen Island and those in the off-island Areas. Our study hopes to provide dedicated support for designing sustainable communities and cities, especially for those in developing countries.

Keywords: big data research; point of interests (POI); sustainability development; spatial accessibility of residential public services; Xiamen City

---

1. Introduction

Dramatic urbanization has resulted in the rapid centralization of the population and uneven distribution of resources that sustain human well-being in urban areas, especially in developing countries [1]. Nowadays, along with a growing awareness of "spatial justice" ("spatial justice" refers to fair treatment and justice being afforded to residents in terms of spatial production and spatial allocation of resources) [2], the suitability of urban human settlements has also attracted considerable attentions in contemporary studies [3]. In urban areas, human settlements are chiefly influenced by the amount and the spatial layout of the surrounding public facilities, which are urban infrastructures distributed in a dotted pattern which provide services to the public [4]. To what extent the local public services could satisfy the residents' demand inevitably influences their sense of belonging and identification to the city [5]. Moreover, eliminating disparities in residents' accessibility to those public products is also central to the Sustainable Development Goals (SDGs) set out by the United

Nations in 2015 [6]. Therefore, sufficiently, ideally, and equitably accessible residential public facilities (services) are critical indicators not only for the residential suitability, but also for the sustainability of communities and cities.

The essential meaning of "residential suitability" is the suitability of an area for residential livability and development. As for the perspective of "Livability", the World Health Organization (WHO) firstly proposed the concept of "Living environment" by summarizing the basic conditions for meeting the living demand of human beings in 1961, which holds "convenience and amenity" as the fundamental guidelines of the residential livability evaluation [7]. In the 1990s, "new-urbanism" emerged. The principle of creating walkable neighborhoods containing a wide range of housing and job types became one of the primary criteria for urban planning and judging the residential livability in the 21st century [8–11]. Since the strategy of "sustainable development" was first proposed in "Our Common Future" by United Nations in 1987 [12], the meaning and necessity of the new development pattern have generally been recognized globally [13]. With a growing number of people living in cities, there has been an ongoing debate about what kind of urban area is suitable for residential development, and particularly, for their sustainable development [14–16]. In 2015, the United Nations adopted a far-reaching and people-centered set of universal and transformative goals and targets for sustainable development by 2030. One particular goal, to "Make cities and human settlements inclusive, safe, resilient and sustainable," also set "to ensure access for all to adequate, safe and affordable housing and basic services "as one of the critical targets for future residential sustainable development [17]. Therefore, the residential suitability of urban human settlement is chiefly influenced by the surrounding public services (facilities) [4]. Numerous studies have been dedicated to urban residential suitability assessments from the view of the service efficiency and the spatial layout of public facilities (services), which also has a profound impact on urban study and planning in China [18,19]. However, most of the studies focused primarily on the suitability of a single type of public facilities, such as education, healthcare, green spaces, etc. [19–29] Composite spatial accessibility assessments for residential suitability have still seldom been published. The reason is that the timeliness and advancement of assessment methods in recent studies, mainly based on traditional social and field surveys, have failed to meet the requirements of contemporary urban studies, which call for full coverage and accurate spatial data for the analysis of larger spatial units and a low updating cost in future studies [30–33].

The technological advancement of big data mining and Geographical Information Systems provide new supports for the data and methods of urban residential suitability studies. Electronic maps of the points of interest (POIs), as a series of point-like data describing the geographical location of urban facilities, have become a research hotspot, and have particularly advanced the study of residents' space-time behavior, urban planning, and public evaluation [34–38]. Meanwhile, increasingly accurate points of interest (POIs) data also provide the possibility for a more comprehensive evaluation of urban residential public services [39–41]. What is more, other plentiful sourced big data also make it possible for realistic assessments to be undertaken on the various types of residential public services. For example, the country's education authorities require that public schools enroll pupils from designated areas in China. A detailed "school district" boundary could bring great benefits to residential education assessment. Moreover, the surrounding natural landscape data also provide the possibility for a comprehensive evaluation of residential leisure services.

Combining network big data and spatial analysis techniques, we developed a composite spatial accessibility assessment method for residential suitability, and demonstrated it in Xiamen City, a rapidly urbanizing city in China, from the views of regional "service supplement" and "residential accessibility" of the communities surrounding public facilities. The assessment involved 1756 urban residential communities and covered five essential types of public services (facilities), i.e., healthcare, commerce, leisure, education, and transportation. We hope this new method could improve the general understanding of urban residential suitability and provide support for sustainable community and city planning.

2. Materials and Methods

2.1. Study Area

Xiamen City is a coastal city located on the southeast coast of China (24°23′ N–24°54′ N, 117°53′ E–118°26′ E). As one of the five fastest-growing special economic zones in China, it serves not only as one of the most prominent international trading ports, but also as the chief economic mainstay of Fujian Province [42]. Benefiting from a renowned reputation due to the coastal water view landscape and fast economic development, Xiamen City has been listed as one of the top 10 most livable cities in China [43]. As for 2016, the total population was 3.92 million, with an urbanization rate of 89%. The completed infrastructure investment was 79.235 billion yuan, which accounted for 36.69% of the total fixed asset investment in Xiamen City [44]. Xiamen City can be divided into two parts: Xiamen Island, which includes Siming and Huli districts, and the off-island area, which consists of 4 districts: Haicang, Jimei, Tong'an, and Xiang'an, from west to east. The Siming and Huli districts are the downtown area. Tong'an is the oldest inhabited area with a long-term regional development history (Table 1). The Haicang, Jimei, and Xiang'an districts are new urban areas for technological and economic development [45]. Until 2016, 1756 residential communities were distributed in Xiamen City. According to the relevant regulation in the real estate industry, 425 of them were defined as first level water view settlements, with a distance between water and community of less than 300 m, 553 of them were secondary water view settlements, with a gap between water and the settlement of about 300–800 m, and the others were ordinary settlements (Figure 1).

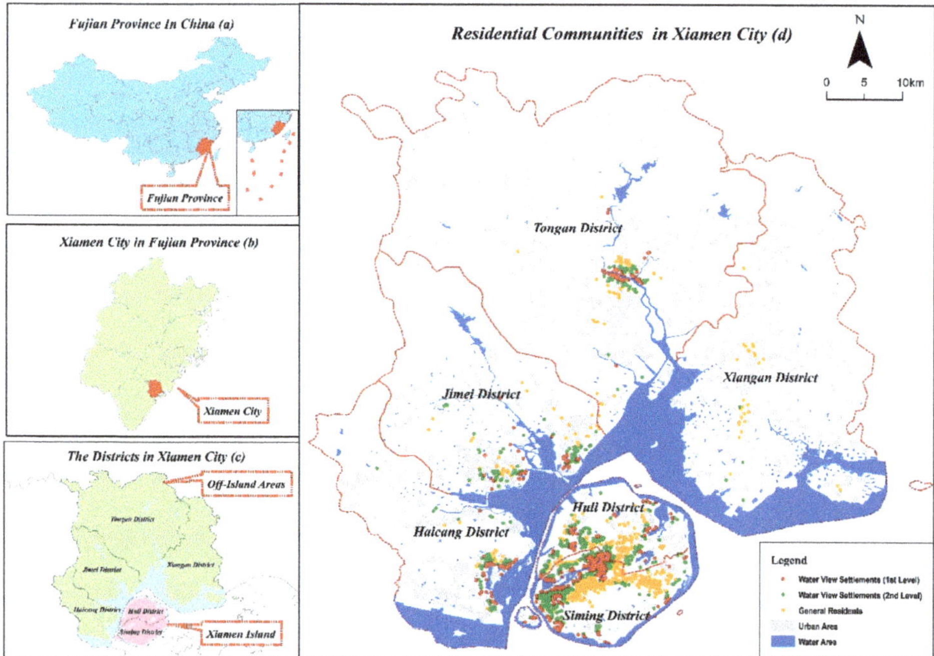

Figure 1. Study area: (**a**) Fujian Province in China, and (**b**) Xiamen City in Fujian Province. (**c**) The six districts of Xiamen City. We separated the regions in Xiamen Island with those in off-island areas using distinct colors and (**d**) the distribution of all the residential communities in Xiamen City.

Table 1. A brief history of administration in Xiamen City.

Year	Downtown	Suburban Area
1911	Siming	—
1950	Siming	Jimei
1957	Siming	Jimei, Tong'an
1987	Siming, Huli	Jimei, Tong'an
2003 until present	Siming, Huli	Jimei, Tong'an, Haicang, Xiang'an

Note: The Siming district was divided from the Tong'an district in 1911. The table neglected some districts that no longer exist, such as the Xinglin and Kaiyuan districts.

2.2. Materials

The spatial datasets of residential communities and public facilities (including roads, hospitals, primary schools, etc.) were derived from the point of interests (POIs) of the Baidu and Gaode Map in 2016. To assess the residential education service, we also mapped the Xiamen education resource division and the school districts according to division regulations from the Xiamen Education Bureau. After projection, spatial adjustment, and field verification, 1756 residential communities and 5953 public facilities points were introduced into the study.

Our study used the point-like data to reflect the residential communities in Xiamen. The centroid points in every residential boundary were chosen as the foothold of our assessment with the assumption that the population is concentrated in these areas. The public facilities were classified into five essential types according to the "Code of urban residential areas planning and design (2016)" (GB50180-93) [46]: commerce, transportation, leisure, healthcare, and education services, and covering 17 kinds of public facilities. Then, we graded them according to local planning and facilities authorities. The scores of each facility were also the general reflections of their service capacities. For example, the process of commerce facilities grading and scoring mainly took the urban planning and facility-scale into consideration. As for the transportation efficiency, we considered that the performance of subway was better than that of the Bus Rapid Transportation system (BRT) in Xiamen City. And the bus stop scores were based on the numbers of routes stopping at the site. The facilities' service boundary was the evaluation standard for grading and scoring the leisure facilities. Moreover, the distance between water and residential communities was the other criterion for the scoring of the settlement's surrounding leisure capacity. Regarding the healthcare facilities, we firstly graded them depending on the hospital level, then separately scored them according to the hospital beds and health care quality. Finally, in terms of the education facilities, as we mainly focused on the primary schools in Xiamen City, we first classified them into several groups based on the level classification of primary school from Xiamen Education Bureau. Then, we marked the service capacity relying on the school's scale and reputation. Details are shown in Table 2. Moreover, to more clearly define the travel behavior of the Xiamen citizens, we also applied the Delphi Methodology in our study. The panel of experts consisted of 18 people drawn from both inside and outside of the relevant field of urban planning, including three academic professors, three urban planners, three government officials, and nine representative Xiamen citizens. After several rounds of expert scoring with the precise knowledge of the study targets and facilities capacities, we finally received the relevant parameters of circle radiuses and weights of distinct demand buffers regarding the diverse types of public services in Xiamen City (Table 3).

Table 2. Service Capacities of Public Facilities and Assessment Standards.

Public Service	Facility Level	Service Capacity Score
	Key Business Circle (Shopping Mall, Supermarket, Local Market and Store)	100–90
Commerce	Non-Key Business Circle (Shopping Mall)	85–80
	Non-Key Business Circle (Supermarket)	75–65

Table 2. *Cont.*

Public Service	Facility Level	Service Capacity Score
Commerce	Non-Key Business Circle (Local Market)	50
	Non-Key Business Circle (Store)	30
Transportation	Subway Hub	100
	Subway Stop	90
	BRT Hub	80
	BRT Stop	75
	Bus Hub	60–10
Leisure	City Level (Historical Sites, Plaza, Park, and Library)	100
	District Level (Historical Sites, Plaza, Park, and Library)	80
	Subdistrict Level (Historical Sites, Plaza, Park, and Library)	60
	Community Level (Historical Sites, Plaza, Park, and Library)	50
	1st Level Water View Settlement	100
	2nd Level Water View Settlement	80
Healthcare	Tertiary	100–75
	Secondary	65–60
	Primary	35–30
	Clinic	25
Education	Provincial key Primary School	100
	City Key Primary School	80
	Key Primary School	70–65
	Ordinary Primary School (Located on Xiamen Island)	50
	Ordinary Primary School (Located on Off-Island Area)	30

Note: The levels of facilities are graded according to the views from local planning and facilities authorities respectively. The scores of each facility are the general reflections of their service capacity.

2.3. Methodology

2.3.1. Location Quotient (LQ)

Location quotient (LQ) can be used to quantify the concentration of a particular industry, cluster, occupation, or demographic group in a region. Here, we used LQ as an indicator to reflect the relative matching degree of regional public facility capacity with the local population scale, indicating the regional disparity of per capita service supplement in each district of Xiamen City.

$$\text{LQ}_{jk} = \left(n_{jk}/p_k \right) / \left(N_j/P \right) \tag{1}$$

where LQ_{jk} is the location quotient of public service j, n_{jk} is the public service j capacity in region k, N_j represents the overall facility capacity in Xiamen City, p_k is the population in region k, and P is Xiamen's total population. The district and city total capacities (n_{jk} and N_j, respectively) for transportation, leisure, and commerce services were represented by their quantity of the public facilities (POI). The n_{jk} and N_j of the healthcare service were estimated by the number of hospital beds. The number of standard hospital beds in various level hospitals was gathered from "The measures for the administration of the hospital grade" released by the National Health Commission, People's Republic of China. The total education service capacity was represented by the enrollment in primary school in the different districts in Xiamen City. Data were sourced from the "Xiamen Special Economic Zone Yearbook (2017)" [44]. If $LQ_{jk} > 1$, the per capita public service j supplement in region k is relatively sufficient compared to the average performance of the whole city, whereas if $LQ_{jk} < 1$, the region k's per capita public service j is in short supply compared with the average level of per capita service supplement in Xiamen City.

Table 3. Multiple demand buffers for transportation, commerce, leisure, and healthcare facilities.

Public Service		Buffer Distance (km)	Weight (λ)
Transportation		0.0–0.5	1.0
		0.5–1.0	0.7
Commerce		0.0–1.0	1.0
		1.0–5.0	0.5
Leisure		0.0–1.0	1.0
		1.0–5.0	0.5
Healthcare		0.0–1.0	1.0
		1.0–3.0	0.5
Traffic Type	**Bus**	**BRT (Rapid Bus Transportation)**	**Subway**
Weight (β)	0.5	0.7	1

2.3.2. Composite Spatial Accessibility Assessment for the Residential Suitability of Public Services

Accessibility assessment was first proposed by Steward and Warntz in 1958 [47]. The "accessibility", defined as the ease to reach the destinations from a given location, has been widely accepted as a suitable method to evaluate the residential suitability of the surrounding public services objectively. And it is initially expressed as a function of spatial isolation and facility attractiveness [48–50].

In practice, how to calculate accessibility is widely debated, and precise definitions of this metric can be arbitrary [6]. Ratio model, proximate-distance model, cumulative-opportunity model, and gravity model (models) are very intuitive and commonly used measures for assessing spatial accessibility [51]. However, when we made a choice among the above models for residential suitability assessment, several challenges had to be taken into consideration. Firstly, spatial isolation cannot merely be defined as the "point (settlements)-to-point (facilities)" distance, known as the Euclidean or Manhattan distance, because of the randomness in people's travel modes [52]. Secondly, the residents' demand for similar service facilities varies at different moments. The optimal public facility, which has the best service capacity, is not always the most sensible choice. Within the range of affordable travel distance, whether residents could make reasonable decisions among similar facilities or not, in accordance with their needs, is a factor that could better reflect the configuration completeness of the residential communities surrounding public facilities. Thirdly, the attractiveness of the facilities to residents has also decayed with increasing spatial isolation [53,54]. Finally, the administrative orders and communities surrounding landscape are also crucial factors affecting service supply mode and resident satisfaction of public services. All in all, distinct methods based on the concept of "accessibility" were necessary when we conducted the residential suitability assessments of different types of public services.

Therefore, in terms of residential transportation, commerce, healthcare, and leisure service, we firstly delineated multiple demand circles around each residential community point. According to the Delphi Methodology, circle radiuses were defined at 500 and 1000 m as the basic demand buffers, meaning that the citizen could access the facilities on foot. Moreover, concerning the residential travel options on public transportation, 3–5 km could also be separately designated as the other demand buffers according to people's practical demands for healthcare, commerce, and leisure services.

Different weights on the multiple demand circles around each residential community point have also been introduced by concerning space isolation effects on facility attractiveness. In our study, facilities in the same demand buffer have equal service attractiveness to the residents. However, the attractiveness of the facilities belonging to different buffers weakens along with increasing spatial isolation. Therefore, we chose the "Binary Discrete" model as our distance-decay function [54–56]. The original weight setting in this model is always by Gaussian function, with the prerequisite of the normal spatial distribution of the facilities, which is hard to satisfy in practice [57]. Therefore, the weights of different buffers in our study have been set depending on expert scoring and suggestions from the facility authorities. Detailed information about the demand buffers could be seen in Table 3. Our study conducted each community residential accessibility assessment of either type of public service by weighted summary of all the demand buffer accessibility scores (Table 3). As for the composition of the accessibility score, in each demand buffer, the facility with the highest service capacity score was chosen as the optimal facility. Its service capacity score was the basic score of the demand buffer. The sum of the other similar facilities capacity scores in the same demand buffer was set as the additional score (Table 2). We also further applied 25-standardization to the original additional score while considering the different significance of the optimal facility and other similar facilities on the satisfaction of residents' demand in each buffer. Finally, each community's accessibility scores were separately standardized on a scale of 100 for easy comparison based on the residential accessibility scores of all the communities in Xiamen City, China (Figure 2).

Considering the impact of surrounding water view on the residential accessibility of leisure service, we conducted the assessment with the multi-ring weighted method and also took the proximity distance from the residential areas to the water as another essential indicator into consideration (Table 2). The three different transportation types, bus, subway, and Bus Rapid Transportation (BRT) in Xiamen City, were used for assessing the residential accessibility of the transportation service. Each was assigned distinct weights according to the residents' preference survey of travel model sourced from the expert scoring and relevant views from the transportation authority in Xiamen City (Table 3).

The primary schools in Xiamen City provide their services based on school district boundaries. To reflect how this feature influence the residential accessibility of education services, we graded the settlement education score according to its school district with the ArcGIS 10.5 spatial overlay analysis (Figure 2).

Finally, we conducted the composite spatial accessibility assessment for residential suitability by summarizing each community's normalized accessibility score of all the essential services. Each essential public service has been assigned with the same weights. For easy comparison, every community's composite accessibility score also experiences 100-standardized with all the residential neighborhoods in Xiamen City. Related equations are as follows:

$$Score_{ij(facilities)} = STD_{100}(\sum_{k=1}^{nj} \lambda_{kj} \times STD_{100}\left(Max.Score_{ij*k} + STD_{25}\left(ADD.Score_{ij*k}\right)\right) \tag{2}$$

$$Score_{i\ (Healthcare)} = Score_{i(Healthcare.facilities)} \tag{3}$$

$$Score_{i\ (Commerce)} = Score_{i\ (Commerce.facilities)} \tag{4}$$

$$Score_{i\ (transportation)} = STD_{100}(\sum_{1}^{type} \beta_{type} \times Score_{i*(transportation.facilities.type)}) \tag{5}$$

$$Score_{i\ (education)} = STD_{100}((Score_{i.school\ district}) \tag{6}$$

$$Score_{i\ (Leisure)} = STD_{100}(\left(Score_{i.Leisure.facilities} + Score_{i.water}\right) \tag{7}$$

$$Score_{i\ (composite)} = STD_{100}\left(\sum_{j}^{5} Score_{ij}\right) \tag{8}$$

$$STD_a = a * \frac{(X - min(X))}{max(X) - min(X)} \qquad (9)$$

where $Score_{ij(facilities)}$ is the accessibility score of residential communities i surrounding public service j depending on the spatial layout of related public facilities. k, λ_{kj} represents each demand buffers and their related weights for public service j separately. In Equation (2), $Max.Score_{ij*k}$ is the optimal facility capacities score in buffer k, and $ADD.Score_{ij*k}$ is the additional score of j kind public facilities in buffer k, which is the total score of the other similar facilities capacities. In Equation (5), the $Score_{i\ (transportation.facilities)}$ is calculated according to the three esential types of transportation system, covering the bus, Bus Rapid Transportation (BRT), and subway respectively. *type* represents the three essential types of transportation system, β_{type} is each transportation types related weights. The $Score_{j.school\ district}$ in Equation (6) is the residential communities i belonging school district service capacity score, and $Score_{i.water}$ in Equation (7) represents the water view service capacity score of residential settlement i. As shown in Equation (8), $Score_{i.composite}$ is the composite spatial accessibility assessment of settlement j with all the essential kinds of public facilities (services), which is the sum of each kind public service accessibility score of residential communities i ($Score_{ij}$). STD_{25} and STD_{100} are the 25 or 100 standardizations on the original scores, respectively. In Equation (9), a = 25 or 100, $min(X)$ and $max(X)$ represent the minimum and maximum original assessment score of residential public service i in all the communities of Xiamen City, China, respectively. In the end, our study also classified the residential accessibility score of public services, from high to low, into three groups by grading the scores with the half standard deviation, labeled as "High", "Medium", and "Low" respectively.

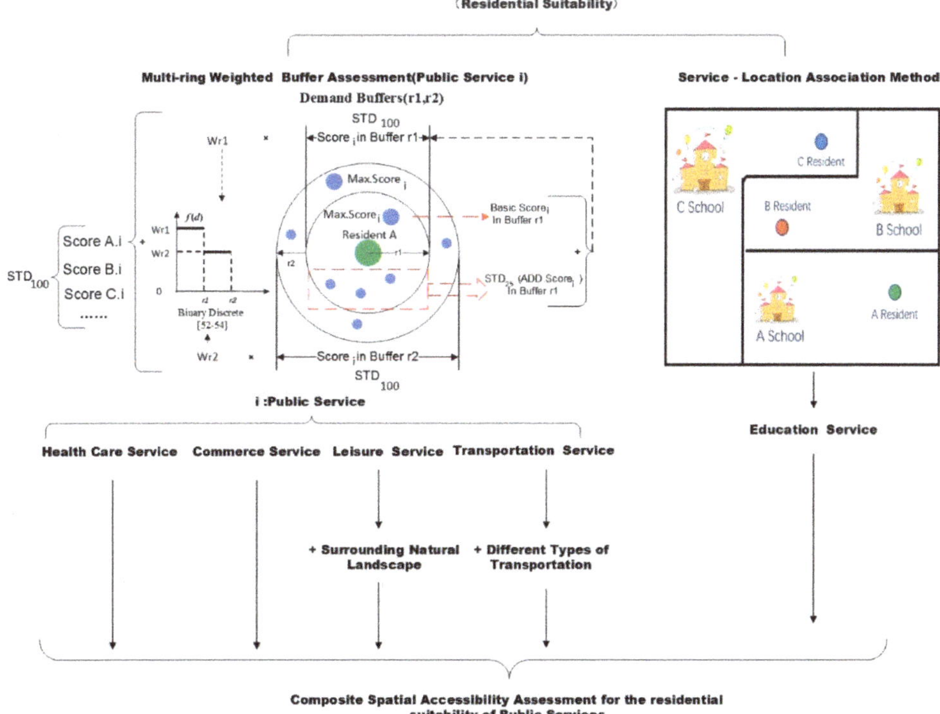

Figure 2. Composite spatial accessibility assessment for the residential suitability of public services.

2.3.3. The Coefficient of Variation (*CV*) and Correlation Analysis

Here, we applied the mean value and the related coefficient of variation (*CV*) to describe the regional facility spatial layout performance on the residential suitability of each type of public service. To identify the allocation pattern of Xiamen urban planning on the residential communities surrounding public facilities, we also conducted a correlation analysis to figure out the mutual relationships among the five essential types of public services.

The *CV* is defined as the ratio of the standard deviation to the mean, and it is widely used to show the extent of variability in relation to the mean value of the population. As a dimensionless indicator, the *CV* has a significant advantage for comparing population variation between data sets with different units or means [58]. Here, we used *CV* as an indicator to reflect the regional disparity in the residential accessibility of public services on different spatial scales.

$$CV_a = \frac{S_a}{X_a} \times 100\% = \frac{1}{X_a} \left[\frac{\sum_{i=1}^{n} \left(X_{ai} - \overline{X_a} \right)^2}{n-1} \right]^{\frac{1}{2}} \times 100\% \tag{10}$$

where S_a is the unbiased estimate of the standard deviation of residential public service evaluation score and $\overline{X_a}$ is the mean of the assessment score.

Correlation analysis measures the variation closeness among multiple variables. The correlation coefficient is the ratio of the multivariate covariance and their standard deviation.

$$Cov_{pq} = \frac{\sum_{j=1}^{N} \left(Score_{pj} - \overline{Score_p} \right) \left(Score_{qj} - \overline{Score_q} \right)}{N-1} \tag{11}$$

$$Corr_{pq} = \frac{Cov_{pq}}{\delta_p \delta_q} \tag{12}$$

where $Corr_{pq}$ is the correlation coefficient of public service p, q in Xiamen City; Cov_{pq} is their covariance; $Score_{pj}$ and $Score_{qj}$ represent the assessment score of the public service p,q for settlement j, respectively; $\overline{Score_p}$ and $\overline{Score_q}$ δ_p, δ_q separately mean the average score and their standard deviation of public service p, q in all the residential communities in Xiamen, respectively. N is the amount of settlements in Xiamen, $N = 1756$.

3. Results and Discussion

3.1. Residential Suitability Assessment on Service Supplement of Public Services in Xiamen City

Most public facilities, accounting for 53.18% of the total public infrastructures in Xiamen City, are concentrated on Xiamen Island. Moreover, the Siming district had the most outstanding performance of public service supplements, with a vast amount (29.73%) of high-quality facilities concentrated in this area. As for the off-island regions, the Tong'an district, an old inhabitant area with long-term regional development, still had advantages regarding the amount of public infrastructure that had accumulated, especially for transportation, education, and healthcare services. However, concerning the number of high-quality facilities, it was not only significantly less than that in the Siming and Huli districts, but also not as adequate as other districts in the off-island area. For example, except for the primary schools, the high-quality facilities in the Jimei district were more sufficient than those in the Tong'an district. Considering the performance in Haicang, Xiang'an districts, distinct disadvantages emerged by comparing their service supplement capacity with those of other regions, regardless of the quantity or the capability of the local public facilities (Table 4).

Table 4. The summary of public facilities in each district of Xiamen City.

Public Service	Facility Level	Siming	Huli	Tong'an	Haicang	Jimei	Xiang'an
Healthcare	**Total**	193	91	147	57	72	123
	Tertiary (HQ)	7	4	1	2	2	2
	Secondary	13	5	3	1	1	0
	Primary	17	12	12	5	11	7
	Clinic	156	70	131	49	58	114
Leisure	**Total**	50	31	12	7	23	9
	Park and historical sites (HQ)	27	14	8	3	15	4
	Plaza	9	5	2	1	2	1
	Others	14	12	2	3	6	4
Commerce	**Total**	1164	893	263	234	347	112
	Shopping mall and Supermarket (HQ)	61	31	8	5	14	6
	Store and local market	1103	862	255	229	333	106
Transportation (Stops)	**Total**	316	342	469	179	324	195
	Subway (HQ)	7	6	0	0	11	0
	BRT	12	8	10	0	13	0
	Bus	297	328	459	179	300	195
Education	**Total**	47	39	80	24	45	65
	Key primary school (HQ)	14	4	3	1	1	0
	Ordinary primary school	33	35	77	23	44	65

Note: The amount of facilities was summarized with the district boundary. "HQ" is the abbreviation of the high-quality public facilities.

Concerning the regional disparity of local service supplement in meeting with the residents' demand, the per capita performance of commerce services in Siming and Huli districts (LQ_{siming} = 1.47 and LQ_{huli} = 1.12) was still significantly better than those in the off-island area ($LQ < 1$) (Figure 3c). However, 52.6% of the population being concentrated on Xiamen Island [44]. Although most of the high-quality transportation and education infrastructures located in Siming and Huli district, the per capita service supplement is still insufficient compared with the average per capita performance in Xiamen City ($LQ < 1$). By contrast, the per capita transportation and education services in the districts outside Xiamen Island were relatively abundant, especially in the Tong'an district ($LQ = 1.83$, $LQ = 1.26$) (Figure 3d,e). Meanwhile, a large amount of healthcare and leisure facilities are concentrated in the Siming district, and the per capita services supplements are more adequate ($LQ > 1$) than those ($LQ < 1$) in the other districts (Figure 3a,b). Despite the LQ of 1.46 for the healthcare service in Xiang'an district (Figure 3a), 92.68% of the healthcare facilities were low-quality, such as clinics, which could not fully satisfy the high-quality healthcare demands of the local residents.

In terms of the per capita service supplement of all the five essential types of public facilities in Xiamen City (Figure 3f), although excellent public resources were mainly concentrated in the Siming and Huli districts, the per capita public service supplement was facing enormous challenges due to a large amount of population crowding in Xiamen Island. The per capita public services supplement in Siming district barely satisfied the average residents' demand in Xiamen City ($LQ = 1.02$), whereas the LQ in Huli district was only 0.84, which had still shown a huge disadvantage by comparing the average level of the per capita service adequacy for the whole city. As for the off-island area, the long-term regional development in Tong'an district has led several relative advantages on the per capita supply supplement of all kinds of public services ($LQ = 1.32$), which was significantly higher than the city's average level. Nevertheless, the high-quality facilities were deficient, as were those in the other off-island districts.

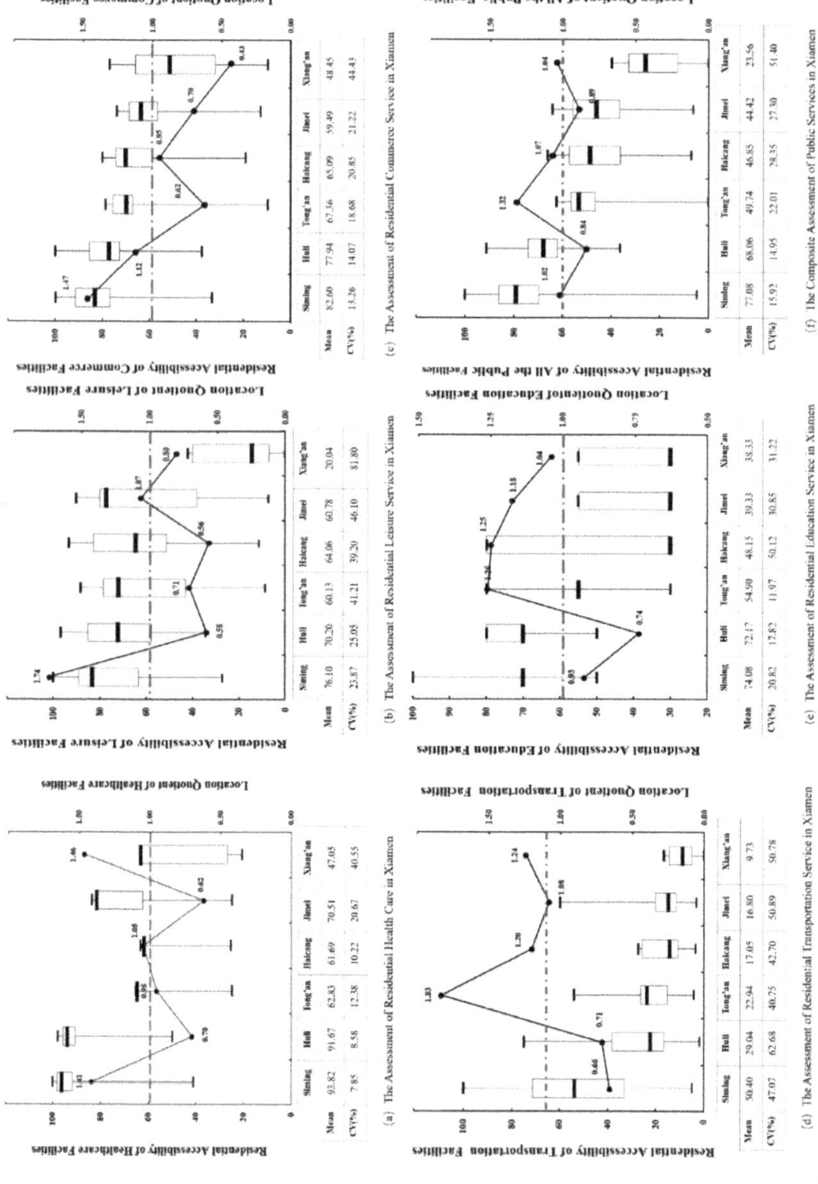

Figure 3. Regional performance of per capita public service supply supplement (LQ) and residential suitability assessment on facility spatial Layout in each district. The tables contain each district's Mean and CV value of the residential accessibility. The assessment results of Healthcare, Leisure, Commerce, Transportation, and Education Service are separately demonstrated in (**a**–**e**). (**f**) is the composite assessment of all the residential public services in Xiamen City.

3.2. Residential Suitability Assessment on the Residential Accessibility of Public Services in Xiamen City

3.2.1. The Regional Residential Suitability Assessment on the Residential Accessibility of Xiamen Public Services

Despite the service supplement, ideally and equitably accessible residential public services are also crucial for residential suitability. Here, we utilized the local mean value and CV of the residential accessibility score as critical indicators to describe the regional residential suitability performance on the spatial layout of diverse types of public services in Xiamen City, China.

As for the healthcare service (Figure 3a), the district's residential suitability in Xiamen City, from high to low, demonstrated a radiation pattern from inside Xiamen Island to the off-island areas. The settlement performances in Siming and Huli were apparently better than those in the other districts, regardless of the local residential accessibility score ($Score_{healthcare.Siming}$ = 93.82 and $Score_{healthcare.Huli}$ = 91.67) or its regional disparities (CV_{Siming} = 7.85%, CV_{Huli} = 8.58%). Amongst the residential public service in the off-island area, the accessibility in Jimei had the highest mean value. However, the regional disparity was apparent (CV_{Jimei} = 20.67%). The performance of residential suitability in Xiang'an district was the worst, with the lowest mean value (47.05) and highest regional disparity (40.55%) in Xiamen City.

The regional residential suitability of commerce and leisure services showed a naturally hierarchical distribution pattern (Figure 3b,c). The settlement performance of Siming and Huli were much better than the other districts on Xiamen Island with high $Score_{ij}$ and low CV_i. Tong'an, Haicang, and Jimei district had similar residential supporting qualities and relatively significant regional disparities. The performance in Xiang'an district was the worst.

In Figure 3d,e, the average residential accessibility of education ($score_{education.Siming}$ = 74.08, score $_{education. Huli}$ = 72.17) and transportation service ($score_{transportatiom.Siming}$ = 50.40, $score_{transportation. Huli}$ = 29.04) in Siming and Huli were much higher than those in the off-island districts. However, the regional disparities were relatively significant based on the uneven distribution of above facilities on Xiamen Island (CV_{Siming} = 47.07%, CV_{Huli} = 62.68% for transportation, and CV_{Siming} = 20.82%, CV_{Huli} = 17.82% for education). In terms of the Tong'an district, although the average residential accessibility scores of the above services were only 54.90 and 22.94 respectively, its regional disparities were the lowest amongst all the districts in Xiamen City (CV_{tongan}=11.97% for education, CV_{tongan}=40.75% for transportation). The performances of the residential suitability of education and transportation services in Haicang, Jimei, and Xiang'an district were quite poor in terms of the low mean value and relatively high regional disparities.

In general, the residential suitability of all public services in Siming and Huli district were the best, with the outstanding mean value (*Composite Score* $_{Siming}$ = 77.08, *Composite Score* $_{Huli}$ = 68.06) and lowest regional disparities (CV_{Siming} = 15.92%, CV_{Huli} = 14.95%) (Figure 3f). Meanwhile, the residential public services in Tong'an, Haicang, and Jimei districts performed similarly in terms of comparable residential accessibility scores and regional disparities. With considerable benefits from the long history of regional development, Tong'an still had some advantages in terms of the reasonable spatial layout of the local public facilities (*Composite Score* $_{Tong'an}$ = 49.74, *CV* $_{Tong'an}$ = 22.01%). However, the performance in Xiang'an district was the worst, with a 23.56 average residential score and 51.40% regional disparity. According to the Urban Master Planning (2011–2020) of Xiamen City, Xiang'an district will be one of the sub-centers of Xiamen City. It is urgent that the quality and quantity of local public facilities in Xiang'an district be improved.

3.2.2. The Overall Residential Suitability Assessment on the Residential Accessibility of Xiamen Public Services

A residential suitability assessment of each type of public service among all the communities in Xiamen City is shown in Figure 4. The spatial layout of residential healthcare and commerce facilities were the most reasonable (Table 5). The average scores for the residential healthcare and commerce accessibility of all the communities in Xiamen were 85.31, 76.24, and the CVs were only 18.56%, 19.17%, respectively. The total proportion of "High" and "Medium" residential communities accounted for 75.17% and 78.25% (Figure 4a,c). In contrast, the residential suitability of the transportation service had the lowest average score (only 36.89) and the highest regional disparity on spatial accessibility (65.15%). Due to the uneven distribution of transportation facilities, 43.68% of the residential communities in Xiamen City were labeled as "Low" (Figure 4d). In Table 6, the correlation coefficient between the healthcare and commerce service was 0.68. Meanwhile, the indexes between transportation & commerce services and transportation & healthcare services were 0.67 and 0.54, respectively, indicating that the above three types of public services have similar spatial patterns in terms of the spatial layout around Xiamen's residential communities. The phenomenon shows that the Xiamen's residential planning has paid considerable attention to the assignment of the above types of public services.

The residential suitability of the leisure services and education performed similarly, with average scores of 70.06 and 66.34 and CVs of 31.71% and 27.86%, respectively (Table 5). Table 6 shows that the correlation coefficients of residential accessibility between the above services and other public services were all less than 0.5, reflecting that the spatial layout of residential communities surrounding education and leisure facilities did not match well with the other public services. As a renowned coastal city, the water-view-oriented leisure facilities played a vital role in the residential suitability of leisure services in Xiamen City. As shown in Figure 4b, most of the residential communities labeled with "High" located in the areas around the Yundang Lake and Gulangyu islet, Moreover, several adjacent water off-island communities also received great benefits on their accessibility of leisure services due to the nearby water bodies. Nowadays, the residential suitability of education services was mainly affected by the "School District." However, the uneven distribution of education resources, high-quality primary schools being mainly concentrated in Siming and Huli district, caused a significant regional difference in the residential suitability of education service in Xiamen City. In Figure 4e, only 18.19% of the residential communities could be labeled as "Low" on their accessibility of education service on Xiamen island, whereas the accessibility of nearly half of the off-island communities, accounting for 48.3% being labeled as "Low", could not fully satisfy the education demands of Xiamen's citizens. Considering the above results, the residential accessibility of education and leisure service depicted apparent regional differences, especially among the communities on and off Xiamen Island, and the neighborhoods near and away from the water bodies.

A composite assessment of residential suitability among all the communities in Xiamen City is shown in Figure 3f. Although only 29.90% of the residential neighborhoods could be graded as "Low," an evident inequality could be witnessed by comparing the residential suitability of all the five essential public services between the communities on Xiamen Island and those in the off-island areas due to the unbalanced regional development in Xiamen City. The urbanization rate in the Siming and Huli districts reached 100% in 2016 [44], with great benefits from regional policy bias and rapid social-economic development. Most of the "High "and "Medium" communities are concentrated on Xiamen Island. In contrast, the residential suitability of all the public services failed to meet the citizen's expectation in the off-island area with the numerous "Low" and "Medium" communities, which is mainly due to the relatively backward regional development. As the 2030 Agenda for Sustainable Development asserts, "no one will be left behind" [8,17]; the apparent inequality could inevitably lead to adverse effects on the sustainable development of Xiamen City, China.

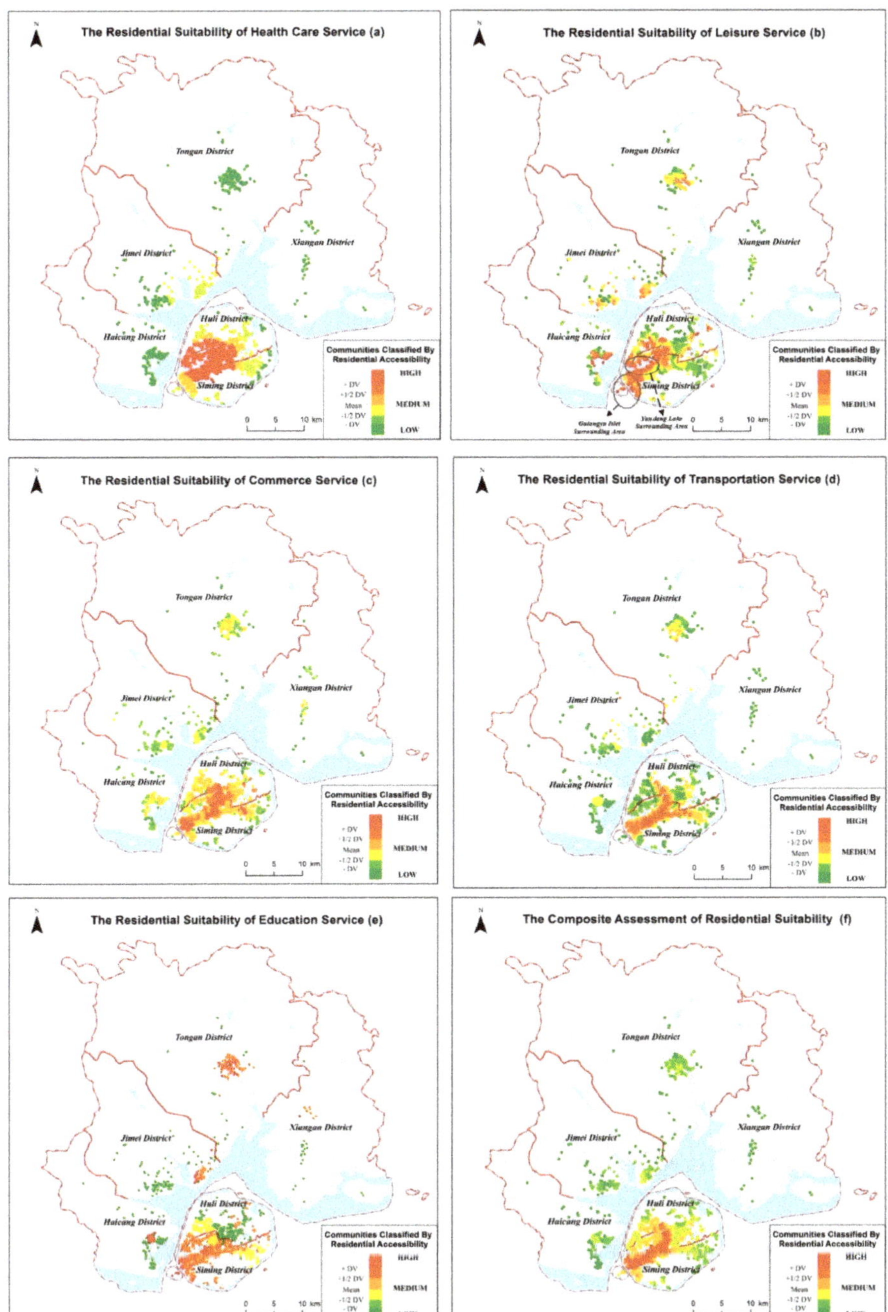

Figure 4. Residential suitability assessment of communities surrounding public services in Xiamen City (Low: $Score < Score_{mean} - 0.5 \times DV$; Medium: $Score \in [Score_{mean} - 0.5 \times DV, Score_{mean} + 0.5 \times DV]$; High: $Score > Score_{mean} + 0.5 \times DV$; "Score" is the abbreviation of residential accessibility score of diverse type of public facilities).

Table 5. Residential suitability on residential accessibility of all the public services in Xiamen City

Area	Service	Mean	CV (%)
	Commerce	76.24	19.17
	Traffic	36.89	65.15
Xiamen City	Leisure	70.06	31.71
	Healthcare	85.31	18.56
	Education	66.34	27.86

Table 6. The spatial correlation of the residential accessibility of each type of public service in Xiamen City.

Service	Commerce	Transportation	Leisure	Healthcare	Education
Commerce	1.00	0.67	0.45	0.68	0.45
Transportation	0.67	1.00	0.30	0.54	0.41
Leisure	0.45	0.30	1.00	0.42	0.29
Healthcare	0.68	0.54	0.42	1.00	0.47
Education	0.45	0.41	0.29	0.47	1.00

4. Conclusions

Residential suitability of urban public services has become a critical issue in many countries. The literature has so far witnessed a steady growth of studies using accessibility to services and facilities as the index of measure. Most of the studies have, nevertheless, targeted "residential suitability" of a single type of public facility. As public facility investigations involve types of public facilities with complex characteristics, existing methods have failed to meet contemporary demand. Moreover, the larger spatial urban study also proposed new requirements on the quality of the spatial data. The application of network big data provided better data resources compared to traditional social and field surveys for its full coverage and accuracy, and the low cost for updating. Meanwhile, the composite spatial accessibility assessment method contributes to a further attempt to integrate the study on residential suitability including various kinds of public facilities (services). In this study, combining network big data and spatial analysis, we conducted a composite spatial accessibility assessment for residential suitability covering five essential types of public services, i.e., healthcare, leisure, commerce, transportation, and education services in the rapidly urbanizing Xiamen City, China, which we hope to provide support for designing a sustainable city and community according to the 2030 Sustainable Development Goals of the United Nations.

As for the service supplement of public services in Xiamen City, although most of the facilities are concentrated in Xiamen Island, the regional per capita public service supplement was facing enormous challenges with the high local population, and the per capita transportation and education service supplements were in short supply compared with the average per capita performance of the whole city. Meanwhile, Tong'an, an old inhabitant district, still had advantages in the amount of public facilities due to its long history of regional development. However, high-quality facilities were insufficient, similar to the situation in other off-island districts.

In terms of the facility spatial layout, the residential suitability of healthcare and commerce services was the most reasonable in Xiamen city, respectively demonstrating radiation and hierarchical patterns from inside Xiamen Island to the off-island area. In contrast, the residential suitability of the transportation service performed the worst. Currently, Xiamen's residential planning has paid considerable attention to the assignment of residential transportation, commerce, and healthcare services, which has a similar allocation pattern in terms of the community's surrounding public facilities, whereas the residential education and leisure services did not match well with the other public services. Inequality is evident through a comparison of the residential composite suitability between the communities on Xiamen Island and those in the off-island area due to the unbalanced regional development in Xiamen City.

Therefore, we recommend (1) Upgrading and balancing the allocation of public facilities, especially those in the off-island districts, and Xiang'an district should be given significant attentions; (2) Reducing the regional differences in the residential accessibility of education and leisure services by redistributing the education resources to the off-island area, expanding the investment in cultural landscapes, and increasing the number of "hanging gardens" and self-service libraries in Xiamen City; and (3) Promoting the transformation of the urban spatial development pattern to "a single-heart multi-cores", which could not only help to ease the supply-demand contradiction of the public services on Xiamen Island, but also provide ways to enhance the sustainable development by narrowing the regional inequality of residential suitability in Xiamen City. Our study calls for more focus on the supplement model of residential public services, including the service supply adequacy, ideally, and equitably-accessible residential public facilities (services), by urban planning and decision-makers, especially for cities in developing countries.

A composite spatial accessibility assessment method for the residential suitability of urban public services in Xiamen City was used as an empirical example. The usefulness of the composite methodology, network big data, and spatial analysis in measuring and analyzing the residential suitability of the urban public services, to some extent, has been verified by the empirical outcomes. However, due to the lack of a publicly-accepted quantitative evaluation system, several parameters in our study were set by the subjective method and only reflected the regional characteristics in Xiamen City. Future studies should focus on developing a universal parameter system on the residential suitability of urban public services. By doing so, residential suitability assessments at various levels could be better discerned.

Author Contributions: Y.Z., G.Z., T.L., X.L., J.L., M.L., H.Y., L.K. worked collectively. Specifically, conceptualization, T.L., and H.Y.; data curation, Y.Z.; formal analysis, Y.Z., X.L., and M.L.; funding acquisition, T.L., and G.Z.; investigation, J.L., and L.K.; methodology, Y.Z. and G.Z.; visualization, Y.Z.; writing—original draft, Y.Z.; writing—review and editing, Y.Z., T.L., G.Z., and H.Y.

Funding: This research was funded by the National Key R&D Program of China (2016YFC05027002018 YFD1100300) and National Natural Science Foundation of China (41771573, 41871167, 41201155, 41771570, 41671444).

References

1. Wei, Y.D.; Ewing, R. *Urban Expansion, Sprawl, and Inequality*; Elsevier: Amsterdam, The Netherlands, 2018.
2. Brennetot, A. In Geographers and Spatial Justice: Genealogy of a Complicated Relationship. *Annales de Géographie* **2011**, *678*, 115–134. [CrossRef]
3. Gunn, A.M. *Habitat: Human Settlements in an Urban Age*; Elsevier: Amsterdam, The Netherlands, 2017.
4. Wu, L.Y. *Introduction to Sciences of Human Settlements*; China Architecture & Building Press: Beijing, China, 2001.
5. Tsou, K.W.; Hung, Y.T.; Chang, Y.L. An accessibility-based integrated measure of relative spatial equity in urban public facilities. *Cities* **2005**, *22*, 424–435. [CrossRef]
6. Weiss, D.; Nelson, A.; Gibson, H.; Temperley, W.; Peedell, S.; Lieber, A.; Hancher, M.; Poyart, E.; Belchior, S.; Fullman, N. A global map of travel time to cities to assess inequalities in accessibility in 2015. *Nature* **2018**, *553*, 333–336. [CrossRef] [PubMed]
7. Higasa, T.; Hibata, Y. *Urban Planning*; Kioritz Corporation Press: Tokyo, Japan, 1977.
8. Boeing, G.; Church, D.; Hubbard, H.; Mickens, J.; Rudis, L. LEED-ND and Livability Revisited. *Berkeley Plan. J.* **2014**, *27*, 31–55.
9. Harvey, D. *Social Justice and the City*; Johns Hopkins University Press: Baltimore, MD, USA, 1973.
10. Calthorpe, P. *The Next American Metropolis: Ecology, Community, and the American Dream*; Princeton Architectural Press: New York, NY, USA, 1993.
11. Dong, J. On the forming of health-oriented urban space. *Mod. Urban Res.* **2009**, *10*, 77–84.

12. Keeble, B.R. The Brundtland report: 'Our common future'. *Med War.* **1988**, *4*, 17–25. [CrossRef]
13. Bruckmeier, K. Sustainability between necessity, contingency, and impossibility. *Sustainability* **2009**, *1*, 1388–1411. [CrossRef]
14. Asami, Y. *Residential Environment: Methods and Theory for Evaluation*; University of Tokyo Press: Tokyo, Japan, 2001.
15. Evans, P.B. *Livable Cities? Urban Struggles for Livelihood and Sustainability*; University of California Press: Berkeley, CA, USA, 2002.
16. Timmer, V.; Seymoar, N.K. *The Livable City: World Urban Forum 2006, Vancouver Working Group Discussion Paper*; International Centre for Sustainable Cities: Vancouver, BC, Canada, 2006.
17. Transforming Our World: The 2030 Agenda for Sustainable Development. Available online: https://sustainabledevelopment.un.org/post2015/transformingourworld (accessed on 8 December 2018).
18. Lineberry, R.L.; Welch, R.E. Who Gets What: Measuring the Distribution of Urban Public Services. *Soc. Sci. Q.* **1974**, *54*, 700–712.
19. McAllister, D.M. Equity and efficiency in public facility location. *Geogr. Ana.* **1976**, *8*, 47–63. [CrossRef]
20. Erkip, F.B. The distribution of urban public services: The case of parks and recreational services in Ankara. *Cities* **1997**, *14*, 353–361. [CrossRef]
21. Kinman, E.L. Evaluating health service equity at a primary care clinic in Chilimarca, Bolivia. *Soc Sci Med.* **1999**, *49*, 663–678. [CrossRef]
22. Kontodimopoulos, N.; Nanos, P.; Niakas, D. Balancing efficiency of health services and equity of access in remote areas in Greece. *Health Policy* **2006**, *76*, 49–57. [CrossRef] [PubMed]
23. Lin, T.; Liu, X.F.; Song, J.C.; Zhang, G.Q.; Jia, Y.Q.; Tu, Z.Z.; Zheng, Z.H.; Liu, C.L. Urban waterlogging risk assessment based on internet open data: A case study in China. *Habitat Int.* **2018**, *71*, 88–96. [CrossRef]
24. Panter, J.; Jones, A.; Hillsdon, M. Equity of access to physical activity facilities in an English city. *Prevent. Med.* **2008**, *46*, 303–307. [CrossRef] [PubMed]
25. Jiang, H.B.; Xu, J.G.; Qi, Y.; Chen, J.T. The quantitative analysis of large-scale supermarkets location based on time accessibility and Gasa rules. *Geogr. Res.* **2010**, *29*, 1056–1068.
26. Ni, J.; Wang, J.; Rui, Y.; Qian, T.; Wang, J. An enhanced variable two-step floating catchment area method for measuring spatial accessibility to residential care facilities in Nanjing. *Int. J. Environ. Res. Public Health* **2015**, *12*, 14490–14504. [CrossRef] [PubMed]
27. Sun, C.G.; Lin, T.; Zhao, Y.; Lin, M.X.; Yu, Z.W. Residential Spatial Differentiation Based on Urban Housing Types—An Empirical Study of Xiamen Island, China. *Sustainability* **2017**, *9*, 1777. [CrossRef]
28. Wang, J.H.; Deng, Y.; Song, C.; Tian, D.J. Measuring time accessibility and its spatial characteristics in the urban areas of Beijing. *J. Geogr. Sci.* **2016**, *26*, 1754–1768. [CrossRef]
29. Wang, S.T.; Zheng, S.Q.; Feng, J. Spatial accessibility of housing to public services and its impact on housing price: A case study of Beijing's inner city. *Prog. Geogr.* **2007**, *26*, 78–85.
30. Becker, R.A.; Caceres, R.; Hanson, K.; Loh, J.M.; Urbanek, S.; Varshavsky, A.; Volinsky, C. A tale of one city: Using cellular network data for urban planning. *IEEE Perv. Comput.* **2011**, *10*, 18–26. [CrossRef]
31. Wu, S.M.; Chen, T.; Wu, Y.J.; Lytras, M. Smart cities in Taiwan: A perspective on big data applications. *Sustainability* **2018**, *10*, 106. [CrossRef]
32. Kharrazi, A.; Qin, H.; Zhang, Y. Urban big data and sustainable development goals: Challenges and opportunities. *Sustainability* **2016**, *8*, 1293. [CrossRef]
33. Kamrowska-Zaluska, D.; Obracht-Prondzyńska, H. The Use of Big Data in Regenerative Planning. *Sustainability* **2018**, *10*, 3668. [CrossRef]
34. Guo, J.; Lyu, Y.Q.; Shen, T.Y. Urban spatial structure based on point pattern analysis: Taking Beijing metropolitan area as a case. *Econ. Geogr.* **2015**, *35*, 68–74.
35. Kwan, M.P. GIS methods in time-geographic research: Geocomputation and geovisualization of human activity patterns. *Geogr. Ann. Ser. B Hum. Geogr.* **2004**, *86*, 267–280. [CrossRef]
36. Stessens, P.; Khan, A.Z.; Huysmans, M.; Canters, F. Analysing urban green space accessibility and quality: A GIS-based model as spatial decision support for urban ecosystem services in Brussels. *Ecosyst. Serv.* **2017**, *28*, 328–340. [CrossRef]
37. Wu, Z.Y.; Zhuo, J. Impact of Urban Built Environment on Urban Short-Distance Taxi Travel: The Case of Shanghai. Available online: http://iopscience.iop.org/article/10.1088/1755-1315/153/6/062019/pdf (accessed on 8 December 2016).

38. Lytras, M.; Visvizi, A. Who uses smart city services and what to make of it: Toward interdisciplinary smart cities research. *Sustainability* **2018**, *10*, 1998. [CrossRef]

39. Cao, X.; Chen, H.; Liang, F.; Wang, W. Measurement and Spatial Differentiation Characteristics of Transit Equity: A Case Study of Guangzhou, China. *Sustainability* **2018**, *10*, 1069. [CrossRef]

40. Tang, F.; Mo, W.; Zhang, X.; Zhou, S. Spatial Distribution of Public Service Facilities Based on POI Data. *Urban. Archit.* **2017**, *27*, 35–39.

41. Yin, C.H.; He, Q.S.; Liu, Y.F.; Chen, W.Q.; Gao, Y. Inequality of public health and its role in spatial accessibility to medical facilities in China. *Appl. Geogr.* **2018**, *92*, 50–62. [CrossRef]

42. Lin, T.; Cao, X.; Huang, N.; Xu, L.; Li, X.; Zhao, Y.; Lin, J. Social cognition of climate change in the coastal community: A case study in Xiamen City, China. *Ocean Manag.* **2018**. [CrossRef]

43. Zhang, W. *China Livable City Report*; China Science Publishing Press: Beijing, China, 2016.

44. Xiamen Statistical Bureau, Xiamen Special Economic Zone Yearbook 2017. Available online: http://www. stats-xm.gov.cn/tjzl/tjsj/tqnj/ (accessed on 28 September 2017).

45. Lin, T.; Li, X.H.; Zhang, G.Q.; Zhao, Q.J.; Cui, S.H. Dynamic analysis of island urban spatial expansion and its determinants: A case study of Xiamen Island. *Acta Geogr. Sin.* **2010**, *65*, 715–726.

46. The Ministry of Housing and Urban-Rural Development of the People's Republic of China. The Code of Urban Residential Areas Planning & Design (2016). Available online: http://www.mohurd.gov.cn/ (accessed on 28 June 2016).

47. Stewart, J.Q.; Warntz, W. Physics of population distribution. *J. Reg Sci.* **1958**, *1*, 99–121. [CrossRef]

48. Su, S.L.; Li, Z.K.; Xu, M.Y.; Cai, Z.L.; Weng, M. A geo-big data approach to intra-urban food deserts: Transit-varying accessibility, social inequalities, and implications for urban planning. *Habitat Int.* **2017**, *64*, 22–40. [CrossRef]

49. Widener, M.J.; Farber, S.; Neutens, T.; Horner, M. Spatiotemporal accessibility to supermarkets using public transit: An interaction potential approach in Cincinnati, Ohio. *J. Transp. Geogr.* **2015**, *42*, 72–83. [CrossRef]

50. Widener, M.J.; Shannon, J. When are food deserts? Integrating time into research on food accessibility. *Health Place* **2014**, *30*, 1–3. [CrossRef]

51. Song, Z.N.; Chen, W.; Zhang, G.X.; Zhang, L. Spatial accessibility to public service facilities and its measurement approaches. *Prog. Geogr.* **2010**, *29*, 1217–1224.

52. Chen, L.; Zhang, W.; Yang, Y. Residents' incongruence between reality and preference of accessibility to urban facilities in Beijing. *Acta Geogr. Sin.* **2013**, *68*, 1071–1081.

53. Wang, F. *Quantitative Methods and Socio-Economic Applications in GIS*; CRC Press: Boca Raton, FL, USA, 2014.

54. Luo, W.; Qi, Y. An enhanced two-step floating catchment area (E2SFCA) method for measuring spatial accessibility to primary care physicians. *Health Place* **2009**, *15*, 1100–1107. [CrossRef]

55. Luo, W.; Whippo, T. Variable catchment sizes for the two-step floating catchment area (2SFCA) method. *Health Place* **2012**, *18*, 789–795. [CrossRef]

56. Wang, F. Measurement, optimization, and impact of healthcare accessibility: A methodological review. *Ann. Assoc. Am. Geogr.* **2012**, *102*, 1104–1112. [CrossRef] [PubMed]

57. TAO, Z.L.; Cheng, Y.; Dai, T.Q.; Li, X. Sensitivity Analysis of Parameters in Measuring Spatial Accessibility to Public Service Facilities. *Mod. Urban Res.* **2017**, *3*, 30–35. [CrossRef]

58. Reed, G.F.; Lynn, F.; Meade, B.D. Use of coefficient of variation in assessing variability of quantitative assays. *Clin. Diagn. Immunol.* **2002**, *9*, 1235–1239. [CrossRef]

 *sustainability*

Article

Destination Image Analytics Through Traveller-Generated Content

Estela Marine-Roig

Serra Hunter Fellow, University of Lleida, 25003 Catalonia, Spain; estela.marine@aegern.udl.cat;
Tel.: +34-973-703-338

Received: 8 May 2019; Accepted: 17 June 2019; Published: 19 June 2019

Abstract: The explosion of content generated by users, in parallel with the spectacular growth of social media and the proliferation of mobile devices, is causing a paradigm shift in research. Surveys or interviews are no longer necessary to obtain users' opinions, because researchers can get this information freely on social media. In the field of tourism, online travel reviews (OTRs) hosted on travel-related websites stand out. The objective of this article is to demonstrate the usefulness of OTRs to analyse the image of a tourist destination. For this, a theoretical and methodological framework is defined, as well as metrics that allow for measuring different aspects (designative, appraisive and prescriptive) of the tourist image. The model is applied to the region of Attica (Greece) through a random sample of 300,000 TripAdvisor OTRs about attractions, activities, restaurants and hotels written in English between 2013 and 2018. The results show trends, preferences, assessments, and opinions from the demand side, which can be useful for destination managers in optimising the distribution of available resources and promoting sustainability.

Keywords: destination image; user-generated content; online travel review; big data analytics; opinion mining; sentiment analysis; resource optimisation; place sustainability; TripAdvisor; Greek Attica

1. Introduction

Nations, states, cities, and regions commit considerable effort and funds to improving their tourist destination image (TDI) and attractiveness [1]. Due to growing competitiveness, promotion strategists need more precise information about the diversity of responses to TDI [2]. The image of a city [3], country [4], region [5], or tourist destination [6–8] has been the subject of constant study for more than half a century in countless scientific publications. This great production on destination image has led to systematic reviews [9–12] and meta-analyses [13,14]. The great success of this concept may be because the authors agree that image—projected and perceived—plays a crucial role in decision-making regarding selecting a holiday destination [7,9,15,16]. Consequently, some authors [17,18] theorised about the image formation process. The agents of image formation can be divided into three groups according to the origin of the sources [18,19]: induced (emanating from the destination promoters), organic (transmitted between individuals) and autonomous (produced independently of the previous ones).

Among the organic agents, along with the experience itself, is the opinion of users and consumers that spreads through word-of-mouth marketing (WoM), in conversations with relatives, friends, colleagues, or acquaintances. From the proliferation of user-generated content (UGC) disseminated through social media, we speak about electronic WoM communication (eWoM). Opinions of other users and consumers transmitted through both WoM and eWoM, have become the main sources of secondary information (not including the primary source of own experience) in the process of procuring goods or contracting services online.

In travel, hospitality and tourism, experiences shared through social media have been increasing [20] as well as consultation (before and during the trip) and consideration of content generated by other travellers. Within traveller-generated content (TGC), some authors have used travel-related forums [21,22], tweets [23], Facebook posts [24], multiple social media [25] and online photographs [26–28] to deduce TDI aspects, but most have focused on travel blogs [29] and online travel reviews (OTR) [30]. It is worth highlighting the transition from travel blogs to OTRs. OTRs have grown dramatically, while many portals that hosted travel blogs have disappeared. For example, the portal TripAdvisor stored 10 million OTRs in 2007 [30], and it has already exceeded 700 million, covering more than 8 million tourist resources worldwide [31]. This abundance of first-hand, spontaneous, disinterested, and freely available online information has led many researchers to choose OTRs as a data source [32,33]. It has gone from analysing a few hundred opinions obtained through expensive surveys to freely dispose of hundreds of thousands of OTRs about places or tourist resources of a destination. For example, TripAdvisor currently stores over 150,000 opinions and 100,000 photographs on the Basilica of the Sagrada Familia in Barcelona.

The figures above and other more spectacular in the field of UGC, and social media (Facebook, Twitter, etc.) gave rise to the link between their analysis and that of big data [34–36]. Therefore, much research has been devoted to social media analytics in general [37–39] and tourism analytics in particular [40–42].

Regarding the percentage of users who consulted the opinions of other travellers spread through WoM and eWoM, the following surveys can be highlighted:

- The European Union surveyed more than 30,000 Europeans from different social and demographic groups [43]. One question was: "Which of the following information sources do you think is most important when you make a decision about your travel plans? (Maximum three answers)". They were, in the first two positions, WoM (recommendations of friends, colleagues or relatives) with 51% followed by eWoM (websites collecting and presenting comments, opinions and ratings from travellers—OTRs) with 34%.

- In another case confined to Britain, more than 11,000 foreign visitors were surveyed [44]. In response to the question: "Thinking about your holiday in Britain, which of the following information sources influenced your choice of destination?", 40% used WoM (conversations with friends or family) and 30% eWoM (OTRs), slightly behind search engines and price comparison portals.

- Recently, results similar to those above were obtained in a survey of more than 2,000 Americans who had travelled for pleasure in the past 12 months [45]. However, eWoM had more weight than WoM. To the question: "In the past 12 months, which of these Internet technologies or services have you used to help plan your leisure travel? (Select all that apply)", 58.2% had used eWoM (TGC) and 45.6% WoM (opinions of friends, colleagues or relatives). Of the travellers who used eWoM, 32.5% consulted OTRs about hotels, 30.8% about restaurants or activities, and 29.6% about destinations.

Previous surveys have highlighted organic information sources, transmitted through WoM and eWoM. These sources are the most consulted by potential tourists. These results are crucial to demonstrate that TGC is an agent of the destination image construction, because travel blogs and OTRs are expressions of the image perceived (and transmitted) by visitors [46]. Besides, TGC has to be consulted by other tourists or prospective tourists to be part of the projected image and close the circle of Figure 1. It is noteworthy that the aforementioned survey on TGC usage [45] coincides with another that was conducted ten years earlier [30], in the sense that the respondents considered extremely important or very important, first "Where to stay"; second, "Where to eat"; and, third, "What to do". Over a few years, TGC and social media have reversed in priority between sources. Late last century, tourist offices, tour operators and travel agencies were protagonists in constructing the projected image [47]; induced sources, especially destination marketing and management organisations (DMO), had a high penetration in the market, while organic sources had minimal [18].

From the perspective of scientific literature, in principle, it was questioned whether prospective travellers intended to use TGC in planning their trips [48]. There were also doubts about the credibility of TGC, especially with hotel OTRs [49]. There are already numerous studies that show the usefulness of TGC [50]. By way of example, the following, based on surveys, can be cited: one, on Turkish users [51], showed the validity of TGC as a source in the search for information related to the trip and, more broadly, in the planning process thereof. Another, on visitors from New Zealand [52], revealed that perceived usefulness and empowerment led to the use of TGC to plan the trip. A third, about Chinese travellers [53], showed that eWoM had a significant utility and influence in the planning and decision making related to the trip. The influence of OTRs in travel decision making has also been demonstrated through artificial intelligence methodologies [54].

Moreover, image is a qualifying and amplifying determinant of destination competitiveness and sustainability [1,55]. DMOs should be caretakers of the image and resources of destinations through policies and incentives that facilitate developing products, desirable from the demand side, but that do not endanger local resources [56]. In this vein, "The very existence of tourism and sustained competitiveness depends on the availability of resources and the degree to which these resources are bundled to meet visitor expectations and needs at the destination" [57] (p. 100). Consequently, DMOs need to know the TDI, from the viewpoint of their visitors, to properly manage available resources. However, literature on this perspective is scarce [58]. These authors assessed the destination image from the tourists' viewpoint through 203 TripAdvisor OTRs. Despite several studies applying big data analytics on OTRs about accommodations, restaurants and tourist attractions, no such studies have integrated the analysis of different types of tourism resources to measure the cognitive, affective and conative components of TDIs.

In response, the chief objective of the study presented here was to propose a theoretical and methodological framework to measure TDIs by analysing TGC. The novelty of the study stems from its integration of big data from OTRs about various kinds of tourism resources (attractions, travel-related activities, hotels and restaurants) to measure three components of TDIs according to common metrics. This proposed model of big data analytics applies to a Greek region, Attica, where the most striking and complete ancient Greek monumental complex is located [59]. To explore which tourist resources are most popular and best valued by tourists, a random sample of 300,000 TripAdvisor OTRs (100,000 from the attractions section, 100,000 from the restaurants section, and 100,000 from the hotels section) are analysed. The results, based on visitors' needs, preferences and opinions, can be useful to DMOs to optimise deployment of available resources and promote sustainability.

2. Image of a Tourist Destination

The images are of paramount importance because they transpose the representation of an area inside the mind of potential tourists and give them a preliminary idea of the destination [60]. The TDI has received numerous definitions throughout its history. The most commonly used terms in its definition have been [61]: impression (45%), perception (27%), belief (18%), idea (18%), and representation (15%). One of the most cited by scientific doctrine is that which says that image can be defined as the sum of beliefs, ideas and impressions that a person has about a tourist destination [15]. Researchers have distinguished projected images from perceived images [62].

Projected images can be conceived as the ideas and representations of destinations that are available for tourists' consideration [63]. The authors agree on the subjectivity of the perceived image. Human behaviour is based more on image than on objective reality because what an individual believes to be true, in fact, is true for him or her [64]. From this perspective, the TDI is defined as the subjective interpretation of reality made by a tourist [65] or a partial, simplified, idiosyncratic, and distorted representation that is not necessarily isomorphic in relation to the real-world environment [66]. In other cases, its complexity is highlighted by defining the TDI as a sum of associations and pieces of information connected to a destination that would include multiple destination components and personal perception [67]. In the following sections, the multiple factors that influence image construction and the components that make it up will be explained.

2.1. Image Building

The image of a town is a multidimensional and complex construct [68]. The image of a pleasure travel destination is a global concept (gestalt). It is a holistic construct that, to a greater or lesser extent, derives from attitudes regarding the perception of the destination's tourist attributes [69].

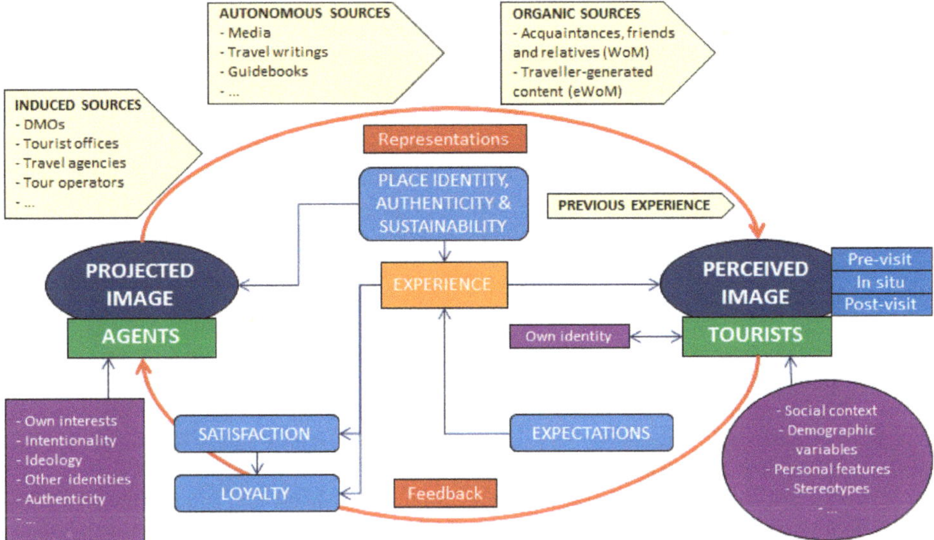

Figure 1. Circle of image construction from a holistic perspective, derived from Marine-Roig [70].

In Figure 1, one can observe agents, constructs, information sources, and major variables involved in the construction of the image. These elements are interrelated in a circle. At opposite points of their diameter, agents project the image and tourists perceive. The perceived image may vary according to the stage of the trip (before, during and after). The representations of the tourist destination are in the arc Agents–Tourists, and the opinions of visitors (feedback) are in the arc Tourists–Agents. The visitor's lived experience is in the centre of the circle. Furthermore, in Figure 1, one can observe some of the variables that explain the subjectivity of the image perceived by tourists and discordance of representations in the projected image by the agents. Next, the most important aspects of the scheme are explained.

- Information sources. The representations come from two sources of information: primary and secondary. Secondary sources are grouped into three types: induced, autonomous and organic [18]. In the organic sources of Gartner, the TGC diffused through eWoM has been added [71] and previous experience has been segregated. The latter is distinguished by being a primary source [72] and enjoying the highest credibility for tourists since it is based on information personally acquired in a previous trip to the area.
- Expectations. The lived experience has as antecedent expectations that the tourist internalized previously. In the pre-visit phase, the image is a set of expectations and perceptions a prospective traveller has about a destination [56]. There are often discrepancies between projected and perceived images. These can be grouped under two concepts [19]: discordance in the representations, when promoters distort reality to suit their interests that may not be coincident, and incongruity in the image when the projected image does not match the current perception of tourists. The contrast between positive image and negative reality often leads to disappointment or anger upon arrival, and false images restrict the learning potential of travel, one of its most

valuable and enduring foundations [73]. The greater the difference between image and reality, that is, between expectation and experience, the more likely it is that tourists will be dissatisfied [74]. For example, when the image perceived in advance is positive and the reality perceived in situ is negative, there is a negative incongruity causing a great dislike [9].

- Place sustainability. Destination competitiveness is illusory without sustainability. From this perspective, the expression 'sustainable competitiveness' is tautological [1]. At the same time, one of the most influential researches in the scientific literature on tourism [56] stated: "Interestingly, the sustainability of local resources becomes one of the most important elements of destination image, as a growing section of the market is not prepared to tolerate over-developed tourism destinations and diverts to more environmentally advanced regions" (p. 101). From another perspective, tourists' perception affects brand image sustainability [75].
- Place identity and authenticity. Identity and authenticity are part of the projected image [70], but can also directly influence the experience through existential authenticity (oriented activity or experienced authenticity) [76]. These authors demonstrated the relationship between image, authenticity, identity, and place attachment.
- Satisfaction and loyalty. Perceived image through experience is a forerunner of tourist satisfaction and loyalty. At the same time, satisfaction is also an antecedent of loyalty. In tourism, loyalty is measured by intentions of future behaviour, specifically by the tourist's predisposition (attitude) to return to the place or recommend it both through WoM and eWoM. Satisfaction and loyalty have their negative side when they become dissatisfaction and disloyalty. Many authors have demonstrated these relationships. For example, a meta-analysis of 66 independent studies revealed that the impact of image on tourist loyalty is significant [13]. Tourism image is a direct antecedent of satisfaction and loyalty [65]; image directly influences satisfaction, and this has a direct and positive impact on loyalty to the destination [77]. Affective image is the main antecedent of loyalty [78]. Last, overall image (cognitive and affective images) indirectly influences tourists' behavioural intentions mediated by their satisfaction [79], which is positively affected by overall image, as are their intentions to recommend the destination [80]. In addition, satisfaction and loyalty are placed on the path that goes from the tourist to the agents, because overall satisfaction positively affects image and loyalty in all models [5].

2.2. Image Components

Image must be thought of as the overall cognitive, affective and evaluative structure of the behaviour unit, or its internal view of itself and its universe [64]. In this line, to analyse the image of a city or tourist destination, the doctrine has mostly used the tripartite cognitive–affective–conative model [18,81–84] inherited from the field of psychology.

Table 1 compares the dominant model of image analysis [84] with another parallel [66] that adds spatial and evaluative dimensions. Both have in common the division of the interaction between the person and the environment in three areas: to have knowledge of something, to feel something about it and, therefore, to do something about it. They also agree in considering three distinct but hierarchically interrelated aspects. In parallel, the first two lead to the overall or composite image. An overall place image is formed because of both cognitive and affective evaluations of that place [17]. A composite place image is subjectively shaped by an interlaced system of both designative and appraisive perceptions [85].

The designative aspect refers to the physical characteristics of the resource such as shape, size, colour, texture, layout, and other details, and the mental map that concerns basic properties such as location, distance, orientation, and other spatial variations. City image is acquired and supported by an underlying network that represents the individual's movement field or activity space [66]. The mental image can also be relatively abstract; for example, the structure is identified as a "restaurant" or "the third building from the corner" [3]. The designative aspect may be less important by itself than the estimative, the meaning attached to—or evoked by—the physical form. Affective meaning is simply the emotional response to the environment that accompanies the perceptive and symbolic meanings. The evaluative dimension is a general opinion or judgement and preference that specifically involves the set of places to assess or classify [66]. The evaluative meaning simply refers to establishing a ranking between the best and the worst.

Table 1. Definition of components or aspects of image.

Rapoport [84] (p. 28)	Pocock and Hudson [66] (p. 30)
Cognitive Involves perceiving, knowing and thinking, the basic processes whereby individuals knows their environment.	**Designative** It is informational in nature, concerned with description and classification—the basic "whatness" and "whereness" of the image.
Affective Involves feelings and emotions about the environment, motivations, desires and values (embodied in the images).	**Appraisive** It is one of appraisal or assessment. It incorporates both evaluation and preference, the former including some general or external standards, the latter reflecting a more personal type of appraisal and affection, the emotional response concerned with feeling, value and meaning attached to the perceived.
Conative Involves acting, doing, striving and thus having an effect on the environment in response to 1 and 2.	**Prescriptive** Relates to predictions and inference of both descriptive and appraisive nature giving the image depth, continuity, pattern, or meaning beyond that justified by the experience of a particular scene alone.

The Pocock and Hudson model and similar ones were adopted in human geography by numerous authors to study human behaviour in relationships of people with the environment [86,87]. In an article on constructing the image of a country [88], the authors compared constructs of Table 1, in terms of country image, from a semantic dimension (meaning): Cognitive component = designative meaning, affective component = appraisive meaning, and conative component = prescriptive meaning, and equating them with the attitude toward the product, from a pragmatic dimension (purpose).

2.3. Proposed Model to Analyse the Image

Figure 2 shows an adaptation of Pocock and Hudson's model (Table 1), with the addition of facilities and temporal dimension to the designative aspect, and the subdivision of the prescriptive aspect in behavioural and attitudinal responses.

- Facilities. Next to the structure and form that characterize physical image, concept has added facilities that cover the relatively abstract mental image [3] of a tourist resource. The visitor identifies a structure such as a museum, aquarium, spa, hotel, restaurant, or other services related to tourism. Not all authors have considered that services are attributes of the image. In a review of 25 articles on TDI [89], among the attributes commonly used in these studies, only 56% considered accommodation, 60% gastronomy, and 32% transportation. However, in one of the most influential papers in the scientific literature on TDI [90], infrastructure and activities (hotels, restaurants, bars, transport, excursions, etc.) are considered as determining dimensions or attributes of the perceived destination image. As has been shown, for example, the positive influence of the gastronomic experiences impacts the destination image and loyalty [91,92].
- Temporal dimension. The image is built and changes over time [93–95]. For example, a Mediterranean seafront does not have the same image in summer as it does in winter; similarly, the image of Japan is different in the season of flowering cherry trees compared to the rest of the year.
- Prescriptive response. The prescriptive response has been specified, dividing it into behavioural and attitudinal to analyse the tourist's actions and loyalty (see Figure 1).

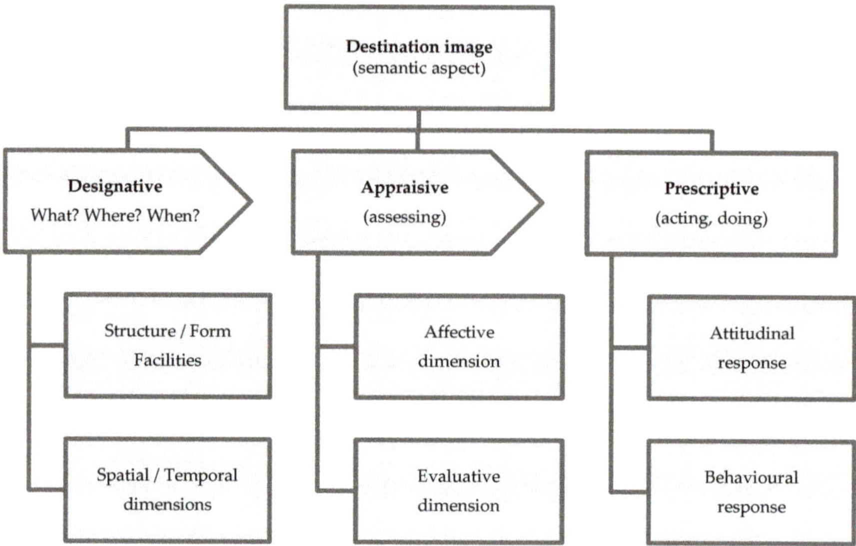

Figure 2. Destination image components, adapted from other authors [66].

2.4. Use of Big Data Analytics in Hospitality and Tourism Research

In a systematic review of literature from the Web of Science and Scopus published between 2000 and 2016, the authors selected 96 articles on big data and tourism, only 17 of which appeared in journals addressing hospitality or tourism [42]. In a similar review of literature from the Web of Science, ScienceDirect, SAGE Journals Online, Emerald Insight, Wiley Online Library and Springer published between 2007 and 2016, the authors examined 144 journal papers and 21 conference papers [41]. In a more recent literature review of work in Scopus published between 1990 and 2017, the authors selected 109 papers for descriptive as well as content analyses [40].

Among research using OTRs as a data source, two systematic literature reviews merit attention; one involved examining 65 papers published between 2000 and 2015 in seven major journals addressing tourism and hospitality [32], whereas the other involved examining 55 papers published between 2008 and 2017 and collected from six popular online databases as well as Google Scholar [33]. Table 2 summarises the research domains of both studies.

Table 2. Industry domains of online reviews of work addressing hospitality and tourism.

Research Domain	Number (Kwok et al. [32])	Percentage	Number (Hlee et al. [33])	Percentage	N	Average
Accommodations	47	72.3	35	63.6	82	68.3%
Restaurants	8	12.3	8	14.5	16	13.3%
Destinations and tourism products	10	15.4	12	21.9	22	18.4%

Other recent studies that have used massive TGC as a data source have been based on reviews of hotels [96,97] or restaurants [98,99]. A study focused on exploring similarities between attractions through 1,695,333 OTRs that highlighted Athens, Cairo and Rome in the category of ancient ruins deserves special attention [100]. As subsequent research has shown, people who use reviews of various tourism products have particular objectives, including opinion mining [101], especially for information about tourist satisfaction [102] and affective image [21]. Recent research based on 25,220 TripAdvisor reviews on things to do in an Italian province [103] is not focused on TDI, but it measures visitor satisfaction through its evaluation that uses between one and five bubbles and implements a content analysis using the commercial application Leximancer.

Among research on TDIs stemming from TGC, the results of which can be compared with those of the study presented here, four studies warrant mention. The first involved collecting and analysing 18,884 travel blogs and OTRs on VirtualTourist and TripAdvisor of the Basilica of La Sagrada Familia in Barcelona [104]; the second did the same with 132,502 travel blogs and OTRs on TravelBlog, VirtualTourist and TripAdvisor addressing Catalonia in general [105]; the third did that with 387,414 OTRs about "Things to Do" in Île de France on TripAdvisor [71]; and the fourth did it with 330,000 OTRs on TripAdvisor regarding Catalan territorial brands [106].

2.5. Online Travel Reviews as a Big Data Source to Analyse the Image

The schema of Figure 2, derived from a theoretical model conceived over 40 years ago [66], is useful for analysing the image perceived by visitors in the time of big data, TGC and social media. The adaptation can be explained by a simple example: A visitor walks through a park and sees a bank with certain characteristics in a particular environment (designative aspect). She/he thinks the bank seems comfortable, and the environment is pleasant (affective dimension). She/he sits on it and writes an online review about the place (behavioural response). She/he rates the place with a high score (evaluative dimension) and recommends the park and the bank (attitudinal response). Moreover, the paratextual elements of the OTR allow situating the experience in space and time (designative aspect).

In addition to the spontaneity of the story, opinion and assessment, OTRs have advantages over information sources based on surveys. For example, to elucidate the prescriptive aspect [13], the respondent had to be asked whether she/he intended to visit the attraction or area (behavioural response), or was asked if she/he thought back to it and recommended it (attitudinal response) to measure loyalty. With the TGC, we know directly what the visitor's behaviour and attitude has been. You can even know if the author had previously visited the area through the paratextual elements of previous OTRs.

The content of OTR can be considered semi-structured information because it houses structured data, but text written by the reviewer does not have a rigid structure that allows a quantitative analysis directly. The web page hosting a review contains three sources of useful data for analysing the image according to the model proposed in Figure 2: textual body of the review, hypertext mark-up language (HTML) metadata, and paratextual elements.

- Textual body. It is the most important part of the OTR. The reviewer recounts his/her experience and gives his/her opinion about the place she/he visited or the tourist resource she/he used. The writing does not contain structured information, except the structure derived from the syntactic grammar rules.
- HTML metadata [71]. Metadata from the web page are intended for reading by Internet browsers and search engines. They give varied information, such as coding and language of the page, but most interesting are those that give information directly related to the OTR, such as title, description, keywords, etc.
- Paratextual elements [107]. The term *paratext* [108] refers to a set of productions (title, preface, author's name, artworks, etc.) that accompany the text of a literary work. This French literary theorist divided the paratext in *peritext* and *epitext* according to the distance of the paratextual elements in relation to the location of the text itself. These productions may be the responsibility of the author, editor/publisher or both. Applying the theory to the case of OTRs, these elements are generated by the web server based on information provided by reviewers and advertisers who market their products or services on the web. Therefore, the title of the OTR, language, date, geographic location, topic or type of resource, the profile of the reviewer with the number of reviews, cities visited, etc., would be peritext. The epitext (related OTRs, contextual advertising, etc.) can be used to follow the path of its links but is not used directly in the analysis of the image.

After seeing the theoretical basis, a case study based on a prominent tourist destination and a popular website dedicated to the promotion and commercialisation of trips is then exposed to empirically demonstrate the usefulness of OTRs as a source of data to analyse the tourist image.

3. Materials and Methods

The proposed methodological framework (Figure 3) is an extension and update of previous research [71] to define and obtain the necessary metrics that allow measuring the image from the proposed theoretical framework (Figure 2). The previous framework was limited to analysing data from OTRs of attractions displayed among results on search engines; however, such information allowed measuring only the cognitive and affective components of TDIs via HTML metadata. By contrast, the current version also allowed analysing the conative component of TDIs in light of all information in OTRs, as well as to include data sources about other tourist resources such as hotels and restaurants.

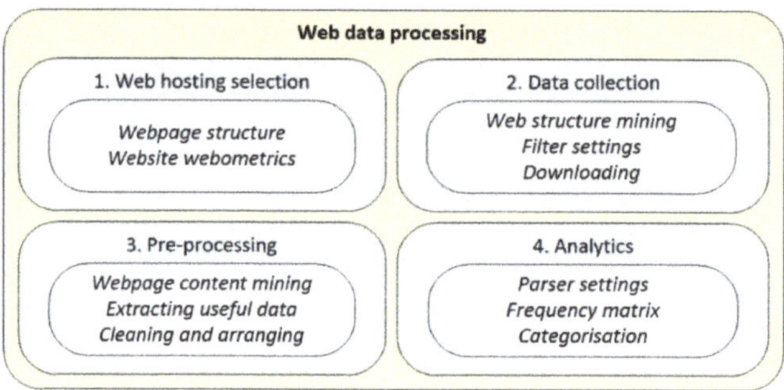

Figure 3. Methodological framework.

3.1. Case Study: Attica

Attica is a historical region of Greece that encompasses the entire metropolitan area of Athens. It is a leading tourist destination with 6.7 million nights spent at tourist accommodation establishments by foreign tourists in 2017 [109]. Its most important tourist resource is the Acropolis, an ancient Greek monumental complex declared a World Heritage Site [59]. Attica is classified by the European Union, as a NUTS-2 region (nomenclature of territorial units for statistics), with the code EL30. To study the spatial dimension of the image, the region has been divided into seven subregions: North Athens (NA), Athens (At), South Athens (SA), East Attica (EA), West Attica (WA), Piraeus (Pi), and islands (Is: mainly Saronic Gulf islands).

3.2. Webhost Selection and Data Collection

Based on the webometrics of popularity, visibility and size [110], TripAdvisor is selected as the most suitable website for the case study. TripAdvisor [111] hosts almost a million reviews and opinions about the Greek region of Attica. Once the filters are established, OTRs of three sections (Things to Do, Restaurants, and Hotels and Places to Stay) are downloaded through a web copy programme.

3.3. Pre-Processing

The relevant information for the case study can be extracted from the OTR webpages, through a search utility that supports regular language expressions (search patterns). The main data sources are the textual and paratextual elements, as well as some HTML metadata contained in the webpage (see Section 2.4). The extracted information must be debugged and arranged for further processing.

The most-representative language of the OTRs posted by foreign visitors is English. To delimit the temporal dimension, OTRs written in English between 2013 and 2018 were selected. As an exploratory study and to facilitate comparison of the metrics between the three segments, a random sample of 100,000 OTRs in each section was extracted (Tables A1–A3, Appendix A). The representativeness of the samples was different in each case (75% attractions, 95% hotels, and 60% restaurants). Figure 4 shows the temporal distribution of the sample.

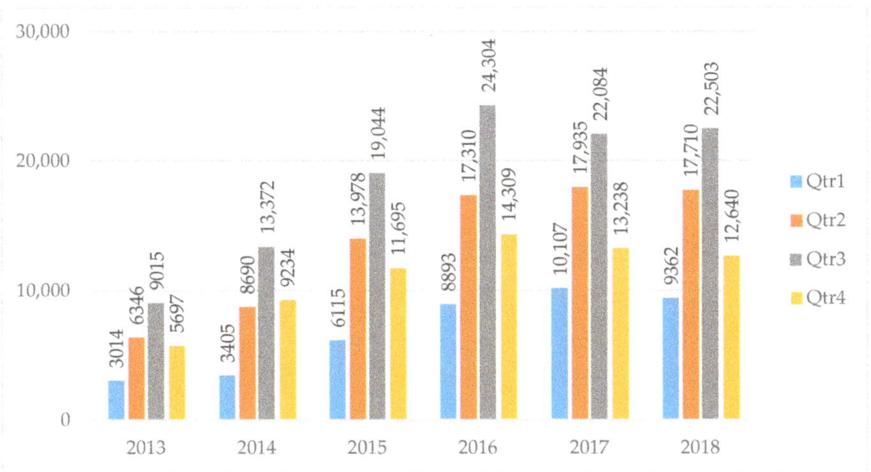

Figure 4. Temporal distribution of 300,000 TripAdvisor online travel reviews on Attica.

3.4. Analytics

Content analysis is a research method for making valid inferences from meaningful matter that contains useful, diverse and unstructured information, through mapping symbolic data into a data matrix suitable for statistical analysis [112]. The most-used techniques are based on a word-frequency count because the most-mentioned words reflect greater interest [113].

Figure 5 shows the algorithm used to generate the word-frequency table based on the text extracted from the OTR webpages. The list of composite words contains groups of two or more words that have a different meaning from words alone (e.g., Temple of Olympian Zeus, must-see, not a must). The blacklist contains words that are not significant in the case study (e.g., determiners, pronouns, adverbs, prepositions, conjunctions). In the case of overlap between words, the algorithm gives priority to composite words. For example, "not so nice" has preference over "not" and "so" (stop words) and over "nice" (keyword). If the overlap is between composite words, the algorithm gives preference by list order. For example, "not do it" has preference over "do it" because it is earlier in the list.

For quantitative and thematic content analysis, categories based on word-frequency tables are constructed. The categories, which must be mutually exclusive and exhaustive, include words or groups of words with similar meaning or connotation [113] excluding polysemous words. To avoid a process of lemmatisation, inflected forms of keywords (e.g., amaze, amazing, amazingly) are included within the categories. The main categories and metrics related to the model in Figure 2 are detailed below.

- Structure/form and facilities. Each touristic resource of TripAdvisor has a code and name. From the outset, TripAdvisor hosts in different sections OTRs about attractions, hotels and restaurants. The attractions are classified according to their type (e.g., monuments and statues, museums), activity (e.g., sightseeing tours, outdoor activities), or service (e.g., transportation, taxis and shuttles). Hotels have a rating of one to five stars. Star ratings indicate the general level of features and amenities to expect. They are provided to TripAdvisor by third-party partners such as Expedia and Giata. Restaurants include bars, cafes and pubs that serve food and are classified by region (e.g., Mediterranean, Asian), country (e.g., Greek, Italian), type of food (e.g., steakhouse, seafood) and other considerations (e.g., fast food, healthy). Metrics derived from the above data provide valuable information about the designative component of the image. Another important metric is popularity, which measures the number of OTRs for each resource sent during the period studied.

- Spatial and temporal dimensions. The resources of TripAdvisor have the name, geographic code and country of the tourist destination (e.g., Spata, Hydra) where they are located. In some cases, they also have the region (e.g., East Attica, Piraeus). With this information, the subregions of Section 3.1 have been delimited. The temporal dimension depends on the date of the experience (perceived image) and the date of publication of the OTR (projected image). Due to the great proliferation of mobile devices, there is little difference between both dates. The analysis has been made based on the publication date because from this moment, the OTR is available online for any user.

- Evaluative dimension. All TripAdvisor OTRs on tourism resources have a score of between one and five bubbles: Excellent (5 bubbles), Good (4 bubbles), Average (3 bubbles), Poor (2 bubbles), and Terrible (1 bubble). The proposed metrics distinguish between the number of positive and negative evaluations: positive scores (*score+*) = 5 bubbles + 4 bubbles, and negative scores (*score−*) = 2 bubbles + 1 bubble Another metric (*average score*) results from calculating the weighted average after converting the bubble ratings to a scale of zero to ten: 5 bubbles = 10; 4 bubbles = 7.5; 3 bubbles = 5; 2 bubbles = 2.5; and 1 bubble = 0.

- Affective dimension. Sentiment analysis tries to deduce from the content of a message the positive or negative polarity of the feelings and moods of the author from, mainly, adjectives that she/he has used. Intensity is difficult to quantify (e.g., amazing, simply amazing, pretty amazing, absolutely amazing). In this case, of study, the affective dimension has been quantified by the number of adjectives and other words that indicate positive or negative feelings or moods. From the table of frequency of words and a lexicon with a list of positive words (e.g., beautiful, happy) and negative words (e.g., crowded, disappointed), two metrics are proposed: positive feelings (*feelings+*) and negative feelings (*feelings−*). Both are calculated by the percentage of positive or negative words in relation to the total number of words (including stop words). The affective dimension can also be classified by topics, building categories of keywords with positive or negative polarities. For example, OTRs of visits to popular attractions often include complaints about crowds, queues and waiting times. In such a case, a category for crowdedness can be constructed to measure the intensity of the sustainability problem. Other categories with negative connotations could be riskiness and dirtiness.

- Attitudinal and behavioural responses. In studies of tourist loyalty, attitudinal loyalty refers to tourists' intention to recommend place or tourist resource, and behavioural loyalty focuses on intention to visit or revisit the place [13]. By analysing OTRs content, it is not necessary to ascertain tourists' intention because you can directly know their behaviour and attitude. Applying the same method of the previous paragraph and a lexicon of positive recommendations (e.g., must-see, cannot miss, recommend) and negative recommendations and warnings (e.g., avoid, do not stay here, would not recommend), two metrics are proposed: positive recommendations (*recommendations+*) and negative recommendations and warnings (*recommendations−*).

The sentiment analysis algorithm goes through the keyword-frequency table and classifies the following locutions from the example OTR (Box 1): lovely (positive feeling); be aware of (warning/negative recommendation); 2 pickpockets (negative feelings); suffering and shame (moods/negative feelings); and worth a visit (positive recommendation). The algorithm classifies the keywords based on the available categories.

```
Load compWord;    // composite word (list of)
Load stopWord;    // non-significant word (list of)
Load textWord;    // text to analyse quantitatively
New   count: 0;   // word counter
New   result;     // table to store frequency of keywords

for each compWord do
{
    if exists compWord in textWord then
    {
        Count occurrences of compWord in textWord;
        Add compWord to result with its frequency;
        Delete occurrences of compWord in textWord;
        Sum frequency to count;
    }
}
for each word in textWord do
{
    if not exists word in stopWord then   // becomes a keyword
    {
        if exists keyword in result increase its frequency;
        else  add keyword to result with frequency 1;
    }
    Increase count;
}
Print result and count;
```

Figure 5. Algorithm to generate the word-frequency table, derived from Marine-Roig [71].

Box 1. TripAdvisor online travel review example (without personal identification data).

> *Reviewer*: A TripAdvisor Member; *from*: New York City, New York; *score*: 4; *date*: 2013-07-24; *attraction*: Plaka; *location*: Athens; *language*: English; *title*: A lovely place to potter, unwind, dine and shop; *text*: The title says it all, but two things to be aware of—pickpockets and graffiti. Apparently, Athens is suffering with Eastern European gangs of pickpockets, like London so keep your hand on your wallet. And there is a lot of graffiti which is quick a shame. None the less worth a visit.

4. Results

The first results of the spatial and temporal dimensions of the image arise directly from the extraction and arrangement of the data (Section 3.3). The metropolis (Athens) accumulates 90% of the OTRs on attractions (Table A1), 72% on hotels (Table A2) and 69% on restaurants (Table A3). The Islands are in second position in relation to the OTRs on hotels and restaurants. In all cases (Figure 4), the third quarter stands out due to the high number of OTRs. In terms of trends, the number of OTRs increased between 2013 and 2016 and decreased in the following years in the case of attractions (Table A1) and hotels (Table A2). On the contrary, OTRs on restaurants continue to grow during 2017 and 2018 (Table A3).

Table A10 shows the 12 most frequent keywords in the text of the OTRs (31.8 million words). Highlights Athens (69,297 occurrences) in attractions, hotel (190,608 occurrences) in hotels, and food (92,552 occurrences) in restaurants. In the three columns, there are keywords related to positive feelings (e.g., good, great). The crowdedness-related keywords crowd/s/ed/ing (7386), overcrowd/ed/ing (186), busy (2623), line/s (3392), queue/s (1139), wait/ed/ing (5243), and await/ing (72), together with ticket/s/ing (8563), and the dirtiness-related keywords abandoned (79), derelict (11), dirty (293), garbage (53), junk (101), rubbish (51), ruined (230), and trash (75), together with graffiti (420), that may indicate sustainability problems, appear in the attractions column. The riskiness-related keywords crime/inal/inally (103), danger/ous (122), pick pocket pickpocket/s (422), robbed robbery (66), steal stole/n (416), and thief/ves (97) also appear in the attractions column.

4.1. Designative Aspect

Table A4 shows the ten most-popular attractions and activities. First, there is the Acropolis [59] and its museum with twice as many reviews as the third classified. Among these top ten, there are three companies dedicated to commercialising transportation and tours. Tours are the most frequent type of activity (Table 3), although it should be remembered that an attraction or activity can be classified according to different concepts (e.g., private tour and transportation companies in the ranks of sixth, ninth and tenth in Table A4).

Table 3. Most frequent types of attractions and activities (Table A4, column Type).

Type	%	Type	%	Type	%
Tours	24.65	Shopping	4.13	Private Tours	2.80
Museums	6.85	Bars & Clubs	3.78	Gift & Speciality Shops	2.56
Sights & Landmarks	6.05	Transportation	3.73	Boat Tours & Water Sports	2.50
Outdoor Activities	5.68	Fun & Games	2.98	City Tours	1.92
Nightlife	4.53	Taxis & Shuttles	2.90	Spas & Wellness	1.71
Bar	4.53	Nature & Parks	2.88	Beaches	1.68

Table A6 shows the ten most-popular hotels. Unlike most popular attractions and restaurants concentrated in Athens, a hotel is located outside of Athens and has the highest number of OTRs. All hotels have four or five TripAdvisor stars, except one that has three stars. Table 4 provides information about the distribution of hotels by star rating (e.g., hotels in the ranks of sixth, seventh and eighth in Table A6). Almost one in three hotels has four stars, followed by three-star (27.09%) and two-star (26.25%) hotels.

Table 4. Frequency (%) of hotels by star rating (Table A6, column Class).

5 Stars	4.5 Stars	4 Stars	3.5 Stars	3 Stars	2.5 Stars	2 Stars	1.5 Star	1 Star
9.99	0.17	30.74	4.57	22.52	1.19	25.06	0.17	5.59

Table A8 shows the ten most-popular restaurants. The first two offer local cuisine (Greek—Mediterranean). Table 5 shows that most restaurants specialise in local cuisine. It must be considered that a restaurant can be classified by different concepts (e.g., restaurants in the ranks of first and second in Table A8).

Table 5. Most frequent specialisation of establishments serving food (Table A8, column Type).

Region	%	Country	%	Structure	%	Kind of Food	%
Mediterranean	17.66	Greek	25.22	Cafe	7.10	Seafood	4.79
European	3.95	Italian	3.57	Bar	6.12	Vegetarian	4.62
American	1.95	Japanese	0.71	Pizza	1.90	Steakhouse	2.68
Asian	1.01	Chinese	0.59	Pub	1.77	Fast food	2.37

4.2. Appraisive and Prescriptive Aspects

Table A5 shows the scores of the ten most-popular attractions or activities. In general, all the scores are very high, but the best scores are for companies dedicated to organising tours. Table A7 shows the scores of the ten most-popular hotels. Only one hotel exceeds the weighted average score of nine points. Table A9 shows the scores of the ten most-popular restaurants. Most of these restaurants have an excellent rating.

Table 6 shows a summary of the results of the sentiment analysis, according to the metrics defined in Section 3.4 (a real-world and comprehensive example related to appraisive and prescriptive stages can be found at the end of Section 3). The evaluative and affective dimensions of the appraisal aspect are both necessary because they can have different values. For example, attractions in Table 6 have the best weighted average of all scores (evaluative dimension) and the highest percentage of keywords related to recommendations and warnings (attitudinal response) but the lowest percentage of keywords related to feelings and moods (affective dimension).

Table 6. Summary of sentiment analysis (percentage) according to the metrics defined in Section 3.4.

Resource	Feelings+	Feelings−	Recommendation+	Recommendation−	Score+	Score−	Average Score
Attractions	3.9687	0.4254	0.5264	0.0709	92.1160	2.4680	87.1465
Hotels	4.6561	0.5883	0.2410	0.0421	80.4760	7.5390	75.6139
Restaurants	5.4675	0.6267	0.4350	0.0457	85.0030	7.3230	81.9132

4.3. Discussion

The seasonality problem shown in Figure 4 is present in other countries of the Mediterranean coast with an increased tourist influx during the summer [105]. Regarding sustainability, the frequency of keywords (Table A10) related to crowdedness in visits to the attractions is much lower than that of other attractions such as the Louvre Museum in Paris [71] or the Basilica of the Sagrada Familia in Barcelona [104]. In all sections (attractions, hotels and restaurants), the spatial dimension shows a high concentration of OTRs in the metropolis and very little in the West Attica subregion. Among the ten most-popular resources, only one hotel is outside Athens in the East Attica subregion. Most restaurants (Table 5) specialise in local cuisine (Greek—Mediterranean). Remarkably, restaurants classified as fast food (2.37%) outweigh those classified as healthy (0.48%). In terms of trends, it is noteworthy that the number of restaurants' OTRs is growing, while it is decreasing in the case of attractions and hotels. Table 6 shows a certain contrast between the evaluative and affective dimensions of the image [71]. As in other cities [106], companies dedicated to organising tours get better scores than attractions declared World Heritage Sites. According to a recent study [100], Attica stands out for its attractions classified as ancient ruins. Overall, the image of Attica is very positive (Table 6), coinciding with a previous study on the TDI of Athens during the recession years [21].

5. Concluding Remarks

The proposed framework allows deducing TDIs from big data extracted from OTRs on sites and tourist resources at destinations. Regarding the case study, the image of Attica has been highly positive, especially in relation to restaurants and attractions. In the random sample of 300,000 OTRs, over 250,000 reviews are for the city of Athens. The two most reviewed restaurants, with a score higher than nine, offer Greek cuisine. Similarly, the Acropolis [59], its museum and the Parthenon stand out in the case of attractions. The hotels have a high level of features and amenities; more than 40% have a rating between four and five stars. In relation to visitors' loyalty, the positive recommendations far outweigh the negative ones. Several private tour and transportation companies are very popular and well valued by reviewers.

5.1. Theoretical Implications

The scientific study of the image of cities and, later, of tourist destinations has continued for more than half a century. After the seminal work of Lynch [3], influential authors in the field have included Hunt [6], Crompton [15], Chon [9], Gartner [18] and Baloglu and McCleary [17]. However, none of those researchers from the last century could have imagined the incredible increase of UGC spread via social media. The dramatic expansion of TGC has induced a paradigm shift in research on travel, tourism and hospitality, and consequently, surveys and in-depth interviews are no longer essential to gathering information on the opinions of visitors about tourist destinations, because it can be obtained for free from social media networks. This TGC constitutes a new and unsolicited organic image-formation agent in Gartner's model [18], with a penetration in the market, through eWoM, higher than that of the induced and autonomous sources.

Most image studies took into consideration the sights for travellers to visit as the main attributes of the destination. Now, as several surveys have shown [30,45], users consult TGC online mainly about hotels and restaurants at the destination. Without undermining the strong influence of tangible heritage on cities' images, both the accommodation sector and the gastronomic image contribute to TDI formation. In this regard, the diagrams in Figures 1 and 2 represent an all-encompassing model to measure TDIs and, indirectly, several aspects related to the sustainability of the destination, as well as the satisfaction and loyalty of visitors. In short, the study presented here involved an attempt to offer an initial integrated framework for analytics on TDIs from a massive amount of TGC, based on the visitor's experience on sightseeing, lodging and dining in the tourist destination, that other researchers can scrutinise, discuss or develop.

Moreover, Figure 1 shows a series of personal and social variables that affect the image perceived by tourists. Big data neutralises this subjective bias because it allows adding the opinions of hundreds of thousands of people, from different countries and cultures, on many places and tourist resources, which collectively constitutes the image as a whole.

5.2. Managerial Implications

Until now, existing studies on OTRs focused on destinations, accommodations, restaurants, or attractions separately. Only 14.5% of the investigations were focused on overall tourism products [33]. In this sense, the proposed metrics have application for OTRs on any tourist place, product or resource. These metrics allow measuring and comparing the image perceived by tourists on two or more resources or groups of tourist resources, places, cities, countries, and regions in certain years or seasons of the year. Based on the paratextual elements, OTRs can be segmented by languages and tourists' nationality [106].

The information obtained with the proposed metrics can be useful for DMOs, because it is based on the opinions and evaluations freely expressed by visitors, which allows deducing their preferences, needs and degrees of satisfaction. The findings can be complemented with results from other sources of big data or with those of conventional approaches that rely on communication-based methods [114]. Knowledge of TDIs perceived by visitors can inform ways to enhance the sustainability of tourist destinations by appropriately distributing available resources at those destinations. The proposed metrics are also useful for extracting business intelligence. For example, the managers of a hotel or a restaurant can compare their results with those of similar properties to gain insights; or they can compare the results before and after making renovations to an establishment. In addition, the crowdedness category can serve to evaluate the success of reforms or changes in the systems of ticket purchases and access to a popular museum, for example.

For policymakers in Attica, the results in the dirtiness and riskiness categories indicate that it would be beneficial to refurbish and clean the urban area of Athens, as well as to improve the area's safety for visitors, especially by controlling the behaviours of graffiti artists and pickpockets. Furthermore, amid the growing popularity of restaurants in the region, most of which offer Greek cuisine, it remains incomprehensible that restaurants classified as fast-food establishments have increased by fivefold compared to those classified as healthy restaurants. Since tourism is highly concentrated in the metropolis, it would be advisable to promote other areas with highly attractive natural resources for tourists, particularly by planning the sustainable tourism exploitation of paradisiacal islands that comprise the region.

5.3. Limitations and Future Research

A limitation of the study was that attractions and restaurants can be classified into several categories on TripAdvisor. In addition, it is virtually impossible to build mutually exclusive and exhaustive categories related to sentiments, crowdedness, dirtiness, riskiness, etc., but correct classification of keywords can be obtained in most cases. Moreover, although the samples were based on a function that generates random numbers of 15 decimals between zero and one, which makes it nearly impossible for repetitions to occur, their representativeness remains questionable. Although that method was performed to facilitate a superficial comparison of results, it also allowed working with all of the available information, while the proposed metrics facilitated the statistical exploitation of the data.

Although the TripAdvisor website has the highest number of OTRs, it would be interesting to contrast results from other online reviews platforms [114] such as Booking, Expedia, Yelp, Ctrip or Airbnb, especially in the accommodation sector.

Funding: The author is pleased to acknowledge the support of the Spanish Ministry of Economy, Industry and Competitiveness (Grant id.: TURCOLAB ECO2017-88984-R).

Conflicts of Interest: The author declares no conflict of interest.

Appendix A

Table A1. Spatial and temporal distribution of 100,000 TripAdvisor OTRs on Attractions.

Resource	Year	Athens	EA	WA	NA	SA	Piraeus	Islands
Attractions	2013	6717	133	7	13	75	98	96
	2014	9447	274	7	20	166	199	292
	2015	15,191	547	19	37	281	300	463
	2016	19,843	714	29	155	416	448	667
	2017	19,654	644	23	143	414	436	602
	2018	19,113	804	29	118	293	394	679

Table A2. Spatial and temporal distribution of 100,000 TripAdvisor OTRs on Hotels.

Resource	Year	Athens	EA	WA	NA	SA	Piraeus	Islands
Hotels	2013	8258	1226	43	223	305	507	1244
	2014	11,019	1401	30	286	472	583	1485
	2015	13,372	1685	59	358	604	608	1669
	2016	14,327	2011	53	432	681	751	1988
	2017	13,427	1630	58	397	606	519	1563
	2018	11,750	1502	47	289	592	429	1511

Table A3. Spatial and temporal distribution of 100,000 TripAdvisor OTRs on Restaurants.

Resource	Year	Athens	EA	WA	NA	SA	Piraeus	Islands
Restaurants	2013	3722	220	5	297	294	103	486
	2014	6455	450	8	525	529	208	845
	2015	11,035	781	26	952	1000	458	1387
	2016	14,948	1423	51	1409	1623	626	2221
	2017	15,866	1568	76	1324	1594	669	2151
	2018	17,019	1718	81	1260	1633	685	2269

Table A4. Top 10 Attractions and activities by number of TripAdvisor OTRs.

Resource Name	Region	Count	Type
Acropolis Museum	Athens	13,258	Museums, History Museums
Acropolis	Athens	12,082	Sights and Landmarks, Historic Sites, Ancient Ruins
Parthenon	Athens	5285	Sights and Landmarks, Points of Interest and Landmarks, Historic Sites, Architectural Buildings, Ancient Ruins

Table A4. *Cont.*

Resource Name	Region	Count	Type
Plaka	Athens	5157	Other, Sights and Landmarks, Neighbourhoods
Archaeological Museum	Athens	2540	Museums, History Museums, Art Museums
Welcome Pickups	Athens	2310	Transportation, Taxis and Shuttles
Panathenaic Stadium	Athens	2130	Sights and Landmarks, Arenas and Stadiums
Mount Lycabettus	Athens	2016	Sights and Landmarks, Lookouts
Private Greece Tours	Athens	1797	Tours, Sightseeing Tours, Day Trips, Multi-day Tours, Private Tours, Archaeology Tours
George's Taxi	Athens	1778	Tours, Transportation, Multi-day Tours, Taxis and Shuttles, City Tours, Sightseeing Tours, Private Tours

Table A5. Reviewer's scores for the top 10 Attractions and activities by number of TripAdvisor OTRs.

Resource Name	Region	Count	5 Bubbles	4 Bubbles	3 Bubbles	2 Bubbles	1 Bubble	Score
Acropolis Museum	Athens	13,258	10,220	2335	560	110	33	9.26
Acropolis	Athens	12,082	9072	2291	569	96	54	9.19
Parthenon	Athens	5285	4089	942	205	35	14	9.28
Plaka	Athens	5157	3104	1562	410	64	17	8.72
Archaeological Museum	Athens	2540	1831	525	143	28	13	9.07
Welcome Pickups	Athens	2310	2144	69	18	13	66	9.56
Panathenaic Stadium	Athens	2130	1253	582	259	32	4	8.58
Mount Lycabettus	Athens	2016	1334	528	121	25	8	8.91
Private Greece Tours	Athens	1797	1752	39	3	1	2	9.92
George's Taxi	Athens	1778	1748	25	3	1	1	9.95

Table A6. Top 10 Hotels by number of TripAdvisor OTRs.

Resource Name, Place	Region	Count	Class
Sofitel Athens Airport, Spata	East Attica	3042	5 stars
Hilton Athens, Athens	Athens	2668	5 stars
Hotel Grande Bretagne, a Luxury Collection Hotel, Athens	Athens	2513	5 stars
The Athens Gate Hotel, Athens	Athens	2404	4 stars
Electra Palace Athens, Athens	Athens	2272	5 stars
Plaka Hotel, Athens	Athens	2072	3 stars
Royal Olympic, Athens	Athens	2031	5 stars
Herodion Hotel, Athens	Athens	1879	4 stars
St. George Lycabettus Lifestyle Hotel, Athens	Athens	1782	5 stars
InterContinental Athenaeum, Athens	Athens	1778	5 stars

Table A7. Reviewer's scores for the top 10 Hotels by number of TripAdvisor OTRs.

Resource Name	Region	Count	5 Bubbles	4 Bubbles	3 Bubbles	2 Bubbles	1 Bubble	Score
Sofitel Athens Airport	EA	3042	1577	941	344	111	69	8.16
Hilton Athens	Athens	2668	1484	779	249	95	61	8.31
Hotel Grande Bretagne	Athens	2513	1938	418	98	31	28	9.19
The Athens Gate Hotel	Athens	2404	1414	811	135	30	14	8.72
Electra Palace Athens	Athens	2272	1411	634	155	49	23	8.70
Plaka Hotel	Athens	2072	1230	709	101	24	8	8.78
Royal Olympic	Athens	2031	681	694	366	178	112	7.04
Herodion Hotel	Athens	1879	1080	644	122	22	11	8.67
St. George Lycabettus Lifestyle Hotel	Athens	1782	633	647	294	146	62	7.30
InterContinental Athenaeum	Athens	1518	683	578	179	51	27	8.03

Table A8. Top 10 Restaurants by number of TripAdvisor OTRs.

Resource Name	Region	Count	Type
Arcadia Restaurant	Athens	1192	Seafood, Mediterranean, Greek
Lithos	Athens	1190	Mediterranean, Greek, Vegetarian Friendly
O Thanasis	Athens	1179	Fast food, Mediterranean, Barbecue
Ta Karamanlidika tou Fani	Athens	1167	Middle Eastern, Mediterranean, Greek
Liondi Traditional Greek Restaurant	Athens	1096	Mediterranean, Greek, Vegetarian Friendly
Oineas Restaurant	Athens	888	Mediterranean, European, Greek
Avocado	Athens	807	Mediterranean, European, Greek
Gods Restaurant	Athens	738	Mediterranean, Greek, Contemporary
Smile Café Restaurant	Athens	719	Mediterranean, Greek, Vegetarian Friendly
Oroscopo	Athens	687	Italian, European, Greek

Table A9. Reviewer's scores for the top 10 Restaurants by number of TripAdvisor OTRs.

Resource Name	Region	Count	5 Bubbles	4 Bubbles	3 Bubbles	2 Bubbles	1 Bubble	Score
Arcadia Restaurant	Athens	1192	916	205	44	15	12	9.19
Lithos	Athens	1190	944	189	39	12	6	9.31
O Thanasis	Athens	1179	606	376	119	49	29	8.14
Ta Karamanlidika tou Fani	Athens	1167	962	159	33	8	5	9.42
Liondi Traditional Greek Restaurant	Athens	1096	916	124	34	15	7	9.40
Oineas Restaurant	Athens	888	712	125	41	7	3	9.32
Avocado	Athens	807	636	129	34	4	4	9.30
Gods Restaurant	Athens	738	421	200	65	22	30	8.25
Smile Café Restaurant	Athens	719	479	153	52	19	16	8.69
Oroscopo	Athens	687	552	107	19	5	4	9.36

Table A10. Content generated by reviewers: 12 most frequent keywords.

	Attractions		Hotels		Restaurants	
	Total: 9,242,536	Unique: 54,409	Total: 14,226,589	Unique: 57,868	Total: 8,332,330	Unique: 54,511
	Keyword	Count	Keyword	Count	Keyword	Count
1	athens	69,297	hotel	19,0608	food	92,552
2	tour	49,202	room	96,628	good	61,763
3	great	40,617	great	71,927	great	53,511
4	acropolis	38,394	good	68,671	place	45,902
5	museum	34,652	staff	67,189	service	45,279
6	time	30,081	breakfast	63,328	restaurant	42,912
7	day	28,535	location	62,276	nice	32,484
8	visit	26,222	athens	60,205	greek	31,969
9	place	22,889	stay	52,209	athens	31,438
10	history	21,540	rooms	48,536	best	25,277
11	amazing	20,890	nice	48,429	excellent	24,016
12	good	20,444	clean	43,029	friendly	23,142

References

1. Ritchie, J.R.B.; Crouch, G.I. The competitive destination: A sustainability perspective. *Tour. Manag.* **2000**, *21*, 1–7.
2. Ahmed, Z.U. The need for the identification of the constituents of a destination's tourist image: A promotion segmentation perspective. *J. Prof. Serv. Mark.* **1996**, *14*, 37–60.
3. Lynch, K. *The Image of the City*; The MIT Press: Cambridge, MA, 1960; ISBN 9780262120043.
4. Roth, K.P.; Diamantopoulos, A. Advancing the country image construct. *J. Bus. Res.* **2009**, *62*, 726–740. [CrossRef]
5. Gim, T.-H. Tourist satisfaction, image, and loyalty from an interregional perspective: An analysis of neighboring areas with distinct characteristics. *Sustainability* **2018**, *10*, 1283. [CrossRef]
6. Hunt, J.D. Image as a factor in tourism development. *J. Travel Res.* **1975**, *13*, 1–7. [CrossRef]

7.   Mayo, E.J. Regional images and regional travel development. In Proceedings of the Travel and Tourism Research Association Fourth Annual Conference, Salt Lake City, Utah, USA, 8–11 September 1973; pp. 211–217.

8.   Gunn, C.A. *Vacationscape: Designing Tourist Regions;* Bureau of Business Research, University of Texas: Austin, TX, USA, 1972; ISBN 978-0877551614.

9.   Chon, K.-S. The role of destination image in tourism: A review and discussion. *Tour. Rev.* **1990**, *45*, 2–9. [CrossRef]

10.   Li, J.; Ali, F.; Kim, W.G. Reexamination of the role of destination image in tourism: An updated literature review. *E-Rev. Tour. Res.* **2015**, *12*, 191–209.

11.   Pike, S. Destination image analysis: A review of 142 papers from 1973–2000. *Tour. Manag.* **2002**, *23*, 541–549. [CrossRef]

12.   Echtner, C.M.; Ritchie, J.R.B. The meaning and measurement of destination image. *J. Tour. Stud.* **1991**, *2*, 2–12.

13.   Zhang, H.; Fu, X.; Cai, L.A.; Lu, L. Destination image and tourist loyalty: A meta-analysis. *Tour. Manag.* **2014**, *40*, 213–223. [CrossRef]

14.   Stepchenkova, S.; Mills, J.E. Destination image: A meta-analysis of 2000–2007 research. *J. Hosp. Mark. Manag.* **2010**, *19*, 575–609. [CrossRef]

15.   Crompton, J.L. An assessment of the image of Mexico as a vacation destination and the influence of geographical location upon that image. *J. Travel Res.* **1979**, *17*, 18–23. [CrossRef]

16.   Goodrich, J.N. The relationship between preferences for and perceptions of vacation destinations: Application of a choice model. *J. Travel Res.* **1978**, *17*, 8–13. [CrossRef]

17.   Baloglu, S.; McCleary, K.W. A model of destination image formation. *Ann. Tour. Res.* **1999**, *26*, 868–897. [CrossRef]

18.   Gartner, W.C. Image formation process. *J. Travel Tour. Mark.* **1993**, *2*, 191–215. [CrossRef]

19.   Marine-Roig, E.; Ferrer-Rosell, B. Measuring the gap between projected and perceived destination images of Catalonia using compositional analysis. *Tour. Manag.* **2018**, *68*, 236–249. [CrossRef]

20.   Sotiriadis, M.D. Sharing tourism experiences in social media: A literature review and a set of suggested business strategies. *Int. J. Contemp. Hosp. Manag.* **2017**, *29*, 179–225. [CrossRef]

21.   Gkritzali, A.; Gritzalis, D.; Stavrou, V. Is Xenios Zeus still alive? Destination image of Athens in the years of recession. *J. Travel Res.* **2018**, *57*, 540–554. [CrossRef]

22.   Garay Tamajón, L.; Cànoves Valiente, G. Barcelona seen through the eyes of TripAdvisor: actors, typologies and components of destination image in social media platforms. *Curr. Issues Tour.* **2017**, *20*, 33–37. [CrossRef]

23.   Jabreel, M.; Moreno, A.; Huertas, A. Semantic comparison of the emotional values communicated by destinations and tourists on social media. *J. Destin. Mark. Manag.* **2017**, *6*, 170–183. [CrossRef]

24.   Lalicic, L.; Huertas, A.; Moreno, A.; Jabreel, M. Which emotional brand values do my followers want to hear about? An investigation of popular European tourist destinations. *Inf. Technol. Tour.* **2019**, *21*, 63–81. [CrossRef]

25.   Huertas, A.; Marine-Roig, E. Differential destination content communication strategies through multiple Social Media. In *Information and Communication Technologies in Tourism 2016;* Springer International Publishing: Cham, Switzerland, 2016; pp. 239–252.

26.   Deng, N.; Liu, J.; Dai, Y.; Li, H. Different cultures, different photos: A comparison of Shanghai's pictorial destination image between East and West. *Tour. Manag. Perspect.* **2019**, *30*, 182–192. [CrossRef]

27.   Paül i Agustí, D. Characterizing the location of tourist images in cities. Differences in user-generated images (Instagram), official tourist brochures and travel guides. *Ann. Tour. Res.* **2018**, *73*, 103–115. [CrossRef]

28.   Paül i Agustí, D. Tourist hot spots in cities with the highest murder rates. *Tour. Geogr.* **2019**. [CrossRef]

29.   Marine-Roig, E. Los "Travel Blogs" como objetos de estudio de la imagen percibida de un destino [Travel blogs as objects of study of the perceived destination image]. In *Turismo y Tecnologías de la Información y las Comunicaciones;* Guevara Plaza, A.J., Aguayo Maldonado, A., Caro Herrero, J.L., Eds.; Facultad de Turismo: Málaga, Spain, 2010; pp. 61–76. ISBN 9788460811152.

30.   Gretzel, U.; Yoo, K.H. Use and impact of online travel reviews. In *Information and Communication Technologies in Tourism 2008;* O'Connor, P., Höpken, W., Gretzel, U., Eds.; Springer Vienna: Vienna, Austria, 2008; pp. 35–46.

31.   TripAdvisor, About us. Available online: https://tripadvisor.mediaroom.com/us-about-us (accessed on 1 January 2019).

32. Kwok, L.; Xie, K.L.; Richards, T. Thematic framework of online review research: A systematic analysis of contemporary literature on seven major hospitality and tourism journals. *Int. J. Contemp. Hosp. Manag.* **2017**, *29*, 307–354. [CrossRef]

33. Hlee, S.; Lee, H.; Koo, C. Hospitality and tourism online review research: A systematic analysis and heuristic-systematic model. *Sustainability* **2018**, *10*, 1141. [CrossRef]

34. Liang, T.-P.; Liu, Y.-H. Research landscape of business intelligence and big data analytics: A bibliometrics study. *Expert Syst. Appl.* **2018**, *111*, 2–10. [CrossRef]

35. Jelvehgaran Esfahani, H.; Tavasoli, K.; Jabbarzadeh, A. Big data and social media: A scientometrics analysis. *Int. J. Data Netw. Sci.* **2019**, 145–164. [CrossRef]

36. Mashingaidze, K.; Backhouse, J. The relationships between definitions of big data, business intelligence and business analytics: a literature review. *Int. J. Bus. Inf. Syst.* **2017**, *26*, 488–505.

37. Rathore, A.K.; Kar, A.K.; Ilavarasan, P.V. Social media analytics: Literature review and directions for future research. *Decis. Anal.* **2017**, *14*, 229–249. [CrossRef]

38. Viñan-Ludeña, M.-S. A systematic literature review on social media analytics and smart tourism. In *Smart Tourism as a Driver for Culture and Sustainability*; Katsoni, V., Segarra-Oña, M., Eds.; Springer: Cham, Switzerland, 2019; pp. 357–374.

39. Pourkhani, A.; Abdipour, K.; Baher, B.; Moslehpour, M. The impact of social media in business growth and performance: A scientometrics analysis. *Int. J. Data Netw. Sci.* **2019**, *3*, 223–244. [CrossRef]

40. Centobelli, P.; Ndou, V. Managing customer knowledge through the use of big data analytics in tourism research. *Curr. Issues Tour.* **2019**. [CrossRef]

41. Li, J.; Xu, L.; Tang, L.; Wang, S.; Li, L. Big data in tourism research: A literature review. *Tour. Manag.* **2018**, *68*, 301–323. [CrossRef]

42. Mariani, M.; Baggio, R.; Fuchs, M.; Höepken, W. Business intelligence and big data in hospitality and tourism: a systematic literature review. *Int. J. Contemp. Hosp. Manag.* **2018**, *30*, 3514–3554. [CrossRef]

43. Eurobarometer. *Flash Eurobarometer 432: Preferences of Europeans Towards Tourism*; European Commission: Brussels, Belgium, 2016.

44. VisitBritain Researching and Planning: Foresight—issue 150. Available online: https://www.visitbritain.org/sites/default/files/vb-corporate/Documents-Library/documents/foresight_150_-_researching_and_planning.pdf (accessed on 8 May 2019).

45. Analysts. *The State of the American Traveler. Destinations Edition*; Destination Analysts: San Francisco, CA, USA, 2018; Volume 27.

46. Marine-Roig, E. *From the Projected to the Transmitted Image: The 2.0 Construction of Tourist Destination Image and Identity in Catalonia*; Rovira i Virgili University: Vila-seca, Catalonia, Spain, 2014.

47. Andreu, L.; Bigné, J.E.; Cooper, C. Projected and perceived image of Spain as a tourist destination for British travellers. *J. Travel Tour. Mark.* **2000**, *9*, 47–67. [CrossRef]

48. Ayeh, J.K.; Au, N.; Law, R. Predicting the intention to use consumer-generated media for travel planning. *Tour. Manag.* **2013**, *35*, 132–143. [CrossRef]

49. Ayeh, J.K.; Au, N.; Law, R. "Do we believe in TripAdvisor?" Examining credibility perceptions and online travelers' attitude toward using user-generated content. *J. Travel Res.* **2013**, *52*, 437–452. [CrossRef]

50. Ukpabi, D.C.; Karjaluoto, H. What drives travelers' adoption of user-generated content? A literature review. *Tour. Manag. Perspect.* **2018**, *28*, 251–273. [CrossRef]

51. Yilmaz, B.S. Turkish tourism consumer's information search behavior: The role of user generated content in travel planning process. *EcoForum* **2017**, *6*, 1–6.

52. Mendes-Filho, L.; Mills, A.M.; Tan, F.B.; Milne, S. Empowering the traveler: an examination of the impact of user-generated content on travel planning. *J. Travel Tour. Mark.* **2018**, *35*, 425–436. [CrossRef]

53. Chong, A.Y.L.; Khong, K.W.; Ma, T.; McCabe, S.; Wang, Y. Analyzing key influences of tourists' acceptance of online reviews in travel decisions. *Internet Res.* **2018**, *28*, 564–586. [CrossRef]

54. Nilashi, M.; Ibrahim, O.; Yadegaridehkordi, E.; Samad, S.; Akbari, E.; Alizadeh, A. Travelers decision making using online review in social network sites: A case on TripAdvisor. *J. Comput. Sci.* **2018**, *28*, 168–179. [CrossRef]

55. Ritchie, J.R.B.; Crouch, G.I. A model of destination competitiveness and sustainability. In *Destination Marketing and Management: Theories and Applications*; Wang, Y., Pizam, A., Eds.; CABI: Wallingford, UK, 2011; pp. 326–339.

56. Buhalis, D. Marketing the competitive destination of the future. *Tour. Manag.* **2000**, *21*, 97–116. [CrossRef]
57. Uysal, M.; Harrill, R.; Woo, E. Destination marketing research: Issues and challenges. In *Destination Marketing and Management: Theories and Applications*; Wang, Y., Pizam, A., Eds.; CABI: Wallingford, UK, 2011; pp. 99–112.
58. Kladou, S.; Mavragani, E. Assessing destination image: An online marketing approach and the case of TripAdvisor. *J. Destin. Mark. Manag.* **2015**, *4*, 187–193. [CrossRef]
59. UNESCO Acropolis, Athens. Available online: https://whc.unesco.org/en/list/404 (accessed on 8 May 2019).
60. Fakeye, P.C.; Crompton, J.L. Image differences between prospective, first-time, and repeat visitors to the Lower Rio Grande Valley. *J. Travel Res.* **1991**, *30*, 10–16. [CrossRef]
61. Lai, K.; Li, X. Tourism destination image: Conceptual problems and definitional solutions. *J. Travel Res.* **2016**, *55*, 1065–1080. [CrossRef]
62. Manheim, J.B.; Albritton, R.B. Changing national images: International public relations and media agenda setting. *Am. Polit. Sci. Rev.* **1983**, *78*, 641–657. [CrossRef]
63. Bramwell, B.; Rawding, L. Tourism marketing images of industrial cities. *Ann. Tour. Res.* **1996**, *23*, 201–221. [CrossRef]
64. Boulding, K.E. *The image: Knowledge in life and society*; University of Michigan Press: Ann Arbor, MI, USA, 1956; ISBN 978-0472060474.
65. Bigné, J.E.; Sánchez, M.I.; Sánchez, J. Tourism image, evaluation variables and after purchase behaviour: inter-relationship. *Tour. Manag.* **2001**, *22*, 607–616. [CrossRef]
66. Pocock, D.; Hudson, R. *Images of the Urban Environment*; Macmillan: London, UK, 1978; ISBN 9780333192115.
67. Murphy, P.; Pritchard, M.P.; Smith, B. The destination product and its impact on traveller perceptions. *Tour. Manag.* **2000**, *21*, 43–52. [CrossRef]
68. Elliot, S.; Papadopoulos, N.; Szamosi, L. Studying place image: an interdisciplinary and holistic approach. *Anatolia* **2013**, *24*, 5–16. [CrossRef]
69. Um, S.; Crompton, J.L. Attitude determinants in tourism destination choice. *Ann. Tour. Res.* **1990**, *17*, 432–448. [CrossRef]
70. Marine-Roig, E. Identity and authenticity in destination image construction. *Anatolia Int. J. Tour. Hosp. Res.* **2015**, *26*, 574–587. [CrossRef]
71. Marine-Roig, E. Measuring destination image through travel reviews in search engines. *Sustainability* **2017**, *9*, 1425. [CrossRef]
72. Phelps, A. Holiday destination image—the problem of assessment. An example developed in Menorca. *Tour. Manag.* **1986**, *7*, 168–180. [CrossRef]
73. Britton, R.A. The image of the Third World in tourism marketing. *Ann. Tour. Res.* **1979**, *6*, 318–329. [CrossRef]
74. Mathieson, A.; Wall, G. *Tourism, Economic, Physical and Social Impacts*; Longman: London, UK, 1982.
75. Wita, N.; Ashton, S.A. Tourist perception toward destination brand image sustainability: Mae Kam Pong Community case study. *J. Int. Thai Tour.* **2019**, *14*, 95–125.
76. Jiang, Y.; Ramkissoon, H.; Mavondo, F.T.; Feng, S. Authenticity: The link between destination image and place attachment. *J. Hosp. Mark. Manag.* **2017**, *26*, 105–124. [CrossRef]
77. Chi, C.G.-Q.; Qu, H. Examining the structural relationships of destination image, tourist satisfaction and destination loyalty: An integrated approach. *Tour. Manag.* **2008**, *29*, 624–636. [CrossRef]
78. Hernández-Lobato, L.; Solis-Radilla, M.M.; Moliner-Tena, M.A.; Sánchez-García, J. Tourism destination image, satisfaction and loyalty: A study in Ixtapa-Zihuatanejo, Mexico. *Tour. Geogr.* **2006**, *8*, 343–358. [CrossRef]
79. Wang, C.; Hsu, M.K. The relationships of destination image, satisfaction, and behavioral intentions: An integrated model. *J. Travel Tour. Mark.* **2010**, *27*, 829–843. [CrossRef]
80. Prayag, G.; Hosany, S.; Muskat, B.; Del Chiappa, G. Understanding the relationships between tourists' emotional experiences, perceived overall image, satisfaction, and intention to recommend. *J. Travel Res.* **2017**, *56*, 41–54. [CrossRef]
81. Agapito, D.; Oom do Valle, P.; da Costa Mendes, J. The cognitive-affective-conative model of destination image: A confirmatory analysis. *J. Travel Tour. Mark.* **2013**, *30*, 471–481. [CrossRef]
82. Han, H.; Kim, Y.; Kim, E.-K. Cognitive, affective, conative, and action loyalty: Testing the impact of inertia. *Int. J. Hosp. Manag.* **2011**, *30*, 1008–1019. [CrossRef]
83. Pike, S.; Ryan, C. Destination positioning analysis through a comparison of cognitive, affective, and conative perceptions. *J. Travel Res.* **2004**, *42*, 333–342. [CrossRef]

84. Rapoport, A. *Human Aspects of Urban Form*; Pergamon Press: Oxford, UK, 1977.
85. Stern, E.; Krakover, S. The formation of a composite urban image. *Geogr. Anal.* **1993**, *25*, 130–146. [CrossRef]
86. Walmsley, D.J.; Lewis, G.J. *People & Environment: Behavioural Approaches in Human Geography*; Pearson Education Ltd.: London, UK, 1984.
87. Wakabayasi, Y. Behavioral studies on environmental perception by Japanese geographers. *Geogr. Rev. Jpn. Ser. B* **1996**, *69*, 83–94. [CrossRef]
88. Brijs, K.; Bloemer, J.; Kasper, H. Country-image discourse model: Unraveling meaning, structure, and function of country images. *J. Bus. Res.* **2011**, *64*, 1259–1269. [CrossRef]
89. Gallarza, M.G.; Gil Saura, I.; Calderón García, H. Destination image. *Ann. Tour. Res.* **2002**, *29*, 56–78. [CrossRef]
90. Beerli, A.; Martín, J.D. Factors influencing destination image. *Ann. Tour. Res.* **2004**, *31*, 657–681. [CrossRef]
91. Folgado-Fernández, J.A.; Hernández-Mogollón, J.M.; Duarte, P. Destination image and loyalty development: the impact of tourists' food experiences at gastronomic events. *Scand. J. Hosp. Tour.* **2017**, *17*, 92–110. [CrossRef]
92. Choe, J.Y.; Kim, S. Effects of tourists' local food consumption value on attitude, food destination image, and behavioral intention. *Int. J. Hosp. Manag.* **2018**, *71*, 1–10. [CrossRef]
93. Gartner, W.C. Temporal influences on image change. *Ann. Tour. Res.* **1986**, *13*, 635–644. [CrossRef]
94. Gartner, W.C.; Hunt, J.D. An analysis of state image change over a twelve-year period (1971–1983). *J. Travel Res.* **1987**, *26*, 15–19. [CrossRef]
95. Chon, K.-S. Tourism destination image modification process. *Tour. Manag.* **1991**, *12*, 68–72. [CrossRef]
96. Gerdt, S.-O.; Wagner, E.; Schewe, G. The relationship between sustainability and customer satisfaction in hospitality: An explorative investigation using eWOM as a data source. *Tour. Manag.* **2019**, *74*, 155–172. [CrossRef]
97. Gunasekar, S.; Sudhakar, S. How user-generated judgments of hotel attributes indicate guest satisfaction. *J. Glob. Sch. Mark. Sci.* **2019**, *29*, 180–195. [CrossRef]
98. Li, H.; Wang, C. (Renee); Meng, F.; Zhang, Z. Making restaurant reviews useful and/or enjoyable? The impacts of temporal, explanatory, and sensory cues. *Int. J. Hosp. Manag.* **2018**. [CrossRef]
99. Vu, H.Q.; Li, G.; Law, R.; Zhang, Y. Exploring tourist dining preferences based on restaurant reviews. *J. Travel Res.* **2019**, *58*, 149–167. [CrossRef]
100. McKenzie, G.; Adams, B. A data-driven approach to exploring similarities of tourist attractions through online reviews. *J. Locat. Based Serv.* **2018**, *12*, 94–118. [CrossRef]
101. Hou, Z.; Cui, F.; Meng, Y.; Lian, T.; Yu, C. Opinion mining from online travel reviews: A comparative analysis of Chinese major OTAs using semantic association analysis. *Tour. Manag.* **2019**, *74*, 276–289. [CrossRef]
102. Song, S.; Kawamura, H.; Uchida, J.; Saito, H. Determining tourist satisfaction from travel reviews. *Inf. Technol. Tour.* **2019**. [CrossRef]
103. Franzoni, S.; Bonera, M. How DMO can measure the experiences of a large territory. *Sustainability* **2019**, *11*, 492. [CrossRef]
104. Marine-Roig, E. Religious tourism versus secular pilgrimage: The basilica of La Sagrada Família. *Int. J. Relig. Tour. Pilgr.* **2015**, *3*, 25–37.
105. Marine-Roig, E.; Anton Clavé, S. A detailed method for destination image analysis using user-generated content. *Inf. Technol. Tour.* **2016**, *15*, 341–364. [CrossRef]
106. Marine-Roig, E.; Mariné Gallisà, E. Imatge de Catalunya percebuda per turistes angloparlants i castellanoparlants (Image of Catalonia perceived by English-speaking and Spanish-speaking tourists). *Doc. Anàlisi Geogràfica* **2018**, *64*, 219–245. [CrossRef]
107. Marine-Roig, E. Online travel reviews: A massive paratextual analysis. In *Analytics in Smart Tourism Design: Concepts and Methods*; Xiang, Z., Fesenmaier, D.R., Eds.; Springer: Heidelberg, Germany, 2017; pp. 179–202. ISBN 978-3-319-44262-4.
108. Genette, G. *Paratexts: Thresholds of interpretation*; Cambridge University Press: New York, NY, USA, 1997; ISBN 0 521 41350.
109. Eurostat Tourism. *Eurostat Regional Yearbook*; Publications Office of the European Union: Luxembourg, 2018; pp. 139–150.
110. Marine-Roig, E. A webometric analysis of travel blogs and review hosting: The case of Catalonia. *J. Travel Tour. Mark.* **2014**, *31*, 381–396. [CrossRef]

111. TripAdvisor Attica, Greece. Available online: https://www.tripadvisor.com/Tourism-g189399-Attica-Vacations.html (accessed on 1 January 2019).
112. Roberts, C.W. Content Analysis. *Int. Encycl. Soc. Behav. Sci.* **2001**, 2697–2702.
113. Stemler, S. An Overview of Content Analysis. Available online: https://pareonline.net/getvn.asp?v=7&n=17 (accessed on 8 May 2019).
114. Xiang, Z.; Du, Q.; Ma, Y.; Fan, W. A comparative analysis of major online review platforms: Implications for social media analytics in hospitality and tourism. *Tour. Manag.* **2017**, *58*, 51–65. [CrossRef]

 *sustainability*

Article

Spatiotemporal Analysis to Observe Gender Based Check-In Behavior by Using Social Media Big Data: A Case Study of Guangzhou, China

Rizwan Muhammad *, Yaolong Zhao * and Fan Liu

School of Geography, South China Normal University, Guangzhou 510631, China; liufan@m.scnu.edu.cn
* Correspondence: rizwan@m.scnu.edu.cn (R.M.); zhaoyaolong@m.scnu.edu.cn (Y.Z.)

Received: 13 February 2019; Accepted: 13 May 2019; Published: 17 May 2019

Abstract: In a location-based social network, users socialize with each other by sharing their current location in the form of "check-in," which allows users to reveal the current places they visit as part of their social interaction. Understanding this human check-in phenomenon in space and time on location based social network (LBSN) datasets, which is also called "check-in behavior," can archive the day-to-day activity patterns, usage behaviors toward social media, and presents spatiotemporal evidence of users' daily routines. It also provides a wide range of opportunities to observe (i.e., mobility, urban activities, defining city boundary, and community problems in a city). In representing human check-in behavior, these LBSN datasets do not reflect the real-world events due to certain statistical biases (i.e., gender prejudice, a low frequency in sampling, and location type prejudice). However, LBSN data is primarily considered a supplement to traditional data sources (i.e., survey, census) and can be used to observe human check-in behavior within a city. Different interpretations are used elusively for the term "check-in behavior," which makes it difficult to identify studies on human check-in behavior based on LBSN using the Weibo dataset. The primary objective of this research is to explore human check-in behavior by male and female users in Guangzhou, China toward using Chinese microblog Sina Weibo (referred to as "Weibo"), which is missing in the existing literature. Kernel density estimation (KDE) is utilized to explore the spatiotemporal distribution geographically and weighted regression (GWR) method was applied to observe the relationship between check-in and districts with a focus on gender during weekdays and weekend. Lastly, the standard deviational ellipse (SDE) analysis is used to systematically analyze the orientation, direction, spatiotemporal expansion trends and the differences in check-in distribution in Guangzhou, China. The results of this study show that LBSN is a reliable source of data to observe human check-in behavior in space and time within a specified geographic area. Furthermore, it shows that female users are more likely to use social media as compared to male users. The human check-in behavior patterns for social media network usage by gender seems to be slightly different during weekdays and weekend.

Keywords: social media big data; lbsn; check-in density; spatiotemporal analysis; KDE; GWR; SDE; Guangzhou

---

1. Introduction

In recent years, social media has dramatically expanded in popularity around the world and became an integral part of the information ecosystem in both application and research perspectives due to its unprecedented reach to masses (i.e., users, consumers, businesses, governments, and nonprofit organizations) [1]. Historically, traditional data sources (i.e., survey, census) [2,3] are analyzed to observe human activity behavior [4,5], lifestyle patterns [6], and gender differences [7], but these data sources are considered to be more expensive both in the collection and the analysis. This, in turn, requires more processing time and results in data sparsity. Policymaking and delivery of services are closely intertwined with city planning and human mobility. However, due to the limitations highlighted above, these traditional methods are considered to be less effective in policy-making and delivery of services [8].

A considerable amount of previous research [9–14] studied the demography of social media users and discussed reasons that influence people to use LBSN. These recent research studies highlighted the motivations for social media network use among both male and female users. Smith [9] argued that female users tend to use online social media to interact with families and friends rather than male users. Muscanell and Guadagno [15] put forward that male users mostly use online social media for making new relationships while female users utilize social media more for the maintenance of the relationship. Moreover, the pattern and motivation to use the social media network by both male and female users seems to be slightly different. Hwang and Choi [16] explored the online usage behavior of Weibo by college students and the motivations of usage by gender. Lastly, it was suggested that online usage behavior of Weibo acts as a platform to search for information on social issues and interests. Rossi and Musolesi [17] proposed methodologies to identify the unique users from check-in data and characterized the users by the spatiotemporal trails from the check-ins made over time and the frequency of visit to specific locations [16,18].

As part of online social interaction in LBSN's [19,20], users [21] can announce their geo-location [22], announce the activity performed [23], and discuss places they visit (referred as "check-in" [24]). By the third quarter of 2017, Weibo [25] amounted up to 376 million monthly active users (MAU), 172 million daily active users (DAU). Among the active users, 93% were accessing Weibo through mobile devices [26,27]. This enormous number of users were attracted worldwide due to fast information sharing and check-in phenomenon [28], which generates high volumes of data (referred to as "Big Data" [29,30]). Irrespective of fundamental limits to demonstrate human check-in behavior [31], i.e., prejudice of gender, frequency sampling prejudice, and location type prejudice. Check-in reveals human check-in behavior in space and time. The motivations for using Weibo may differ between male and female users. Statistics show that 50.10% of Weibo users are male, 49.90% are female [32], and it is considered one of the most popular social media platforms in China [33] due to the unavailability of Facebook and Twitter. According to the China Internet Network Information Center (CNNIC) [34], 72% of the total Sina Weibo users were 20 to 35 years of age. Among them, the majority of users are in their 20s and constitute the heaviest users [16,35].

Currently, LBSN data are obtainable at a relatively cheap cost with information such as timestamps, location, and gender [36], and can be analyzed to perceive human check-in behavior as equated to the previously stated traditional datasets. Intrinsically, LBSN data offers new dimensions to help and create new techniques and methods to observe human check-in behaviors [37] and differences in gender. In the current study, we explore the LBSN data to observe human check-in behavior and intensity of check-ins during the period within a city at an individual level. Moving toward this direction, the research presented in the current study aims to investigate the spatiotemporal information related to the check-in to identify and determine human check-in behavior. The simple hypothesis is that people follow a typical daily routine: e.g., go to work, eat at some preferred restaurant, go shopping, and go back home.

Consequently, if we have enough data to observe distinctive human behaviors, such knowledge can be analyzed to understand human behavior by using LBSN check-in information as a proxy measure. Recent research [38–42] explored the LBSN datasets to examine people's daily check-in behavior and mobility patterns in different cities rather than Guangzhou, China. However, most of the existing literature focused on Facebook and/or Twitter rather than Weibo. Therefore, this study will also serve to fill a research gap by focusing on the most popular Chinese local social network site, Sina Weibo, and study area as Guangzhou, China. Moreover, previous studies [16,18] explored gender-based check-in behavior analysis in Sina Weibo and suggested that women are more likely to use Sina Weibo to provide help and information to others. However, studies to date have not fully investigated gender-based check-in behavior analysis in Sina Weibo usage especially in Guangzhou, which also motivated the current study. The primary objective and contribution of the paper are twofold and can be summarized as follows.

1. The primary objective of the research is to characterize behavioral differences between male and female using the "check-in" function of the Sina Weibo (launched by Sina Corporation [43] in 2009).
2. The main contribution of our work consists of examining the check-in density by using KDE. The GWR method was applied to observe the relationship between check-in and districts with a focus on gender during weekdays and weekend. Lastly, the standard deviational ellipse (SDE) analysis is used to analyze the orientation, direction, spatiotemporal expansion trends and the differences of check-in behavior by male and female in Guangzhou, China, which was missing in the existing literature regarding gender-based check-in behavior analysis.

Moreover, this line of research can help improve our understanding of human check-in behavior and consider LBSN data (a source of big data) as a supplement to than a substitute of traditional data sources while taking a decision on policy making [44–46] associated with urban planning [47,48] and city functionalities [49].

The organization of the rest of the paper is as follows. Section 2 presents the literature review. Section 3 defines the study area and dataset. Section 4 presents the methodology. Section 5 presents the results and discussion for the experimental results performed on the dataset. Lastly, Section 6 concludes the paper and proposes some further research issues.

2. Literature Review

The research on spatial analysis has significantly progressed toward observing human behavior, which has long been constrained by traditional data sources with improved abilities to capture, analyze, and process LBSN data [50,51]. The terminology "social network site" (SNS) [52] denotes to web-based services [53] and is a social structure made up of individuals connected by one or more specific types of interdependency, such as friendship, common interests, and shared knowledge [54]. It allows users to (1) construct a profile, (2) articulate users' social links, and (3) track and view shared social ties within the system [55–57]. Moreover, it reflects the real-life social networks among people through online platforms such as a website, providing ways for users to share ideas, activities, events, and interests over the Internet.

SNSs first emerged in the mid-1990s [56,58–60] as a simple mode of communication to interact with people over the Internet by using personal computers only [61]. Recent technological advancements of "smart" mobile devices empowered users in a variety of ways in existing social networks by adding location dimension and providing a potential benefit to access social network accounts on personal computers along with mobile devices [21]. Primarily, desktop computers were the modes to use, connect, and share information on SNSs [62], but, with the introduction of smartphones, the access to SNSs became convenient to use, connect, and share information with their "friends" [63] on the move [64,65]. With this rapid development of mobile phone technology, users can easily communicate and share information (i.e., text, audio, and video) progressively by using the geo-location [66]. The development of LBSNs progressed with the integration of communication technologies [67], which, in turn, provide fast sharing of information about what, where, why, and with whom users share information. LBSNs include geographic services (i.e., geo-location) and capabilities (geo-tagging) to assist in exploring social dynamics and make it an essential type of social networking [20,68], which allows the sharing of users' current geolocation and discovering their friends' location, which, in turn, raises users' privacy concerns [69,70]. Privacy in LBSN primarily depends on legislative and business-oriented actors involved in data sharing even though privacy is not an individual issue. Yet, some of the personal data is shared unintentionally or willingly by the user [71–74]. Sometimes, the location is deliberately shared by users for the sake of benefits (i.e., customers can enter competitions, donate to charities, or earn additional loyalty credit) in exchange of information, branded hashtags, check-ins, or experiences on LBSN [75].

Various studies based on LBSN datasets to observe human check-in behavior under domains like privacy [73,76,77], gender differences [78], geographic spaces [56], urban emotions [79], activity location choice, lifestyle patterns [6,80–82], and operations and production management [83] have been conducted. Li and Chen [63] studied location sharing by the users in the real world, and presented data analysis results over user profiles, update activities, mobility characteristics, social graphs, and attribute correlations. Benevenuto et al. [84] analyzed the frequency and duration of social network connectivity, as well as the users, conduct of different activities on these sites differentiated by types and sequences. Chang and Sun [85] analyzed the LBSN dataset to point out the influence of factors where users check-in, including historical check-ins, similarity to historical places, where their friends check-in, time of day, and demographics. Lei et al. [86] spatiotemporally analyzed the LBSN dataset to observe the human dynamics regarding differences in gender, behavior in check-in, and online time duration in Beijing's Olympic Village. Moreover, it argued that female users are more likely to interact in social media in comparison to male users. Hu and Zhang [87] utilized clustered spatiotemporal data and suggested a selection method. Moreover, exploratory spatial data analysis (ESDA) is performed to acquire the datasets with the prospects of quick grouping by mining the Weibo check-in data. Saleem et al. [88] explored the prominent locations and introduced a method of location influence with the ability to reach out geographically by using LBSN data. Furthermore, a memory-efficient algorithm was proposed, which resulted in efficient and scalable diverse sets of locations with a broad geographical spread. In addition, previous research [89–92] focused on observing human mobility patterns and analyzing check-in data for location prediction and venue tagging in the city by using LBSN datasets. While References [37,93] mainly focused on examining the factors that can predict the uses and patterns of using LBSN.

Many applications utilized the concept of automatic venue tagging to observe spatial differences [94,95]. While Gao and Liu [96] argued that temporal features and ranking of a user's geo-location history are considered to be irrelevant with the integration of human mobility in LBSN. Yang et al. [97] explored check-in behavior and mobility patterns by analyzing the spatiotemporal distribution of geotagged social media data messages and activity patterns. Moreover, References [92,98,99] analyzed the large LBSN datasets to study the variation of urban spaces and observed the spatial characteristics of the social networks, which may arise in LBSN users. Muscanell and Guadagno [15] examined the impact of gender and personality on the use of Facebook and MySpace and reported that male users use social media for relationship formation while female users use social media for relationship maintenance. Moreover, female users are reported low in agreeableness while using instant messaging more often than male users is high in agreeableness, whereas male users are reported low in openness while playing more online games compared to female users are high in openness. Rzeszewski and Beluch [100] addressed the gap (representation and representativeness) in data by investigating the LBSN users, based on the spatiotemporal distribution of the content produced (demographics of the user population). While Guan et al. [101] studied the concentration and significance of users' thoughts on Sina Weibo and Feng et al. [102] analyzed China's city network based on users' friend relationships and check-in behavior on Sina Weibo.

LBSN datasets have been exploited in various research studies for the urban development and its environmental hazards [103], expansion and exploration [104–106], travel and activity patterns [107,108], and disaster management [109–111], emergency mapping [112], Special Event Population [113], and urban sustainability [114]. Hong [115] highlighted various factors to observe the payment patterns and willingness of buyers by utilizing the LBSN dataset. Mazumdar et al. [116] proposed a prediction model, which gathers surreptitiously visited locations from an available user trajectory. Moreover, the relationship between a user's checked-in data for predicting the unchecked or hidden locations was investigated. Dokuz and Celik [117] proposed a method to discover the user's historical data and measures based on communally important locations for each user's (individual's) preferences.

Furthermore, an algorithm was proposed that was compared with a naïve alternative using real-life Twitter dataset. Fiorio et al. [118] developed a methodology for parsing the population-level migration signal from individual-level point-in-time data using flexible time-scales. Moreover, a stochastic model was proposed for simulating patterns in digital trace data and test it against three datasets: geo-tagged Tweets and Gowalla check-ins. Wu et al. [119] analyzed the impact on housing prices when neighborhood land uses are mixed. By using geographic information system data, three quantitative measures of the land-use mix were created, and these measures were computed for various neighborhoods in Beijing's central city. The research base on check-in behavior analytics is useful to know about gender-based human check-in behavior, but, under the scope of the current study, the connection with other indicators of gender equality [120,121] are not considered.

3. Dataset and Study Area

The dataset mined in this study was obtained from Sina Weibo. It covered the Guangzhou area for the period between January and May 2016, which contains 852,560 check-ins from 20,634 users. Guangzhou is considered to be one of the most attractive destinations in China due to its heterogeneous population and job opportunities regarding demographic characteristics [122], socioeconomic status [123,124], and place of origin [28]. Guangzhou, China (longitude from 112°57′ to 114°3′E and latitude from 22°26′ to 23°56′N [125]) is located on the south coast of Pearl River Delta (PRD) with a 14.5 million population [126] and had a total area of 7434.4 km^2 [127]. In 2015, Guangzhou was divided into 11 districts (Baiyun, Conghua, Haizhu, Huadu, Huangpu, Liwan, Nansha, Panyu, and Zengcheng) [128]. Six of the districts (Baiyun, Huangpu, Haizhu, Liwan, Tianhe, and Yuexiu) are denoted as the center of the city [129,130], as shown in Figure 1.

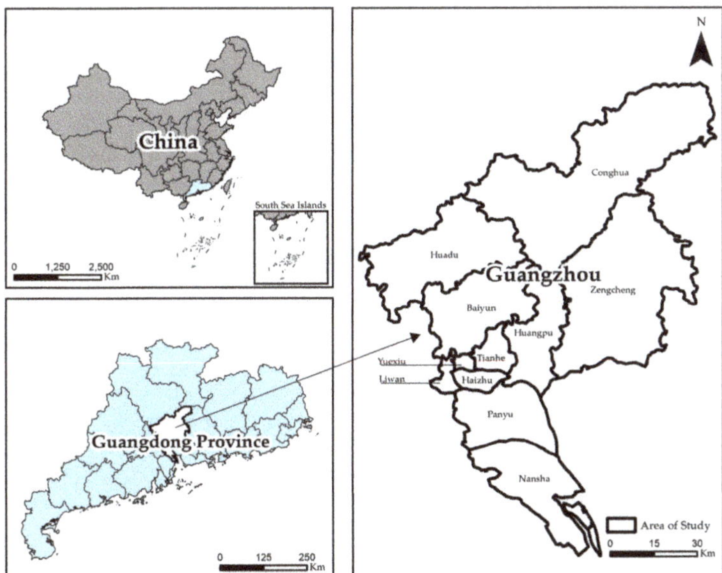

Figure 1. Map of Guangzhou, China.

The Weibo dataset used in the current study contains information like the unique id of user id, time, and date of the check-in. Additionally, geo-location (longitude and latitude), venue type, venue category, and gender collected via the web or mobile applications [131]. Therefore, it is assumed that the LBSN dataset archives the day-to-day activity patterns, usage behaviors toward social media, and presents spatiotemporal evidence, which is related to the daily routines of users [114]. A typical Weibo "check-in" is represented as: check-in (1305141104 006810) = {5503767214, ####, 1305141104 006810, Fri Apr 22 09:37:03 +0800 2016, m, 113.854085, 23.527322}. Where 1305141104 006810 denotes "status_id," 5503767214 denotes "user_id," #### denotes the "user_name," Fri Apr 22 09:37:03 +0800 2016 denotes "day, month, date, time and year," m denotes "gender" and 113.854085, 23.527322 denotes geo-location.

4. Methodology

In the current study, we analyzed the Weibo based geo-location dataset (Jan-May 2016) from Guangzhou, China. Figure 2 presents the check-in behavior analytics framework, where the LBSN data analysis methodology involves the two stages: collection, storage, and analysis of LBSN data. The download of Weibo data is the significant step of Weibo data collection and storage stage. To collect check-in data, we implemented a multi-threaded crawler to access the Weibo API. In turn, the crawler collects the check-in data filtered by gender, and the results are processed with entries that have geolocation. The outcome is in single JSON (JavaScript Object Notation) file by utilizing a python-based Weibo API (an open interface of Sina Weibo) [132,133], which is considered an extensively used data format [134,135]. To be adequately analyzed, the dataset is converted into a distinct file in the CSV (Comma-Separated Values) format so that the check-ins could be listed regarding their publishing time. However, the critical task in the data analysis stage is to mine and investigate the features of LBSN data. Moreover, during data pre-processing, invalid records are excluded by considering four criteria points: (a) availability of information i.e. user id, date, time, gender, geo-location, (b) location of the records must be in Guangzhou, China, (c) the range of record is within the date and time, and, (d) as a minimum, each user checked-in twice a month. After pre-processing (noises, void records, and bogus users) of 903,008 anonymized check-in records, 852,560 check-in records associated with the geographical area are picked up between January to May, 2016. Lastly, the task in the data insight stage is to analyze and investigate the features of LBSN check-in data by considering location, time, and gender and visualize data by using ArcGIS [136] to produce density maps [137] and trends [138,139].

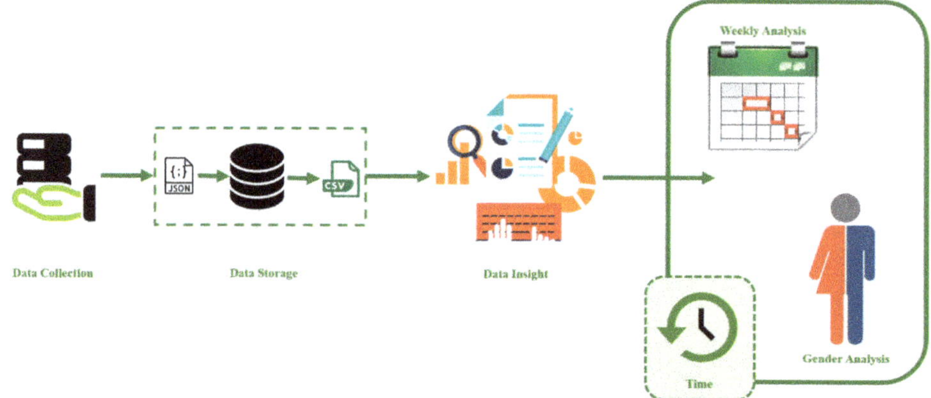

Figure 2. Check-in behavior analytics framework.

4.1. Kernel Density Estimation

In order to detect hot-spots and observe gender differences in check-in behavior, we estimated the density function of check-in using a kernel density estimation (KDE) [140–142]. KDE is considered a popular spatiotemporal investigation practice that is used to observe the features of location (i.e., destination, time) comparative to each other. KDE is an evolving spatiotemporal means that has earlier been used [143–146] to examine several characteristics of the social media (but not limited to LBSN) data analytics such as users' online activity and movement patterns [38], check-in behavior [147], city boundary definitions [148,149], and point-of-interest recommendations [150]. Moreover, it examines the diffusion of destinations in neighborhoods, allows investigators to see where destinations are densely distributed, and where they are more intensely dispersed. Lastly, it attempts to produce a smooth density surface of spatial point events in the geographic space [151].

The goal is to produce a smooth density surface that signifies the density of the point group. The algorithm is functioned by setting the search scope (window). The central grid of the window gives the weight of each grid unit to an outward grid, according to the principle of anti-distance weight. Moreover, in the window, the weights and density values are the sum of kernel density value that belongs to the central grid.

To measure the density of historical check-ins at point "x," let $f_i(x)$ be a density function at geo-location "x."

$$f_i(x) = k(x,h) = \frac{1}{xh} \sum_{j=1}^{X} K(\frac{\|x - x_j\|}{h})$$ (1)

where "x" represents the geo-location (longitude and latitude) of check-in dataset "$1 < i < n$" at which density estimation with bandwidth "h" is calculated. In KDE, bandwidth is considered an important parameter. If the bandwidth is too large, then the point density surface will become too smooth, while too small will change point density distribution abruptly. Therefore, the optimal bandwidth is determined by repeatedly setting the bandwidth and comparing the smoothness of the point density surface. Bandwidth "h" is dependent on the resulting density estimate "$f_i(x)$". "X" is the total number of check-ins in the dataset, "j" points to a signal geo-location, "K" is a standard normal density function, "$\|.\|$" denotes the Euclidean norm [152–154], and "x_j" is the geo-location of check-in "j."

The log-probability data-driven option is used to assess the value of bandwidth "h" in constructing the density estimate below.

$$L(h) = \frac{1}{x_t} \sum_{j=1}^{x_t} log\, f_i(x^j | X, h)$$ (2)

where the "x_t" events "x^j" are data points in the dataset X. Higher value of $L(h)$ is ideal since it shows that the higher probability is being allocated to new but invisible data. Hence, a simple method for bandwidth selection is to perform a grid-search on "h" using a validation set.

4.2. Geographically Weighted Regression (GWR)

GWR is a spatial regression technique that considers spatial nonstationarity and allows local parameters to be estimated. It is considered an extension of a traditional linear regression framework, and is, accordingly, easy for the specification. Unlike the complex mechanism of the Bayesian spatial model, the GWR method is easier for researchers to understand and is widely used in a practical application. In particular, in the GWR models, the coefficients of variables can be visualized in an easily identifiable manner, which could provide insightful suggestions for city planners and check-in behavior analysis [155]. A typical GWR model takes the following form.

$$y_i = \beta_0(u_i, v_i) + \sum_k \beta_k(u_i, v_i) x_{ik} + \varepsilon_i$$ (3)

where (u_i, v_i) represents the geo-coordinates (longitude, latitude) of observation I, $\beta_0\,(u_i, v_i)$ represents the intercept value, $\beta_k\,(u_i, v_i)$ represents the estimated parameter for the kth variable of observation I, and ε_i represents the error term.

This means that the estimated coefficients are allowed to vary in space. One assumption of GWR is that the observed data near the observation i have more influence in estimating $\beta_k\,(u_i, v_i)$ than the data farther from i. The parameter $\beta_k\,(u_i, v_i)$ is estimated below.

$$\beta_k(u_i, v_i) = (X^T W(u_i, v_i) X)^{-1} X^T W(u_i, v_i) Y.$$ (4)

where the weighting matrix $W(u_i, v_i)$ is a diagonal matrix, and the off-diagonal elements are all zero. The estimation of GWR is part depend on the bandwidth selection for observation i neighbors. For areas with more data points, the bandwidth of the kernel will be lower, while, for areas with few data points, the bandwidth of the kernel will be larger. In the current study, an appropriate bandwidth is selected based on the minimum Akaike information criterion for the GWR model (AIC) [156].

4.3. Standard Deviational Ellipse (SDE) Analysis

The standard deviation ellipse (SDE) [157] analysis is often used to depict the spatial characteristics of a geographical entity, such as central tendency, dispersion, and directional trends. SDE not only is an abstract expression for individual spatial distribution, but it also builds more comprehensive and realistic models of human mobility and online behavior [158]. It is quite effective for a discrete description of anisotropic events in the spatial point pattern analysis, which has been widely used in extensive research such as urban structure analysis [159]. This useful tool is chosen in this study to analyze the check-in behavior at a more detailed level.

There are four parameters of SDE that include the ellipse center, major axis, minor axis, and azimuth. The major and minor axes of the SDE are calculated according to Equation (5), and their proportional relations denote the degree of flattening the SDE. The rotating azimuth is calculated according to Equation (6), which reflects the main trend directions [160,161]. The standard deviations of the major and minor axes of the SDE are calculated according to Equation (7). The major and minor axes of the SDE form the spatial region of the check-in distribution, and the direction of the major axis is defined as the dominant direction of the variation trend [162]. The ellipse center of SDE $(\overline{X}_w, \overline{Y}_w)$ is calculated below.

$$\begin{cases} \overline{X}_w = \dfrac{\sum_{i=1}^{n} w_i x_i}{\sum_{i=1}^{n} w_i} \\ \overline{Y}_w = \dfrac{\sum_{i=1}^{n} w_i y_i}{\sum_{i=1}^{n} w_i} \end{cases} \tag{5}$$

The azimuth α of SDE is calculated using the equation below.

$$\tan \alpha = \frac{\left(\sum_{i=1}^{n} w_i^2 \tilde{x}_i^2 - \sum_{i=1}^{n} w_i^2 \overline{y}_i^2\right) + \sqrt{\left(\sum_{i=1}^{n} w_i^2 \tilde{x}_i^2 - \sum_{i=1}^{n} w_i^2 \overline{y}_i^2\right)^2 + 4 \sum_{i=1}^{n} w_i^2 \tilde{x}_i^2 \overline{y}_i^2}}{2 \sum_{i=1}^{n} w_i^2 \tilde{x}_i \overline{y}_i} \tag{6}$$

The standard deviations of the ellipse σ_x and σ_y in the x and y directions are calculated using the formulas below.

$$\begin{cases} \sigma_x = \sqrt{\dfrac{\sum_{i=1}^{n} \left(w_i \tilde{x}_i \cos \alpha - w_i \overline{y}_i \sin \alpha\right)}{\sum_{i=1}^{n} w_i^2}} \\ \sigma_y = \sqrt{\dfrac{\sum_{i=1}^{n} \left(w_i \tilde{x}_i \sin \alpha - w_i \overline{y}_i \cos \alpha\right)}{\sum_{i=1}^{n} w_i^2}} \end{cases} \tag{7}$$

where (x_i, y_i) in Equations (5)–(7) denote the deviation between coordinates of an element and the geometric center coordinates of an element set, which represents the spatial location of the object. In addition, w_i is the corresponding weight and $\left(\tilde{x}_i, \overline{y}_i\right)$ denote the coordinates deviation from the spatial location of each object to the ellipse center of SDE $(\overline{X}_w, \overline{Y}_w)$.

5. Results and Discussion

5.1. Density Variations and Distribution of Check-Ins

In the current study, the geo-location-based check-in dataset, which comes from Weibo, is utilized and analyzed the check-in density by using KDE. Figure 3a shows the main highways, water channels, and vegetation of Guangzhou, where Figure 3b presents the overall check-in density in Guangzhou. However, by comparing Figure 3a,b, it can be observed that the city center has a relatively dense check-in distribution. Generally, it is considered that more and more people prefer to live near the communities where they have easy access to services, i.e., transport, healthcare, and entertainment [163,164]. Even from the current study, high check-in density is found nearby subways and highways. Historically, the density of the Guangzhou city population rose abruptly, and the attraction of the port mainly influenced population distribution. After the reforms in China, suburban areas were covered by mega-factories, which attracted more people from the areas outside of Guangzhou for the development of mega-factories. However, a small proportion of these people preferred to live in suburban areas while the majority preferred to reside within the city center [165–167]. Hence, the result of the current study also shows the same pattern of population density in Guangzhou.

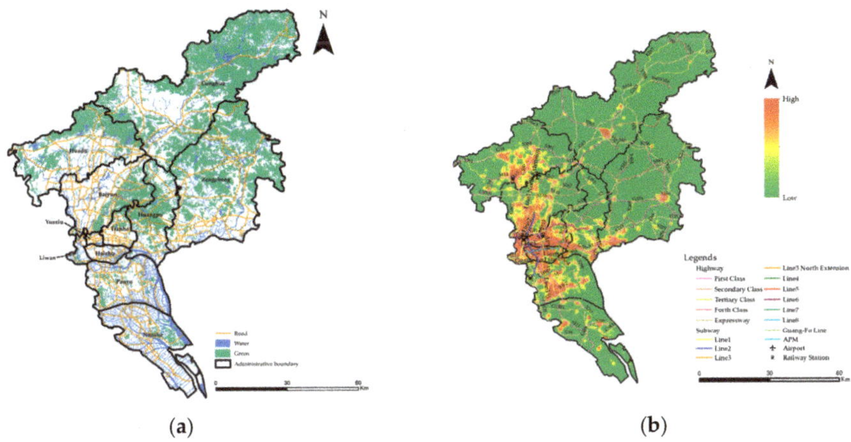

(a) (b)

Figure 3. (**a**) Land use map of Guangzhou and (**b**) kernel density of check-in in Guangzhou.

Gender-based (male and female) analysis on the LBSN data set is performed to investigate the check-in behavior in Guangzhou. Figure 4a,b represents the check-in trend during a weekend and illustrates that, from the dataset, the female users are observed to be more inclined towards the use of Weibo in comparison to the male users during a week in Guangzhou. Moreover, it is also observed that the check-in frequency of the male starts to increase during Friday and almost matches females during Saturday and Sunday. Furthermore, Figure 4 shows that female users are more likely to use Weibo than males. Figure 5 demonstrates that the pattern is similar in all districts.

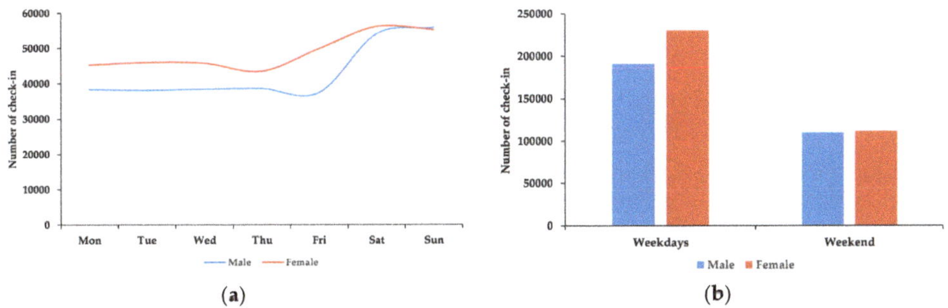

(a) (b)

Figure 4. (**a**) Weekly check-in trend. (**b**) Weekdays and weekend check-in distributions.

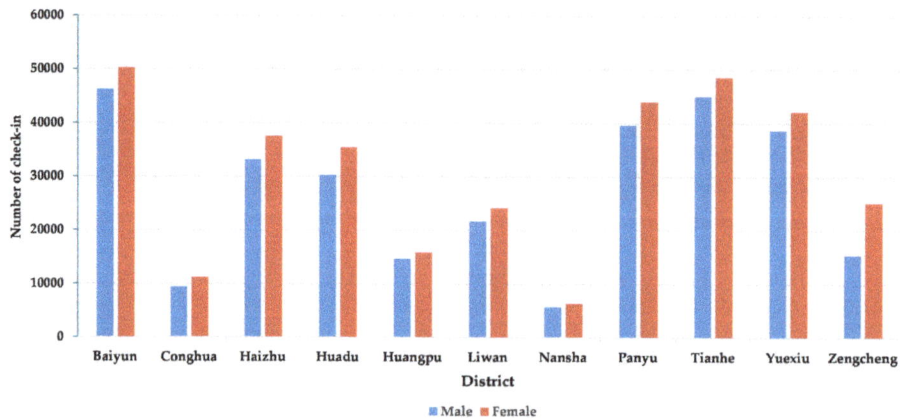

Figure 5. Gender wise check-in distribution in Guangzhou.

To study the check-in trend in Guangzhou, it is vital to observe the gender-based check-in trend through weekdays and weekend. In Figure 6a,b, the growing activity trend can be witnessed through the weekdays from 05:30–09:30 and 17:00–22:30. However, an increasing check-in trend was also observed throughout the weekend from 08:00–22:30. Furthermore, it can also be observed that the frequency of use from female users is a bit steady with a minor rise during the weekend in comparison to male users. Comparatively, the check-in frequency of male users differs a lot during weekdays and weekends. Furthermore, it is observed that, during the whole week, check-in frequency increases from 20:00–23:59. Additionally, during the weekend, the check-in frequency of male users peaks between 19:00–23:00 as compared to female users.

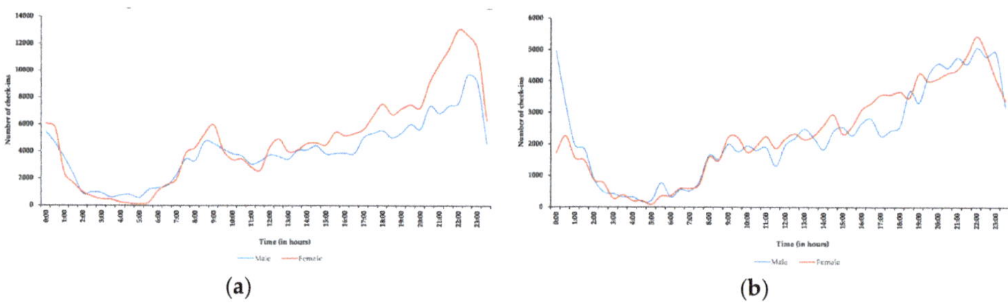

(a)   **(b)**

Figure 6. Temporal trend of check-in during (**a**) weekdays and (**b**) weekend.

Figure 7 illustrates the check-in distribution in all districts of Guangzhou. In comparison, Baiyun, Panyu, Tianhe, and Yuexiu districts (which are considered the business center of Guangzhou) have denser check-in frequency. However, from Figure 8, we can observe more check-ins during Saturday as compared to Sunday in Huadu, Huangpu, and Zengcheng districts when compared to other districts.

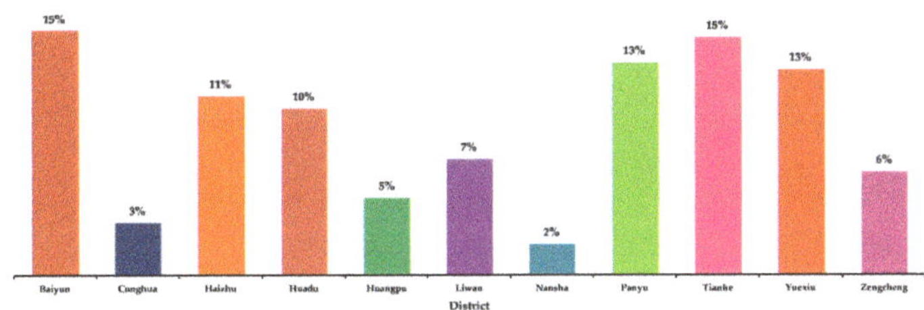

Figure 7. Check-in distribution in districts of Guangzhou.

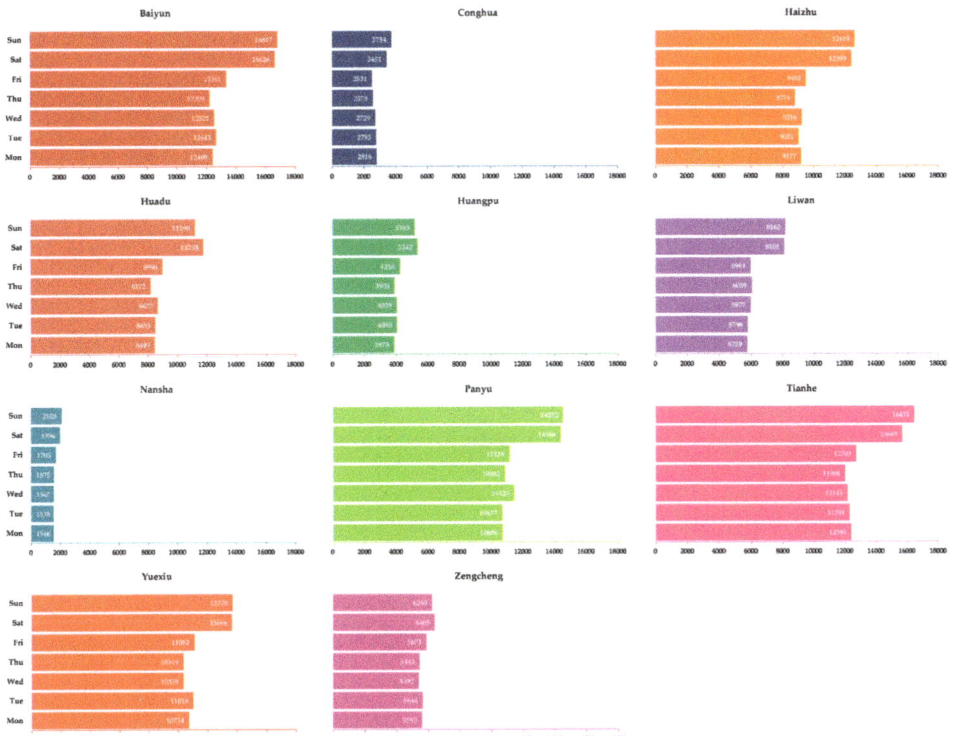

Figure 8. Weekly check-in distribution of check-ins in districts of Guangzhou.

Figure 9a,b show the gender wise check-in distribution in Guangzhou during weekdays and weekend. Meanwhile, more check-ins are made by female users during the weekdays, as shown in Figure 9a. Surprisingly, more check-ins are made by male users in most of the districts (Baiyun, Haizhu, Huangpu, Liwan, Nansha, Tianhe, and Yuexiu) in Guangzhou as compared to female users during the weekend.

Figure 10a,b represent the gender-based weekly check-in trend in all districts of Guangzhou. The difference of check-in behavior by male and female users during Saturday and Sunday can be observed clearly and is mainly due to the change in check-in behavior by male users. Hence, the results of Figure 10a,b also justify the results of Figure 9a,b.

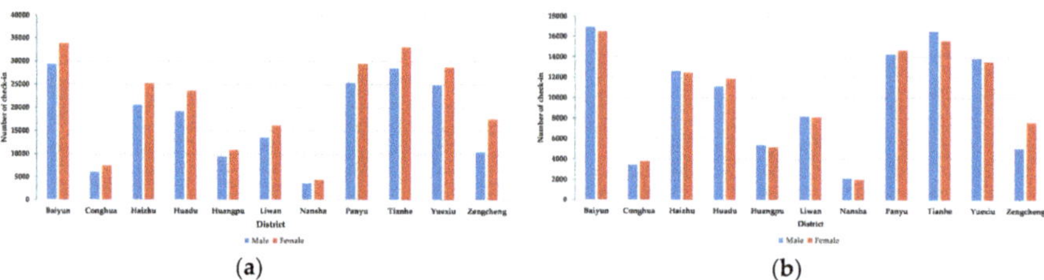

Figure 9. Gender wise check-in distribution in Guangzhou during (**a**) weekdays (**b**) and weekend.

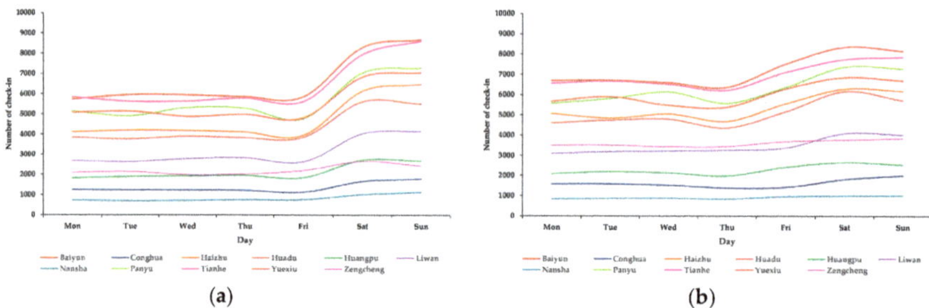

Figure 10. Gender wise weekly check-in distribution in Guangzhou (**a**) male and (**b**) female.

Furthermore, we analyzed the Weibo dataset to observe a gender-based check-in trend during 24 hours in all districts of Guangzhou and is presented in Figure 11a,b,c. Figure 11a presents the day-to-day check-in trend in Guangzhou. An increase in check-in trend is observed from 07:00 AM to 09:00 AM. However, a decrease in the check-in trend is observed from 09:00 to 11:30. Furthermore, It can also be observed that the check-in trend starts to rise again at 11:30 AM and continues to rise until 22:30, as shown in Figure 11b,c.

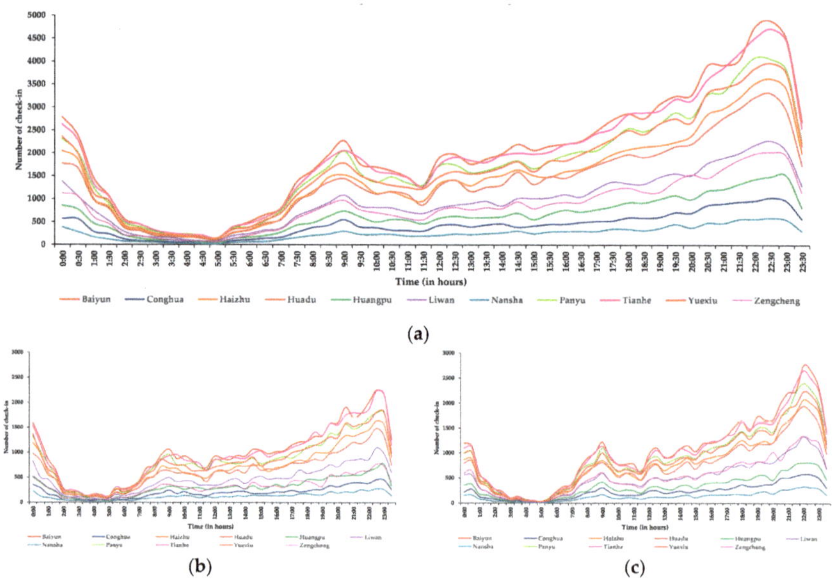

Figure 11. Temporal trend of check-in in Guangzhou (**a**) average daily (**b**), average daily male (**c**), and average daily female.

Average density distribution of check-ins made in Guangzhou was calculated using the KDE method, and the density maps are visualized using ArcGIS in both space and time. Figure 12 reveals the spatiotemporal dynamics of Guangzhou, and it reveals that the center of the Guangzhou city has a higher density of check-ins. Besides higher density of check-ins, it can also be observed that most of the check-ins are made near the borders of the district that are mostly near highways and the subway.

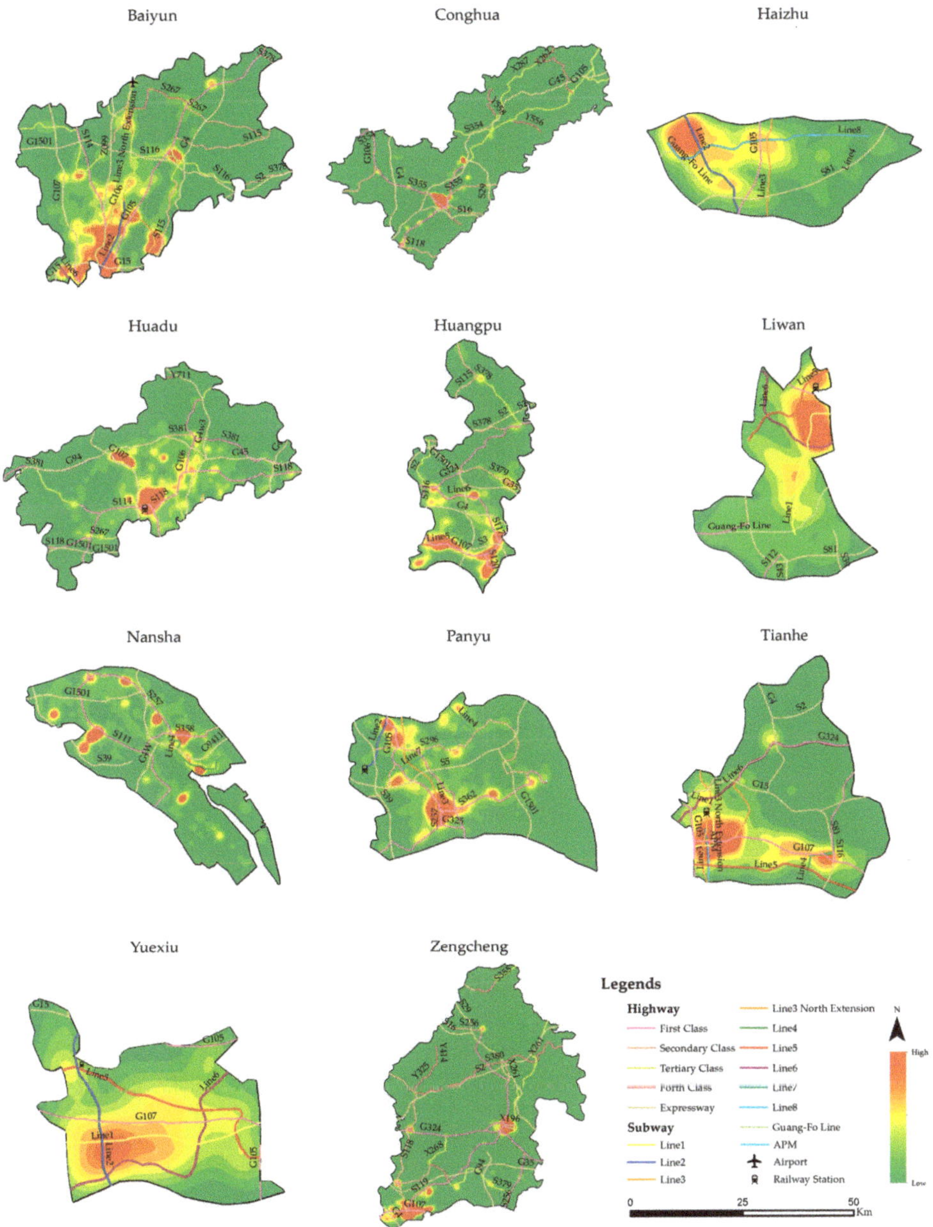

Figure 12. Kernel densities of check-in in the districts of Guangzhou.

In summary, the KDE results reveal that the check-in behavior patterns in Guangzhou are as follows: (1) in terms of the spatial distribution, main agglomeration areas with high frequencies and densities are evident in the center of the city (Baiyun, Haizhu, Liwan, Tianhe, and Yuexiu). (2) From a temporal perspective, people's activity frequencies are almost zero between midnight and early morning (00:00-05:00) and relatively low during working times (09:30-11:30) and relatively high during leisure times and at dinner time. The information from KDE can facilitate studying the dynamic evolution of check-in across both space and time. Additionally, the KDE results verify that the check-in behavior varies at fine temporal (i.e., a day) and spatial (i.e., a city) scales. The results also show that the check-in data can reflect more refined phenomena and results than traditional data with fine time and spatial granularity.

5.2. Geographically Weighted Regression (GWR)

The GWR results for male and female during weekdays and weekends are presented in Table 1. All selected variables are significant at the 1% level. The R^2 values are 0.9288 and 0.9157, which denote that the selected variables can explain 92.8% and 91.5% of the variation in the check-in on weekdays and weekend, with optimal spatiotemporal bandwidths of 0.300 and 0.301 for weekdays and weekend, respectively. The variation trends for male and female during weekdays vary due to the higher number of check-ins by a female as compared to males and can be observed in Figure 14. The variation trends for males and females during the weekend are roughly the same due to the increased number of check-ins by the male as compared to weekdays and can be observed in Figure 14.

Table 1. Estimated GWR parameters for weekdays and weekends by males and females.

	Weekdays					Weekend				
	Min	Mean	Max	SD	*p*-Value	Min	Mean	Max	SD	*p*-Value
Male	0	25.737	1266.668	92.991	0.000 ***	0	15.133	1206.001	58.904	0.000 ***
Female	0	31.078	1437.804	112.434	0.001 ***	0	15.117	670.302	54.261	0.000 ***
	Diagnostic information:					Diagnostic information:				
	Moran's I			0.2082		Moran's I			0.1999	
	R^2			0.9288		R^2			0.9157	
	AIC			10010.62		AIC			10002.32	
	Bandwidth			0.3000		Bandwidth			0.3010	

*** represents a significance level of 1%.

Table 2 presents the GWR results for weekdays and weekends in all districts of Guangzhou. All selected variables are significant at the 1% level. The R^2 values are 0.9203 and 0.9212, denoting that the selected variables can explain 92% and 92.1% of the variation in the check-in on weekdays and weekends in all districts of Guangzhou, with optimal spatiotemporal bandwidths of 0.331 and 0.320 for weekdays and weekends, respectively. The variation trends during weekdays and weekends are very different in most of the districts due to the higher number of check-ins in the city center during weekdays. There is a slight increase in activity in the suburban districts (i.e., Conghua, Nansha, and Zengcheng) during the weekend as compared to weekdays.

Table 3 presents the GWR results for males and females in all districts of Guangzhou during weekdays. All selected variables are significant at the 1% level. The R^2 values are 0.9024 and 0.9186, which denote that the selected variables can explain 90% and 92% of the variation in the check-in for male and female in all districts of Guangzhou during weekdays, with optimal spatiotemporal bandwidths of 0.3554 and 0.3138, respectively. The variation trends during weekdays by males and females are quite different in most of the districts due to the higher number of check-ins by females in all districts of Guangzhou.

Table 2. Estimated GWR parameters for weekdays and weekends in districts of Guangzhou.

	Weekdays					Weekend				
	Min	Mean	Max	SD	*p*-Value	Min	Mean	Max	SD	*p*-Value
Baiyun	0	68.688	1500.464	142.125	0.010 ***	0	74.817	1571.392	159.727	0.000 ***
Conghua	0	4.744	1372.073	35.885	0.000 ***	0	5.652	1727.263	45.294	0.000 ***
Haizhu	0	367.824	1799.977	380.681	0.001 ***	0	418.028	2112.234	440.147	0.010 ***
Huadu	0	31.090	1310.871	89.061	0.000 ***	0	36.435	1580.608	105.450	0.000 ***
Huangpu	0	31.306	703.263	67.124	0.000 ***	0	33.854	825.094	75.191	0.000 ***
Liwan	0	329.503	2092.561	413.853	0.010 ***	0	368.540	2470.301	478.330	0.001 ***
Nansha	0	15.854	712.007	42.482	0.000 ***	0	17.645	876.957	50.111	0.010 ***
Panyu	0	67.906	1363.761	130.817	0.001 ***	0	75.283	1593.028	150.951	0.001 ***
Tianhe	0	334.986	3318.354	477.392	0.001 ***	0	361.432	3778.293	519.751	0.000 ***
Yuexiu	0	1145.853	3011.301	664.719	0.001 ***	0	1246.494	3353.170	748.794	0.001 ***
Zengcheng	0	9.496	1089.747	51.324	0.000 ***	0	15.428	1734.161	83.137	0.000 ***
	Diagnostic information:					Diagnostic information:				
	Moran's I		0.0628			Moran's I		0.0600		
	R^2		0.9203			R^2		0.9212		
	AIC		10128.92			AIC		1010.515		
	Bandwidth		0.3317			Bandwidth		0.3204		

*** represents a significance level of 1%.

Table 3. Estimated GWR parameters for males and females in districts of Guangzhou during weekdays.

	Male					Female				
	Min	Mean	Max	SD	*p*-Value	Min	Mean	Max	SD	*p*-Value
Baiyun	0	43.528	937.598	88.674	0.001 ***	0	50.339	1076.319	107.096	0.000 ***
Conghua	0	3.024	902.968	23.323	0.000 ***	0	3.745	1121.849	29.513	0.000 ***
Haizhu	0	227.986	1108.790	236.494	0.001 ***	0	280.048	1416.303	295.593	0.001 ***
Huadu	0	19.686	793.073	54.664	0.000 ***	0	24.262	1046.881	69.608	0.000 ***
Huangpu	0	19.937	423.187	42.100	0.000 ***	0	22.920	534.047	50.288	0.000 ***
Liwan	0	206.122	1395.648	258.472	0.001 ***	0	245.638	1677.278	319.213	0.001 ***
Nansha	0	9.984	444.819	26.769	0.000 ***	0	11.908	575.094	32.835	0.010 ***
Panyu	0	43.531	824.870	81.660	0.000 ***	0	50.323	1030.004	99.977	0.010 ***
Tianhe	0	211.895	2112.353	304.614	0.010 ***	0	245.488	2567.303	353.042	0.000 ***
Yuexiu	0	734.656	1907.466	429.445	0.000 ***	0	845.272	2254.169	504.612	0.010 ***
Zengcheng	0	6.380	762.068	34.471	0.000 ***	0	10.765	1222.445	57.740	0.001 ***
	Diagnostic information:					Diagnostic information:				
	Moran's I		0.0665			Moran's I		0.0529		
	R^2		0.9024			R^2		0.9186		
	AIC		11334.555			AIC		11275.415		
	Bandwidth		0.3554			Bandwidth		0.3138		

*** represents a significance level of 1%.

Table 4 presents the GWR results for males and females in all districts of Guangzhou during the weekend. All selected variables are significant at the 1% level. The R^2 values are 0.9264 and 0.9020, which denote that the selected variables can explain 93% and 90% of the variation in the check-in for males and females in all districts of Guangzhou during weekdays, with optimal spatiotemporal bandwidths of 0.3315 and 0.3449, respectively. The variation trends during the weekend by males and females are roughly the same due to the higher number of check-ins by the males as compared to females.

Moreover, Figure 13 shows the average spatial change tendencies of the parameter estimates for check-ins, which are pivotal and focus factors in this article. This article also manually sets zero as a threshold to distinguish between the positive and negative effects. This shows the general tendencies and nuances of check-ins by male and female users. The effects of check-ins by the female users in the city center are greater than 0.5 StdResid as compared to male. The greater StdResid means that check-in frequency is high among the female users.

Table 4. Estimated GWR parameters for males and females in districts of Guangzhou during the weekend.

	Male					Female				
	Min	Mean	Max	SD	*p*-Value	Min	Mean	Max	SD	*p*-Value
Baiyun	0	25.160	562.867	53.821	0.010 ***	0	24.478	500.145	52.878	0.000 ***
Conghua	0	1.720	483.899	12.687	0.000 ***	0	1.908	605.415	15.849	0.000 ***
Haizhu	0	139.838	691.187	145.229	0.001 ***	0	137.981	695.931	145.194	0.001 ***
Huadu	0	11.405	517.797	34.904	0.000 ***	0	12.173	533.726	36.229	0.000 ***
Huangpu	0	11.369	290.520	25.540	0.000 ***	0	10.935	291.047	25.160	0.000 ***
Liwan	0	123.381	751.988	156.679	0.010 ***	0	122.901	806.843	159.533	0.010 ***
Nansha	0	5.871	267.188	16.019	0.010 ***	0	5.736	301.862	17.533	0.000 ***
Panyu	0	24.375	538.892	49.567	0.001 ***	0	24.960	563.024	51.292	0.000 ***
Tianhe	0	123.091	1206.001	173.388	0.000 ***	0	115.943	1210.991	167.107	0.001 ***
Yuexiu	0	411.197	1103.834	237.592	0.001 ***	0	401.222	1108.737	245.131	0.001 ***
Zengcheng	0	3.115	327.679	17.024	0.000 ***	0	4.663	511.716	25.604	0.000 ***
	Diagnostic information:					Diagnostic information:				
	Moran's I		0.0367			Moran's I		0.0580		
	R^2		0.9264			R^2		0.9020		
	AIC		10511.428			AIC		10050.409		
	Bandwidth		0.3315			Bandwidth		0.3449		

*** represents a significance level of 1%.

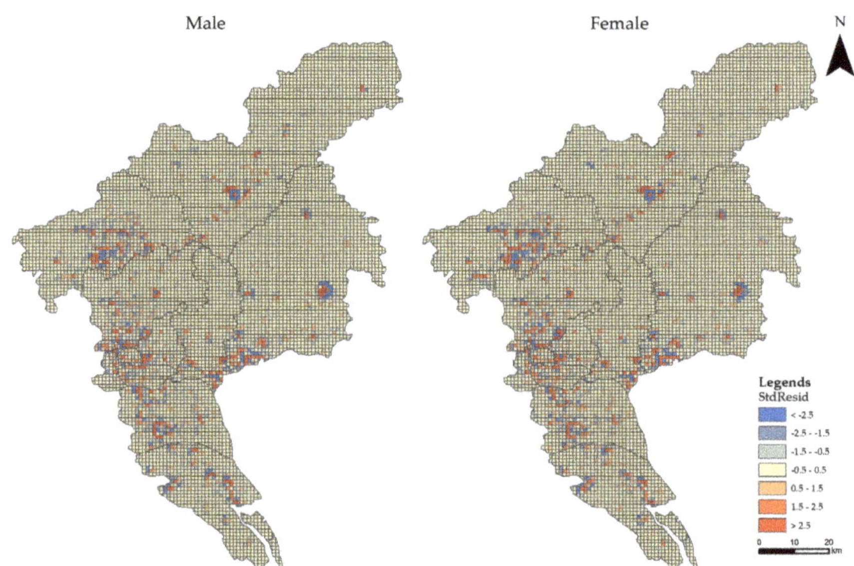

Figure 13. Spatial distribution of check-ins by males and females in Guangzhou.

Figure 14 shows the average spatial change tendencies of the parameter estimates for check-ins by males and females during weekdays and weekends. The effects of check-ins during weekdays in Guangzhou are mostly greater than 0.5 StdResid by females. However, the effects of check-ins during the weekend in Guangzhou are almost the same by male and female users. Lastly, urban areas that are also considered the city center, including Baiyun, Haizhu, Liwan, Tianhe, and Yuexiu, all have positive effects on check-in frequency. The other districts have mostly negative effects, which have few settlements.

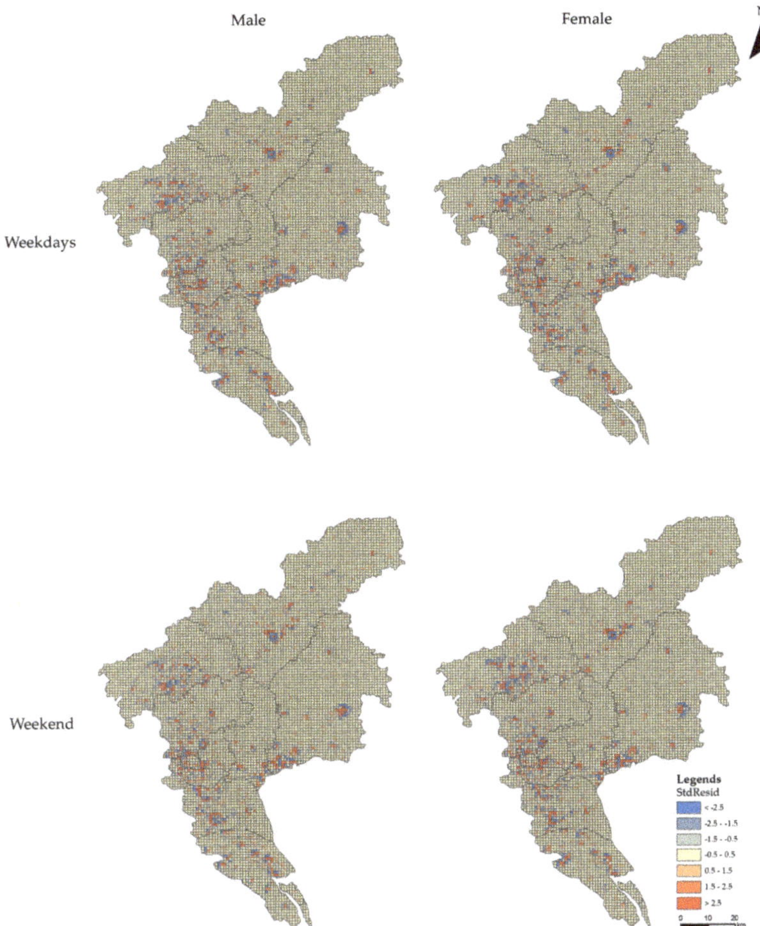

Figure 14. Spatial distribution of check-ins by males and females during weekdays and weekends in Guangzhou.

5.3. Standard Deviational Ellipse (SDE) Analysis

The overall spatial pattern changes of check-ins by males and females across Guangzhou were determined through a standard deviational ellipse analysis and is plotted in Figure 15, which shows the overall spatial pattern of check-in within the city by males and females. In this case, the major axis indicates the direction and the minor axis indicates the range of the data distribution.

The trajectory of the center in Figure 15 shows a linear movement by the female users in a clockwise direction toward the north-east by rotating azimuth 3°, and the movement of the male users is counterclockwise when rotating azimuth 179°. However, it can be observed from Table 5 that the value of the minor axis for the female users is 40.86 km with an area of 6343.12 km², which results in the larger eccentricity of the SDE. The larger eccentricity of the SDE represents that the check-ins for females are more scattered in the city as compared to males. It also represents that the check-in distribution is mostly close toward the city center by the male users.

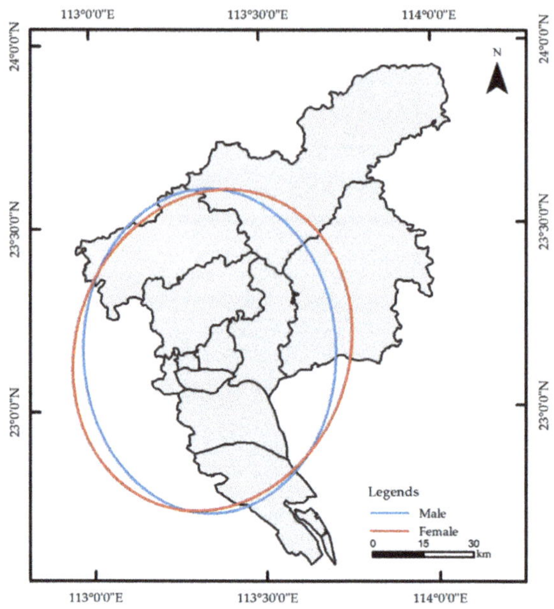

Figure 15. SDEs for check-in distribution by males and females in Guangzhou.

Table 5. SDE features on the major axis, minor axis, area, azimuth, and flattering in Guangzhou.

	Major Axis (km)	Minor Axis (km)	Area (km^2)	Azimuth°	Flattering
Male	49.11	37.75	5823.19	179	1.301
Female	49.41	40.86	6343.12	3	1.209

Figure 16 shows the SDEs for the overall spatial pattern of check-in within the city by males and females during weekdays and weekends in Guangzhou. During weekdays, a linear movement by the female users in a clockwise direction toward the north-east by rotating azimuth 5° and the movement of the male users is counterclockwise by rotating azimuth 179°. However, during the weekend, a linear movement is taken by the female users in a clockwise direction toward the north-east by 1° and the movement of the male users is counterclockwise when rotating azimuth 174°. Moreover, it can be observed from Table 6 that the value of the minor axis for females during weekdays is 38.14 km with an area of 6381.89 km^2, which results in the larger eccentricity of the SDE as compared to males. The larger eccentricity of the SDE represents that, during weekdays, the check-ins for females are more scattered in the city as compared to males. It also represents that, during the check-in, distribution is mostly close toward the city center by the male users as compared to the female users. However, during the weekend, a linear movement by the female users in a counterclockwise direction toward the north-west by rotating azimuth 1° as compared to rotating azimuth 5° during weekdays and the movement of the male users is also counterclockwise toward the north-west when rotating azimuth by 174° as compared to rotating azimuth 179° during weekdays. This change in a pattern during the weekend indicates that the check-in distribution by both males and females shifts to the city center.

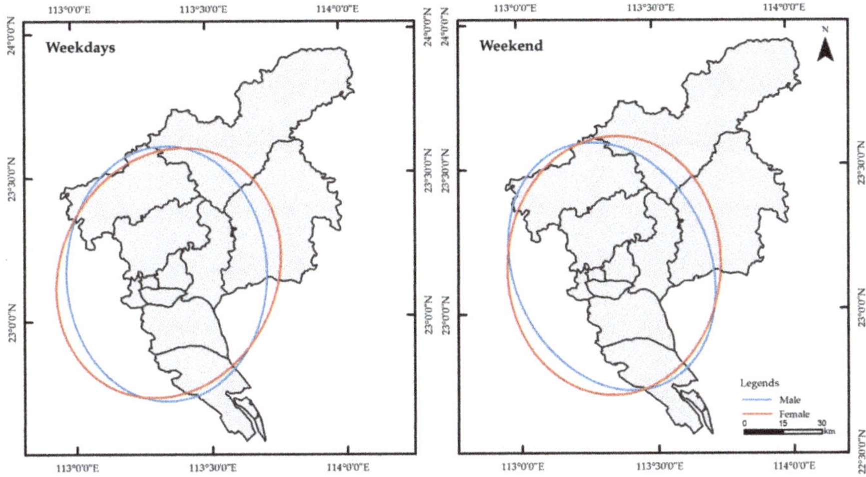

Figure 16. SDEs for check-in distribution by males and females during weekdays and weekends in Guangzhou.

Table 6. SDE features on the major axis, minor axis, area, azimuth, and flattering during weekdays and weekends in Guangzhou.

		Major Axis (km)	Minor Axis (km)	Area (km^2)	Azimuth$^\circ$	Flattering
Weekdays	Male	49.15	41.20	5888.83	179	1.1931
	Female	49.30	38.14	6381.89	5	1.2927
Weekends	Male	49.03	40.13	5705.79	174	1.2218
	Female	49.64	37.04	6258.46	1	1.3403

Moreover, it can be observed from Table 6 that the value of the minor axis for females during weekdays is 37.04 km with an area of 6258.46 km^2, which results in the larger eccentricity of the SDE as compared to males. The larger eccentricity of the SDE represents that, during the weekend, the check-in distribution for females are also more scattered in the city as compared to males, but, during the weekend, the spatial pattern of check-ins is mostly scattered toward the city center.

Figure 17 shows the district-wise SDEs for the overall spatial pattern of check-ins within the city by males and females in all (eleven) districts of Guangzhou. It can be observed that the overall spatial pattern of check-ins is almost the same by males and females in Baiyun, Huadu, Liwan, and Yuexiu districts. However, in the Conghua district, a linear movement is taken by the female users in a clockwise direction toward the north-east by rotating azimuth 50° and the movement of the male users is taken as the clockwise direction toward the north-east by rotating azimuth 51°. It can be observed in Table 7 that the value of the minor axis for the female users is 16.94 km with an area of 430.80 km^2 and a minor axis for the male users is 18.50 km with an area of 487.26 km^2, which results in the larger eccentricity of the SDE as compared to females. The spatial pattern of check-ins by the male users is more scattered within the district as compared to females. In the Haizhu district, a linear movement by the female users is taken in a clockwise direction toward the south-east by rotating azimuth 98° and the movement of the male users is taken in the clockwise direction toward the south-east when rotating azimuth by 99°. The value of the minor axis for the female users is 5.58 km with an area of 35.30 km^2 and the minor axis for the male users is 4.62 km with an area of 29.54 km^2, which results in the larger eccentricity of the SDE by the female users as compared to males. The spatial pattern of check-in by the female users is more scattered within the district as compared to males.

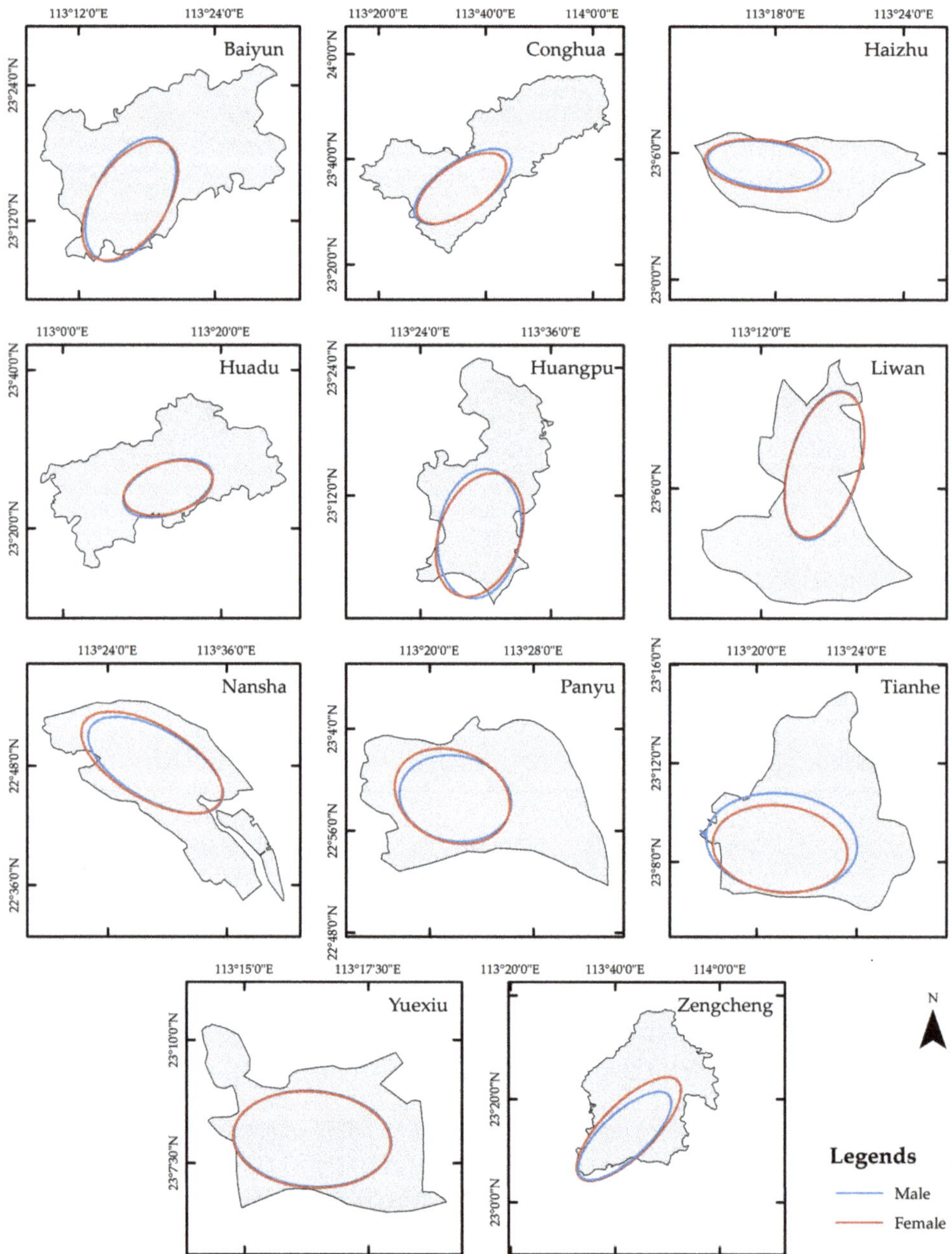

Figure 17. SDEs for check-in distribution by male and female users in districts of Guangzhou.

Table 7. SDE features on the major axis, minor axis, area, azimuth, and flattering in districts of Guangzhou.

		Major Axis (km)	Minor Axis (km)	Area (km^2)	Azimuth$^\circ$	Flattering
Baiyun	Male	10.76	5.68	191.87	29	1.8951
	Female	10.75	5.62	189.80	31	1.9130
Conghua	Male	8.39	18.50	487.26	51	2.2056
	Female	8.09	16.94	430.80	50	2.0933
Haizhu	Male	2.04	4.62	29.54	99	2.2675
	Female	2.01	5.58	35.30	98	2.7727
Huadu	Male	5.96	10.15	190.00	72	1.7041
	Female	5.90	9.91	183.82	71	1.6793
Huangpu	Male	11.26	6.45	227.97	6	1.7464
	Female	11.06	6.38	221.64	9	1.7328
Liwan	Male	4.07	1.74	22.21	13	2.3412
	Female	3.99	1.72	21.53	14	2.3191
Nansha	Male	6.33	13.18	262.13	122	2.0810
	Female	6.30	12.98	290.50	122	2.0612
Panyu	Male	6.42	7.93	143.03	120	1.2347
	Female	6.43	7.96	160.78	121	1.2389
Tianhe	Male	3.26	5.58	59.99	97	1.7093
	Female	3.24	4.62	47.06	98	1.4253
Yuexiu	Male	1.79	2.72	15.32	94	1.5149
	Female	1.81	2.72	15.44	94	1.5073
Zengcheng	Male	20.49	7.85	505.09	42	2.6103
	Female	20.37	11.81	661.52	37	1.7252

In the Huangpu district, a linear movement by the female users is taken in a clockwise direction toward the north-east by rotating azimuth 9° and the movement of the male users is taken in the clockwise direction toward the north-east when rotating azimuth by 6°. The value of the minor axis for the female users is 6.38 km with an area of 221.64 km^2 and the minor axis for the male users is 6.45 km with an area of 227.97 km^2, which results in almost the same eccentricity of the SDE by both males and females. Moreover, the overall spatial pattern of check-in is almost the same by males and females in Nansha and Panyu districts, where the only difference of more scattered distribution of check-ins is conducted by females as compared to males. In the Tianhe district, a linear movement by the female users in a clockwise direction toward the north-east direction by rotating azimuth 98° and the movement of the male users is in the clockwise direction toward the north-east when rotating azimuth by 97°. The value of the minor axis for the female users is 4.62 km with an area of 47.06 km^2 and the minor axis for the male users is 5.58 km with an area of 59.99 km^2, which results in the larger eccentricity of the SDE by the male users. The spatial pattern of check-ins by the male users is more scattered within the district as compared to females. In the Zengcheng district, a linear movement by the female users in a clockwise direction is conducted toward the north-east by rotating azimuth 37° and the movement of the male users is conducted in the clockwise direction toward the north-east when rotating azimuth by 42°. The value of the minor axis for the female users is 11.81 km with an area of 661.52 km^2 and the minor axis for the male users is 7.85 km with an area of 505.09 km^2, which results in the larger eccentricity of the SDE by the female users. The spatial pattern of check-in by the female users is more scattered within the district as compared to the male users.

Furthermore, considering the eleven districts in the study area, we plotted the trajectory of SDEs center for each district, as shown in Figure 18. The geometry of trajectories indicate different spatial patterns of check-in behavior by males and females in each district. In general, the spatial aggregating degree of trajectories in the district indicates the check-in distribution in that district and a certain degree of temporal autocorrelation. The apparent features of change in trajectories for all check-ins are presented in Figure 18. It indicates that a larger difference in trajectories is observed in the Conghua, Haizhu, Huangpu, Panyu, Tianhe, and Zengcheng districts. That may be attributed to Conghua and Zengcheng having a larger area, which accounts for the larger extent of the trajectory and indicates more dispersing in check-ins as compared to the Baiyun, Huadu, Liwan, Nansha, and Yuexiu districts having a comparatively smaller area.

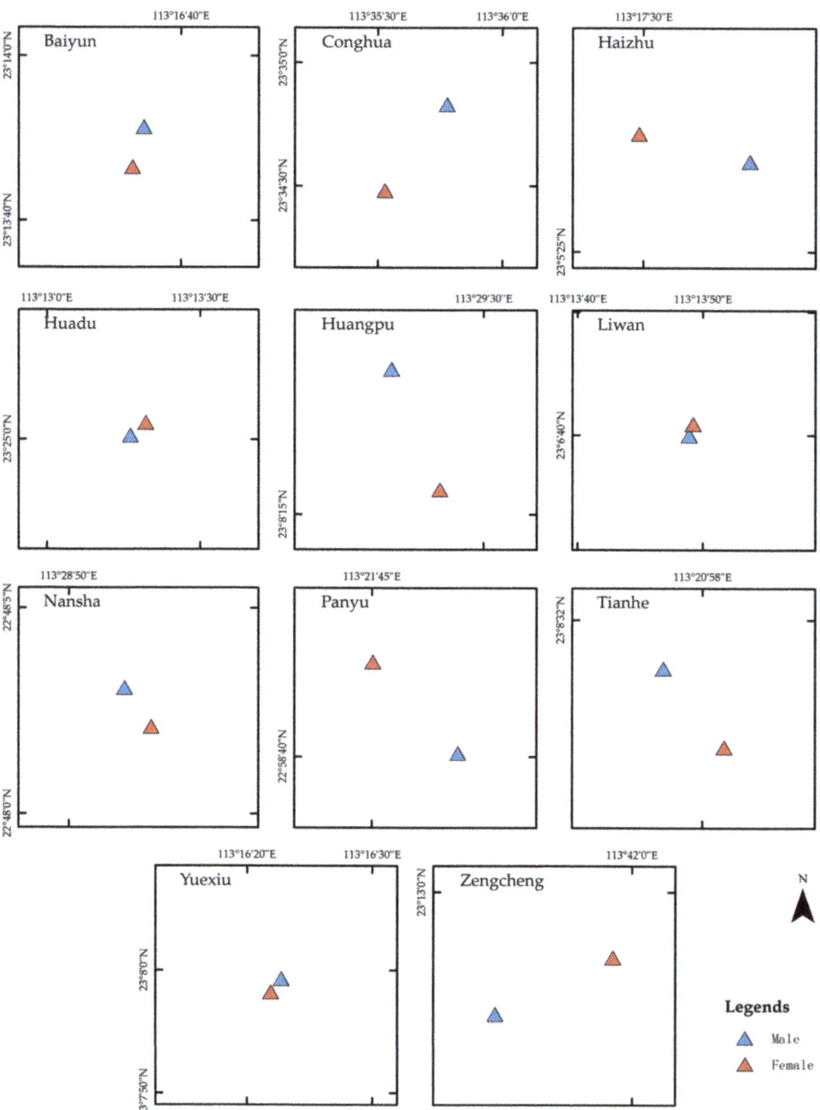

Figure 18. The trajectory of SDEs center by male users and female users in districts of Guangzhou.

Although check-in data are very raw and imprecise for movement of both males and females from one destination to another and, thus, check-in data cannot precisely indicate the nature of the movement of users among the districts, it can still be helpful to understand the check-in behavior.

6. Conclusions

The current study utilized density maps and trends to present the spatiotemporal investigation of gender-based human check-in behavior to explore hourly and daily check-in patterns, as well as patterns during weekdays and weekends. Results show that, in almost all districts of Guangzhou, females are more inclined toward using Weibo as compared to males during the weekdays. However, during the weekend, almost the same check-in trend is observed by both males and females. Furthermore, the center of the city has a comparatively high density of check-in near the subway and highways.

With a supplement to the prior research on check-in behavior, we also consider LBSN data as a supplement rather than a substitute for traditional data sources to observe (i.e., human mobility, activity analysis, and defining city boundary and social issues in a city). Additionally, compared to other traditional data sources, the LBSN dataset has some advantages (low cost and high spatial precision) and disadvantages (i.e., the gender prejudice, a low frequency in sampling, and location type prejudice).

Lastly, based on the results, we consider the LBSN dataset as a novel source of big data with the potential to offer a new viewpoint as an add-on to observe the gender-based check-in density in space and time. The information from KDE can facilitate studying the dynamic evolution of check-in across both space and time. Additionally, the KDE results verify that the check-in behavior varies at fine temporal (i.e., a day) and spatial (i.e., a city) scales. The results also show that the check-in data can reflect more refined phenomena and results other than traditional data with fine time and spatial granularity. Despite the difference of methodologies being used with different types of LBSN and datasets, both early studies [16,18] and the current study based on gender-based check-in behavior on LBSN, draw a similar conclusion that the female users are more likely to use LBSN than male users. Additionally, it can be helpful for policymakers to define policies regarding the supply of services (i.e., transport, health, and entertainment) by highlighting the check-in hotspots in the city. The SDEs indicate the difference of trajectories by males and females in the district of Guangzhou and the trajectory of the SDEs center by males and females in districts of Guangzhou represents the differences in check-in behavior patterns in those districts.

In the future, we will tend to study the use of LBSN data to explore the activities associated with the check-ins and study the motivation toward those activities. Moreover, we will also tend to examine the aspects that bring change in the check-in behavior toward those activities in space and time.

Author Contributions: R.M. and Y.Z. conceived and designed the research; R.M. and F.L. performed the simulations; R.M. and Y.Z. wrote the article; R.M. and Y.Z. proof read the article for language editing. All authors read and approved the final manuscript.

Funding: This work is supported by the National Natural Science Foundation of China (No. 41871292), the Science and Technology Program of Guangzhou, China (No: 201803030034, 201802030008), and the Project of the Science and Technology Foundation of Guangdong Province (No. 2018B020207002).

Conflicts of Interest: The authors declare no conflicts of interest.

References

1. Zeng, D.; Chen, H.; Lusch, R.; Li, S.-H. Social media analytics and intelligence. *IEEE Intell. Syst.* **2010**, *25*, 13–16. [CrossRef]
2. Moser, C.A.; Kalton, G. *Survey Methods in Social Investigation*; Routledge: London, UK, 2017.
3. Edwards, A.; Housley, W.; Williams, M.; Sloan, L.; Williams, M. Digital social research, social media and the sociological imagination: Surrogacy, augmentation and re-orientation. *Int. J. Soc. Res. Methodol.* **2013**, *16*, 245–260. [CrossRef]

4. Schatzki, T.R.; Schatzki, T.R. *Social Practices: A Wittgensteinian Approach to Human Activity and the Social*; Cambridge University Press: Cambridge, UK, 1996.

5. Eagle, N.; Pentland, A. Reality mining: Sensing complex social systems. *Pers. Ubiquitous Comput.* **2006**, *10*, 255–268. [CrossRef]

6. Hasan, S.; Ukkusuri, S.V.; Zhan, X. Understanding social influence in activity location choice and lifestyle patterns using geolocation data from social media. *Front. ICT* **2016**, *3*, 10. [CrossRef]

7. Lenhart, A.; Purcell, K.; Smith, A.; Zickuhr, K. *Social Media & Mobile Internet Use among Teens and Young Adults. Millennials*; Pew Internet & American Life Project: Washington, DC, USA, 2010.

8. Kheiri, A.; Karimipour, F.; Forghani, M. Intra-urban movement flow estimation using location based social networking data. *Int. Arch. Photogramm. Remote Sens. Spat. Inf. Sci.* **2015**, *40*, 781. [CrossRef]

9. Smith, A. *Why Americans Use Social Media*; Pew Internet & American Life Project: Washington, DC, USA, 2011; pp. 1–11.

10. Zhang, L.; Pentina, I. Motivations and usage patterns of weibo. *Cyberpsychol. Behav. Soc. Netw.* **2012**, *15*, 312–317. [CrossRef]

11. Li-Barber, K.T. Self-disclosure and student satisfaction with facebook. *Comput. Hum. Behav.* **2012**, *28*, 624–630.

12. Pentina, I.; Basmanova, O.; Zhang, L. A cross-national study of twitter users' motivations and continuance intentions. *J. Mark. Commun.* **2016**, *22*, 36–55. [CrossRef]

13. Shao, W.; Ross, M.; Grace, D. Developing a motivation-based segmentation typology of facebook users. *Mark. Intell. Plan.* **2015**, *33*, 1071–1086. [CrossRef]

14. Kim, H.-S. A study on use motivation of sns and communication behavior. *J. Korea Acad. -Ind. Coop. Soc.* **2012**, *13*, 548–553.

15. Muscanell, N.L.; Guadagno, R.E. Make new friends or keep the old: Gender and personality differences in social networking use. *Comput. Hum. Behav.* **2012**, *28*, 107–112. [CrossRef]

16. Hwang, H.S.; Choi, E.K. Exploring gender differences in motivations for using sina weibo. *Ksii Trans. Internet Inf. Syst. (TIIS)* **2016**, *10*, 1429–1441.

17. Rossi, L.; Musolesi, M. It's the way you check-in. In Proceedings of the Second Edition of the ACM Conference on Online Social Networks, COSN'14, Dublin, Ireland, 1–2 October 2014; ACM: New York, NY, USA, 2014; pp. 215–226.

18. Chun, M.-h. The affective/cognitive involvement and satisfaction according to the usage motivations of social network services. *Manag. Inf. Syst. Rev.* **2012**, *31*. [CrossRef]

19. Traynor, D.; Curran, K. Location-based social networks. In *Mobile Services Industries, Technologies, and Applications in the Global Economy*; IGI Global: Hershey, PA, USA, 2013; pp. 243–253.

20. Symeonidis, P.; Ntempos, D.; Manolopoulos, Y. Location-based social networks. In *Recommender Systems for Location-Based Social Networks*; Springer: Berlin, Germany, 2014; pp. 35–48.

21. Zheng, Y. Location-based social networks: Users. In *Computing with Spatial Trajectories*; Zheng, Y., Zhou, X., Eds.; Springer: New York, NY, USA, 2011; pp. 243–276.

22. Valverde-Rebaza, J.C.; Roche, M.; Poncelet, P.; de Andrade Lopes, A. The role of location and social strength for friendship prediction in location-based social networks. *Inf. Process. Manag.* **2018**, *54*, 475–489. [CrossRef]

23. Reddy, R.; Kim, R.S.-H. Social Discovery of User Activity for Media Content. Available online: https://patents.google.com/patent/US8661550B2/en (accessed on 1 January 2019).

24. Lu, E.H.-C.; Chen, C.-Y.; Tseng, V.S. Personalized trip recommendation with multiple constraints by mining user check-in behaviors. In Proceedings of the 20th International Conference on Advances in Geographic Information Systems, Redondo Beach, CA, USA, 6–9 November 2012; ACM: New York, NY, USA, 2012; pp. 209–218.

25. Weibo. Available online: http://www.weibo.com (accessed on 2 April 2019).

26. Sina Weibo q4 2017 Financial Report. Available online: http://ir.weibo.com/financial-information/quarterly-results (accessed on 7 January 2019).

27. The 41st Statistical Report on Internet Development in China. Available online: https://cnnic.com.cn/IDR/ReportDownloads/201807/P020180711391069195909.pdf (accessed on 7 January 2019).

28. Liu, Y.; Zhang, F.; Wu, F.; Liu, Y.; Li, Z. The subjective wellbeing of migrants in guangzhou, china: The impacts of the social and physical environment. *Cities* **2017**, *60*, 333–342. [CrossRef]

29. De Mauro, A.; Greco, M.; Grimaldi, M. A formal definition of big data based on its essential features. *Libr. Rev.* **2016**, *65*, 122–135. [CrossRef]

30. Gandomi, A.; Haider, M. Beyond the hype: Big data concepts, methods, and analytics. *Int. J. Inf. Manag.* **2015**, *35*, 137–144. [CrossRef]

31. Tsou, M.-H. Research challenges and opportunities in mapping social media and big data. *Cartogr. Geogr. Inf. Sci.* **2015**, *42*, 70–74. [CrossRef]

32. Sabrina. Sina Weibo User Demographics Analysis in 2013. Available online: https://www.chinainternetwatch.com/5568/what-weibo-can-tell-you-about-chinese-netizens-part-1/ (accessed on 15 March 2019).

33. Daily, C. Special: Micro Blog's Macro Impact. *China Daily*, 2 March 2011.

34. Statistical Report on Internet Development in China. Available online: https://cnnic.com.cn/IDR/ReportDownloads/201411/P020141102574314897888.pdf (accessed on 17 March 2019).

35. Deng, S.; Yang, L.; Zhang, Y. Social q&as or library virtual reference service: What is behind the choices of chinese information seekers? *Libri* **2014**, *64*, 327–340.

36. Miller, H.J.; Goodchild, M.F. Data-driven geography. *GeoJournal* **2015**, *80*, 449–461. [CrossRef]

37. Luo, F.; Cao, G.; Mulligan, K.; Li, X. Explore spatiotemporal and demographic characteristics of human mobility via twitter: A case study of chicago. *Appl. Geogr.* **2016**, *70*, 11–25. [CrossRef]

38. Hasan, S.; Zhan, X.; Ukkusuri, S.V. Understanding urban human activity and mobility patterns using large-scale location-based data from online social media. In Proceedings of the 2nd ACM SIGKDD International Workshop on urban Computing, Chicago, IL, USA, 11 August 2013; ACM: New York, NY, USA, 2013; p. 6.

39. Bao, M.; Yang, N.; Zhou, L.; Lao, Y.; Zhang, Y.; Tian, Y. The spatial analysis of weibo check-in data: The case study of wuhan. In *Geo-Informatics in Resource Management and Sustainable Ecosystem*; Springer: New York, NY, USA, 2013; pp. 480–491.

40. Cao, J.; Hu, Q.; Li, Q. A study of users' movements based on check-in data in location-based social networks. In Proceedings of the International Symposium on Web and Wireless Geographical Information Systems, Seoul, Korea, 29–30 May 2014; Springer: New York, NY, USA, 2014; pp. 54–66.

41. Feng, Y.; Yu, Z.; Lu, X.; Tian, J. Understanding human dynamics of check-in behavior in lbsns. In Proceedings of the 2013 IEEE International Conference on Green Computing and Communications and IEEE Internet of Things and IEEE Cyber, Physical and Social Computing, Beijing, China, 20–23 August 2013; IEEE: New York, NY, USA, 2013; pp. 923–930.

42. Preoţiuc-Pietro, D.; Cohn, T. Mining user behaviours: A study of check-in patterns in location based social networks. In Proceedings of the 5th Annual ACM Web Science Conference, Paris, France, 2–4 May 2013; ACM: New York, NY, USA, 2013; pp. 306–315.

43. Sina Corporation. Available online: http://www.sina.com.cn/ (accessed on 2 April 2019).

44. Charalabidis, Y.; Loukis, E. Participative public policy making through multiple social media platforms utilization. *Int. J. Electron. Gov. Res. (IJEGR)* **2012**, *8*, 78–97. [CrossRef]

45. Rost, M.; Barkhuus, L.; Cramer, H.; Brown, B. Representation and communication: Challenges in interpreting large social media datasets. In Proceedings of the 2013 Conference on Computer Supported Cooperative Work, San Antonio, TX, USA, 23–27 February 2013; ACM: New York, NY, USA, 2013; pp. 357–362.

46. He, W.; Wang, F.-K.; Akula, V. Managing extracted knowledge from big social media data for business decision making. *J. Knowl. Manag.* **2017**, *21*, 275–294. [CrossRef]

47. Afzalan, N.; Evans-Cowley, J. Planning and social media: Facebook for planning at the neighbourhood scale. *Plan. Pract. Res.* **2015**, *30*, 270–285. [CrossRef]

48. Gao, X.; Yu, W.; Rong, Y.; Zhang, S. Ontology-based social media analysis for urban planning. In Proceedings of the 2017 IEEE 41st Annual Computer Software and Applications Conference (COMPSAC), Turin, Italy, 4–8 July 2017; IEEE: New York, NY, USA, 2017; pp. 888–896.

49. Shelton, T.; Poorthuis, A.; Zook, M. Social media and the city: Rethinking urban socio-spatial inequality using user-generated geographic information. *Landsc. Urban Plan.* **2015**, *142*, 198–211. [CrossRef]

50. Reed, P.J.; Khan, M.R.; Blumenstock, J. Observing gender dynamics and disparities with mobile phone metadata. In Proceedings of the Eighth International Conference on Information and Communication Technologies and Development, Ann Arbor, MI, USA, 3–6 June 2016; ACM: New York, NY, USA, 2016; p. 48.

51. Abbasi, M.-A.; Chai, S.-K.; Liu, H.; Sagoo, K. Real-world behavior analysis through a social media lens. In Proceedings of the International Conference on Social Computing, Behavioral-Cultural Modeling, and Prediction, College Park, MD, USA, 3–5 April 2012; Springer: Berlin, Germany, 2012; pp. 18–26.

52. Chen, H.; Beaudoin, C.E. An empirical study of a social network site: Exploring the effects of social capital and information disclosure. *Telemat. Inform.* **2016**, *33*, 432–435. [CrossRef]
53. Alonso, G.; Casati, F.; Kuno, H.; Machiraju, V. Web services. In *Web Services*; Springer: New York, NY, USA, 2004; pp. 123–149.
54. Bendler, J.; Brandt, T.; Neumann, D. Does social media reflect metropolitan attractiveness? Behavioral information from twitter activity in urban areas. In *Analytics and Data Science*; Springer: New York, NY, USA, 2018; pp. 119–142.
55. Ellison, N.B.; Steinfield, C.; Lampe, C. The benefits of facebook "friends:" social capital and college students' use of online social network sites. *J. Comput. Mediat. Commun.* **2007**, *12*, 1143–1168. [CrossRef]
56. Boyd, D.M.; Ellison, N.B. Social network sites: Definition, history, and scholarship. *J. Comput. Mediat. Commun.* **2007**, *13*, 210–230. [CrossRef]
57. Chen, S.; Shao, B.; Zhi, K. Predictors of chinese users' location disclosure behavior: An empirical study on wechat. *Information* **2018**, *9*, 219. [CrossRef]
58. Lacohée, H.; Wakeford, N.; Pearson, I. A social history of the mobile telephone with a view of its future. *Bt Technol. J.* **2003**, *21*, 203–211. [CrossRef]
59. Bellavista, P.; Küpper, A.; Helal, S. Location-based services: Back to the future. *IEEE Pervasive Comput.* **2008**, *7*, 85–89. [CrossRef]
60. Kwon, O.; Wen, Y. An empirical study of the factors affecting social network service use. *Comput. Hum. Behav.* **2010**, *26*, 254–263. [CrossRef]
61. Erl, T.; Khattak, W.; Buhler, P. *Big Data Fundamentals*; Prentice Hall: Englewood Cliffs, NJ, USA, 2016.
62. Vastardis, N.; Kun, Y. Mobile social networks: Architectures, social properties, and key research challenges. *IEEE Commun. Surv. Tutor.* **2013**, *15*, 1355–1371. [CrossRef]
63. Li, N.; Chen, G. Analysis of a location-based social network. In Proceedings of the 2009 International Conference on Computational Science and Engineering, Vancouver, BC, Canada, 29–31 August 2009; IEEE: New York, NY, USA, 2009; pp. 263–270.
64. Ahmed, A.M.; Tie, Q.; Feng, X.; Jedari, B.; Abolfazli, S. Event-based mobile social networks: Services, technologies, and applications. *IEEE Access* **2014**, *2*, 500–513. [CrossRef]
65. Grabowicz, P.A.; Ramasco, J.J.; Goncalves, B.; Eguiluz, V.M. Entangling mobility and interactions in social media. *PLoS ONE* **2014**, *9*, e92196. [CrossRef]
66. Andreassen, C.S. Online social network site addiction: A comprehensive review. *Curr. Addict. Rep.* **2015**, *2*, 175–184. [CrossRef]
67. Borgatti, S.P.; Everett, M.G.; Johnson, J.C. *Analyzing Social Networks*; Sage: New York, NY, USA, 2018.
68. Bao, J.; Zheng, Y.; Wilkie, D.; Mokbel, M. Recommendations in location-based social networks: A survey. *GeoInformatica* **2015**, *19*, 525–565. [CrossRef]
69. Furini, M.; Tamanini, V. Location privacy and public metadata in social media platforms: Attitudes, behaviors and opinions. *Multimed. Tools Appl.* **2015**, *74*, 9795–9825. [CrossRef]
70. Li, H.; Zhu, H.; Du, S.; Liang, X.; Shen, X.S. Privacy leakage of location sharing in mobile social networks: Attacks and defense. *IEEE Trans. Dependable Secur. Comput.* **2018**, *15*, 646–660. [CrossRef]
71. Kumar, S.; Saravanakumar, K.; Deepa, K. On privacy and security in social media—A comprehensive study. *Procedia Comput. Sci.* **2016**, *78*, 114–119.
72. Ruths, D.; Pfeffer, J. Social sciences. Social media for large studies of behavior. *Science* **2014**, *346*, 1063–1064. [CrossRef]
73. Vicente, C.R.; Freni, D.; Bettini, C.; Jensen, C.S. Location-related privacy in geo-social networks. *IEEE Internet Comput.* **2011**, *15*, 20–27. [CrossRef]
74. Li, J.; Yan, H.; Liu, Z.; Chen, X.; Huang, X.; Wong, D.S. Location-sharing systems with enhanced privacy in mobile online social networks. *IEEE Syst. J.* **2017**, *11*, 439–448. [CrossRef]
75. Fuchs, C. *Social Media: A Critical Introduction*; Sage: New York, NY, USA, 2017.
76. Benson, V.; Saridakis, G.; Tennakoon, H. Information disclosure of social media users: Does control over personal information, user awareness and security notices matter? *Inf. Technol. People* **2015**, *28*, 426–441. [CrossRef]
77. Strater, K.; Richter, H. Examining privacy and disclosure in a social networking community. In Proceedings of the 3rd Symposium on Usable Privacy and Security, Pittsburgh, PA, USA, 18–20 July 2007; ACM: New York, NY, USA, 2007; pp. 157–158.

78. Stefanone, M.A.; Huang, Y.C.; Lackaff, D. Negotiating social belonging: Online, offline, and in-between. In Proceedings of the 2011 44th Hawaii International Conference on System Sciences, Kauai, HI, USA, 4–7 January 2011; pp. 1–10.
79. Resch, B.; Summa, A.; Sagl, G.; Zeile, P.; Exner, J.-P. Urban emotions—Geo-semantic emotion extraction from technical sensors, human sensors and crowdsourced data. In *Progress in Location-Based Services 2014*; Springer: New York, NY, USA, 2015; pp. 199–212.
80. Varughese, B.M.; Joseph, M.S.; Thomas, P.E.; Sherly, K. Analyzing the behavior of youth to sociality using social media mining. In Proceedings of the 2017 International Conference on Intelligent Computing and Control Systems (ICICCS), Madurai, India, 15–16 June 2017; IEEE: New York, NY, USA, 2017; pp. 1231–1235.
81. Yuan, Y.; Wang, X. Exploring the effectiveness of location-based social media in modeling user activity space: A case study of weibo. *Trans. GIS* **2018**, *22*, 930–957. [CrossRef]
82. Shaw, S.L.; Sui, D. Giscience for human dynamics research in a changing world. *Trans. GIS* **2018**, *22*, 891–899. [CrossRef]
83. Chan, H.K.; Lacka, E.; Yee, R.W.; Lim, M.K. The role of social media data in operations and production management. *Int. J. Prod. Res.* **2017**, *55*, 5027–5036. [CrossRef]
84. Benevenuto, F.; Rodrigues, T.; Cha, M.; Almeida, V. Characterizing user behavior in online social networks. In Proceedings of the 9th ACM SIGCOMM Conference on Internet Measurement, Chicago, IL, USA, 4–6 November 2009; ACM: New York, NY, USA, 2009; pp. 49–62.
85. Chang, J.; Sun, E. Location 3: How users share and respond to location-based data on social networking sites. In Proceedings of the Fifth International AAAI Conference on Weblogs and Social Media, Barcelona, Spain, 17–21 July 2011; AAAI Press: Menlo Park, CA, USA, 2011; pp. 74–80.
86. Lei, C.; Zhang, A.; Qi, Q.; Su, H.; Wang, J. Spatial-temporal analysis of human dynamics on urban land use patterns using social media data by gender. *ISPRS Int. J. Geo-Inf.* **2018**, *7*, 358. [CrossRef]
87. Hu, Q.; Zhang, Y. An effective selecting approach for social media big data analysis—Taking commercial hotspot exploration with weibo check-in data as an example. In Proceedings of the 2018 IEEE 3rd International Conference on Big Data Analysis (ICBDA), Shanghai, China, 9–12 March 2018; IEEE: New York, NY, USA, 2018; pp. 28–32.
88. Saleem, M.A.; Kumar, R.; Calders, T.; Xie, X.; Pedersen, T.B. Location influence in location-based social networks. In Proceedings of the Tenth ACM International Conference on Web Search and Data Mining, Cambridge, UK, 6–10 February 2017; ACM: New York, NY, USA, 2017; pp. 621–630.
89. Humphreys, L. Mobile social networks and urban public space. *New Media Soc.* **2010**, *12*, 763–778. [CrossRef]
90. Roche, S. Geographic information science i. *Prog. Hum. Geogr.* **2014**, *38*, 703–711. [CrossRef]
91. Anthopoulos, L.G.; Vakali, A. Urban Planning and Smart Cities: Interrelations and Reciprocities. In *The Future Internet Assembly*; Springer: New York, NY, USA, 2012; pp. 178–189.
92. Scellato, S.; Noulas, A.; Lambiotte, R.; Mascolo, C. Socio-spatial properties of online location-based social networks. *ICWSM* **2011**, *11*, 329–336.
93. Li, N.; Chen, G. Sharing location in online social networks. *IEEE Netw.* **2010**, *24*, 20–25. [CrossRef]
94. Chen, C.; Ma, J.; Susilo, Y.; Liu, Y.; Wang, M. The promises of big data and small data for travel behavior (aka human mobility) analysis. *Transp. Res. Part C Emerg. Technol.* **2016**, *68*, 285–299. [CrossRef]
95. Hesse, B.W.; Moser, R.P.; Riley, W.T. From big data to knowledge in the social sciences. *Ann. Am. Acad. Pol. Soc. Sci.* **2015**, *659*, 16–32. [CrossRef]
96. Gao, H.; Liu, H. Mining human mobility in location-based social networks. *Synth. Lect. Data Min. Knowl. Discov.* **2015**, *7*, 1–115. [CrossRef]
97. Yang, C.; Xiao, M.; Ding, X.; Tian, W.; Zhai, Y.; Chen, J.; Liu, L.; Ye, X. Exploring human mobility patterns using geo-tagged social media data at the group level. *J. Spat. Sci.* **2018**, 1–18. [CrossRef]
98. Cheng, Y.; Park, J.; Sandhu, R. Relationship-based access control for online social networks: Beyond user-to-user relationships. In Proceedings of the 2012 International Conference on Privacy, Security, Risk and Trust (PASSAT), Amsterdam, The Netherlands, 3–6 September 2012; IEEE: New York, NY, USA, 2012; pp. 646–655.
99. Kaltenbrunner, A.; Scellato, S.; Volkovich, Y.; Laniado, D.; Currie, D.; Jutemar, E.J.; Mascolo, C. Far from the eyes, close on the web: Impact of geographic distance on online social interactions. In Proceedings of the 2012 ACM Workshop on Workshop on Online Social Networks, Helsinki, Finland, 17 August 2012; ACM: New York, NY, USA, 2012; pp. 19–24.

100. Rzeszewski, M.; Beluch, L. Spatial characteristics of twitter users—Toward the understanding of geosocial media production. *ISPRS Int. J. Geo-Inf.* **2017**, *6*, 236. [CrossRef]
101. Guan, W.; Gao, H.; Yang, M.; Li, Y.; Ma, H.; Qian, W.; Cao, Z.; Yang, X. Analyzing user behavior of the micro-blogging website sina weibo during hot social events. *Phys. A Stat. Mech. Its Appl.* **2014**, *395*, 340–351. [CrossRef]
102. Feng, Z.; Bo, W.; Yingxue, C. Research on china's city network based on users' friend relationships in online social networks: A case study of sina weibo. *GeoJournal* **2016**, *81*, 937–946. [CrossRef]
103. Cui, L.; Shi, J. Urbanization and its environmental effects in shanghai, china. *Urban Clim.* **2012**, *2*, 1–15. [CrossRef]
104. Han, B.; Cook, P.; Baldwin, T. Geolocation prediction in social media data by finding location indicative words. *Proc. COLING 2012* **2012**, *2012*, 1045–1062.
105. Schoen, H.; Gayo-Avello, P.T.M.D.; Gayo-Avello, D.; Takis Metaxas, P.; Mustafaraj, E.; Strohmaier, M.; Gloor, P. The power of prediction with social media. *Internet Res.* **2013**, *23*, 528–543. [CrossRef]
106. Backstrom, L.; Sun, E.; Marlow, C. Find me if you can. In Proceedings of the 19th International Conference on World Wide Web—WWW '10, Raleigh, NC, USA, 26–30 April 2010; ACM: New York, NY, USA, 2010; pp. 61–70.
107. Sun, Y.; Li, M. Investigation of travel and activity patterns using location-based social network data: A case study of active mobile social media users. *ISPRS Int. J. Geo-Inf.* **2015**, *4*, 1512–1529. [CrossRef]
108. Gu, Z.; Zhang, Y.; Chen, Y.; Chang, X. Analysis of attraction features of tourism destinations in a mega-city based on check-in data mining: A case study of shenzhen, china. *ISPRS Int. J. Geo-Inf.* **2016**, *5*, 210. [CrossRef]
109. Yin, J.; Lampert, A.; Cameron, M.; Robinson, B.; Power, R. Using social media to enhance emergency situation awareness. *IEEE Intell. Syst.* **2012**, *27*, 52–59. [CrossRef]
110. Yates, D.; Paquette, S. Emergency knowledge management and social media technologies: A case study of the 2010 haitian earthquake. *Int. J. Inf. Manag.* **2011**, *31*, 6–13. [CrossRef]
111. Cervone, G.; Schnebele, E.; Waters, N.; Moccaldi, M.; Sicignano, R. Using social media and satellite data for damage assessment in urban areas during emergencies. In *Seeing Cities through Big Data*; Springer: Berlin, Germany, 2017; pp. 443–457.
112. Pernici, B.; Francalanci, C.; Scalia, G.; Corsi, M.; Grandoni, D.; Biscardi, M.A. Geolocating social media posts for emergency mapping. *arXiv* **2018**, arXiv:1801.06861.
113. Sims, K.M.; Weber, E.M.; Bhaduri, B.L.; Thakur, G.S.; Resseguie, D.R. Application of social media data to high-resolution mapping of a special event population. In *Advances in Geocomputation*; Springer: Berlin, Germany, 2017; pp. 67–74.
114. Wang, Y.; Wang, T.; Ye, X.; Zhu, J.; Lee, J. Using social media for emergency response and urban sustainability: A case study of the 2012 beijing rainstorm. *Sustainability* **2015**, *8*, 25. [CrossRef]
115. Hong, I. Spatial analysis of location-based social networks in seoul, korea. *J. Geogr. Inf. Syst.* **2015**, *7*, 259. [CrossRef]
116. Mazumdar, P.; Patra, B.K.; Babu, K.S.; Lock, R. Hidden location prediction using check-in patterns in location-based social networks. *Knowl. Inf. Syst.* **2018**, *57*, 571–601. [CrossRef]
117. Dokuz, A.S.; Celik, M. Discovering socially important locations of social media users. *Expert Syst. Appl.* **2017**, *86*, 113–124. [CrossRef]
118. Fiorio, L.; Zagheni, E.; Abel, G.; Hill, J.; Pestre, G.; Letouzé, E.; Cai, J. Understanding patterns of human mobility at different time scales. *PAA 2018 Annual Meeting*. April 26–28, Denver. 2017, pp. 1–24. Available online: https://paa.confex.com/paa/2018/meetingapp.cgi/Paper/21412 (accessed on 2 April 2019).
119. Wu, C.; Ye, X.; Ren, F.; Du, Q. Check-in behaviour and spatio-temporal vibrancy: An exploratory analysis in shenzhen, china. *Cities* **2018**, *77*, 104–116. [CrossRef]
120. Vikat, A.; Jones, C. *Indicators of Gender Equality*; United Nations Economic Commission for Europe: Geneva, Switzerland, 2014; pp. 1–130.
121. O'Dorchai, S.; Meulders, D.; Crippa, F.; Margherita, A. *She Figures 2009–Statistics and Indicators on Gender Equality in Science*; Publications Office of the European Union: Copenhagen, Denmark, 2009.
122. Li, S.-M.; Siu, Y.-M. Residential mobility and urban restructuring under market transition: A study of guangzhou, china. *Prof. Geogr.* **2001**, *53*, 219–229. [CrossRef]
123. Dinsa, G.D.; Goryakin, Y.; Fumagalli, E.; Suhrcke, M. Obesity and socioeconomic status in developing countries: A systematic review. *Obes. Rev.* **2012**, *13*, 1067–1079. [CrossRef]

124. McMurray, R.G.; Harrell, J.S.; Deng, S.; Bradley, C.B.; Cox, L.M.; Bangdiwala, S.I. The influence of physical activity, socioeconomic status, and ethnicity on the weight status of adolescents. *Obes. Res.* **2000**, *8*, 130–139. [CrossRef]

125. Yang, Y.; Li, D.; Mu, D. Levels, seasonal variations and sources of organochlorine pesticides in ambient air of guangzhou, china. *Atmos. Environ.* **2008**, *42*, 677–687. [CrossRef]

126. Dong, Z.; Liu, J.; Sha, S.; Li, X.; Dong, J. Regional disparity of real estate investment in china: Characteristics and empirical study in the context of population aging. *Eurasia J. Math. Sci. Technol. Educ.* **2017**, *13*, 7799–7811. [CrossRef]

127. Wan, Q.; Ma, G.; Li, J.; Wang, X.; Fan, J.; Li, Q.; Lu, W. A comparison of gps-tec with iri-tec at low latitudes in china in 2006. *Adv. Space Res.* **2017**, *60*, 250–256. [CrossRef]

128. Xiong, X.; Jin, C.; Chen, H.; Luo, L. Using the fusion proximal area method and gravity method to identify areas with physician shortages. *PLoS ONE* **2016**, *11*, e0163504. [CrossRef]

129. Shen, J.; Kee, G. Shanghai: Urban development and regional integration through mega projects. In *Development and Planning in Seven Major Coastal Cities in Southern and Eastern China*; Springer: New York, NY, USA, 2017; pp. 119–151.

130. Shen, J.; Kee, G. *Development and Planning in Seven Major Coastal Cities in Southern and Eastern China*; Springer: New York, NY, USA, 2017; Volume 120.

131. Campagna, M. Social media geographic information: Why social is special when it goes spatial. In *European Handbook of Crowdsourced Geographic Information*; Ubiquity Press: London, UK, 2016; pp. 45–54.

132. Littman, J.; Chudnov, D.; Kerchner, D.; Peterson, C.; Tan, Y.; Trent, R.; Vij, R.; Wrubel, L. Api-based social media collecting as a form of web archiving. *Int. J. Digit. Libr.* **2018**, *19*, 21–38. [CrossRef]

133. Weibo api. Available online: http://open.weibo.com/wiki/API (accessed on 9 December 2018).

134. Fernandes, R.; D'Souza, R. Analysis of product twitter data though opinion mining. In Proceedings of the 2016 IEEE Annual India Conference (INDICON), Bangalore, India, 16–18 Decmeber 2016; IEEE: New York, NY, USA, 2016; pp. 1–5.

135. Batrinca, B.; Treleaven, P.C. Social media analytics: A survey of techniques, tools and platforms. *Ai Soc.* **2015**, *30*, 89–116. [CrossRef]

136. McCoy, J.; Johnston, K. *Using Arcgis Spatial Analyst*; Esri Redlands: Redlands, CA, USA, 2001.

137. Shen, Y.; Karimi, K. Urban function connectivity: Characterisation of functional urban streets with social media check-in data. *Cities* **2016**, *55*, 9–21. [CrossRef]

138. Schwartz, R.; Halegoua, G.R. The spatial self: Location-based identity performance on social media. *New Media Soc.* **2015**, *17*, 1643–1660. [CrossRef]

139. Johnson, I.L.; Sengupta, S.; Schöning, J.; Hecht, B. The geography and importance of localness in geotagged social media. In Proceedings of the 2016 CHI Conference on Human Factors in Computing Systems, San Jose, CA, USA, 7–12 May 2016; ACM: New York, NY, USA, 2016; pp. 515–526.

140. Silverman, B.W. Algorithm as 176: Kernel density estimation using the fast fourier transform. *J. R. Stat. Soc. Ser. C (Appl. Stat.)* **1982**, *31*, 93–99. [CrossRef]

141. Bowman, A.W. An alternative method of cross-validation for the smoothing of density estimates. *Biometrika* **1984**, *71*, 353–360. [CrossRef]

142. Silverman, B.W. *Density Estimation for Statistics and Data Analysis*; Routledge: New York, NY, USA, 2018.

143. Wu, C.; Ye, X.; Ren, F.; Wan, Y.; Ning, P.; Du, Q. Spatial and social media data analytics of housing prices in shenzhen, china. *PLoS ONE* **2016**, *11*, e0164553. [CrossRef]

144. Lichman, M.; Smyth, P. Modeling human location data with mixtures of kernel densities. In Proceedings of the 20th ACM SIGKDD International Conference on Knowledge Discovery and Data Mining—KDD '14, New York, NY, USA, 24–27 August 2014; ACM: New York, NY, USA, 2014; pp. 35–44.

145. Silverman, B.W. *Density Estimation for Statistics and Data Analysis*; CRC Press: New York, NY, USA, 1986; Volume 26.

146. King, T.L.; Bentley, R.J.; Thornton, L.E.; Kavanagh, A.M. Using kernel density estimation to understand the influence of neighbourhood destinations on bmi. *BMJ Open* **2016**, *6*, e008878. [CrossRef]

147. Li, L.; Goodchild, M.F.; Xu, B. Spatial, temporal, and socioeconomic patterns in the use of twitter and flickr. *Cartogr. Geogr. Inf. Sci.* **2013**, *40*, 61–77. [CrossRef]

148. Sun, Y.; Fan, H.; Li, M.; Zipf, A. Identifying the city center using human travel flows generated from location-based social networking data. *Environ. Plan. B Plan. Des.* **2016**, *43*, 480–498. [CrossRef]

149. Yuan, N.J.; Zheng, Y.; Xie, X.; Wang, Y.; Zheng, K.; Xiong, H. Discovering urban functional zones using latent activity trajectories. *IEEE Trans. Knowl. Data Eng.* **2015**, *27*, 712–725. [CrossRef]

150. Li, H.; Ge, Y.; Hong, R.; Zhu, H. Point-of-interest recommendations: Learning potential check-ins from friends. In Proceedings of the 22nd ACM SIGKDD International Conference on Knowledge Discovery and Data Mining, San Jose, CA, USA, 13–17 August 2016; ACM: New York, NY, USA, 2016; pp. 975–984.

151. Xie, Z.; Yan, J. Kernel density estimation of traffic accidents in a network space. *Comput. Environ. Urban Syst.* **2008**, *32*, 396–406. [CrossRef]

152. Pardoux, E.; Peng, S. Adapted solution of a backward stochastic differential equation. *Syst. Control Lett.* **1990**, *14*, 55–61. [CrossRef]

153. Hahn, K.; Pflug, P. On a minimal complex norm that extends the real euclidean norm. *Mon. Für Math.* **1988**, *105*, 107–112. [CrossRef]

154. Sun, J.; Wang, G.; Cheng, X.; Fu, Y. Mining affective text to improve social media item recommendation. *Inf. Process. Manag.* **2015**, *51*, 444–457. [CrossRef]

155. Bao, J.; Liu, P.; Yu, H.; Xu, C. Incorporating twitter-based human activity information in spatial analysis of crashes in urban areas. *Accid Anal Prev* **2017**, *106*, 358–369. [CrossRef]

156. Fotheringham, A.S.; Brunsdon, C.; Charlton, M. *Geographically Weighted Regression*; John Wiley & Sons, Limited West Atrium: New York, NY, USA, 2013.

157. Baojun, W.; Bin, S.; Inyang, H.I. Gis-based quantitative analysis of orientation anisotropy of contaminant barrier particles using standard deviational ellipse. *Soil Sediment Contam.* **2008**, *17*, 437–447. [CrossRef]

158. Loo, B.P.; Lam, W. A multilevel investigation of differential individual mobility of working couples with children: A case study of hong kong. *Transp. A Transp. Sci.* **2013**, *9*, 629–652. [CrossRef]

159. Liu, Y.; Wang, F.; Xiao, Y.; Gao, S. Urban land uses and traffic 'source-sink areas': Evidence from gps-enabled taxi data in shanghai. *Landsc. Urban Plan.* **2012**, *106*, 73–87. [CrossRef]

160. Peng, J.; Chen, S.; Lü, H.; Liu, Y.; Wu, J. Spatiotemporal patterns of remotely sensed pm2. 5 concentration in china from 1999 to 2011. *Remote Sens. Environ.* **2016**, *174*, 109–121. [CrossRef]

161. Shi, Y.; Matsunaga, T.; Yamaguchi, Y.; Li, Z.; Gu, X.; Chen, X. Long-term trends and spatial patterns of satellite-retrieved pm2.5 concentrations in south and southeast asia from 1999 to 2014. *Sci. Total Environ.* **2018**, *615*, 177–186. [CrossRef]

162. Lefever, D.W. Measuring geographic concentration by means of the standard deviational ellipse. *Am. J. Sociol.* **1926**, *32*, 88–94. [CrossRef]

163. Nara, A.; Tsou, M.-H.; Yang, J.-A.; Huang, C.-C. The opportunities and challenges with social media and big data for research in human dynamics. In *Human Dynamics Research in Smart and Connected Communities*; Springer: Berlin, Germany, 2018; pp. 223–234.

164. Nikolaidou, A.; Papaioannou, P. Utilizing social media in transport planning and public transit quality: Survey of literature. *J. Transp. Eng. Part A Syst.* **2018**, *144*, 04018007. [CrossRef]

165. Zhou, C.; Xu, X.; Sylvia, S. Population distribution and its change in guangzhou city. *Chin. Geogr. Sci.* **1998**, *8*, 193–203. [CrossRef]

166. *National Bureau of Statistics of China*; National Bureau of Statistics of China: Beijing, China, 2016.

167. *Guangzhou Statistical Yearbook 2013*; China Statistics Press: Beijing, China, 2013.

 *sustainability*

Article

Sales Prediction by Integrating the Heat and Sentiments of Product Dimensions

Xiaozhong Lyu [1,*], **Cuiqing Jiang** [1,2,*], **Yong Ding** [1,2], **Zhao Wang** [1] **and Yao Liu** [1]

[1]   School of Management, Hefei University of Technology, Hefei 230009, China; dingyong@hfut.edu.cn (Y.D.); xcwangzhao@163.com (Z.W.); liuyaoemail@foxmail.com (Y.L.)

[2]   Key Laboratory of Process Optimization and Intelligent Decision Making of Ministry of Education, Hefei 230009, China

*   Correspondence: adolflv@mail.hfut.edu.cn (X.L.); jiangcuiq2017@163.com (C.J.)

Received: 20 January 2019; Accepted: 7 February 2019; Published: 11 February 2019

Abstract: Online word-of-mouth (eWOM) disseminated on social media contains a considerable amount of important information that can predict sales. However, the accuracy of sales prediction models using big data on eWOM is still unsatisfactory. We argue that eWOM contains the heat and sentiments of product dimensions, which can improve the accuracy of prediction models based on multiattribute attitude theory. In this paper, we propose a dynamic topic analysis (DTA) framework to extract the heat and sentiments of product dimensions from big data on eWOM. Ultimately, we propose an autoregressive heat-sentiment (ARHS) model that integrates the heat and sentiments of dimensions into the benchmark predictive model to forecast daily sales. We conduct an empirical study of the movie industry and confirm that the ARHS model is better than other models in predicting movie box-office revenues. The robustness check with regard to predicting opening-week revenues based on a back-propagation neural network also suggests that the heat and sentiments of dimensions can improve the accuracy of sales predictions when the machine-learning method is used.

Keywords: big data; sales prediction; online word-of-mouth; dynamic topic model; product attributes; back-propagation neural network

---

1. Introduction

Social media are forms of electronic communication (such as Facebook, WeChat, and IMDb.com) through which people create online communities to share information, ideas, personal messages, etc. Consumers increasingly use online word-of-mouth (eWOM) on social media for decision support before making purchases [1]. Therefore, social media marketing has recently appeared as an interdisciplinary and cross-functional concept that uses social media to achieve organizational goals by creating value for stakeholders [2]. Sales prediction is a foundation for social media marketing. Highly accurate and timely sales predictions can allow firms to reduce their profit losses and to improve their market performance [3]. Due to the superiority of big data on social media, sales predictions are being produced more than ever before to increase their accuracy and to enable them to support real time marketing strategies for online retailers and enterprises. However, the accuracy of these models is still unsatisfactory. We need to extract more predictive information from high-frequency social media data to improve the accuracy of sales predictions.

High-frequency big data, such as eWOM on social media [4] and Google search index (GSI) data on the Google search engine [5], contain timely information and can improve the accuracy of sales predictions [6]. However, the accuracy of sales predictions is still unsatisfactory for irregular or nonseasonal sales trends [7,8]. Based on multiattribute attitude theory [9,10], we argue that the heat and sentiments of product dimensions discussed in eWOM, which previous predictive models do not consider, can improve the accuracy of sales predictions. These factors have effects on product sales [11,12]. Therefore, this paper proposes a framework to simultaneously extract the heat and sentiments of product dimensions from eWOM and to then integrate them into a sales prediction model.

We chose the movie industry as our research context. We obtain reviews from IMDb.com, online search data from Google.com, and film-related data from BoxOfficeMojo.com. Finally, we construct a large dataset including data on films, Google Trends, and 349,269 reviews of 122 movies.

To extract the heat and sentiments of product dimensions, in this study, we developed a dynamic topic analysis (DTA) framework that integrates machine-learning techniques and lexicon-based methods. The framework has two major functions. First, DTA captures key product dimensions from eWOM without manual annotation. Second, DTA simultaneously extracts the heat and sentiments of the extracted dimensions. Next, we integrate the heat and sentiments of the dimensions to construct a new sales prediction model, called the autoregressive heat-sentiment (ARHS) model, to dynamically predict sales. We focused on the three most important dimensions discussed in movie eWOM: the *star*, the *genre*, and the *plot*. We found that the proposed ARHS model has better accuracy than previous models in predicting movie box-office revenues. Furthermore, the ARHS model can predict sales of all kinds of products if the products have multiple attributes and sufficient eWOM. The robustness check with regard to forecasting opening-week revenues using a back-propagation (BP) neural network demonstrates that the predictive model integrating the heat and sentiments of dimensions is more accurate.

2. Literature Review

eWOM influences consumer purchase intentions by changing the preferences for alternatives and in turn influences product sales based on information theory [13,14]. We introduce multiattribute attitude theory in this research domain.

2.1. eWOM's Effect on Sales

Some research on eWOM has shown mixed findings regarding the direct effects of eWOM on product sales [15,16]. Other research shows the moderating effects of rating variance [17], review helpfulness [18], and the features of reviewers [19], products [20,21], and social media platforms [22–24]. In this paper, we focus on the direct effects of eWOM.

The volume of eWOM represents the popularity (overall heat) of products supplied by reviewers, such as the number of online reviews. Previous studies have found mixed results regarding the effects of eWOM volume on sales [25,26]. Many studies have found that eWOM volume positively affects sales [25,27–29], whereas several other studies have not found a significant effect [30,31]. Xu [32] found that more information could even reduce sales under certain conditions. Therefore, under some conditions, volume cannot be used to predict sales. The multiattribute attitude model demonstrates that only the most important attributes that reflect consumers' perceptual dimensions can influence consumer purchasing decisions [33]. In this paper, we divide the overall heat of products into the heat of their important attributes. Previous research has proven that the heat of key product dimensions can influence product sales [11]. In this paper, we demonstrate that the heat of dimensions has predictive power in predicting movie box-office revenues.

The valence of eWOM can be the average rating on the rating scale (e.g., 1–5), or it can be binary (positive and negative). It can also be regarded as the overall sentiment of eWOM [34,35]. Most studies have reported a significant, positive effect of valence [27,36], but other studies have not found a significant effect [29,37]. The overall sentiment of eWOM represents the emotion conveyed by reviewers to consumers. However, the overall sentiment represents the aggregation of the sentiments of all attributes discussed in eWOM, which include irrelevant and abundant attributes. Perhaps for this reason, prior studies have found that the overall sentiment of eWOM has no effect on movie box-office revenues [38,39]. Chen and Xie [40] demonstrate that eWOM provides product-dimension preference information that helps consumers find products that match their needs. Potential consumers will change their purchase intentions regarding a product after perceiving the sentiments of important product dimensions from online reviews [12]. We argue that analyzing the eWOM sentiments of key product dimensions can provide new insights for sales prediction and overcome the weakness of overall sentiment.

2.2. eWOM-Based and GSI-Based Sales Prediction

Online search data include indexes (from zero to 100) of the frequency of the object searched in an online search engine, such as Google.com. This type of data has been used to predict movie box-office revenues [41]. Bughin [42] finds that the valence of eWOM influences sales more than Google Trends. Geva et al. [5] find that adding Google search data to models based on the more commonly used eWOM data improves the accuracy of sale predictions for search products. Regarding the different natures of search products and experience products, the effect of online information is always different in these two kinds of products [27]. In this paper, we aim to demonstrate whether the model used for search products is valid for experience products. Geva et al. [5] also found that for search products, Google search index (GSI) models based on inexpensive Google Trends provide accuracy that is at least comparable to that of eWOM-based prediction models. These studies have proven that both online search data and eWOM have powerful predictive ability. To date, however, the predictive ability of the heat and sentiments of product dimensions has not been researched. This study attempts to improve the prediction accuracy of movie box-office revenues by proposing a comprehensive model that simultaneously integrates the heat and sentiments of product dimensions.

3. Materials and Methods

3.1. Research Framework

Figure 1 shows the framework of our study; it will help researchers develop a sales prediction model for products with abundant eWOM and multiple attributes. First, based on the findings that eWOM volume [25,27–29] and ratings [27,36] positively affect sales, we developed the eWOM model by integrating eWOM variables into the autoregressive (AR) model. Additionally, we developed the GSI model by integrating Google Trends into the AR model based on findings in the literature [41]. We then integrated Google Trends into the eWOM model following the method of [5] and named this benchmark model the autoregressive online (ARO) model.

The above models consider volume (heat) and valence as a whole. Multiattribute attitude theory decomposes a consumer's overall attitude toward a product into smaller components [9,10]. These components, which are the most important attributes, reflect only consumers' perceptual dimensions rather than product characteristics that are directly controllable and measurable by marketing managers [33]. Multiattribute attitude theory shows that only the importance and valence of important product attributes can predict consumer purchase predispositions. Volume and sentiment, which represent the sum of the heat and sentiments of all attributes implied in eWOM, may have little effect on sales because of the offset effects of irrelevant and redundant attributes. This may be why some research finds that the volume and valence of eWOM have no effect on sales [15,38,39]. In previous studies, attributes were generated based on expert judgment and in-depth interviews. Additionally, the measures of attribute importance and valence in the attitude model were obtained by surveying a sample of respondents [43]. Recent research shows that topic models can be used to generate important attributes that reflect consumers' perceptual dimensions, and the heat and sentiment of these product dimensions can be used as a proxy for attribute importance and valence [44–46]. Therefore, the heat and sentiments of dimensions are more suitable than eWOM volume and valence for predicting sales.

Thus, according to the discussion above, we hypothesize the following:

H1. *The heat and sentiments of dimensions have predictive power for sales and can improve the performance of the benchmark model.*

Finally, to verify the hypothesis, we used DTA to extract the heat and sentiments of product dimensions from eWOM and integrated them into the benchmark model to determine whether the new model, the ARHS model, has better prediction accuracy.

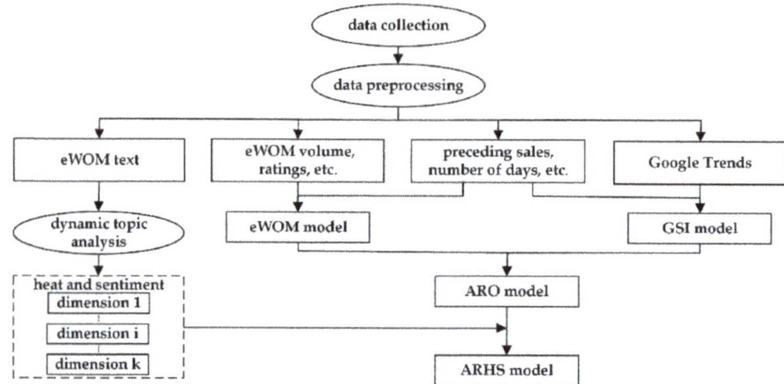

Figure 1. Framework for constructing the autoregressive heat-sentiment (ARHS) model.

3.2. Data and Variables

3.2.1. Data Collection

This paper focuses on online reviews of movies because within the film industry, online reviews are more popular than other types of eWOM [29]. In this study, IMDb.com, Google.com, and BoxOfficeMojo.com are the data sources. We examined movie reviews on IMDb.com, the most popular and authoritative information source for movies worldwide, for approximately seven weeks after movie releases. We then collected daily data on box-office revenues, budgets, and distributors as well as other movie information from BoxOfficeMojo.com. Our unit of time was one day; however, we aggregated the reviews published before the release day into one time window. The weekly Google Trends before a movie release and the daily Google Trends one day before and 49 days after a movie release were obtained and reviewed. Because the unit of time was one day, we had a sufficiently long study period with enough observations to provide credible results. The final dataset included Google Trends and eWOM information for 50 consecutive time windows and revenue information for 49 time windows. All movies were released in the US from 2010 to 2016.

After filtering out movies with fewer than 100 reviews by the end of the data period, we identified 349,269 reviews for 122 sample movies. We chose a threshold of 100 reviews to ensure that we have sufficient reviews to train the topic model used in the DTA. As shown in Table 1, our sample movies exhibited great diversity in terms of film distributor, movie genre, release month, and Motion Picture Association of America (MPAA) rating. Table 2 indicates that the total domestic gross and production budgets of the movies are right-skewed; that is, all but a few movies have low box-office revenues and product budgets.

Tables 3 and 4 list the definitions of and statistics regarding eWOM, Google Trends, and our film-related variables. First, we measured the eWOM volume and valence, which are represented by $v_{t,1}$ and $v_{t,2}$, respectively. Volume is the log-transformation of the daily number of reviews. We added one to the daily number of reviews to ensure that the log-transformation result was not negative [47]. Valence is the mean of the daily review ratings, reflecting the overall sentiment of reviewers with regard to a specific movie [28]. If there were no reviews on one day, then we used the average valence of the preceding days as a proxy [48]. Second, we used the variable $v_{t,3}$ to denote the number of days since the movie release to consider the time effect. Third, we set the dummy variable $v_{t,4}$ to one if the day was on the weekend and zero otherwise to consider the seasonal effect. Fourth, the variable $v_{t,5}$ represents the number of cinemas at which a film was being shown [17]. Finally, we used the

Google Trends of movie names, and the initial trends range from 0 to 100 in terms of online search data. Table 4 shows that sales, volume ($v_{t,1}$), and theaters ($v_{t,5}$) have right-skewed distributions and that the skewness of volume and sales is very large. This result means that very few movies had high box-office revenues or high heat and that most movies had low box-office revenues or low heat. The distributions of valence ($v_{t,2}$) are relatively evenly distributed.

Table 1. The diversity of movies.

Distributor	Freq.	Genre	Freq.	Release Month	Freq.	MPAA Ratings	Freq.
Warner Bros.	18	Drama	38	January	10	R	57
Lionsgate	16	Comedy	37	February	11	PG-13	50
Paramount	12	Thriller	14	March	12	PG	14
Weinstein	10	Action	13	April	7	NC-17	1
Fox	10	Sci-Fi	10	May	10	Total	122
Sony	9	Horror	9	June	6		
Universal	7	Animation	8	July	7		
Open Road Films	7	Crime	6	August	11		
Focus Features	6	Fantasy	5	September	11		
Roadside Attractions	6	Adventure	3	October	12		
FilmDistrict	4	Sports	2	November	11		
Relativity	4	Music	2	December	14		
Buena Vista	4	Romance	2				
CBS Films	2	Documentary	1				
Bleecker Street	2	War	1				
TriStar	2						
A24	1						
Radius-TWC	1						
Rogue Pictures	1						

Table 2. The distribution of movie gross and budgets.

Domestic Gross (Million)	Freq.	Production Budget (Million)	Freq.
≤ 25	40	≤ 25	59
25–50	32	25–50	30
50–75	21	50–75	11
75–100	10	75–100	7
100–125	9	100–125	3
125–150	1	125–150	6
150–175	3	150–175	1
175–200	3	175–200	5
200–225	1	Total	122
225–250	1		
250–275	1		

Table 3. Key variables for each movie: numerical.

Variable	Description (for Each Movie)	Measure and Data Sources
Sales	Daily domestic box-office revenues	Dollars (log-transformation); BoxOfficeMojo.com
$v_{t,1}$	Daily number of reviews	Number (log-transformation); IMDb.com
$v_{t,2}$	Daily valence of reviews	Average of daily ratings (0–10); IMDb.com
$v_{t,3}$	Days from initial release	Number (1–49)
$v_{t,4}$	Whether the day is on the weekend	1 = the day is on the weekend (Fri, Sat, and Sun), 0 = others
$v_{t,5}$	Daily number of cinemas	Number (log-transformation); BoxOfficeMojo.com
$v_{t,6}$	Daily Google Trends of movie name	Number (0–100); Google.com

Table 4. Summary statistics of the key variables.

Variable	Mean	Median	Maximum	Minimum	Std. Dev.	Skewness	Kurtosis
Sales	1,039,207	263,875	35,167,017	10	2,204,483.7	5.351	46.855
$v_{t,1}$	22.11526	11	506	0	38.754737	3.956	26.541
$v_{t,2}$	3.801627	4	10	0	3.6083277	0.162	1.410
$v_{t,5}$	1483.412	1195	4324	1	1264.8718	0.343	1.634
$v_{t,6}$	33.49281	28	100	2	21.922873	1.101	3.815

Figure 2a shows the relationship between Google Trends and the box-office revenues of the movie Gravity. Figure 2b shows the relationship between eWOM volume and the box-office revenues of the movie Gravity. We observe that both the eWOM and GSI data have high correlations with movie box-office revenues.

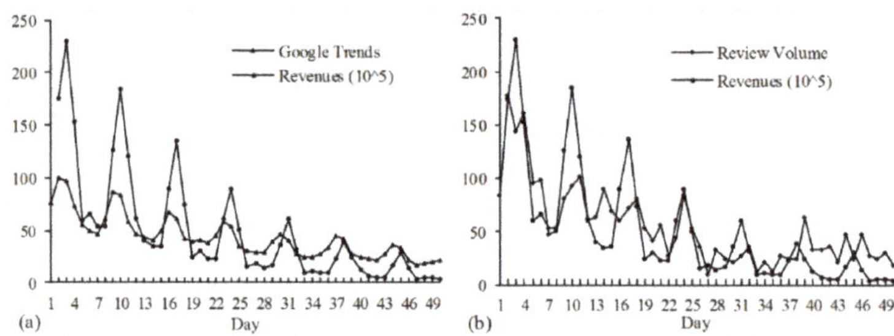

Figure 2. The relationship between online information and the box-office revenues of the movie Gravity. (a) The relationship between Google Trends and box-office revenues; (b) the relationship between the number of reviews and box-office revenues.

3.2.2. Dynamic Topic Analysis

For 122 movies, we constructed a DTA framework by integrating the dynamic topic model (DTM) [49], the lexicon-based method [50], and the Stanford natural language processing (NLP) technique [51] to derive the heat and sentiments of dimensions from online reviews. We obtained 122 daily documents by integrating hundreds of daily reviews for each movie into one document. Finally, the daily documents compiled over 50 days constitute our review corpus, which contains 349,269 reviews. Figure 3 shows the structure of the corpus.

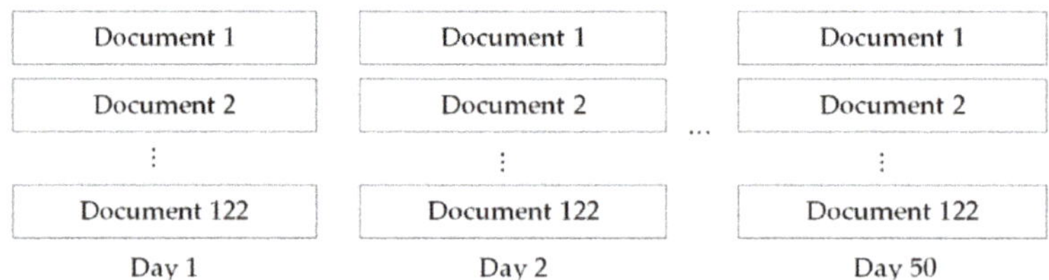

Figure 3. The structure of our review set.

We preprocessed each document by using the steps used in a study by Guo et al. [45]. First, we eliminated non-English words and spelling errors, such as web sites, punctuation marks, and numbers. We then used the Stanford NLP package for word text tokenization, part-of-speech tagging, and word stemming. Finally, each document became a word-of-bag.

To extract key product dimensions from a large corpus of text data in an effective manner, previous studies have used the latent Dirichlet allocation (LDA) model [44,45]. The DTM is more suitable for extracting key product dimensions from our structured review set [49] and is an extended

LDA method [52]. The DTM can quickly identify a conglomeration of connected topics from a very large number of documents over different time windows, which LDA alone cannot do.

As a machine-learning method, the DTM is highly efficient in handling online big data. We used the DTM to extract key product dimensions, the heat of these dimensions, words that represent each dimension and the changes in these factors over different time windows. The DTM assumes that a review comprises a sequence of N words, $d = (w_1, w_2, \ldots, w_N)$, D reviews form a review set, $C_t = [d_1, d_2, \ldots, d_D]$, and T review sets form a corpus over T time windows, $C = \{C_1, C_2, \ldots, C_T\}$. The DTM also assumes that reviewers share K dimensions across the corpus over the T time windows. In each time window, the DTM assumes that reviewers express their experience with a product or service over K dimensions. For instance, a reviewer may comment about a movie in a review by focusing on three dimensions with different heat and sentiments: 30% and 4.9 for movie stars, 40% and 3.4 for the story plot, and 30% and 2.1 for the background music. Thirty percent is the dimension heat of movie stars, which means that one-third of the review is about movie stars; additionally, 4.9 is the sentiment strength of movie stars, which means that the reviewer has a strong sentiment toward movie stars.

Comparing the perplexity of the DTM and the semantics of the dimensions when using different values of K, we determine the optimal number of key product dimensions [11]. Ultimately, we find three movie dimensions that can perfectly represent the review corpus. The formula for the perplexity of the DTM for the document set on day t is as follows:

$$perplexity(C_t) = exp\left(-\frac{\sum_{d=1}^{D} \sum_{n=1}^{N_d} \log \sum_{k=1}^{K} p(W_{d,n} = w | Z_{d,n} = k) p(Z_{d,n} = k | d)}{\sum_{d=1}^{D} N_d} \right) \tag{1}$$

where C_t is the document set on day t; D is the number of documents on day t; N_d is the number of words in document d; K is the number of dimensions; $p(W_{d,n} = w | Z_{d,n} = k)$ is the heat of word w in dimension k; $p(Z_{d,n} = k | d)$ is the heat of dimension k in document d. DTM learning with Gibbs sampling can simultaneously generate the heat of the words in each dimension and the heat of the dimensions in each document. Readers can refer to [49] for details on the DTM. Let $\vartheta_{i,t}$ be the heat of the kth dimension of the ith movie on day t. $\vartheta_{i,t}$ can be calculated as follows:

$$\vartheta_{i,t,k} = \frac{\sum_{d=1}^{D_{i,t}} p(Z = k | t, d, i)}{D_{i,t}} \tag{2}$$

where $p(Z = k | t, d, i)$ is the heat of dimension k in document d of movie i, and $D_{i,t}$ is the number of documents for movie i on day t. In our research context, $D_{i,t}$ equals one.

We name the three dimensions *plot*, *star*, and *genre*, following the method of Guo et al. [45]; these dimensions have been regarded as the three most important attributes of movies [7,53]. Table 5 shows the changes in the dimension *plot* in different time windows.

Table 5. The change in words and the weight of the dimension *plot*.

plot	weight	*plot*	weight	*plot*	weight
story	0.9%	plot	0.5%	plot	0.5%
plot	0.4%	story	0.4%	story	0.4%
book	0.1%	book	0.3%	book	0.4%
horror	0.3%	horror	0.3%	horror	0.3%
dark	0.2%	dark	0.3%	dark	0.2%
original	0.3%	original	0.2%	original	0.2%
scary	0.2%	scary	0.2%	scary	0.2%
real	0.2%	maze	0.2%	maze	0.2%
pretty	0.2%	pretty	0.2%	pretty	0.2%
action	0.2%	love	0.2%	house	0.2%

The heat of a dimension refers to the proportion of reviewers' discussion concerned with the dimension of a product in eWOM. For example, the heat of the dimension *plot* denotes the proportion of consumers' discussion concerned with *plot*-related information in reviews. Figure 4 shows the changes in the heat of the three movie dimensions over 50 days. Consumers talk more about movie stars and the story plot in the early days after a movie's release than they do at the end of the release.

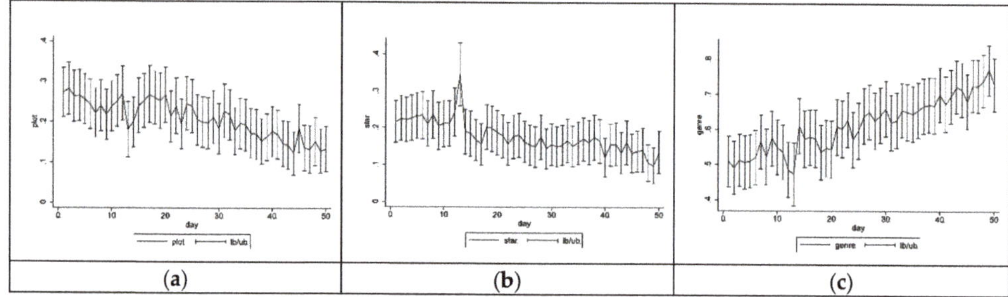

Figure 4. Average heat of the three dimensions for 122 movies. (**a**) The average heat of the dimension *plot*; (**b**) the average heat of dimension *star*; (**c**) the average heat of dimension *genre*.

We then used the sentiment lexicon and syntax relation to calculate the sentiments of dimensions. Lexicon-based methods that use a publicly recognized sentiment lexicon are more objective and suitable for big data sentiment analysis than machine-learning-based methods that require expert annotations because expert annotation has a high cost and there are artificial deviations. Most studies on dimension sentiment analysis divide dimensions into positive and negative classes [54], and sentiment analysis methods are different based on different applications. We calculated the sentiment strength of each dimension that can forecast movie box-office revenues. We extracted the syntactic relations between the dimension words and sentiment words in the daily review sentences using the Stanford NLP package, and we obtained the sentiments of the dimension words based on the extracted relations. Table 6 shows the main sentiment mining rules used in our framework.

Table 6. The main rules for mining the sentiments of dimension words.

Syntax Relations	Examples	Word Sentiments
Nominal subject	The *plot* is *boring*.	*Plot*: 3.0
Adjectival modifier	She is a *good actor*.	*Actor*: 3.8612
Direct object	I *enjoy* 3D.	*3D*: 3.9782
Open clausal complement	I think the actor *enjoys acting*.	*Acting*: 3.9782
Adverb modifier	Tom *performed earnestly*.	*Perform*: 3.5
Relative clause modifier	I saw an *actor* who people *dislike*.	*Actor*: 3.5417

Finally, we calculated the average daily sentiment strength of the dimensions for each movie. Let $s_{i,n,d}$ be the sentiment value for the nth dimension word at the ith time (location) in document d for one movie. The sentiment of the kth dimension for one movie on the tth day can then be formulated as follows:

$$\theta_{t,k} = \frac{1}{N} \sum_{n=1}^{N} \frac{1}{D} \sum_{d=1}^{D} \frac{1}{I} \sum_{i=1}^{I} s_{i,n,d}. \tag{3}$$

Intuitively, $\theta_{t,k}$ represents the average strength of the sentiment of the kth dimension. Figure 5 shows the average sentiments of the dimension *plot* for 122 movies. Using Figures 4 and 5, we can easily monitor consumer feedbacks (heat and sentiments) on product dimensions over time.

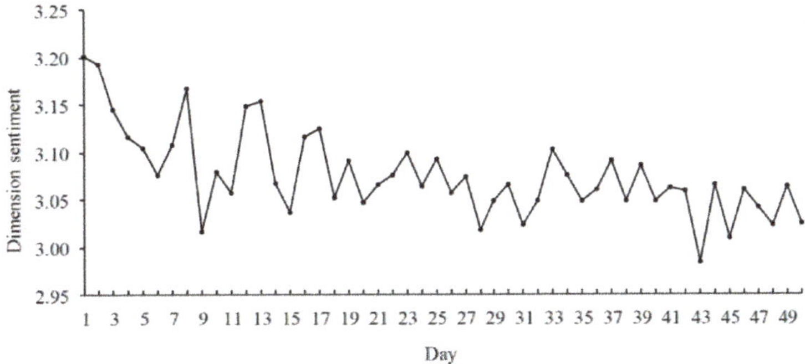

Figure 5. Average sentiments of the dimension *plot* for 122 movies.

In Table 7, we describe the key variables of the dimensions.

Table 7. Key variables for each movie: dimensions.

Variable	Description	Measures
$\vartheta_{t,1}$	The heat of the dimension *plot* on day t	Probabilistic
$\vartheta_{t,2}$	The heat of the dimension *star* on day t	Probabilistic
$\vartheta_{t,3}$	The heat of the dimension *genre* on day t	Probabilistic
$\theta_{t,1}$	The sentiment of the dimension *plot* on day t	Numerical value
$\theta_{t,2}$	The sentiment of the dimension *star* on day t	Numerical value
$\theta_{t,3}$	The sentiment of the dimension *genre* on day t	Numerical value

Table 8 shows the summary statistics of the variables. The heat of the dimensions ($\vartheta_{t,i}$) is between zero and one. The median of the sentiments of the dimensions ($\theta_{t,i}$) is three.

Table 8. Summary statistics of the dimension variables.

Variable	Mean	Median	Maximum	Minimum	Std. Dev.	Skewness	Kurtosis
$\vartheta_{t,1}$	0.26932	0.00971	0.9999957	1.93×10^{-6}	0.4063216	1.082	2.284
$\vartheta_{t,2}$	0.13574	0.00971	0.9999957	2.16×10^{-6}	0.3078812	2.184	5.992
$\vartheta_{t,3}$	0.59495	0.95943	0.9999949	1.43×10^{-6}	0.4544953	−0.426	1.250
$\theta_{t,1}$	3.07341	3	4.83333	0.130435	0.3685301	−3.347	28.987
$\theta_{t,2}$	3.09445	3	4.90476	0.130435	0.3519009	−2.837	28.437
$\theta_{t,3}$	3.08610	3	4.60417	0.130435	0.299457	−3.282	35.261

3.3. Predictive Model

We used the first 40 days of data to train the predictive model and the last 9 days of data to test the trained model. The regressive model can have better forecasting performance than the machine-learning models when the amount of relevant information is sufficient and when the variation in box-office revenues is small [55]. However, if there is not enough information, then the machine-learning model can help improve the forecasting accuracy by more thoroughly utilizing the limited information given. According to the sufficient predictors discussed in Section 3.2 and the relatively stable revenues in the test period, the proposed approach that we constructed was based on the autoregressive model because the regressive model is the most efficient predictive model [7]. We also needed to address some methodological concerns. First, we log-transformed some skewed variables to give them similar normal distributions. Second, we used the variance inflation factor (VIF) to assess multivariate multicollinearity. The VIF values were lower than the threshold of five; thus, multicollinearity was not a serious issue [56].

3.3.1. Autoregressive Model

We started with an AR model as our base model to forecast movie box-office revenues. We used this AR model with the parameter p to model the relationship between preceding box-office revenues and current box-office revenues as follows:

$$\log(Sales_t) = \alpha + \sum_{i=1}^{p} \varphi_i \log(Sales_{t-i}) + \epsilon_t \tag{4}$$

where $\varphi_1, \varphi_2, \ldots, \varphi_p$, are the parameters to be estimated, α is the effect of the combination of time-invariant variables, such as the production budgets and genres of movies, and ϵ_t is an error term. The AR model uses only preceding sales to predict current or future sales.

3.3.2. ARO Model

In addition to preceding box-office revenues, online information, such as Google Trends and eWOM volume, might greatly influence box-office revenues. According to the discussion above, we propose a predictive model by integrating online information into the AR model. This model includes all the variables of previous GSI models and eWOM models. Our ARO model is similar to that proposed in [5], and it can be formulated as follows:

$$\log(Sales_t) = \alpha + \sum_{i=1}^{p} \varphi_i \log(Sales_{t-i}) + \sum_{i=0}^{q} \sum_{j=1}^{J} \rho_{i,j} v_{t-i,j} + \epsilon_t \tag{5}$$

where $v_{t,j}$ represents the jth online information variable on day t. We determined p and q by comparing model accuracy when using different values of p and q. φ_i and $\rho_{i,j}$ are parameters that need to be estimated. The parameter q specifies the lags of the preceding days of the online information variables; J indicates the number of these variables. The ARO model uses preceding sales, Google Trends, the eWOM variables and other predictors in Table 4 to predict current and future sales.

3.3.3. The ARHS Model

According to previous studies, the heat and sentiments of product dimensions are very important for sales [11,12]; thus, it is desirable to integrate the heat and sentiments of movie dimensions into predictive models to achieve better accuracy. In this section, we extend the ARO model to the ARHS model. We formulate the ARHS model as follows:

$$\log(Sales_t) = \alpha + \sum_{i=1}^{p} \varphi_i \log(Sales_{t-i}) + \sum_{i=0}^{q} \sum_{j=1}^{J} \rho_{i,j} v_{t-i,j} + \sum_{i=0}^{\gamma} \sum_{k=0}^{K} \omega_{i,k} \vartheta_{t-i,k} + \sum_{i=0}^{\delta} \sum_{k=0}^{K} \mu_{i,k} \theta_{t-i,k} + \epsilon_t \tag{6}$$

where p, q, γ, and δ are user-defined parameters, ϵ_t is an error term, and φ_i, $\rho_{i,j}$, $\omega_{i,k}$, and $\mu_{i,k}$ are parameters that need to be estimated. $\vartheta_{t,k}$ and $\theta_{t,k}$ are the heat and sentiments, respectively, of the kth dimension at time t, and are obtained by using DTA. p, q, γ, and δ specify how far the model "looks back" into the past, whereas J and K specify how many related variables we would like to consider. J and K are fitted as described in Section 3.1. We used the least squares method to train all the models. The ARHS model extends the ARO model by integrating the preceding heat and sentiments of the movie dimensions into the ARO model.

4. Results

In this section, we compare the ARHS model with the AR model, the eWOM-based model, the GSI-based model, and the ARO model to validate its effectiveness.

In this paper, we use the mean absolute percentage error (MAPE) to measure the performance of the predictive models:

$$MAPE = \frac{1}{n} \sum_{i=1}^{n} \frac{|Pred_i - True_i|}{True_i} \times 100\% \tag{7}$$

where n is the number of predictions made on the test data, $Pred_i$ is the predicted box-office revenues, and $True_i$ represents the true value of the box-office revenues. In statistics, $MAPE$ is a suitable measure of accuracy for time-series-value predictions. We can compare the error of the fitted time series because it is a percentage error. All the $MAPE$ results reported herein are the mean value of the independent runs of 122 movies on different days. This metric is robust to comparing the performance of the sales prediction models [5,57]. For brevity, we removed the percent sign (%) of the $MAPE$ value from Figures 6–9.

4.1. Performance of the Parameters in the ARHS Model

In the ARHS model, the parameters p, q, γ, and δ provide the flexibility to fine tune the model for optimal performance. We now explore how the choices of these parameter values affect prediction accuracy.

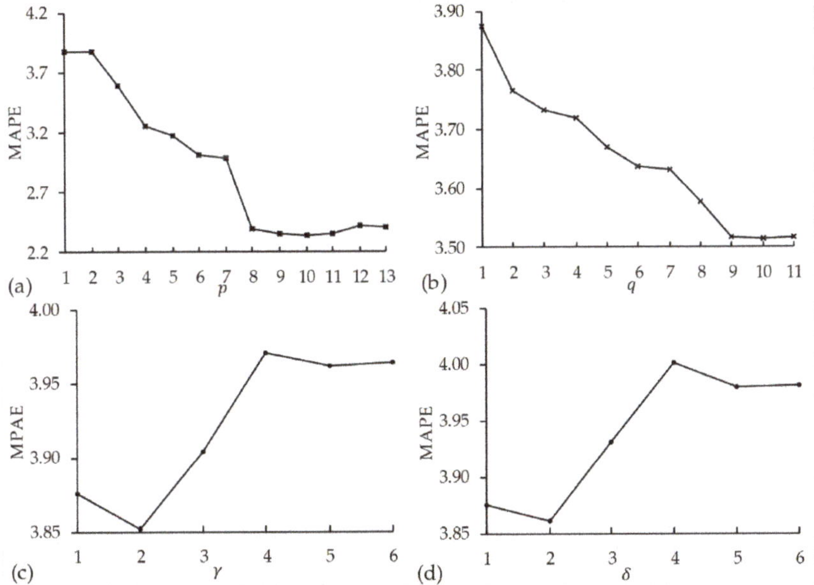

Figure 6. The effects of the parameters on prediction accuracy. (**a**) Effects of p; (**b**) effects of q; (**c**) effects of γ; (**d**) effects of δ.

First, we varied p with fixed values of parameters q, γ and δ ($q = \gamma = \delta = 1$) to study how preceding box-office revenues affect the prediction accuracy of the ARHS model. As shown in Figure 6a, the model achieves its best prediction accuracy when $p = 10$. The change in accuracy is minor after $p = 8$, and the accuracy even decreases after $p = 11$. These findings suggest that p should be large enough to factor in all significant influences of preceding box-office revenues but that it should not be so large that it lets irrelevant preceding box-office revenues reduce prediction accuracy.

Next, using a fixed value of p, γ, and δ ($p = \gamma = \delta = 1$), we varied the value of q from 1 to 11 to study its effect on prediction accuracy. Figure 6b shows that the model achieves its best performance when $q = 10$. However, the accuracy is basically the same after $q = 9$, which means that numerical online information will affect box-office revenues over the following nine days. Based on the above results, we suggest that the predictive power of numerical online information for box-office revenues lasts slightly longer than the preceding box-office revenues.

By using fixed values for p, q, and δ ($p = q = \delta = 1$), we varied γ from 1 to 6 to study the prediction accuracy of the ARHS model. As shown in Figure 6c, the ARHS model achieves the best prediction accuracy at $\gamma = 2$, which implies that the effect of the heat of dimensions captured from the text of eWOM lasts two days.

We also varied δ from 1 to 6, using fixed values for p, q, and γ ($p = q = \gamma = 1$). As shown in Figure 6d, the ARHS model achieves the highest accuracy at $\delta = 2$, which implies that the effects of the sentiments of dimensions on box-office revenues also last two days.

From the results above, we conclude that the product-dimension information captured from online comments has a shorter effect on box-office revenues than numerical online information. We think the reason for this result is that consumers look through the text of eWOM posted only in recent days but glance at the numerical information of eWOM posted over a longer period of time before they decide to see a movie. The optimal parameter values of the ARHS model should be simultaneously searched (p, q from 1 to 12; δ, γ from 1 to 6). After comparing 4356 experimental results, we found that the optimal parameter values are $p = 8$, $q = 9$, and $\delta = \gamma = 2$.

4.2. Comparison of the Predictive Models

To verify the superiority of the ARHS model, we compared its performance with that of the other models.

First, we compared the ARHS model ($q = 9$, $\delta = \gamma = 2$) with the AR model. As shown in Figure 7, the ARHS model consistently outperforms the AR model as p ranges from 1 to 10. We observe that the ARHS model has much higher accuracy when p is small, which implies that the eWOM of a movie has more predictive power when we know little about the preceding box-office revenues. When $p = 4$, our proposed sales prediction model improves the MAPE of the AR model by 27.65%. When the lag of sales is 8, the improvement of the MAPE is the smallest, 2.69%. These improvements suggest that the ARHS model has higher accuracy.

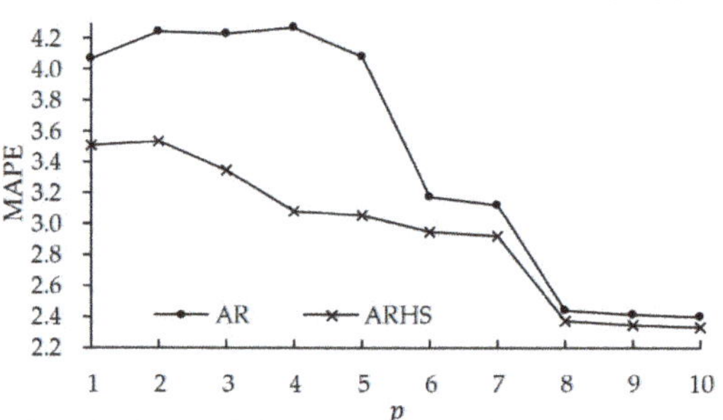

Figure 7. Comparison with the autoregressive prediction model.

We then conducted experiments to compare the ARHS ($\delta = q = \gamma = 1$) model with the eWOM model, the GSI model [41], and the ARO model. Both our study and previous studies prove that these models are better than the AR model. As shown in Figure 8a, the eWOM model and the GSI model have nearly the same accuracy performance with regard to forecasting the sales of experience products. As shown in Figure 8b–d, the ARHS model always outperforms the eWOM model, the GSI model, and the ARO model when p ranges from 1 to 6. Thus, the ARHS model is the best among these models, which supports our hypothesis. The effects of the eWOM text on box-office revenues decrease over time, and our test occurs at the end of the release period of the movie. Therefore, compared with the eWOM model, the GSI model, and the ARO model, the ARHS model improves the *MAPE*, but not much. We argue that the improvement in accuracy of the ARHS model will be higher earlier after a movie is released. Because of the high gross of movies, a very small improvement in forecasting accuracy might result in a difference of millions of dollars. Therefore, the ARHS model should be meaningful to movie marketers and theater managers.

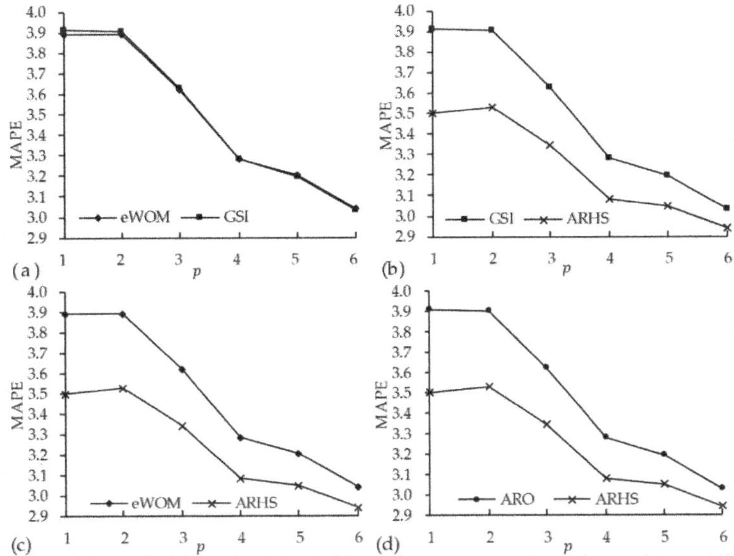

Figure 8. Comparisons of model accuracy. (**a**) Comparison of the eWOM model and the GSI model; (**b**) comparison of the GSI model and the ARHS model; (**c**) comparison of the eWOM model and the ARHS model; (**d**) comparison of the ARO model and the ARHS model.

To verify the time robustness of the ARHS model, we compared its accuracy for different predictive periods. We use the first 20, 30, and 40 days of data as training data and the following 9 days of data as test data. Figure 9 shows the results. The prediction accuracy increases when p ($0 < p < 8$) increases, and it barely changes after $p \geq 8$. The prediction accuracy for 21–29 days is always higher than that for 31–39 days, and the prediction accuracy for 31–39 days is higher than that for 41–49 days. This means that the prediction performance of the ARHS model is higher in the initial stage of a movie's release and that earlier is better. Therefore, we conclude that the heat and sentiments of dimensions have greater predictive power in the early days after a movie's release.

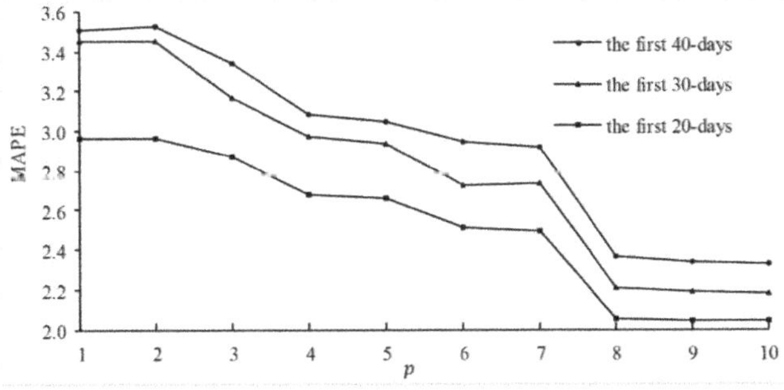

Figure 9. Comparison of different prediction intervals.

4.3. Robustness of the Predictive Power of the Heat and Sentiments of Dimensions

To further verify the predictive power of the heat and sentiments of dimensions, we conduct a robustness check with regard to predicting the opening-week revenues of movies, which determines the gross of movies, by using a BP neural network with 10-fold cross validation. We filtered out the movies that did not have any online reviews before being released. The final dataset comprises 111 movies with 14,328 online reviews, Google Trends and film-related factors; however, it does not include preceding revenues.

We calculated the overall sentiment of the reviews using the previous method [12]. To verify the forecasting performance of the extracted dimension-specific information, we included the predicted factors in the ARO model without volume in the BP neural network as a benchmark model [41]. We then constructed a new predictive model by integrating the volume and overall sentiment into the benchmark model, which we call the volume–overall–sentiment (VOS) model. Finally, we constructed the dimension-heat-sentiment (DHS) model by integrating the heat and sentiments of dimensions. We used the root mean squared error (*RMSE*) and *MAPE* to measure the prediction accuracy of the BP neural network:

$$RMSE = \sqrt{\frac{1}{n} \sum_{n=1}^{n} (y_i - \hat{y}_i)^2} \tag{8}$$

where n is the sample size, y_i is the actual box-office revenues, and $\hat{y}_i$ represents the predicted box-office revenues.

Figure 10 shows the *RMSE* and *MAPE* of these models for predicting the opening-week box-office revenues. We found that the VOS model is always better than the benchmark model and that the DHS model is better than the VOS for opening-week revenue prediction. Therefore, the heat and sentiments of dimensions have better prediction performance for movie opening-week revenues when the machine-learning method is used.

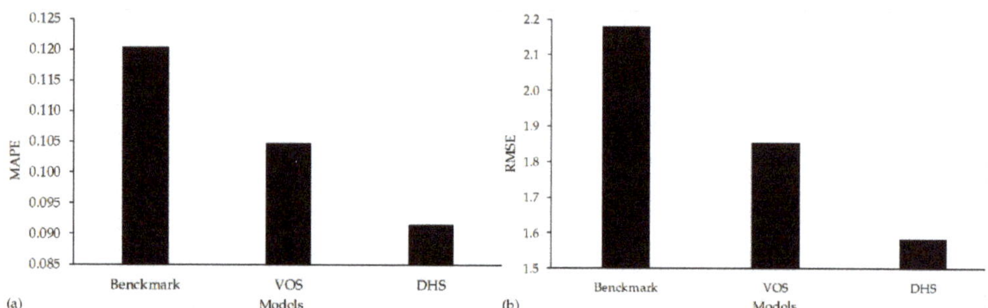

Figure 10. Accuracy of models predicting opening-week revenues. (**a**) *MAPE* of the three models; (**b**) *RMSE* of the three models.

5. Conclusion and Discussion

Previous research demonstrates that eWOM text implies the heat and sentiments of product dimensions that influence product sales [11,12,43]. Thus, we propose a method called DTA to extract the heat and sentiments of product dimensions from big data on eWOM. Previous research has proven that the multiattribute attitude model with three attributes is sufficiently concise and effective [43,58]. Based on our DTA results, we obtained the dynamic heat and sentiments of three key movie dimensions: the *plot*, the *genre*, and the *star*. These dimensions have also been regarded as important movie dimensions in previous studies, but the heat and sentiments of these dimensions had not been investigated [7,53]. To improve accuracy, we propose the ARHS model by integrating the heat and sentiments of dimensions into a prediction model for movie daily ticket sales. This model's performance was compared with that of other predictive models, and the results indicate that the

ARHS model is more accurate than the benchmark model [5], which supports our hypothesis. We also found that the ARHS model performs much better in the early stage of product release. The robustness check with regard to predicting opening-week revenues using the machine-learning method also demonstrates that the heat and sentiments of dimensions have more predictive power.

Our research has some theoretical implications. First, our research extends the use of multiattribute attitude theory to a big data environment using DTA, which is a framework based on a machine-learning method, syntactic method, and lexicon-based method. In previous research, attributes used to predict consumer purchase predispositions were generated based on expert judgment and in-depth interviews. Additionally, the measures of attribute importance and valence were obtained by surveying a sample of respondents [43]. However, individual deviations, the limited number of survey samples and halo effects may bias the results [59], and these methods often have high manpower and time costs that are not suitable for big data. Therefore, we propose DTA to extract the heat and sentiments of the most important dimensions from eWOM big data as a proxy for attribute importance and valence in social media marketing research, making it possible to avoid the issues of the methods used in previous studies. Second, we demonstrate that multiattribute attitude theory [9,10] is valid for forecasting sales in a big data environment. Previous research used eWOM volume and sentiment as a whole to predict sales, with limited performance. Our research shows that the heat and sentiment of dimensions extracted from social media based on multiattribute attitude theory can improve the predictive performance of the benchmark model. Third, compared with the results regarding search products [5], our study proves that integrating social media data and online search data together can improve the performance of sales predictions for experience products. Fourth, our research demonstrates that the heat and sentiments of dimensions implied in social media improve the accuracy of predictive models. Therefore, related research, such as studies that predict election outcomes, stock prices, and internet security, can use the heat and sentiments of dimensions to construct predictive models.

Additionally, our paper has some practical implications. First, using the DTA, we can extract the heat and sentiments of dimensions in a timely way and can then monitor consumer feedbacks on brand dimensions by analyzing the dynamic heat and sentiments of brand dimensions over time. There may be an initial collaborative attack on a brand dimension when the heat of the brand dimension suddenly increases and the sentiment of the brand dimension decreases suddenly and simultaneously. At that point, we can provide a warning of "collaborative brand attacks" to enterprises. The stakeholders can then confirm it in social media and carry out certain activities to stop the attack if the information is accurate. Second, we find that the predictive power of the volume and ratings of eWOM lasts longer than that of eWOM text: the heat and sentiments of product dimensions. Therefore, managers should pay attention to the volume and ratings of eWOM over a long period and pay attention to the text of new eWOM only. Third, the proposed ARHS model has better predictive performance for sales. Therefore, marketing managers can provide an early warning with regard to a sales explosion or collapse and more accurately determine whether and when to carry out promotion activities using our method. In the movie industry, marketers can provide an early warning with regard to ticket sales before a movie is released, and theaters can adjust the number of screens for different movies based on accurate daily box-office predictions. Fourth, we can determine the strength and weakness of relevant product attributes by comparing the DTA results of different products. For example, if one dimension of a product has the lowest sentiment among all competitive products, this dimension can be the weakness of the product. Therefore, this method can also be used to suggest specific changes in product design and the marketing support for the product.

This paper also has some limitations. Our research can only benefit the social media marketing of products and brands with multiple attributes and abundant social media data. We predict only box-office revenues in America to demonstrate the predictive power of the heat and sentiments of dimensions. To improve our theory, we plan to conduct further research that forecasts the sales of other products in different regions. Additionally, in this paper, we use only one type of eWOM and social media data. We should use multiple types of eWOM and data in other forms of media in future research to obtain results that are more robust. Future research can be conducted to examine how dimension-specific information influences product sales differently, which can help to identify the different economic effects of each product attribute.

Author Contributions: Conceptualization, X.L. and C.J.; methodology, X.L.; software, X.L.; validation, X.L., Y.D., and Z.W.; formal analysis, X.L. and Y.L.; investigation, X.L.; resources, X.L.; data curation, X.L.; writing—original draft preparation, X.L.; writing—review and editing, X.L. and C.J.; visualization, X.L.; supervision, X.L., Z.W., and Y.L.; project administration, C.J. and Y.D.; funding acquisition, C.J.

Funding: This research was funded by the National Natural Science Foundation of China (NSFC), key grant number 71731005 and grant number 71571059.

References

1. Wang, X.; Yu, C.; Wei, Y. Social media peer communication and impacts on purchase intentions: A consumer socialization framework. *J. Interact. Mark.* **2012**, *26*, 198–208. [CrossRef]
2. Felix, R.; Rauschnabel, P.A.; Hinsch, C. Elements of strategic social media marketing: A holistic framework. *J. Bus. Res.* **2017**, *70*, 118–126. [CrossRef]
3. Amornpetchkul, T.; Duenyas, I.; Şahin, Ö. Mechanisms to induce buyer forecasting: Do suppliers always benefit from better forecasting? *Prod. Oper. Manag.* **2015**, *24*, 1724–1749. [CrossRef]
4. Hu, N.; Koh, N.S.; Reddy, S.K. Ratings lead you to the product, reviews help you clinch it? The mediating role of online review sentiments on product sales. *Decis. Support Syst.* **2014**, *57*, 42–53. [CrossRef]
5. Geva, T.; Oestreicher-Singer, G.; Efron, N.; Shimshoni, Y. Using forum and search data for sales prediction of high-involvement products. *MIS Q.* **2017**, *41*, 65–82. [CrossRef]
6. Chern, C.C.; Wei, C.P.; Shen, F.Y.; Fan, Y.N. A sales forecasting model for consumer products based on the influence of online word-of-mouth. *Inf. Syst. e-Bus. Manag.* **2015**, *13*, 445–473. [CrossRef]
7. Ghiassi, M.; Lio, D.; Moon, B. Pre-production forecasting of movie revenues with a dynamic artificial neural network. *Expert Syst. Appl.* **2015**, *42*, 3176–3193. [CrossRef]
8. See-To, E.W.K.; Ngai, E.W.T. Customer reviews for demand distribution and sales nowcasting: A big data approach. *Ann. Oper. Res.* **2018**, *270*, 415–431. [CrossRef]
9. Fishbein, M. An investigation of the relationships between beliefs about an object and the attitude toward that object. *Hum. Relations* **1963**, *16*, 233–239. [CrossRef]
10. Rosenberg, M.J. Cognitive structure and attitudinal affect. *J. Abnorm. Psychol.* **1956**, *53*, 367–372. [CrossRef]
11. Li, X.; Wu, C.; Mai, F. The effect of online reviews on product sales: A joint sentiment-topic analysis. *Inf. Manag.* **2018**. [CrossRef]
12. Liang, T.P.; Li, X.; Yang, C.T.; Wang, M. What in consumer reviews affects the sales of mobile apps: A multifacet sentiment analysis approach. *Int. J. Electron. Commer.* **2015**, *20*, 236–260. [CrossRef]
13. Siering, M.; Muntermann, J.; Rajagopalan, B. Explaining and predicting online review helpfulness: The role of content and reviewer-related signals. *Decis. Support Syst.* **2018**, *108*, 1–12. [CrossRef]
14. Yang, D.; Zhang, L.; Wu, Y.; Guo, S.; Zhang, H.; Xiao, L. A sustainability analysis on retailer's sales effort in a closed-loop supply chain. *Sustainability* **2019**, *11*, 18. [CrossRef]
15. Floyd, K.; Freling, R.; Alhoqail, S.; Cho, H.Y.; Freling, T. How online product reviews affect retail sales: A meta-analysis. *J. Retail.* **2014**, *90*, 217–232. [CrossRef]
16. Ghose, A.; Ipeirotis, P.G.; Li, B. Designing ranking systems for hotels on travel search engines by mining user-generated and crowd-sourced content. *Mark. Sci.* **2012**, *31*, 493–520. [CrossRef]
17. Wang, F.; Liu, X.; Fang, E. User reviews variance, critic reviews variance, and product sales: An exploration of customer breadth and depth effects. *J. Retail.* **2015**, *91*, 372–389. [CrossRef]
18. Chen, P.-Y.; Dhanasobhon, S.; Smith, M.D. All reviews are not created equal: The disaggregate impact of reviews and reviewers at Amazon.com. 2006. Available online: http://archive.nyu.edu/bitstream/2451/14961/2/USEDBOOK19.pdf (accessed on 20 July 2006).

19. Baker, A.M.; Donthu, N.; Kumar, V. Investigating how word-of-mouth conversations about brands influence purchase and retransmission intentions. *J. Mark. Res.* **2016**, *53*, 225–239. [CrossRef]
20. Gu, B.; Tang, Q.; Whinston, A.B. The influence of online word-of-mouth on long tail formation. *Decis. Support Syst.* **2013**, *56*, 474–481. [CrossRef]
21. Kostyra, D.S.; Reiner, J.; Natter, M.; Klapper, D. Decomposing the effects of online customer reviews on brand, price, and product attributes. *Int. J. Res. Mark.* **2016**, *33*, 11–26. [CrossRef]
22. Hafeez, H.A.; Manzoor, A.; Salman, F. Impact of social networking sites on consumer purchase intention: An analysis of restaurants in Karachi. *J. Bus. Strateg.* **2017**, *11*, 1–20. [CrossRef]
23. Yan, Q.; Wu, S.; Wang, L.; Wu, P.; Chen, H.; Wei, G. E-WOM from e-commerce websites and social media: Which will consumers adopt? *Electron. Commer. Res. Appl.* **2016**, *17*, 62–73. [CrossRef]
24. Parboteeah, V.; Valacich, J.; Wells, J. Online impulse buying: understanding the interplay between consumer impulsiveness and website quality. *J. Assoc. Inf. Syst.* **2011**, *12*, 32.
25. Dewan, S.; Ramaprasad, J. Social media, traditional media, and music sales. *MIS Q.* **2014**, *38*, 101–121. [CrossRef]
26. Allan J., K.; Philip J., K. WOM and social media: Presaging future directions for research and practice. *J. Mark. Commun.* **2014**, *20*, 5–20.
27. Cui, G.; Lui, H.; Guo, X. The effect of online consumer reviews on new product sales. *Int. J. Electron. Commer.* **2012**, *17*, 39–57. [CrossRef]
28. Chevalier, J.A.; Mayzlin, D. The effect of word of mouth on sales: Online book reviews. *J. Mark. Res.* **2006**, *43*, 345–354. [CrossRef]
29. Duan, W.; Gu, B.; Whinston, A.B. Do online reviews matter?—An empirical investigation of panel data. *Decis. Support Syst.* **2008**, *45*, 1007–1016. [CrossRef]
30. Clemons, E.K.; Gao, G.G.; Hitt, L.M. When online reviews meet the hyperdifferentiation: A study of craft beer industry. *J. Manag. Inf. Syst.* **2006**, *23*, 149–171. [CrossRef]
31. Chintagunta, P.K.; Gopinath, S.; Venkataraman, S. The effects of online user reviews on movie box-office performance: Accounting for sequential rollout and aggregation across local markets. *Mark. Sci.* **2010**, *29*, 944–957. [CrossRef]
32. Xu, H. Is More Information Better? An Economic Analysis of Group-Buying Platforms. *J. Assoc. Inf. Syst.* **2018**, *19*, 1130–1144. [CrossRef]
33. Hansen, F. Consumer choice behavior: An experimental approach. *J. Mark. Res.* **1969**, *6*, 436–443. [CrossRef]
34. Relling, M.; Schnittka, O.; Sattler, H.; Johnen, M. Each can help or hurt: Negative and positive word of mouth in social network brand communities. *Int. J. Res. Mark.* **2016**, *33*, 42–58. [CrossRef]
35. Sweeney, J.C.; Soutar, G.N.; Mazzarol, T. Word of mouth: Measuring the power of individual messages. *Eur. J. Mark.* **2012**, *46*, 237–257. [CrossRef]
36. Lin, Z.; Wang, Q. E-commerce product networks, word-of-mouth convergence, and product sales. *J. Assoc. Inf. Syst.* **2018**, *18*, 848–871. [CrossRef]
37. Amblee, N.; Bui, T. Harnessing the influence of social proof in online shopping: The effect of electronic word of mouth on sales of digital microproducts. *Int. J. Electron. Commer.* **2011**, *16*, 91–114. [CrossRef]
38. Liu, Y. Word of mouth for movies: Its dynamics and impact on box office revenue. *J. Mark.* **2006**, *70*, 74–89. [CrossRef]
39. Zhang, Z.; Li, X.; Chen, Y. Deciphering word-of-mouth in social media: Text-based metrics of consumer reviews. *ACM Trans. Manag. Inf. Syst.* **2012**, *3*, 1–23. [CrossRef]
40. Chen, Y.; Xie, J. Online consumer review: Word-of-mouth as a new element of marketing communication mix. *Manag. Sci.* **2008**, *54*, 477–491. [CrossRef]
41. Lee, C.; Jung, M. Predicting movie incomes using search engine query data. In Proceedings of the International Conference on Artificial Intelligence and Pattern Recognition, Kuala Lumpur, Malaysia, 17–19 November 2014; pp. 45–49.
42. Bughin, J. Google searches and twitter mood: nowcasting telecom sales performance. *NETNOMICS Econ. Res. Electron. Netw.* **2015**, *16*, 87–105. [CrossRef]
43. Wilkie, W.L.; Pessemier, E.A. Issues in marketing's use of multi-attribute attitude models. *J. Mark. Res.* **1973**, *10*, 428. [CrossRef]
44. Tirunillai, S.; Tellis, G.J. Mining marketing meaning from online chatter: Strategic brand analysis of big data using latent Dirichlet allocation. *J. Mark. Res.* **2014**, *51*, 463–479. [CrossRef]

45. Guo, Y.; Barnes, S.J.; Jia, Q. Mining meaning from online ratings and reviews: Tourist satisfaction analysis using latent Dirichlet allocation. *Tour. Manag.* **2017**, *59*, 467–483. [CrossRef]
46. Cheung, C.M.K.; Thadani, D.R. The impact of electronic word-of-mouth communication: A literature analysis and integrative model. *Decis. Support Syst.* **2012**, *54*, 461–470. [CrossRef]
47. Duan, W.; Gu, B.; Whinston, A.B. The dynamics of online word-of-mouth and product sales-An empirical investigation of the movie industry. *J. Retail.* **2008**, *84*, 233–242. [CrossRef]
48. Hur, M.; Kang, P.; Cho, S. Box-office forecasting based on sentiments of movie reviews and independent subspace method. *Inf. Sci.* **2016**, *372*, 608–624. [CrossRef]
49. Blei, D.M.; Lafferty, J.D. Dynamic topic models. In Proceedings of the 23rd International Conference on Machine Learning, Pittsburgh, PA, USA, 25–29 June 2006; pp. 113–120.
50. Guerini, M.; Gatti, L.; Turchi, M. Sentiment analysis: How to derive prior polarities from SentiWordNet. In Proceedings of the 2013 Conference on Empirical Methods in Natural Language Processing, Seattle, WA, USA, 18–21 October 2013; pp. 1259–1269.
51. Socher, R.; Perelygin, A.; Wu, J.; Chuang, J.; Manning, C.; Ng, A.; Christopher, P. Recursive deep models for semantic compositionality over a sentiment treebank. In Proceedings of the 2013 Conference on Empirical Methods in Natural Language Processing, Seattle, WA, USA, 18–21 October 2013; pp. 1631–1642.
52. Blei, D.M.; Ng, A.Y.; Jordan, M.I. Latent Dirichlet allocation. *J. Mach. Learn. Res.* **2003**, *3*, 993–1022.
53. Lash, M.T.; Zhao, K. Early predictions of movie success: The who, what, and when of profitability. *J. Manag. Inf. Syst.* **2016**, *33*, 874–903. [CrossRef]
54. Schouten, K.; Frasincar, F. Survey on aspect-level sentiment analysis. *IEEE Trans. Knowl. Data Eng.* **2016**, *28*, 813–830. [CrossRef]
55. Moon, S.; Bergey, P.K.; Iacobucci, D. Dynamic effects among movie ratings, movie revenues, and viewer satisfaction. *J. Mark.* **2010**, *74*, 108–121. [CrossRef]
56. Kleinbaum, D.; Kupper, L.; Nizam, A.; Rosenberg, E. *Applied Regresion Analysis and Other Multivariable Methods*; Duxbury Press: Pacific Grove, CA, USA, 2013.
57. Yu, X.; Liu, Y.; Huang, X.; An, A. Mining online reviews for predicting sales performance: A case study in the movie domain. *IEEE Trans. Knowl. Data Eng.* **2012**, *24*, 720–734. [CrossRef]
58. Kraft, F.B.; Granbois, D.H.; Summers, J.O. Brand evaluation and brand choice: A longitudinal study. *J. Mark. Res.* **1973**, *10*, 235–241. [CrossRef]
59. Lehmann, D.R. Television show preference: Application of a choice model. *J. Mark. Res.* **1971**, *8*, 47. [CrossRef]

 *sustainability*

Article

Assessing Technology Platforms for Sustainability with Web Data Mining Techniques

Desamparados Blazquez, Josep Domenech * and Jose-Maria Garcia-Alvarez-Coque

Department of Economics and Social Sciences, Universitat Politècnica de València, 46022 València, Spain;
mdeblzso@upvnet.upv.es (D.B.); jmgarcia@upvnet.upv.es (J.-M.G.-A.-C.)
* Correspondence: jdomenech@upvnet.upv.es

Received: 12 November 2018; Accepted: 27 November 2018; Published: 29 November 2018

Abstract: Public policies have encouraged the proliferation of technology platforms that support the transition towards sustainable agriculture and the development of innovations in the food system. Provided the difficulty associated with assessing the outputs and outcomes of technology platforms, this work proposes a practical assessment method based on the retrieval and analysis of online documents related to the technology platforms. Concretely, the method consists of applying web scraping techniques to retrieve documents related to a technology platform from the Internet and then applying web data-mining techniques to automatically classify these documents into the functions that the platform should fulfill, which are described from the viewpoint of co-evolution of innovation. Data are automatically processed to obtain a variety of metrics, which are applied to measure the impact of European Technology Platforms (ETPs) on promoting an organic food paradigm. This method provides time-series data that helps to follow the evolution of the different functions of the platform and to describe its lifecycle. It has been applied to one platform taken as a case study, TP Organics, which represents a key platform for stakeholders that promote organic farming and agroecology as core components of an ambitious program for sustainable agriculture. The obtained online-based measures have been proven to assess the global evolution of the platform, its dissemination through the European Union (EU) Member States, and the evolution of the different functions expected to be fulfilled by it regarding the diffusion and promotion of innovations in organic agriculture.

Keywords: technology platforms; sustainable agri-food systems; innovation in sustainable agriculture; online data; data mining; TP organics

---

1. Introduction

The agricultural research agenda has been subject to major changes in recent years, resulting in a diverse range of innovations. The present stream not only promotes innovations to obtain more food and agricultural raw materials, but also to alleviate rural poverty, improve diets and health, and allow a transition towards sustainable systems. Such a move also includes sustainable agriculture, which involves the adoption of practices that enhance the environmental quality and the natural base of agriculture. Progress towards a sustainable paradigm requires new ways of organizing research, new ways of establishing priorities and flexible ways of evaluating results and impacts [1].

Research policies, their priorities, and implementation have been developed for supporting innovation in the agricultural systems through technological platforms (TPs), which facilitate industry-based partnerships and networking. Specifically, in 2004, the European Commission invited the industry to establish European Technology Platforms (ETPs), especially to define research agendas that would attract private investment. The ETP became an industry-led stakeholder *fora* to develop research and innovation agendas and roadmaps for action at the EU and national levels [2,3].

The ETPs came to the agri-food sector to involve all relevant stakeholders in the development of common visions that could meet the needs and benefits of society. For the agro-food-forestry-biotechnology sectors, the ETPs were initiated mainly by industrial organizations led by multinational companies, with the support and proposals of scientific organizations. The concept quickly extended from capital-intensive industries to ecological and social sustainability [4,5]. The technology platforms continue to be a tool within European policies and long-term strategies to promote innovation, extended to social and environmental benefits related to public concerns. The ETP concept has developed in sectoral forms, represented by platforms such as EATIP (European Aquaculture Technology and Innovation Platform) in the aquaculture sector or TP Organics in the agroecological sector; in horizontal forms such as the ETP Nanofutures, a multi-sectorial cross-ETP devoted to connect ETPs that require nanotechnologies in their industrial sectors; and has also been accompanied by other initiatives such as the ERANETs and the European Innovation Partnership for agricultural productivity and sustainability (EIP-Agri), which are instruments of the European Commission designed to support public-public and private-public partnerships in their preparation, coordination, and funding [6].

Previous ETP assessments have proven the success of ETPs in developing joint visions, setting strategic research agendas and contributing to the definition of the research priorities [7]. However, the evaluation exercises frequently result in costly efforts given the lack of indicators available to monitor the impact of ETPs, in particular in emerging areas, such as the innovation for sustainable systems. Since most ETPs have been set up recently, it is difficult to assess, in a quantified way, their results and impacts, which are typically revealed after a certain period of time. We acknowledge that evaluation is a comprehensive exercise that attempts to answer questions which depend on the policy objectives of the platform and that are normally supported by surveys and interviews.

Meanwhile, the transformation and digitization of society have resulted in the emergence of social and economic behavior on the internet as a digital footprint. Certainly, this includes the activity of the ETPs and the actors involved in them. The digital activity of individuals and companies is closely related to their underlying behavior, as previous research showed [8–10]. For this reason, it is currently being studied by researchers and official institutions as a source of information to monitor and understand the underlying socio-economic behavior. See, for instance, the efforts by the European statistical system with the ESSnet Big Data project [11] or the academic literature reviewed by Blazquez and Domenech [12].

The high availability of online data, together with the advances in Big Data research allows us to envision novel monitoring and assessment methods with lower operational costs as they are not based on surveys and interviews. Current Big Data technology enables us to design assessments systems that include the automatic retrieval of content generated by some economic agents and then apply document analysis and text mining to classify the documents and generate activity indicators. Some examples include evaluating the export orientation of companies [10], the diffusion of technology [13], or job skills demand in the labor market [14].

The objective of this paper is to measure the activity and impact of an ETP using the information publicly available about the platform on the Internet as the main source. With a view to the transition of food systems, we depart from a conceptual framework about the functions an ETP needs to meet and propose a method to retrieve the documents relevant to the ETP and classify them according to the main function they are related to. This would allow us to provide insights on how the activity of the ETP is being carried out and understand it as part of an ETP lifecycle. Our method also enables a regional analysis by mapping the activities of an ETP to the European countries.

To provide a method validation, we focused the method implementation on a specific ETP that has a wide scope as it faces technological, societal, and environmental challenges. We explored TP Organics as a platform that promotes an ambitious transformation of agricultural systems through organic farming and agroecology. TP Organics, according to its website (http://tporganics.eu/about-us/.) aims at strengthening "research & innovation for organics and other agroecological approaches that contribute to sustainable food and farming systems". It is broad in scope as it combines large companies, small and medium enterprises, researchers, farmers, consumers, and civil society organizations active in the organic value chain from production, input, and supply to food processing, marketing, and consumption. TP Organic, as a technological platform, promotes holistic and long-term agroecological solutions and the "co-creation of knowledge, combining science with the traditional, practical, and local knowledge of producers"

In summary, taking TP Organics as a pilot case, the main contributions of this paper are: (i) formulate a framework for ETP monitoring based on the functions that they seek to meet, (ii) devise a method for online data collection that provides a comprehensive monitoring of the defined functions; (iii) apply the framework and method to an ETP of a broad scope such as TP Organics; (iv) provide a first evaluation of the lifecycle of TP Organics and its functions.

The remainder of the paper is structured as follows. Section 2 defines a conceptual framework based on the functions of innovation intermediaries. Section 3 describes the case study, TP Organics, which is a functional ETP for developing new concepts and stakeholder networking towards sustainable food systems. Section 4 proposes a practical methodology for the digital monitoring of TP functions that aims to monitor the differential performance of innovation functions in TPs from a co-evolutionary perspective that reflects interactions between societal subsystems [15]. Section 5 describes the results obtained after applying the proposed methodology to TP Organics. Finally, Section 6 draws some concluding remarks.

2. Analytical Background

We consider TP (or ETP in its EU version) as a specific form of innovation intermediary, which is a broad concept that can be applied to TP but also to more flexible configurations such as research and innovation networks. For the transition of food systems, an innovation platform has certain characteristics: (i) it combines multiple actors; (ii) it performs activities around agricultural innovation challenges at different levels in agricultural systems (value chain, sector, country, village, etc.).

The TPs can be observed as intermediaries that connect different actors within the innovation systems in order to encourage co-evolution [16]. Howells [17] defines innovation intermediaries as "organization[s] or bod[ies] that act as agent[s] or broker[s] in any aspect of the innovation process between two or more parties".

TPs and other innovation platforms could be considered single innovation intermediaries that coordinate the platform, but it may be more realistic to see a TP as a space for several intermediaries to achieve the connection between stakeholders and thus facilitate the transition of innovation systems [18–21]. A TP plays a role of an innovation intermediary or of an ecosystem of individuals and organizations that perform intermediary functions from a bottom-up perspective [22,23]. TPs form a space of nested systems of intermediaries that fulfill complementary functions that foster co-evolution in innovation systems [16].

Kivimaa et al. [24] point out the lack of literature on intermediate activities and on the type of contribution of the different types of intermediaries to better support the acceleration of innovation processes. TPs could be considered as a type of knowledge-intensive business service (KIBS) in multilevel contexts [25–28]. Seen as KIBS, TPs provide a space for complex operations in which the generation and dissemination of knowledge take place through an intense interaction between actors in which human capital plays a critical role.

Whereas functions of innovation intermediaries could be embedded into a broader perspective of innovation system functions [29,30], we propose to assess TP functions in a perspective which looks at the ways that innovation intermediaries may support innovation. Indeed, much of the literature on innovation intermediaries has focused on their functions, such as the studies of Bessant et al. [31] and Howells [17] that go beyond the coordination and brokering relations in complex multi-actor configurations in the innovation systems.

Kilelu et al. [16,20] established a classification of intermediary functions applicable to the interpretation of innovation platforms. These functions can be adapted to the case of TPs by applying some modifications to the definitions proposed in those studies. As a result, these functions are as described below:

- Function F1: Demand articulation. It involves activities that identify innovation challenges and opportunities, such as information gathering, needs assessment, and strategic planning.
- Function F2: Institutional support. This function promotes institutional change by facilitating changes in regulations, working on attitudes and practice, and also linking science, policy, and practice.
- Function F3: Capacity building. It promotes incubating and strengthening organizational forms and networks. This includes objectives regarding organization development, as well as training and competence building activities.
- Function F4: Network brokering. This involves establishing key relationships among different actors to form partnerships and promote cooperation.
- Function F5: Innovation process management. This function promotes the interaction among different actors for research and innovation activities by aligning agendas and mediating relationships.
- Function F6: Knowledge brokering. It identifies knowledge needs and facilitates knowledge transfer from different sources by articulating, communicating, and disseminating knowledge and technology.

These functions are perfectly valid to assess ETPs; although, ETPs were originally designed as an arena to guide the research and innovation agenda at the EU level, for example, for H2020 projects. Such functions require monitoring efforts and we intend to provide a tool for this based on online data. Once indicators for each function are defined, a further step will be to use the measured indicators to assess the contribution of a TP to the transition of innovation systems. The monitoring of functions can allow us to determine their relevance in terms of impact but it is not possible to assure that a function with more events has a higher economic or social impact than a function with a lower number of events. However, it is possible to build synthetic indicators that will make it possible to explore the dynamics of functions and check possible patterns of interaction between functions.

Whereas the monitoring of the functions is not precise enough to fully unravel co-evolution processes as in previous studies (e.g., Kilelu et al. [16]), comprehensive data mining can provide the basis for monitoring TPs functions. This opens the door to a further understanding of the dynamics of TPs.

The co-evolutionary nature of innovation covers different dimensions including technological change but also social and institutional changes [32,33]. A basic operationalization of the co-evolutionary approach would attach actors' practices to hardware (technological innovation), software (culture, thinking, and learning) and orgware (organizations and institutions). In a broad sense, we could attach functions F1 ("Demand articulation") and F5 ("Innovation process management") to hardware, function F4 ("Network brokering") and F6 ("Knowledge brokering") to software, and F2 ("Institutional support") and F3 (Capacity building) to orgware. Therefore, we could carry out a first assessment on how the alignment of functions takes place in a co-evolution process of innovation systems.

3. TP Organics as a Technology Platform

Organic agriculture is characterized by promoting the efficient use of nutrients by keeping its cycles short and as closed as possible [34,35]. The organic sector has generated a wide range of innovations and has developed new ideas that have been put into practice in farms and businesses across the EU and the world [36,37]. Many organic farms and agro-food companies have become living creative laboratories for smart and green innovations. The organic sector has already generated a plethora of useful new practices for sustainable agriculture both within and outside the organic sector, with a promising future [38]. Controversy on the role of organic agriculture to feed the world continues [39]. However, our intention is to illustrate how online data collection can be used to monitor TP Organics as an innovation platform.

The structure of TP Organics has been summarized in Figure 1. Its core is formed by the secretariat and the steering committee, which is integrated by Bionext (a national technology platform of the Netherlands), Naturland e.V. (an association of farmers), FiBL (the Research Institute of Organics Agriculture), IFOAM EU (an international not-for-profit organization) and ACT Alliance EU (a network of 14 faith-based development agencies), as relevant representatives working towards a sustainable organic system. The rest of the platform is formed by 30 umbrella organizations and international networks, 36 firms, national technology platforms from seven Member States and 27 supporting members which may also act as financial supporters. Also, TP Organics cooperate with 10 initiatives; six at EU level and four at an international level.

We consider that the TP Organics platform is an interesting case to test our monitoring methodology due to several reasons. In the first place, it is a relatively recent ETP, which responds to an increasing demand for facing research challenges of the transition towards sustainable innovation systems. Secondly, the ETP was the result of a bottom-up process, bringing together farmers, processors, retailers, and scientists in a reflection on the future scenarios of agriculture and food systems. Third, TP Organics integrates not only technologies but also many social aspects of innovation, including viable concepts for the empowerment of rural economies, the sustainability of food systems towards eco-sustainable methods and the production of high-quality food as a basis to improve the quality of life and health. The performance of organic systems versus conventional systems of agriculture has been evaluated by the integration of several dimensions that consider production methods, environment, farmers' livelihood, consumer health, and food access [40]. Additionally, and connected with the previous points, TP Organics is a good example of a double way of governance of innovation systems from a multilevel perspective. TP Organics offers a space for an ecosystem of intermediaries that facilitate the interconnection between innovative practices at the niche level and more structural changes at the regime level with respect to beliefs, general rules, routines and standardized ways of doing things, policy paradigms, and social expectations and norms [41,42]. Therefore, it is a good example that brings together the six functions of the innovation platforms mentioned in the previous section. Finally, TP Organics has a very active corporate website, which is a sign of its offline and online activity.

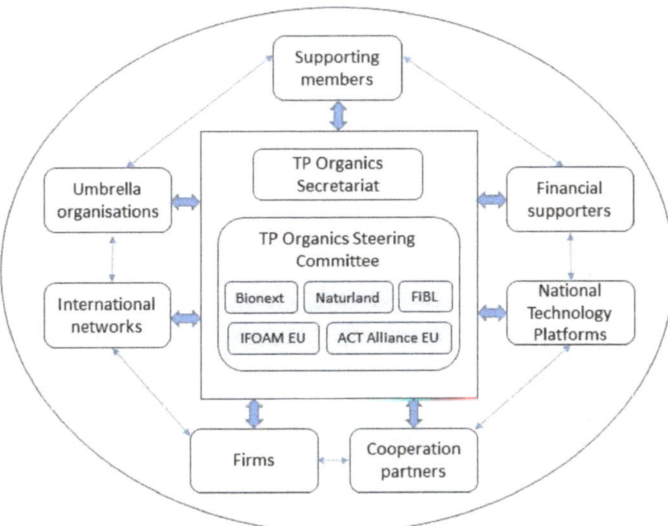

Figure 1. Structure of TP Organics in November 2018.

Even without the initial recognition of the European Commission, promoters of organic products sought the support of interested stakeholders, including an international association (IFOAM), the relevant commercial actors throughout the agri-food value chain, as well as environmental non-governmental organizations (NGOs). In 2008, they published a Vision for an Organic Food and Agriculture Research Agenda until 2025 [43]. As a result of this process, TP Organics was designed with a focus on sustainable food systems and public goods and officially launched in the fall of 2008. This was followed by a Strategic Research Agenda, which linked the term "innovation" with public goods, efficiency, farmers' knowledge, learning, and competitive advantage. In 2010, an Implementation Action Plan was published describing what is required for developing key research topics: eco-functional intensification; the economy of low use of external inputs; animal health systems. All this evolves with a focus on resilience and sustainability, from agricultural diversity to the diversity of natural foods and the creation of innovation centers in agricultural communities. In 2013, the promoters of TP Organics and other international actors launched a global TP to establish the Technological Innovation Platform of IFOAM—Organics International (TIPI) as a global, informal network to mobilize different organizations working on organic research topics [44]. The timeline of TP Organics is summarized in Figure 2.

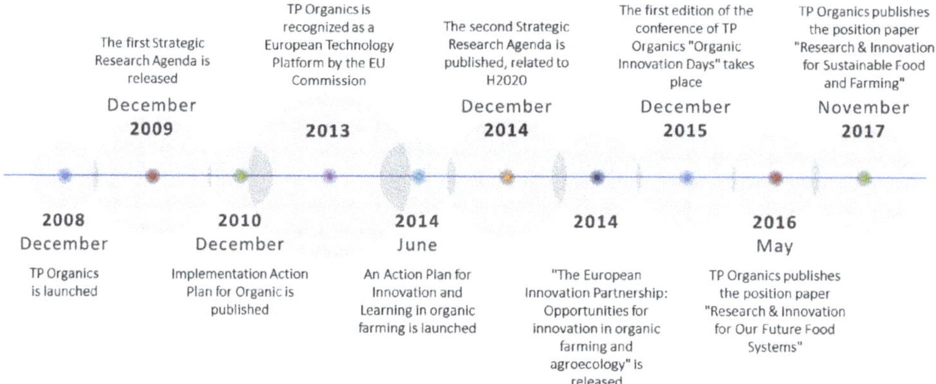

Figure 2. Timeline of TP Organics.

4. Methods

The assessment of the activities and impact of the TP organics platform was conducted by applying a combination of web data mining techniques. In particular, a web scraping system was employed to prepare a corpus of TP organics-related documents published online and content mining was applied to analyze the retrieved texts.

The corpus was created by introducing the term "TP Organics" (and the variant "TPOrganics") in Google and retrieving all documents in the result list. Since our objective is to understand the evolution of the ETP, only PDF files were kept because the time and date of their creation can be reliably determined, unlike HTML web pages. This way, the web mined content is put in the context of the date in which it was created. A deduplication process to remove multiple copies of the same document was also performed to avoid double counting. The web scraping process resulted in the retrieval of 197 different documents related to TP Organics.

To understand the ETP functions involved in the contents of each document, we compiled a set of keywords describing each of the ETP functions introduced in Section 2. For instance, finding "research agenda" in a document suggests that it is related to the demand articulation (F1), while "workshop" suggests capacity building (F3). The complete list of the keywords we used to describe each function can be found in Appendix A. Similar methods for classifying documents were successfully validated in previous works [13,45,46].

Once the keywords for each function are defined, content mining techniques were applied to the documents in the corpus. Particularly, documents were transformed to compute the frequency of the n-gram terms (i.e., the keywords) describing the ETP functions. Also, in order to allow for regionalizing the activity of the ETP, the frequency of the co-occurrence of each keyword with each EU country's name was also accounted for. To do so, a script that extracts the text from the documents and computes the frequencies of keyword occurrence and co-occurrence was developed in bash. Its execution resulted in a dataset of about 1 million rows, one for each triplet of document, keyword, and country. The dataset also included the creation date timestamp of each document, as extracted from the PDF metadata.

The activity reflected in a document typically involves more than one ETP function, as they are not mutually exclusive. To deal with this, we classified the documents in the function whose keywords were more intensively found in them.

The combination of the PDF creation timestamp with the keyword count allows us to create a powerful tool to monitor the evolution through time of the contents related to the ETP. This is similar to what Google Trends presents for the user searches in the Google Search Engine.

5. Results and Discussion

This section presents the online-based measures related to the assessment of TP Organics that were obtained after applying the methods described above.

5.1. Expansion of the ETP

The first measure is reflected in Figure 3. It shows the evolution over time of the quantity of online documents that mention "TP Organics". As can be seen, there is a clear increasing tendency, where the number of created documents generally grow during the platform lifetime. This sustained increase in the number of documents related to TP Organics suggests that the activity of the platform has also increased during these years. This constitutes a first approach to assess the activity of the ETP. The increasing tendency observed is in line with the growth in the production and consumption of organic food in the EU, as Figure 4 shows. From 2009 to 2016 (the last year for which data are available), the area dedicated to organic cultivation increased by 43%, from 8.3 million hectares (ha) to almost 12 million ha. In the same period, the retail sales of organic food increased from 16.9 billion euros to 30.7 billion euros, which represents a growth of 82%.

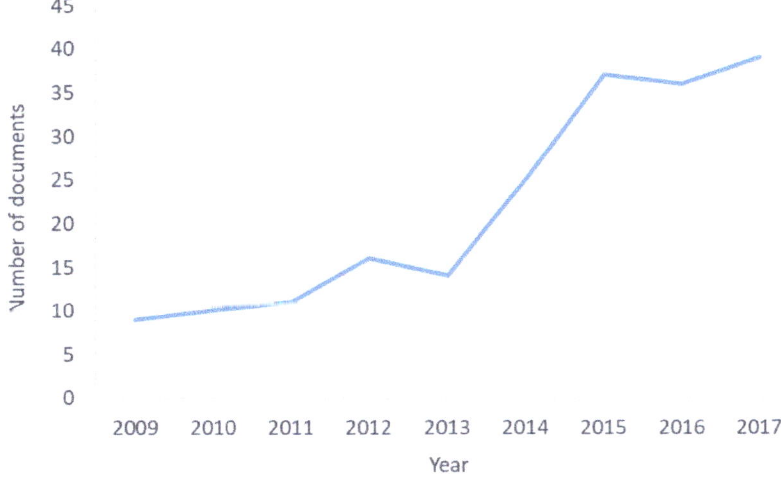

Figure 3. Evolution of the number of documents related to TP Organics created each year.

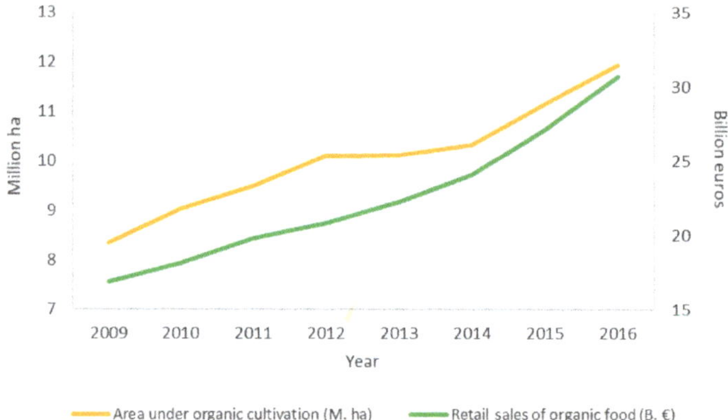

Figure 4. Evolution of the organic market in the EU. Sources: FiBL and Eurostat.

A more profound analysis of the pattern shown in Figure 3 provides some relevant information. Particularly, one can observe two periods in which the platform activity only increases slightly (2009 to 2013, and 2015 to 2017), while there is a sharp increment between years 2013 and 2015, a period in which the number of new documents related to TP Organics more than doubled. This period in which the expansion of the online activity of TP Organics is more pronounced took place at the same time that this platform was officially recognized as a European Technology Platform by the European Commission [7]. This was also influenced by the launch of its Second Research and Strategic Agenda under a larger and more mature structure than when the first one was launched.

5.2. Evolution in the Functions of the ETP

The documents related to TP Organics were classified into each of the six functions according to their content. To do so, each document was assigned to the predominant function, which is the one with the highest proportion of present keywords among the total number of keywords defining the function. Figure 5 shows the evolution of the number of documents that predominantly reflect each function. This chart reveals interesting insights into the activity of TP Organics. First, the variety of functions carried out by the platform increases gradually over time. As one can see, just two functions are predominant in the documents created in the year 2009. Another function appears as predominant in 2010, and the same happens in 2011 and 2012. This trend continues and it is only since 2015 that the six identified functions of the ETP are found to be predominant in the content of the documents related to the platform.

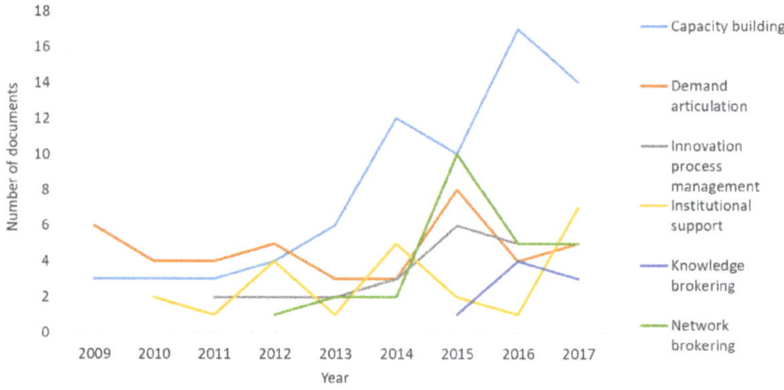

Figure 5. Number of documents that predominantly reflects each function.

Second, this pattern can be seen as a reflection of the ETP lifecycle. During the first few years from its conception, the platform is commencing its activity and trying to acquire members and resources, so it is located at the first part of the learning curve. Therefore, it is focused only on few functions, especially on those that are more related to determining and settling the strategic perspectives of the platform, i.e., "demand articulation" and "capacity building". The first is focused on articulating a research and innovation strategy involving the different actors of the system by means of developing a Strategic Research Agenda, which is a main objective of ETPs, while the second is focused on detecting knowledge and technology needs and promoting learning among its members and related community on organic food and sustainable agri food systems.

After the initial stage, the scope of the ETP increases together with its knowledge and resources. Thus, the number of functions it can fulfill or support intensively is larger. In this regard, the platform incorporates a third function ("institutional support") in its second year. This would indicate that the ETP was consolidated as a political tool. The third year incorporates the function "innovation process management", which consists in coordinating the interaction among different actors of the system regarding research and innovation activities, particularly the proposal of projects under the topic of organic food and sustainable agri-food systems under schemes of European funding (FP7, H2020, and FP9). It is reasonable that this function appears in this moment, as it is not possible to identify the key research and development challenges and coordinate different actors to work towards them at least until a Strategic Research Agenda is settled ("demand articulation"), the most important knowledge and technology needs are detected ("capacity building") and the political influence/role of the platform is developed ("institutional support"). However, "network brokering," which is the next function to appear, comes unexpectedly late despite being one of the basic activities for which European Technology Platforms were established [47].

It is reasonable that the "knowledge brokering" function is the last one to appear, as it is only possible to spread results of the ETP or inform about the organization and participation in relevant activities after some years of work. However, we should point out that the frequency of documents in which this function is predominant is not as important as expected. The reasons behind this should be explored in future works. Despite this, the way in which the different functions appeared over time is generally consistent with what was expected.

Finally, the order in which the different functions appear over time reflect that, regarding the co-evolutionary approach to innovation, the hardware (technological innovation) and orgware (organizations and institutions) dimensions are represented since the establishment of the platform, while the software dimension (culture, thinking, and learning) is not represented until the platform has been working for some years, when the "network brokering" function appears.

The different stages that TP Organics went through are graphically represented in Figure 6. These stages are summarized into first, the establishment of the platform; second, its growth; and finally, its maturity.

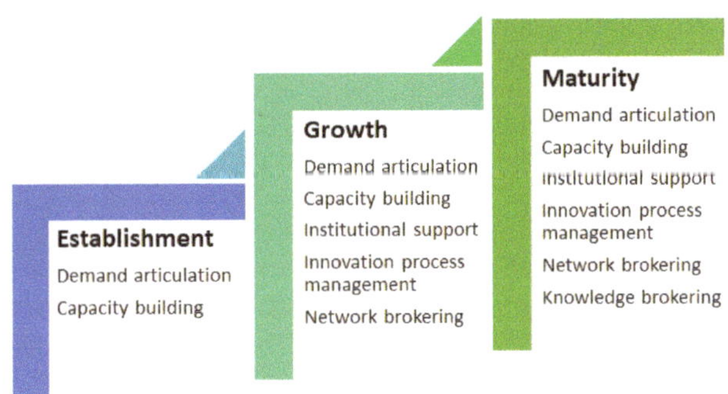

Figure 6. Stages of TP Organics as a technology platform.

5.3. Geographical Penetration of the ETP

Figure 7 includes two heat maps that represent the frequency of documents mentioning each country in the European Union. The heat map on the left represents the frequency of documents from the years 2009 to 2013; while on the right, the period considered was from the years 2014 to 2017. The values are standardized to a scale of 0 to 1, where 0 represents no documents available and 1 represents the highest frequency in the sample.

As can be seen, the activity of TP Organics has generally increased in the second time period with respect to the first in all countries. This is consistent with the previous results and in line with the ETP lifecycle: in the initial years, the platform has to grow, acquire members and position itself in the sector, so that its impact grows over time and is expected to be higher, as it is, once it has been working some years. It can also be seen that the leading countries, in which the presence of TP Organics is higher, are France and Germany, followed by Belgium and Italy. In the rest of the countries, the impact is lower. This is in line with the origin of the main members of the platform, whose headquarters are established in those countries (for instance, IFOAM and Act Alliance EU are located in Belgium, while Naturland e.V. is located in Germany and FiBL has some offices in France, Belgium, and Germany).

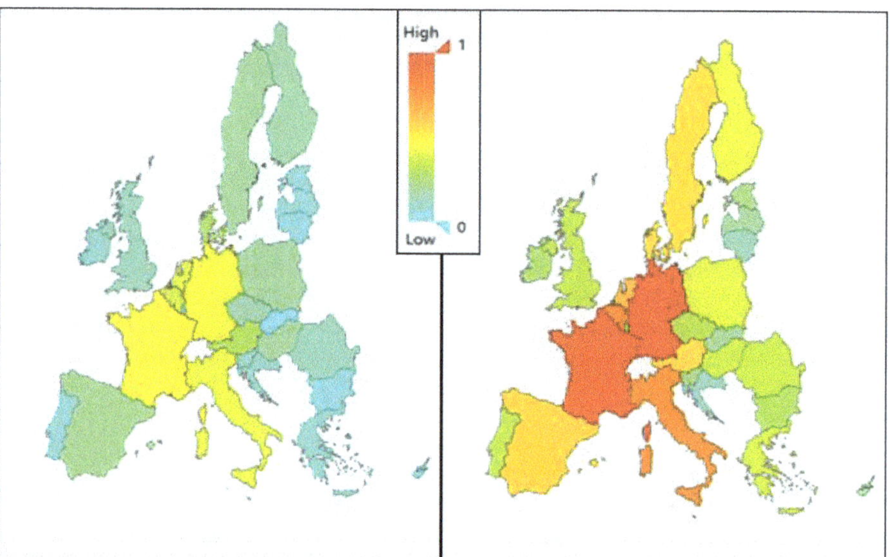

Figure 7. Documents mentioning each EU country for periods 2009–2013 (**Left**) and 2014–2017 (**Right**). Colder colors (blue, green) represent low quantity of documents, while warmer colors (yellow, orange, and red) represent higher quantity of documents.

In general, the impact of TP Organics in the European Union increases over time, although there are slight differences in the evolution when looking at each particular country. The geographical spread of TP Organics tends to follow the extension of the total organic area in the EU Member States (see Figure 8). Although, the platform's incidence tends to concentrate on Western Europe and with some leading countries in organic areas, such as Italy and Spain, not appearing as leading Member States in the TP. The dominant role of France and Germany is maintained in both periods, while in the case of Sweden and Finland, in which the impact of the ETP is similar in the first period, the evolution over time is different as in the second period the impact in Sweden is higher than in Finland (as the orange color reflects by contrast to the greener tonality).

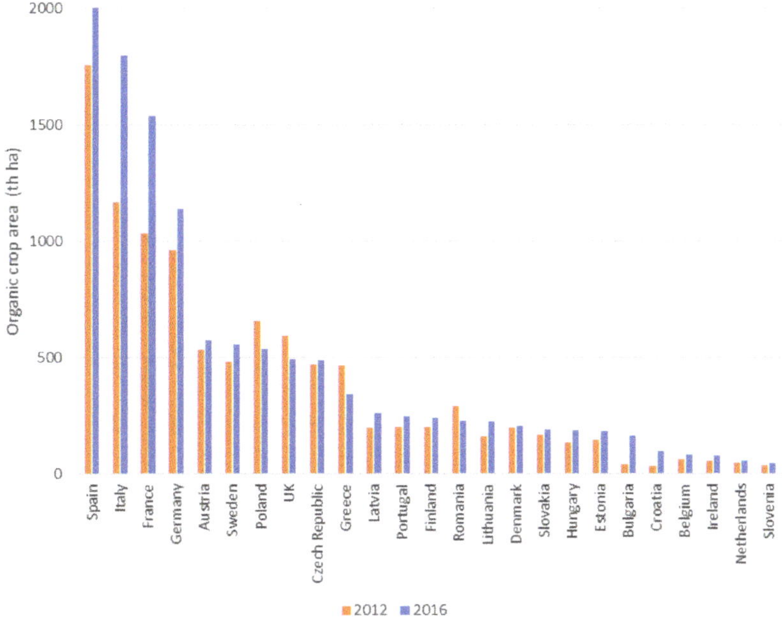

Figure 8. Organic crop area by EU country. Source: Eurostat.

6. Concluding Remarks

The technique shown in the present paper makes uses of online data collection to monitor and evaluate the performance of technology platforms. The method focused on European Technology Platforms, in particular, onTP Organics, a platform that promotes knowledge related to the future challenges of sustainable agriculture. We have not participated in the controversy on the future role of organic agriculture but we recognize that it offers a technology package, which surely has significant room for development, from the viewpoint of yields and environmental benefits per unit of product. Whatever the benefits organic agriculture show in the present and in the future, a great deal of knowledge and partnership will be required through the contribution of technology platforms.

While standardized methods to evaluate technology platforms are not generally agreed, the framework presented here is based on functions and on a co-evolutionary approach that, combined with web data mining techniques, provide the basis for a timely assessment of technology platforms.

The method exploits a number of possibilities that, in the explorative approximation followed in this research, open the chance of monitoring the TP activities, their results, their geographical dissemination, and their development through time. In the case of TP Organics, a lifecycle has been suggested with three different stages (establishment, growth, and maturity) with a varying relevance of the six innovation functions defined. With the proposed method, we are able to assess the impact of the platform regarding each of its functions, as well as over time and per country. The method makes use of publicly accessible and dated information and allows for a promptly update, very suitable to assess the technological platforms by replacing or complementing other methodologies of evaluation based on direct surveys.

Three areas of research will continue the present effort. The first area is the use of online data processing not only for monitoring but also for forecasting. This direction can facilitate policy analysis of the present trends and can help to foresee future directions of innovation activities and policy-making, whether applied to one technology platform or to a group of them.

A second direction is the use of the proposed monitoring technique to compare the evolution of different TP that could enrich the scope of innovation policies applied to sustainability.

We acknowledge the scaling up effort that will imply the definition of a standardized and common methodology to describe functions for various TP. If there are several technological solutions to the problems of sustainability in the agri-food system, we would like to assess the comparative performance of the related TPs.

A third direction is the reformulation of TP functions, not according to those adapted to innovation intermediaries (as was the approach followed in the present study) but to specific topics related to the main subject matter of the selected TP. Thus, it is ensured that TP Organics and other platforms for sustainability fulfill sets of functions related to climate adaptation and mitigation, resilience, vulnerability of agricultural systems, and other functions that will require a definition in terms of keywords and indicators to further evaluate TP performance.

Our work is not without limitations. On the one hand, our analysis only focused on PDF documents, thus leaving a high quantity of information out of reach of the contributed method just because it was published in other formats. On the other hand, the results are subject to certain English bias, as the list of keywords is mainly composed of terms in this language.

Although the results presented here are encouraging, there is still room for improvement. There are alternative ways to solve the problem of documents involving several ETP functions that could be explored in the future. This could be addressed by employing more advanced techniques, such as fuzzy clustering or with a term frequency–inverse document frequency (tf-idf) scheme. A more precise document classification would allow for a deeper understanding of the evolution of the platform.

In summary, this article has introduced a new tool for TP assessment with a wide range of possible applications for policy monitoring and evaluation.

Author Contributions: Conceptualization, J.-M.G.-A.-C.; methodology, D.B. and J.D.; software, J.D.; validation, D.B. and J.D.; formal analysis, D.B. and J.D.; investigation, D.B.; resources, J.D.; data curation, D.B. and J.D.; writing—original draft preparation, D.B., J.D. and J.-M.G.-A.-C.; writing—review and editing, D.B., J.D. and J.-M.G.-A.-C.; visualization, D.B. and J.D.; project administration, J.D. and J.-M.G.-A.-C.; funding acquisition, D.B., J.D. and J.-M.G.-A.-C.

Funding: This research was partially supported by the Spanish Ministry of Science, Innovation and Universities with grant AGL2015-65897-C3-3-R; and by the Spanish Ministry of Education with grant FPU14/02386.

Conflicts of Interest: The authors declare no conflict of interest.

Appendix A

Table A1. Keywords selected to represent each TP function.

TP Function	Keywords
Demand articulation	agenda, research agenda, innovation agenda, strategic agenda, strategic research agenda, sra
Institutional support	regulation/s, regulated, recommendation/s, policy maker/s, policy making, legislation, legislative, council, european commission
Capacity building	workshop/s, conference/s, forum, congress, organic innovation days symposium, science day, training
Network brokering	eatip, eposs, etpgah, eurobotics, fabre tp, fabre, food for life, manufuture, wsstp, cooperation

Table A1. *Cont.*

TP Function	Keywords
Innovation process management	work programme, h2020, horizon 2020, deliverable, fp7, fp 7, seventh framework programme, societal challenge 2 project, funds, funded, funding, era net, cofund, multi actor project, budget, eip agri, eip, jpi, face, facce jpi, era net core, organic core, core organic I, core organic II, core organic plus, cost action, research action/s, research project/s, operational group, focus group, university, universities, business, businesses, enterprise/s, company, companies, firm/s, umbrella organization/s, SME/s, federation/s, association/s, research institute/s, research group/s, research center/s, research centre/s, ngo/s, non-governmental organization/s, farmer/s, farm/s
Knowledge brokering	newsletter/s, article/s, research paper/s, report/s, book/s, paper/s, leaflet/s, brochure/s, dossier/s, publication/s, portfolio/s, journal/s, transfer, transfers, transferred, press, newspaper/s, magazine/s, ok net arable, patent/s, patented, pct, wipo, epo

References

1. Sayer, J.; Cassman, K.G. Agricultural innovation to protect the environment. *Proc. Natl. Acad. Sci. USA* **2013**, *110*, 8345–8348. [CrossRef] [PubMed]

2. European Commission. *Technology Platforms: From Definition to Implementation of a Common Research Agenda*; Office for Official Publications of the European Communities: Luxembourg, 2008; ISBN 92-894-8191-9.

3. Könnölä, T.; Salo, A.; Brummer, V. Foresight for European coordination: Developing national priorities for the forest-based sector technology platform. *Int. J. Technol. Manag.* **2011**, *54*, 438–459. [CrossRef]

4. Levidow, L.; Birch, K.; Papaioannou, T. EU agri-innovation policy: Two contending visions of the bio-economy. *Crit. Policy Stud.* **2012**, *6*, 40–65. [CrossRef]

5. Levidow, L.; Birch, K.; Papaioannou, T. Divergent paradigms of european agro-food innovation. *Sci. Technol. Hum. Values* **2013**, *38*, 94–125. [CrossRef]

6. ERA-NET in Horizon 2020. Available online: http://ec.europa.eu/research/era/era-net-in-horizon-2020_en.html (accessed on 22 September 2018).

7. European Commission. *Strategy for European Technology Platforms: ETP 2020*; Commission Staff Working Document, SWD(2013) 272 Final; European Commission: Brussels, Belgium, 2013.

8. Choi, H.; Varian, H. Predicting the Present with Google Trends. *Econ. Rec.* **2012**, *88*, 2–9. [CrossRef]

9. Vicente, M.R.; López-Menéndez, A.J.; Pérez, R. Forecasting unemployment with internet search data: Does it help to improve predictions when job destruction is skyrocketing? *Technol. Forecast. Soc. Chang.* **2015**, *92*, 132–139. [CrossRef]

10. Blazquez, D.; Domenech, J. Web data mining for monitoring business export orientation. *Technol. Econ. Dev. Econ.* **2018**, *24*, 406–428. [CrossRef]

11. ESSNET Big Data. Available online: https://webgate.ec.europa.eu/fpfis/mwikis/essnetbigdata/index.php/ESSnet_Big_Data (accessed on 8 November 2018).

12. Blazquez, D.; Domenech, J. Big Data sources and methods for social and economic analyses. *Technol. Forecast. Soc. Chang.* **2018**, *130*, 99–113. [CrossRef]

13. Arora, S.K.; Youtie, J.; Shapira, P.; Gao, L.; Ma, T. Entry strategies in an emerging technology: A pilot web-based study of graphene firms. *Scientometrics* **2013**, *95*, 1189–1207. [CrossRef]

14. De Mauro, A.; Greco, M.; Grimaldi, M.; Ritala, P. Human resources for Big Data professions: A systematic classification of job roles and required skill sets. *Inf. Process. Manag.* **2018**, *54*, 807–817. [CrossRef]

15. Darnhofer, I. Socio-technical transitions in farming: Key concepts. In *Transition Pathways towards Sustainability in Agriculture: Case Studies from Europe*; Sutherland, L.-A., Darnhofer, I., Wilson, G., Zagata, L., Eds.; CABI: Oxfordshire, UK, 2015; pp. 17–31.

16. Kilelu, C.W.; Klerkx, L.; Leeuwis, C. Unravelling the role of innovation platforms in supporting co-evolution of innovation: Contributions and tensions in a smallholder dairy development programme. *Agric. Syst.* **2013**, *118*, 65–77. [CrossRef]

17. Howells, J. Intermediation and the role of intermediaries in innovation. *Res. Policy* **2006**, *35*, 715–728. [CrossRef]

18. Klerkx, L.; Leeuwis, C. Establishment and embedding of innovation brokers at different innovation system levels: Insights from the Dutch agricultural sector. *Technol. Forecast. Soc. Chang.* **2009**, *76*, 849–860. [CrossRef]

19. Klerkx, L.; Aarts, N.; Leeuwis, C. Adaptive management in agricultural innovation systems: The interactions between innovation networks and their environment. *Agric. Syst.* **2010**, *103*, 390–400. [CrossRef]

20. Kilelu, C.W.; Klerkx, L.; Leeuwis, C.; Hall, A. Beyond knowledge brokering: An exploratory study on innovation intermediaries in an evolving smallholder agricultural system in Kenya. *Knowl. Manag. Dev. J.* **2011**, *7*, 84–108. [CrossRef]

21. Eastwood, C.R.; Chapman, D.F.; Paine, M.S. Networks of practice for co-construction of agricultural decision support systems: Case studies of precision dairy farms in Australia. *Agric. Syst.* **2012**, *108*, 10–18. [CrossRef]

22. Boon, W.P.; Moors, E.H.; Kuhlmann, S.; Smits, R.E. Demand articulation in emerging technologies: Intermediary user organisations as co-producers? *Res. Policy* **2011**, *40*, 242–252. [CrossRef]

23. Hakkarainen, L.; Hyysalo, S. The evolution of intermediary activities: Broadening the concept of facilitation in living labs. *Technol. Innov. Manag. Rev.* **2016**, *6*, 45–58. [CrossRef]

24. Kivimaa, P.; Boon, W.; Hyysalo, S.; Klerkx, L. *Towards a Typology of Intermediaries in Transitions: A Systematic Review*; Ciarli, T., Rotolo, D., Eds.; SPRU Working Paper Series; Science Policy Research Unit: Brighton, UK, 2017.

25. Alvesson, M. *Management of Knowledge-Intensive Companies*; Walter de Gruyter: Berlin, Germany; New York, NY, USA, 1995; ISBN 978-3-11-090056-9.

26. Miles, I. Knowledge intensive business services: Prospects and policies. *Foresight* **2005**, *7*, 39–63. [CrossRef]

27. Strambach, S. Knowledge-Intensive Business Services (KIBS) as drivers of multilevel knowledge dynamics. *Int. J. Serv. Technol. Manag.* **2008**, *10*, 152–174. [CrossRef]

28. Mas-Verdú, F.; Wensley, A.; Alba, M.; Álvarez-Coque, J.M.G. How much does KIBS contribute to the generation and diffusion of innovation? *Serv. Bus.* **2011**, *5*, 195. [CrossRef]

29. Hekkert, M.P.; Suurs, R.A.; Negro, S.O.; Kuhlmann, S.; Smits, R.E. Functions of innovation systems: A new approach for analysing technological change. *Technol. Forecast. Soc. Chang.* **2007**, *74*, 413–432. [CrossRef]

30. Hekkert, M.P.; Negro, S.O. Functions of innovation systems as a framework to understand sustainable technological change: Empirical evidence for earlier claims. *Technol. Forecast. Soc. Chang.* **2008**, *76*, 584–594. [CrossRef]

31. Bessant, J.; Rush, H. Building bridges for innovation: The role of consultants in technology transfer. *Res. Policy* **1995**, *24*, 97–114. [CrossRef]

32. Biggs, S.D. A multiple source of innovation model of agricultural research and technology promotion. *World Dev.* **1990**, *18*, 1481–1499. [CrossRef]

33. Smits, R.E.; Kuhlmann, S.; Shapira, P. *The Theory and Practice of Innovation Policy*; Edward Elgar Publishing: Cheltenham, UK, 2010; ISBN 9781845428488.

34. Padel, S.; Röcklinsberg, H.; Schmid, O. The implementation of organic principles and values in the European Regulation for organic food. *Food Policy* **2009**, *34*, 245–251. [CrossRef]

35. Robertson, G.P.; Gross, K.L.; Hamilton, S.K.; Landis, D.A.; Schmidt, T.M.; Snapp, S.S.; Swinton, S.M. Farming for ecosystem services: An ecological approach to production agriculture. *BioScience* **2014**, *64*, 404–415. [CrossRef] [PubMed]

36. Willer, H.; Lernoud, J. *The World of Organic Agriculture. Statistics and Emerging Trends 2016*; Research Institute of Organic Agriculture FiBL/IFOAM Organics International/Medienhaus Plump: Bonn, Germany, 2016; ISBN 978-3-03736-307-2.

37. Reganold, J.P.; Wachter, J.M. Organic agriculture in the twenty-first century. *Nat. Plants* **2016**, *2*, 15221. [CrossRef] [PubMed]

38. Rahmann, G.; Ardakani, M.R.; Bàrberi, P.; Boehm, H.; Canali, S.; Chander, M.; David, W.; Dengel, L.; Erisman, J.W.; Galvis-Martinez, A.C.; et al. Organic Agriculture 3.0 is innovation with research. *Org. Agric.* **2017**, *7*, 169–197. [CrossRef]

39. Muller, A.; Schader, C.; Scialabba, N.E.H.; Brüggemann, J.; Isensee, A.; Erb, K.H.; Smith, P.; Klocke, P.; Leiber, F.; Stolze, M.; et al. Strategies for feeding the world more sustainably with organic agriculture. *Nat. Commun.* **2017**, *8*, 1290. [CrossRef] [PubMed]

40. Seufert, V.; Ramankutty, N.; Foley, J.A. Comparing the yields of organic and conventional agriculture. *Nature* **2012**, *485*, 229–235. [CrossRef] [PubMed]

41. Geels, F. The multi-level perspective on sustainability transitions: Responses to seven criticism. *Environ. Innov. Soc. Transitions* **2011**, *1*, 24–40. [CrossRef]

42. Elzen, B.; Barbier, M.; Cerf, M.; Grin, J. Stimulating transitions towards sustainable farming systems. In *Farming Systems Research into the 21st Century: The New Dynamic*; Darnhofer, I., Gibbon, D., Dedieu, B., Eds.; Springer: Dordrecht, The Netherlands, 2012; pp. 431–455. ISBN 9789400745025.

43. Niggli, U.; Slabe, A.; Schmid, O.; Halberg, N.; Schlüter, M. *Vision for an Organic Food and Farming Research Agenda 2025. Organic Knowledge for the Future*; TP Organics: Brussels, Belgium, 2008; Available online: http://orgprints.org/13439/ (accessed on 18 September 2018).

44. Niggli, U.; Willer, H.; Baker, B. *A Global Vision and Strategy for Organic Farming Research*; TP Organics: Brussels, Belgium, 2016; Available online: http://orgprints.org/31340/ (accessed on 18 September 2018).

45. Blazquez, D.; Domenech, J.; Gil, J.A.; Pont, A. Business Export Orientation Detection through Web Content Analysis. In *Web Information Systems Engineering—WISE 2014, Proceedings of the 15th International Conference, Thessaloniki, Greece, 12–14 October 2014, Lecture Notes in Computer Science*; Springer: Cham, Switzerland, 2014; pp. 435–444.

46. Blazquez, D.; Domenech, J.; Gil, J.A.; Pont, A. Monitoring e-commerce adoption from online data. *Knowl. Inf. Syst.* **2018**, in press. [CrossRef]

47. ERA LEARN—Other ERA relevant Partnership Initiatives. Available online: https://www.era-learn.eu/p2p-in-a-nutshell/type-of-networks/other-era-relevant-partnership-initiatives#ETPs (accessed on 8 November 2018).

 *sustainability* 

Article

Semantic Network Analysis of Legacy News Media Perception in South Korea: The Case of PyeongChang 2018

Sung-Won Yoon [1] and Sae Won Chung [2,*]

[1]   Division of Business Administration, The University of Suwon, Hwaseong-si 18323, Korea; syoon@suwon.ac.kr
[2]   Graduate School of International Studies, Korea University, Seoul 02841, Korea
*   Correspondence: swchung37@korea.ar.kr; Tel.: +82-2-3290-5323

Received: 5 October 2018; Accepted: 26 October 2018; Published: 2 November 2018

Abstract: This paper aims at exploring how conservative and liberal newspapers in South Korea framed PyeongChang 2018 directly. Our research questions addressed four points: first, different attitudes of conservative and liberal newspapers in the PyeongChang news reporting; second, their success and failure in influencing public opinion; third, South Koreans' perceptions on PyeongChang 2018; and fourth, South Korean public reliance on the newspapers. To investigate the framing differences, we employed a big data analytic method (automated semantic network analysis) with NodeXL (analytic software). Conclusively, we were able to find out four main findings. First, the conservative media showed pessimistic attitudes to the Olympics, and the liberal media did conversely. Second, despite the conservative media's resourcefulness, they could not succeed in influencing public opinion. Third, the conservative media perceived the Olympics as an undesirable event, but the liberal media did the Olympics as a significant event for further peace promotion. Fourth, the conservative media's framings did not considerably influence upon the public opinion. As a conclusion, the public are no longer passive recipients of the messages from the media. Instead, they tend to selectively accept the information from the media based on 'collective intelligence'. This trend provides a significant implication for enhancing the sustainability of the media environment in South Korea.

Keywords: big data analytic methods; semantic network analysis; framings; NodeXL

1. Introduction

The XXIII Winter Olympics held from 9 to 25 February 2018 in PyeongChang County, South Korea drew a considerable amount of attention from global media. The Olympics, commonly known as PyeongChang 2018, was a highly striking event not simply because it accommodated the largest number of participating countries, people, and games in the Olympic history. It was because of the 'peaceful mood' created between two Koreas as realised in introducing a joint-marching in the opening ceremony, establishing a unified Korean female ice hockey team, and inviting North Korean delegates to the opening and closing ceremonies. It was thus even expected that North Korean and US representatives might meet for talks during or after the Olympics. By enhancing the expectation on the peace settlement in the Korean peninsula, such peaceful mood resulted in official talks between North and South Koreas, and also between Pyongyang and Washington in the end. In this regard, the Olympics has created a significant image as a "Peace Olympics". However, such image bears vast disparities between major media in South Korea in accordance with their political orientations. On the one hand, the conservative media framed this event 'negatively' by projecting the images of this event as 'Pyongyang Olympics' or an inappropriate event. On the other hand, the liberal media framed this event 'positively' by giving positive evaluations enforcing its peacemaking aspects. This paper aims at exploring how differently these two types of Korean media framed this event. This study employs automated semantic network analysis, which is one of alternative content analysis methods [1]. This method is useful in representing the content of the message as a network of objects, such as positive and negative accounts of a certain object [1]. The schematic representation of semantic network analysis is presented in Figure 1.

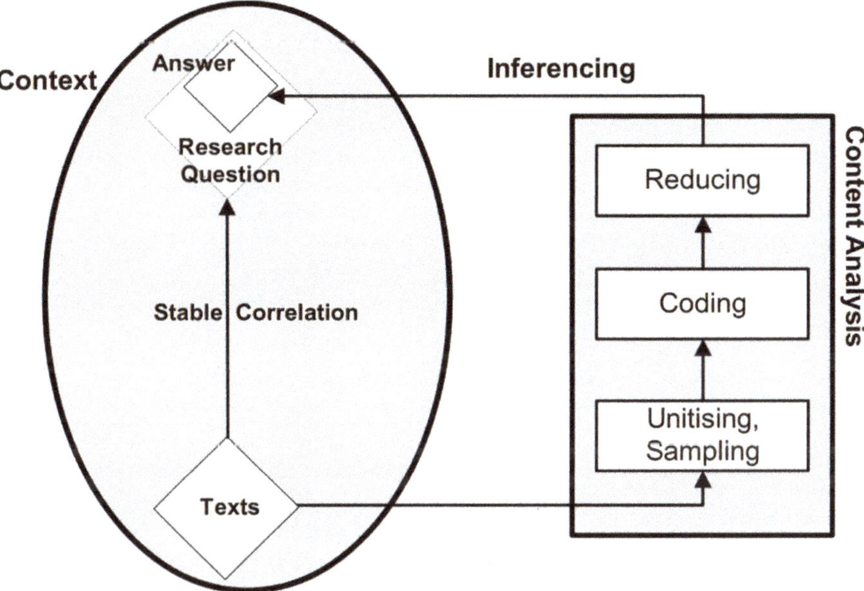

Figure 1. Semantic network analysis [1] (p. 17).

The contribution of this study lies in providing a new interpretation to PyeongChang 2018 based on semantic networks extracted from news media texts. Empirical studies on Pyeongchang 2018 have touched upon event preparation or civic participation. Especially, in 2011, when the hosting the PyeongChang Olympics was confirmed, most studies were conducted in order to prepare a plan for regional development by analysing the hosting environment of the PyeongChang Olympics [2–6]. Alternatively, in the preparation of hosting the PyeongChang Olympics, the regional residents' willingness to cooperate can have a positive influence in terms of regional development and the diffusion of sport culture [7]. However, there is no study on the media attitudes and their subsequent formation of public opinion about this event. In this vein, this study can be timely and significant to some extent. By conducting semantic network analysis, this study can suggest another novel way of interpreting this event. Semantic network analysis seeks to explore what are the keywords in the media discourse (*vertices*) and their relations among themselves (*edges*) [8]. Semantic network analysis is a standardised method of extracting essential representations from huge amount of unstructured data [8]. Hence, the research outcomes of this study can suggest extended interpretation of this event in social context and structure (Figure 2).

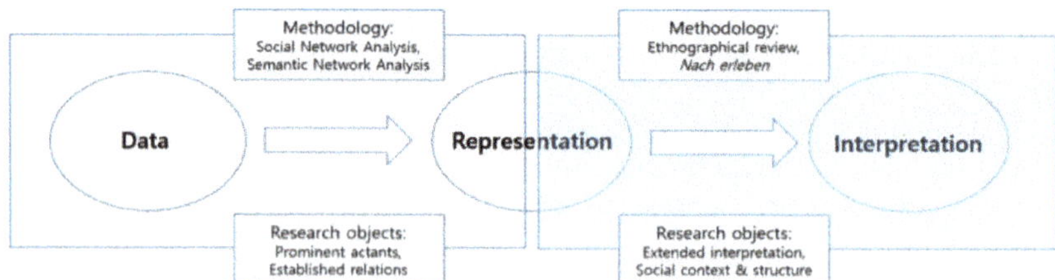

Figure 2. The role of semantic network analysis in interpretive sociology [8] (p. 11).

This paper consists of five sections. First, it reviews empirical literatures on press partisanship in South Korea. Second, a theoretical framework based on two relevant theories—*framing theory* and *graph theory*—is presented. Then, the third section concerns our big data analytic methods—automated semantic network analysis. The fourth section elaborates our data profiles, such as the volume of data, data collection period and the analysis process. Fifth, this paper presents our research findings based on our methods. Finally, the differences between the research findings and the public perceptions based on public opinion surveys on PyeongChang 2018 are discussed.

In investigating media framings about PyeongChang 2018, the following four research questions were addressed in Table 1.

Table 1. Research questions of this study.

Research Question 1 (Q1)	How have the conservative and the liberal newspapers shown their attitudes to PyeongChang 2018 in their news framings? Optimistically or pessimistically?
Research Question 2 (Q2)	Which type of newspapers was successful in influencing public opinion about PyeongChang 2018? Conservative or liberal?
Research Question 3 (Q3)	How did the South Korean public perceive PyeongChang 2018? Positively or negatively?
Research Question 4 (Q4)	Why did the South Korean public buy more of a particular type of newspapers (conservative or liberal)?

Based on these research questions, we drew out four hypotheses. They are as follows:

Hypothesis 1 (H1). *The conservative media would have applied pessimistic framings to PyeongChang 2018 by selecting its negative aspects which reflect the conservative media's perceptions of this event. Conversely, the liberal newspapers would have taken opposite stances by choosing positive aspects of this event.*

Hypothesis 2 (H2). *The conservative media would have been more successful than the liberal ones in influencing a public opinion. It is because the conservative newspapers have larger readerships than the liberal ones. The conservative newspapers can disseminate their intended messages to their readers, and the public is more likely to accept the framings created by the conservative media.*

Hypothesis 3 (H3). *Considering the framing exposure by the conservative media, the South Korean public would have perceived PyeongChang 2018 Olympics negatively rather than positively.*

Hypothesis 4 (H4). *The South Korean public would have been influenced by the conservative media's framings. It is because the conservative newspapers are more resourceful than the liberal ones in terms of their provocative framings out of careful lexical choices.*

2. Literature Review

Partisanship of South Korean Press and Framing

Much has been studied extensively on the impact of press partisanship on media framing. Press partisanship refers to a consistent position and attitudes toward the social structure (such as power structure, economic distribution structure, and so on) which is formed through the consideration about the self's position of his or her values, experience and society, and public issues which influenced such social structure [9] (pp. 113–114). Press partisanship can be understood in the journalism as maintaining neutral attitudes [9] (pp. 113–114). Kahn and Kenney investigated US media editorials and other news items during 60 Senatorial election campaign periods to probe press partisanship. They suggested that the editorials and news articles are closely related with newspaper company owners' and newsmakers' political orientations [10]. In addition, their beliefs and political orientations are found in their biased framings in reporting their news articles [10]. Shoemaker and Reese also pointed out that most of the linguistic and visual information media delivers are generally influenced by the factors such as the preferences of the news producers or the news company not to mention the dominant ideology of the society [11]. Press partisanship could cause a social division and conflict in this regard [12].

In South Korea, press partisanship has became an issue since Kim Dae-jung's administration [13,14]. Press partisanship in South Korea is considered to have been intensified due to three causations [9] (pp. 124–128). First, it resulted from a series of hegemony struggles from a power shift. Right after gaining its power from the conservative government, Kim Dae-jung's liberal government conducted media reformation. By strengthening the standard of social transparency and procedural justification, this reformation put those who were privileged from the former governments into a critical situation. In this vein, such change gave the newspaper owners and social elites a feeling of crisis because they were the privileged members of the stable regime until the power shifted. Consequently, this transformation has strengthened press partisanship in South Korea.

Second, it arose from political parties' undemocratic nature. At the time of political democratisation in South Korea, political parties were not able to resolve newly emerging social conflicts. Instead of political parties, it was media that played a mediating role. This is believed to have strengthened press partisanship in South Korea. However, as media tried to play mediating or intervening roles instead of their original role of a messenger, they have often either exaggerated or concealed certain facts in order to represent a particular stakeholder's interests [15].

Third, press partisanship derived from the particular environment of the Korean newspaper industry. The latter has been more interested in extending their realm of political influence than securing loyal readership by producing instructive articles. This seems to have been caused by the excessively competitive nature of the media industry [9] (pp. 124–128).

Regarding the political landscape of South Korean media, there are two opposing camps—conservative camp (i.e., *Chosun Ilbo*, *JoongAng Ilbo* and *DongA Ilbo*, for instance), and liberal camp (i.e., *The Hankyoreh* and *The Kyunghyang Shinmun*, for instance). Both *Ilbo* and *Shinmun* mean 'newspapers' in Korean. Among these conservative newspapers, *Chosun Ilbo* and *DongA Ilbo* were established in the 1920s, in the middle of the Japanese colonial times. Since then, they have become representative conservative media. *JoongAng Ilbo* was established by Samsung in 1965 and became independent from it in 1999. The liberal newspapers were established much later than the conservative newspapers since South Korea's democratisation in 1987. *The Hankyoreh* was established in 1988. *The Kyunghyang Shinmun* became a liberal newspaper when its ownership was changed from Hanhwa (one of the conglomerates in Korea) to the employees of the newspaper. South Korean press partisanship became more evident when *The Hankyoreh* was established as the first liberal newspaper in South Korea. Regarding political orientation, three abovementioned conservative newspapers possess 70% of Korean readership; liberal newspaper possesses less than 10 percent of it [9] (p. 108).

Then, what kind of influence has Korean press partisanship made upon the news production process? Several case studies have been conducted on a range of issues. Most representative cases include media coverage of the North Korean crisis, North–South conflict, and security of the Korean peninsula [16–20]. Notably, Kim and Roh analysed the editorials of *Chosun Ilbo*, *DongA Ilbo*, *The Hankyoreh* and *The Kyunghyang Shinmun* published from 2008 to February 2010. They argued that the apparent differences were found between different political orientations in reporting North Korean issues. They analysed the editorials' political ideology by focusing on their respective views on 'Progressive Party', 'Korea-US relations', and 'National Security Law' [20]. In each theme, different newspapers framed the issues differently based on their political orientations. In dealing with South Korea's aid to North Korea, *DongA Ilbo*, and *Chosun Ilbo* covered the issue with conservative framings and *The Hankyoreh* did with liberal framings [19].

In fact, press partisanship may not be an issue to be criticized unless the media do not seriously distort the truth [9] (p. 114). One might plausibly request public media organisation to discard their partisanship and keep their neutrality. However, it is almost impossible to request private media to give up their own partisanship [9] (p. 114). Basically, such diverse partisanship of media could be helpful to provide their diverse perspectives to the readers, as is a positive function of media [9] (p. 114). However, South Korean press partisanship was prevalent not only in 'ideological' issues, but in any controversial issues related to the government's policies as well. Some of the exemplary cases include the coverages on the corruptions of presidential families [21], the evaluations on the current government and president [22], South Korea's import of US beef [23], the prospects on possible economic crisis in South Korea [24], the introduction of comprehensive real estate tax and the abolition of patriarchal family system [19], and so on. Furthermore, partisanship is also strikingly captured in a wide range of policy areas such as political, diplomatic, economic, legal, educational, labour, taxation and welfare, and so forth.

According to Shoemaker and Reese's model, the government can be the external factor which influences framing media contents [11]. In other words, when the media have a political orientation different from the current government's, they are very likely to face a series of conflicts. The incumbent government may pressurise them to delete or amend some contents of news articles, whilst the latter may negatively depict the former's performances. Even though reporting on the government is the media's official and formal task, their news collecting activities are normally performed based on their private relationships [25]. Hence, irrespective of the presence of the government's pressure, there may be some sorts of self-controls by journalists or media company who were aware of the aftermath their candid coverage may cause [25]. In addition, media coverage is highly influenced not only by the media's political orientations but also their political compatibility with the government's [26]. Table 2 sums up a media-government relationship matrix based on their compatibility of their political orientations.

Table 2. Media–government relationship [26] (p. 164).

Media-Government Relationship		Political Orientations of the Regime	
		Conservative	Liberal
Media	Conservative	I. Compatible	II. Incompatible
	Liberal	III. Incompatible	IV. Compatible

Media-government relationships can be interpreted as either 'compatible' (I and IV) or 'incompatible' (II and III). When they have a compatible relationship, the media will show supportive attitudes towards the government's policies. Otherwise, the media will show negative and opposing attitudes toward the government's policies.

In this context, this study aims to investigate comprehensively how the newspapers in two opposing camps (conservative vis-à-vis liberal) in news framing. By looking into the case of

PyeongChang 2018, we would like to address how the newspapers in each camp differentiated their attitudes in their news coverage.

3. Theoretical Framework

3.1. Framing

Framing is "to select some aspects of a perceived reality and make them more salient in a communicating text" [27] (p. 52). Framing's distinctive aspects come out of its applicability effects [28]. This effect becomes valid when the audiences accept the message exposed by the media [28]. For instance, if the news message suggested a framing about the connection between tax policy and unemployment rates, the audience might consider about the applicability of unemployment rates to tax issues [29]. In other words, issue salience from the media has a certain amount of impact upon the audience's process of cognition process. Based on this theoretical base, this paper explores how representative media in South Korea framed PyeongChang 2018 in connection to the peace and security issue of the Korean peninsula. Then, the media framings are composed with the South Korea's public opinion to the issue. D'Angelo's study presents the algorithm of frame construction flow, framing effects flow and frame definition flow (Figure 3) [30]. This study suggests the importance of diversified methods of framing analysis [30]. In this regard, automated semantic network analysis can be one of the helpful approaches showing how the framings were constructed and connected to each other from the working out the constellation of the groups of the salient words in the news texts.

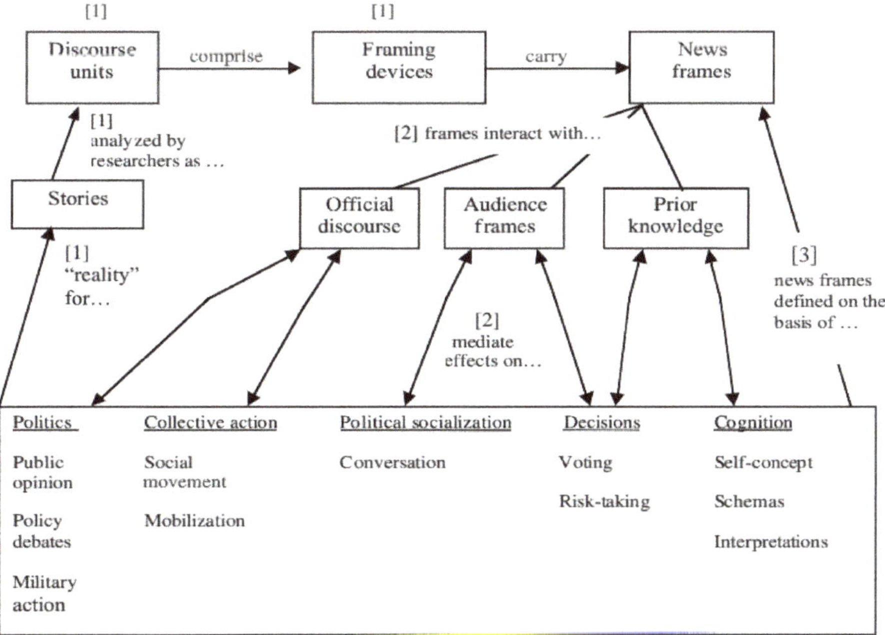

Note. [1] = Frame construction flow; [2] = Framing effects flow; [3] = Frame definition flow.

Figure 3. A model of the news framing process [30] (p. 880).

The framings exist in "located lexical choices of codes" [31] (p. 59), and can be retrieved from the most frequently mentioned certain words (or phrases) within the news texts. In recent years, there have been several accounts that point to the exploring reference words which are closely related to thematic framings. Fillmore and Atkins explored the risk frame by indicating the associate words such as danger, peril, hazard, venture, gamble, risky, perilously and so on [32] (pp. 79–80). Nerlich and

Koteyko explored associate frames regarding carbon reduction activism in the UK by focusing on the lexical devices. They coded "carbon debt" into the finance framing, "carbon living" into the lifestyle framing, "carbon sinner" into the attitude framing [33] (p. 211). Degani focused on exploring framings by focusing on the association of lexical choices within Obama's speeches [34]. The results of these studies indicate the framing constitutes a group of reference lexical items. However, these studies do not present the overview of such connectedness. To show such connectedness, some scholars employed *semantic network analysis* using a computer-assisted method. Such analysis is useful regarding indicating an overview of framings based on clear visualisation established by analytic software. After addressing *graph theory*, this paper will introduce and review automated semantic network analysis.

3.2. Graph Theory

Graph theory is the study of graphs. A graph consists of points (nodes or vertices) and lines (edges or relations). This theory originated from Euler's analysis of his Königsberg (now Kaliningrad Russia) bridge problem in 1736 [35]. The problem was "can one cross all seven bridges and never cross the same one twice [35]?" A background map is provided in Figure 4 below.

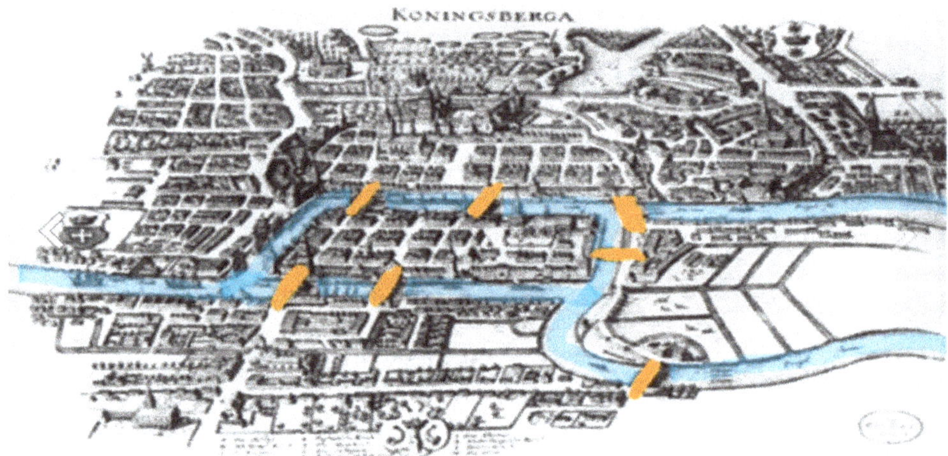

Figure 4. A map of 18th-century Königsberg, with its seven bridges, highlighted [35].

To solve this problem, Euler applied the vertices for the four landmasses and the edges for the seven walking paths. The visualised Euler's proof is presented in Figure 5 below. This proof is the first recorded case of using a graph to solve a mathematical problem. According to Barabási, there are two important implications. First, graph representation makes some problems simpler and more tractable. Second, path existence does not depend on our ingenuity but the graph's properties [36]. Due to such implications, *graph theory* played a vital role in providing the means of visualisation.

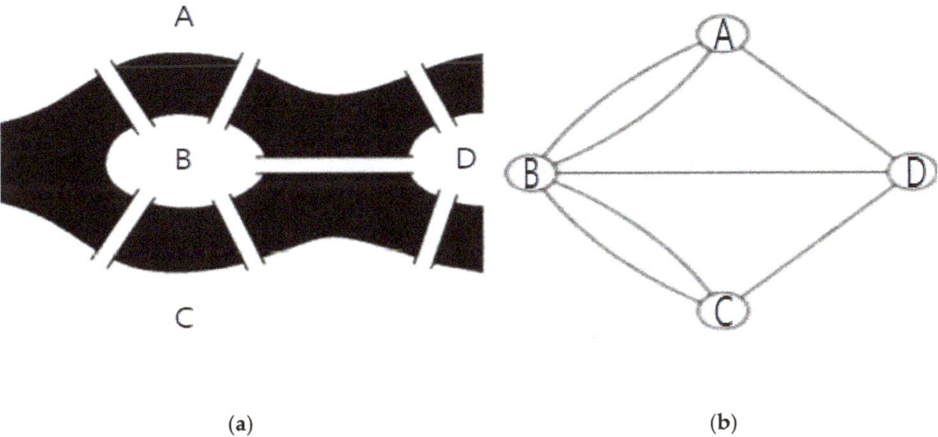

(a)   (b)

Figure 5. (a) a simplified depiction of the pattern of the rivers and bridges in the Königsberg bridge problem [35]; (b) the corresponding network of vertices and edges [35].

In addition, *graph theory* has been applied to the diverse ranges of academic disciplines. These applications were useful when someone wants to explore a complex interrelation in a network format. The most recent application of the graph theory to network analysis is topological data analysis. The most recent study conducted topological data analysis to analyse social media to predict the popularity of images in the data [37]. In addition, their study investigates the feasibility of topological data analysis for social media mining. In addition, their study showed that the shape of data can produce meaningful information [37]. This approach can be useful in analysing visual images in the media, but, as mentioned in the Introduction, our study concerned more about text as a unit of analysis. In this regard, we apply the semantic network analysis in order to discern South Korean conservative and liberal newsmakers' perceptions of PyeongChang 2018. According to de Saussure, understanding text can be formalised as a system of signs [38]. To be more specific, the text consists of a system of multiple subordinate units, such as sentences, phrases or words. A semantic network refers to a group of different words being connected to each other [39]. In this approach, the words can be represented as vertices or nodes. Word relations can be represented as undirected edges [40]. Based on de Saussure's approach, Drieger employed a semantic network model which considers the text as a network of connected words [41]. Based on the theoretical model of the semantic network, this study will explore the word connections from the texts embedded in the conventional news media—leading conservative and liberal newspapers in South Korea.

3.3. Algorithm of Theoretical Framework

The study aims at probing how the South Korean conservative and liberal media framed PyeongChang 2018 differently. The problem of this research can be a negative impact resulted by this contrasting news frames from the two types of newspapers. To address this problem, this study employed two theories—framing theory and graph theory. To formulate the theoretical framework, research problem and its further consequences, this study suggests the algorithm in Figure 6.

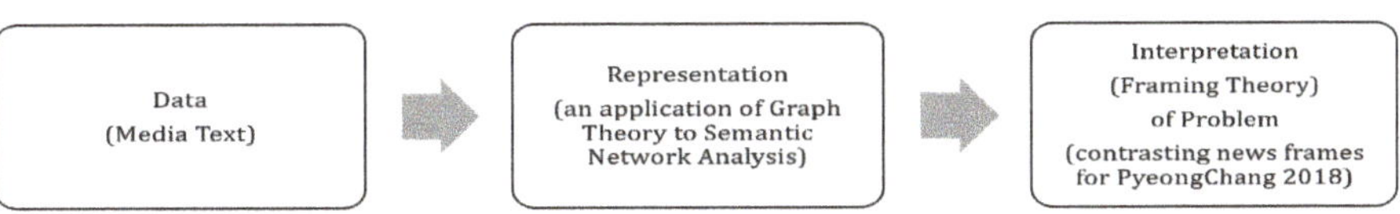

Figure 6. Overview of the proposed framework.

4. Methods

4.1. Semantic Network Analysis

Recently, *semantic network analysis* has been widely accepted as a helpful method exploring word associations in the texts. *Semantic network analysis* is a form of content analysis that identified the network of associations between concepts expressed in the text [42,43]. The core of *semantic network analysis* is "collecting a small set of texts from 'critical junctures' in the life of 'a social movement for study'" [44] (p. 229). Doerfel and Barnett conducted a *semantic network analysis* to probe the structure of the International Communication Association (ICA) [45]. They extracted the words from the titles of the papers presented to ICA. *Semantic network analysis* identifies a new cluster of concepts, so it enables us to explore the meaningful ideas from the texts [43]. Analysing a meaningful cluster of concepts corresponds well with framing analysis. *Semantic network analysis* addresses the issue saliences mentioned by Entman's study on frames [46].

Since the importance of big data research, automated content analytic methods (such as text mining, natural language processing, knowledge representation network analysis and visual analytics) drew a significant amount of attention [41]. Among these possible options suggested, we focused on automated semantic network analysis focusing on exploring framings. Schultz et al. explored associate frames about the 2010 BP Oil Spill Crisis to explore the interplay of public relations and news in a crisis. They collected a considerable amount of news articles to examine the frame difference between UK and US news items in the BP crisis [47]. Their results contributed to analysing the interplay between public relations and news in crisis situations [47]. Motta and Baden aimed to investigate the dynamic of frames for finding out their evolutionary factors. Their works are useful to overlook the trend of dynamic transformation regarding the forceful discourses, such as emergence, evolution, consolidation or crisis [48]. David et al. examined the associated media framings of the population issue in the Philippines to reveal the strategic framings [49]. Jiang et al. aimed to investigate how American and Chinese media framed the Arab Spring in 2010. Their results presented the competition between the framings of two main global actors with different political, cultural frames [46]. Based on the aforementioned empirical studies, we believe that an application of *semantic network analysis* is helpful to explore three aspects. The first one is frame difference between conservative and liberal media news items in PyeongChang 2018. This aspect could show how different types of newspapers perceived PyeongChang 2018. The second one is frame dynamics of such framing over the two months of this event. It suggests how these newspapers took either constant or different attitudes to PyeongChang 2018. The third one is strategic frames by these media camps. It reveals that these newspapers' emphasise PyeongChang 2018 in terms of their public relations.

4.2. Data

In line with empirical studies, this study conducts automated semantic network analysis by using NodeXL (software for automatic text analysis). The data collection for this study consists of three steps. First, this study selected 'the PyeongChang Winter Olympics' as a keyword for exploring the relevant articles published in conservative and liberal newspapers. Second, this study collected the articles over a set period of time (from January to February 2018) for capturing the moment of most heated debates on PyeongChang 2018. Third, this study conducted a keyword search by employing the search engine in the websites of these newspapers—*Chosun Ilbo* [50], *JoongAng Ilbo* [51], *The Hankyoreh* [52], and *The Kyunghyang Shinmun* [53]. A total of 445 articles were collected from two conservative newspapers in South Korea (*Chosun Ilbo* and *JoongAng Ilbo*), 102 articles from two liberal newspapers in South Korea (*The Hankyoreh* and *The Kyunghyang Shinmun*). The summarised overview of the number of collected articles are presented in Table 3.

Table 3. Overview units of analysis.

Political Orientation	Newspapers	Number of Articles
Conservative	*Chosun Ilbo*	376
	JoongAng Ilbo	69
Liberal	*The Hankyoreh*	35
	The Kyunghyang Shinmun	67

The data analysis of this study consists of three steps. NodeXL was a software for extracting, analysing and visualising the semantic networks from the collected data. First, the collected data were stored by each month as two separate worksheets for the purpose of comparative analysis. Second, this study used 'words and word pairs' option (one of the graph metrics of NodeXL) to work out the vertices and edges of the semantic networks. Finally, NodeXL was employed for the visualisation of the semantic networks.

This study chose four newspapers containing two conservative ones and two liberal ones. The sampling criteria of this study is mainly based on the number of paid circulation, which refers to the number of subscribers. Table 4 shows the number of the circulation number in 2017. For the conservative media, *Chosun Ilbo*, *DongA Ilbo* and *JoongAng Ilbo* were the major three newspapers in South Korea. Among these three, we chose *Chosun Ilbo* and *JoongAng Ilbo* due to their availability of the search engines. For the liberal media, *The Hankyoreh* and *The Kyunghyang Shinmun* were the two major liberal newspapers.

Table 4. Top ten daily newspapers based on circulation (in 2017) [54].

Ranking	Newspaper Names	Paid Circulation	Remarks
1	*Chosun Ilbo*	1,254,297	Nationwide Daily Newspaper
2	*DongA Ilbo*	729,414	Nationwide Daily Newspaper
3	*JoongAng Ilbo*	719,931	Nationwide Daily Newspaper
4	*The Maeil Business Newspaper*	550,536	Economic Newspaper
5	*The Korea Economic Daily*	352,999	Economic Newspaper
6	*The Nongmin Shinmun (published three times a week)*	287,884	Newspaper for Farmers
7	*The Hankyoreh*	202,484	Nationwide Daily Newspaper
8	*The Kyunghyang Shinmun*	165,133	Nationwide Daily Newspaper
9	*Munhwa Ilbo*	163,090	Nationwide Daily Newspaper
10	*The Hankook Ilbo*	159,859	Nationwide Daily Newspaper

The research period is from January to February 2018. This study set this period for two reasons. First, PyeongChang 2018 was held from 9–25 February 2018. Second, January 2018 was the preceding month of this event. In South Korea, there has been a number of debates emerged during this month. Figure 7 shows that there has been a sharp increase in terms of the numbers of related-news items published from January 2018. Even though the data collection period can be short, it can be beneficial to capture such heated debates on PyeongChang 2018.

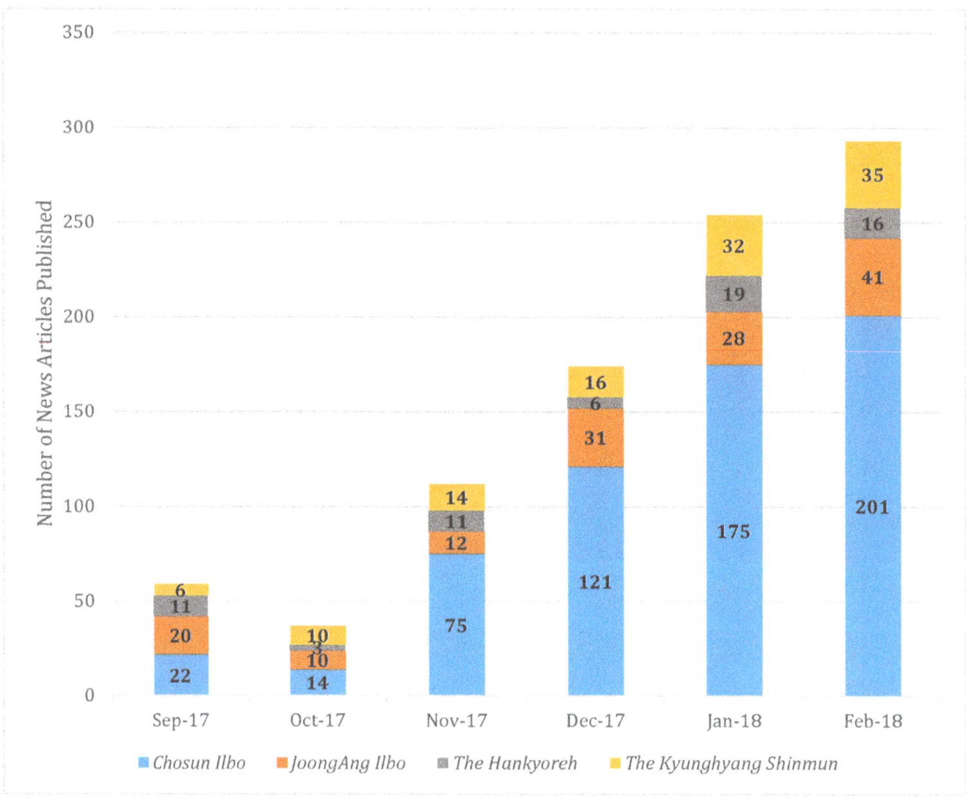

Figure 7. Growth in numbers of media coverage on PyeongChang 2018 from September 2017 to February 2018 (focusing on four selected newspapers).

5. Results

First of all, it is useful to look at the number of articles these two groups of newspapers published (Figure 4). In January, a total of 254 articles were released on PyeongChang 2018 (203 by conservative and 51 by liberal media). In February, the Olympic month, a total of 293 articles (15% more than the previous month) were published (conservative: 242, liberal: 51). Given the quantity of article published (Figure 8), we were able to find out three findings. First, the conservative newspapers' editorials seemed to pay more interests towards PyeongChang 2018 than the liberal ones. Second, the conservative newspapers' editorials seem to have temporarily allocated more resources for covering PyeongChang 2018. Korean newspapers generally have a flexible resource management system to flexibly cover any issues if necessary [55]. In other words, when the agents want to cover a certain issue more, the editorial board invests more resource in reporting PyeongChang 2018. Third, based on their huge circulation (Table 3), the conservative newspapers should have attempted to influence public opinion by approaching the subscribers.

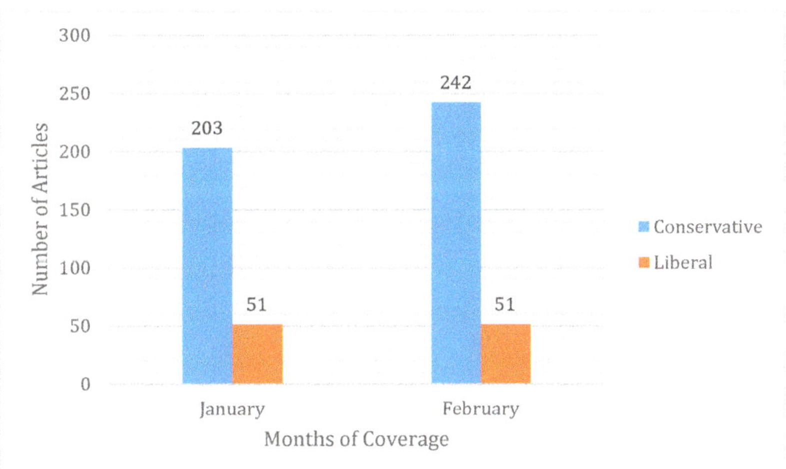

Figure 8. Composition of the articles over the period (January to February 2018).

To compare coverage frequency by period and by newspaper, a weekly publication dynamics was illustrated as in Figure 9. For the conservative newspapers, two peaks were found—the last week of January (24–31 January) and the second week of February (8–14 February). During the period of the first peak, the most news items were dedicated to covering the readiness of PyeongChang 2018 organisation committee, and South Korean athletes' expected results during the Olympics. During the second peak, the newspapers mainly covered the issues of forming a unified Korean female ice hockey team and North Korea's participation in the PyeongChang Olympics. For the liberal newspapers, the second week of February (8–14 February) was the only peak. North Korea's participation in PyeongChang 2018 was the central theme for the coverage during this period.

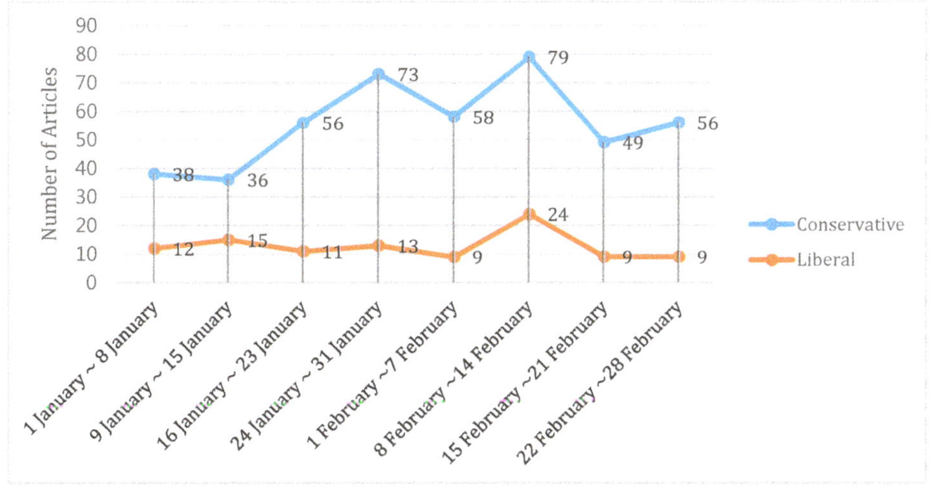

Figure 9. Weekly dynamics of articles published about PyeongChang 2018.

Although two groups of newspapers had the same peak, their responses were different. Looking at the articles' sentiment about PyeongChang 2018 (Figures 10 and 11), each group of newspapers has different evaluations. In Figure 10, conservative newspapers negatively evaluated the PyeongChang

Olympics. Such a negative sentiment usually becomes evident in the case of covering North Korea's participation in PyeongChang 2018. They viewed North Korea as a violator of peacebuilding of the Korean peninsula. Conversely, liberal newspapers positively evaluated the PyeongChang Olympics. They considered North Korea's participation can be a momentum for peacebuilding of the Korean peninsula. In Figure 11, both conservative and liberal newspapers reduced the articles with either positive or negative sentiments. Newspapers of these groups focused on covering PyeongChang 2018 as an event per se. They covered the achievements of South Korean team's achievements such as medal-winning in the skeleton, short track and speed skating and outstanding achievement in figure skating. From the results of Figures 10 and 11, three implications can be pointed out. First, both conservative and liberal newspapers are more likely to give their positive and negative evaluations regarding the PyeongChang 2018 in January. It means that these newspapers tried to clinch their argument. Second, the conservative newspapers generally employed more negative framings for reporting PyeongChang 2018, and the liberal newspapers did positive framings for reporting the Olympics. Third, the conservative newspapers are more likely to attempt to politicize PyeongChang 2018 in South Korea.

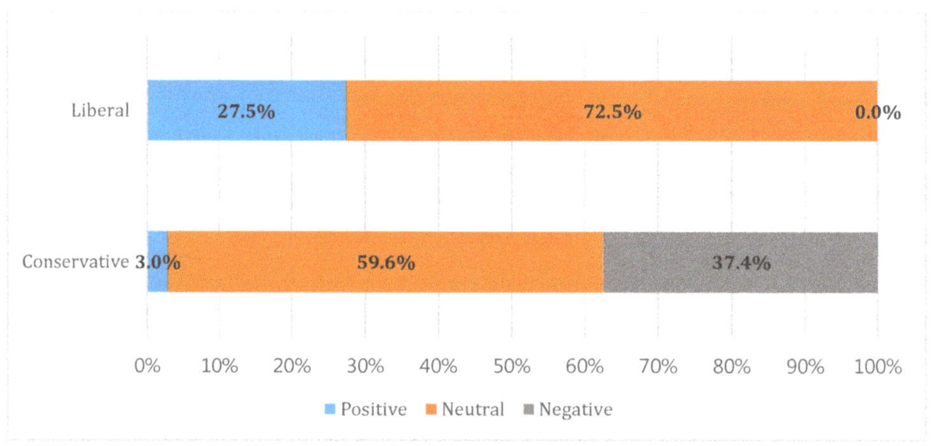

Figure 10. Journalists' evaluations regarding PyeongChang 2018 published in January.

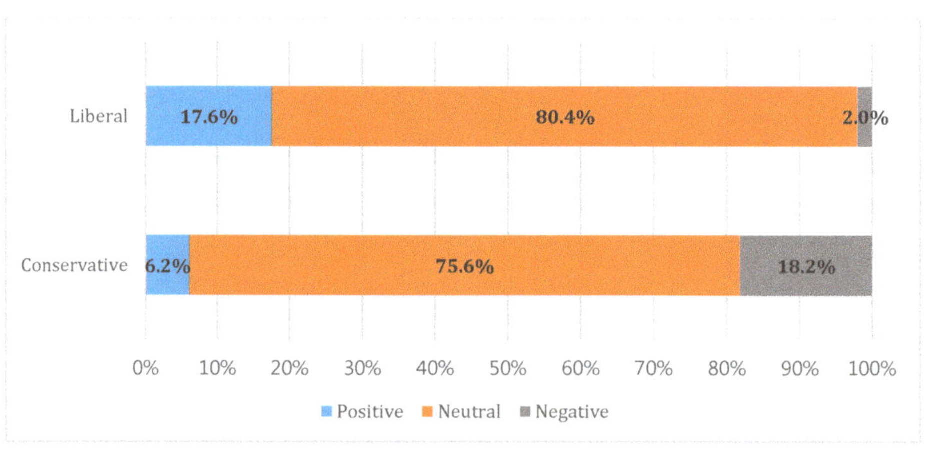

Figure 11. Journalists' evaluations regarding PyeongChang 2018 published in February.

Figure 12 shows the semantic network of conservative newspapers' media framings, and Figure 13 does that of liberal newspapers in January 2018. Table 4 shows how the betweenness centrality of significant vertices ('keywords') are different between the conservative and the liberal newspapers. The values in the table are significant because they identify the core concepts within their news texts. It means that, when a vertex records a high value in betweenness centrality, this vertex is highly connected to other vertices within the semantic networks. To extend journalist evaluations about PyeongChang 2018, this study employed automated semantic network analysis for presenting salient framings of conservative newspapers and liberal newspapers. With this regard, NodeXL enables this study to visualise the results.

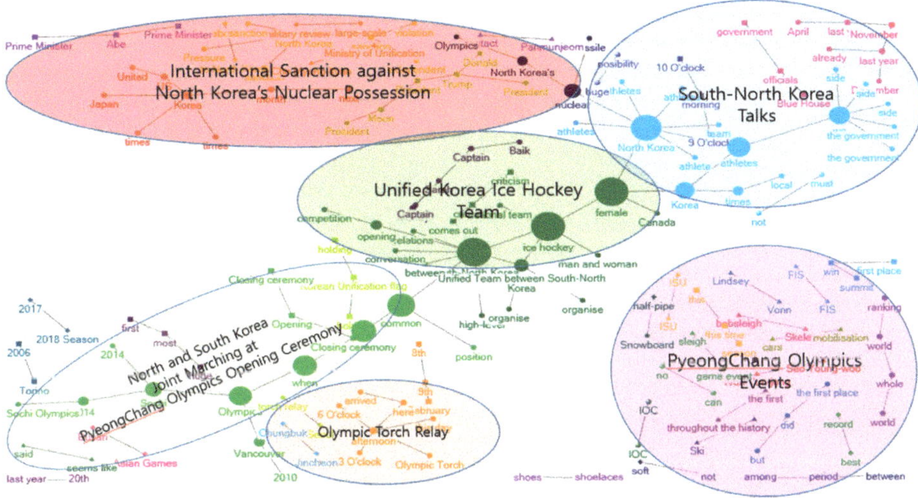

Figure 12. Semantic network of conservative media in January 2018.

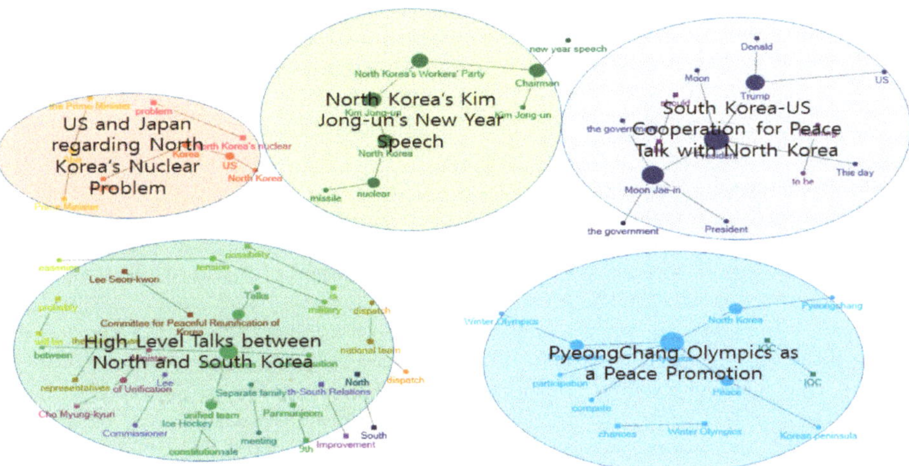

Figure 13. Semantic network of liberal media in January 2018.

In Figure 12, based on the size of vertices, there are four framings are worth further investigation—"international sanction against North Korea's nuclear possession," "South-North Korea talks", "Unified Korea ice hockey team", and "North and South Korea joint marching at PyeongChang

Olympics opening ceremony." For the first framing, "North Korea" and "nuclear" were two biggest vertices (also presented in Table 4). It means that the conservative newspapers stressed these words when they publish articles connected to this framing. Considering the published articles further, *Chosun Ilbo* is likely to criticise South Korean liberal government in line with global sanctions against North Korea. For instance, "critics warned South Korea might violate UN resolution regarding North Korea when the South Korean government supports the North Korean athletes and cheering groups participating in the 2018 PyeongChang Olympics" [56]. Another example is "when the South Korean government supports North Korean team's participation, it is highly possible to violate the international sanctions against North Korea [...]. The US government voice concern about the importance of keeping such international sanctions" [57]. For the second framing, "North Korea", "athletes", "(South) Korea" and "(South Korean) government" were emphasised in terms of the size of vertices (also in Table 4). The conservative newspapers tried to convey their putative North Korea's ulterior motives to their readers. For example, "we all know why North Korea suddenly wanted to participate in the PyeongChang Olympics. North Korea wanted to utilise this chance for creating a gap between the South Korean government, who wishes to North–South talks, and the US government, who wished to dismantle North Korea's nuclear weapons" [58]; "North Korea wanted to use North–South talks related with PyeongChang Olympics as one of instruments to achieve its two strategic objectives—'sticking to nuclear violence' and 'overcoming international sanctions'—which were adopted and proclaimed by North Korea's Workers' Party during their New Year's congress" [59]. In the third framing, "North Korea", "ice hockey", and "female" were emphasised in the texts of the conservative media (also in Table 4). Connected to this framing, "common", "closing ceremony", "Olympics" were stressed in the fourth framing (also in Table 4). The conservative newspapers warned a unified Korean ice hockey team and North–South Korea joint marching can mar South Korea's national identity and legitimacy. *Chosun Ilbo* pointed out Korean unification flag replaced *Taegeukgi* (the flag of South Korea) at the opening ceremony of PyeongChang 2018. For example, "North and South Koreas will establish a unified Korean ice hockey team and joint marching at the opening ceremony [...]. The South Korean government said 'we might hold Korean unification flag' at the ceremony [...]. This means we will not be able to use the Republic of Korea during the ceremony [...]. This is unacceptable" [60].

In Figure 13, four framings are worth further investigation—"North Korea's Kim Jong-un's News Year Speech", "South Korea-US cooperation for peace talk with North Korea", "high level talks between North and South Korea", and "PyeongChang Olympics as a peace promotion". In the first framing, the liberal newspapers attempt to check out possibility of North–South Korea relations improvement. "North Korea's Workers' Party", "North Korea", and "Kim Jong-un" were the key words within this framing (also in Table 4). From the articles in the liberal newspapers, these keywords were used to address the North Korea's willingness to talk with its international counterparts. In the related article by *The Hankyoreh*, "Chairman Kim Jong-un mentioned the chance of North Korea's participation to PyeongChang 2018 in his New Year speech [...]. Regarding the possibility of North–South talks, there were some mixed reactions to his speech [...]. The Minjoo Party of Korea, the ruling party, optimistically viewed his speech as a further peace movement [...]. The Liberty Korean party, the opposition party, interpreted his speech as North Korea's two-facedness [61]." *The Hankyoreh* assessed practicality of North Korea's sending a delegation over to PyeongChang 2018 [62]. In the second framing, "Moon Jae-in", and "Trump" were the keywords (in Table 4). The liberal newspapers addressed that South Korean and the US leaders showed the supportive attitudes towards North–South talks after PyeongChang Olympics. US President Trump mentioned that "(he) supports the North–South talks, over the phone [...] Within the US, some pointed out North Korea's willingness to talk could be its camouflage of peace promotion [...]. However, the US should establish a momentum for the talks with North Korea by utilising this situation wisely. The aim of Trump administration's sanctions against North Korea is its denuclearisation. In order to achieve such objective, they should sit at the negotiation table [63]." In the third framing, the liberal newspapers suggested the implication of North–South high level talks as an improvement of inter-Korean relations. For example, North–South high level talks was the opportunity to suggest an implication of their thawing relations [64]. In the fourth framing, "Olympics", "participation", and "peace" were main vertices. In this regard, the liberal newspapers underscored the South Korean government's effort of using PyeongChang 2018 as a peace promotion of the Korean peninsula. For instance, the South Korean government has emphasized the peace promotion of the Korean peninsula by North Korea's participation in PyeongChang 2018. [65] Kim Hyun, the spokesman of the Minjoo Party of Korea, mentioned that "Moon Jae-in administration's effort to turn PyeongChang 2018 into peace Olympics is complying with the Special Act on PyeongChang Olympics [   ]" [66]. Park Joo sun, the Second Deputy Speaker of the National Assembly, asserted that "any trials to make PyeongChang 2018 as a political battle should be stopped" [66].

Revisiting Figures 12 and 13 and Table 5, three contrasting points between the conservative and liberal newspapers were found. First, these newspapers showed contrasting interpretation around North Korea's participation in PyeongChang 2018. On the one hand, the conservative newspapers perceived North Korea's participation as an illegitimate issue. On the other hand, the liberal newspapers did the participation as an implication for further peace promotion. Second, these newspapers showed the contrasting expectation to North Korea. The conservative newspapers viewed North Korea's changes with scepticism. On the contrary, the liberal newspapers cautiously suggested the possible change of North Korea regarding its peace promoting efforts in the Korean peninsula. Finally, these newspapers had opposing understanding regard prerequisite to the peacebuilding of the Korean peninsula. The conservative newspapers perceived North Korea's disestablishment of nuclear weapons as an ultimate precondition. The liberal ones perceived talks with North Korea as an effective way of promoting peace in the Korean peninsula.

Table 5. Comparison between Top 20 Vertices by betweenness centrality in January 2018.

Conservative		Liberal	
Top Vertices	**Betweenness Centrality**	**Top Vertices**	**Betweenness Centrality**
North and South Korea	556	(South Korean) President	27
Ice Hockey	549	Olympics	25
Female	539	Moon Jae-in	21
North Korea	445	Trump	15
Unification	359	North and South Korea	13
Entrance	296	(North Korean) Workers' Party	12
Moment	266	Kim Jong-un	12
Athletes	249	Chairman	11
Competition	242	North Korea	10
We	210	Peace	7
Korea	177	Participation	7
Nuclear	128	North Korea's	7
Sochi	128	Nuclear	6
Unified Team	87	Unified Team	5
Held	44	High-level Talks	5
Vancouver	44	South Korea	2
2014	44	The US	2
North Korea	44	Tension	1
Perspective	44	(South Korean) Representatives	1
One	19	(South Korean) Ministry of Unification	1

In Figure 14, the framings related to North Korea were salient in the texts of the conservative newspapers. They mainly addressed the events of PyeongChang 2018—South Korean female curling team and unified Korean female ice hockey team. Table 5 indicates that "female curling team" bears the high value of betweenness centrality in the conservative newspapers. In Figure 11, we were able to find out liberal newspapers focused on covering the PyeongChang 2018 as a peace-promotion event. Table 5 shows that "ice hockey team" bears the high value of betweenness centrality in the liberal newspapers.

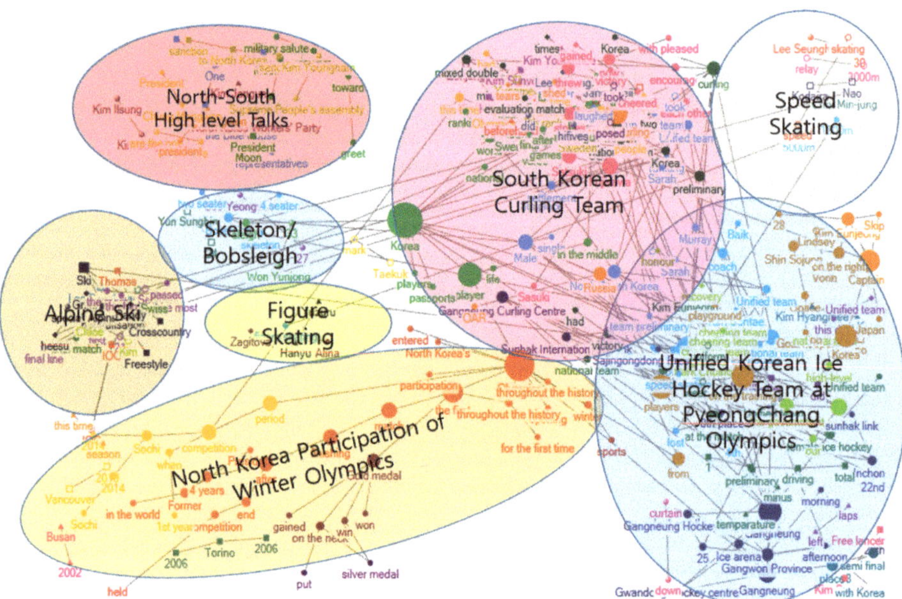

Figure 14. Semantic network of conservative media in February 2018.

From Figure 14 and Table 6, the conservative newspapers employed the words (vertices) related to the South Korean team's achievement such as "victory", "pleased", "laughed", and "cheered". In addition, they employed negative words such as "lost" and "minus" when they cover the issues about unified female ice hockey team (Figures 14 and 15 and Table 6). Other than these events, the conservative newspapers paid their attention to cover South Korean teams in other games such as alpine ski, figure skating, speed skating and skeleton/bobsleigh. They employed the wordings of their achievements such as "gold medal", and "silver medal" (Figure 14 and Table 6). In Figure 14 and Table 5, the conservative newspapers were likely to emphasise South Korean teams' achievements in PyeongChang 2018 rather than a unified Korean ice hockey team. From Figure 15 and Table 5, the liberal newspapers focused on covering unified female ice hockey team and North Korea's *Samjiyon* Orchestra's performance in South Korea during PyeongChang 2018. Considering other minor framings, the liberal newspapers covered the articles on South Korean team at the Olympics, Russia's participation in the name of OAR (Olympic Athletes from Russia), a novel related to PyeongChang 2018 (*Happy Days of Grump*, written by a Finnish writer Tuomas Kyro) and the investigations into Woo Byung-woo's (a senior presidential secretary during the ousted Park's government) dispatching a spy to Kim Jin-sun (the former chief of the PyeongChang Winter Olympics Organising Committee).

Table 6. Comparison between Top 20 words by betweenness centrality in February 2018.

Conservative		Liberal	
Top Vertices	**Betweenness Centrality**	**Top Vertices**	**Betweenness Centrality**
Korea	8888.6829	Held	29
Olympic Games	6428.7636	Saimdang Hall	25
Female	4203.9763	Gangneung Art Centre	24
Athletes	4074.0719	Gangwon Province	21
Of athletes	3749.5833	Female	17
During the training	3380	Afternoon	16
Female curling	3374.2258	Ice Hockey	10
Held	3351.7491	8th of February	9
South Korea's	3230	At Ice Arena	9
Claim	3165	An Eccentric Main	6
The first	2994	Former	6
At the competition	2118.7878	Pence (US Vice President)	6
Gangneung	2007.9263	North and South Korea	6
Players	1966.9673	Single Skating	6
Fujisawa	1628	Old man	6
Competitions	1620	Grump	4
Male	1616.4879	Whole	4
Match	1561.5119	Winter Olympics	2
Completed	1448	Short Track	2
South Korean National Team	1438.7903	Past	1

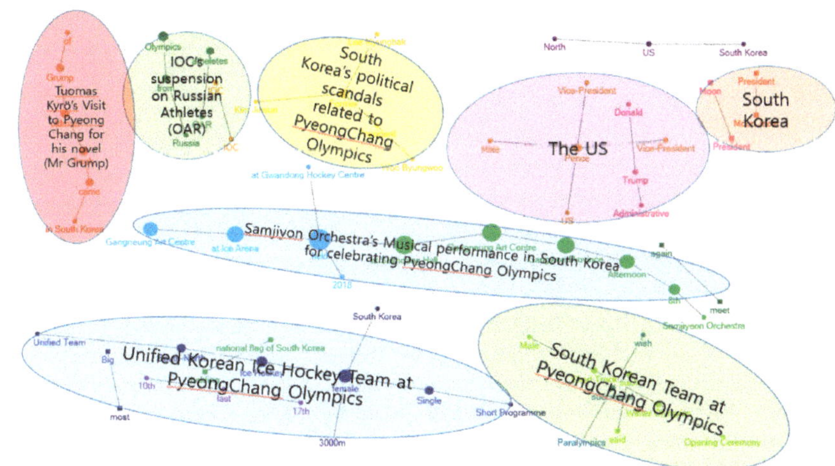

Figure 15. Semantic network of liberal media in February 2018.

Overviewing the salient framings formed by two different newspaper camps, we were able to find out three findings. First, the conservative newspapers evaluated the PyeongChang Olympics as a negative event while the liberal newspapers did oppositely. Second, the conservative newspapers featured more diverse framings about the South Korean team's achievements than the liberal ones. Third, the conservative newspapers were more resourceful than the liberal ones regarding particular emphasis on the framings. Arguably, based on these findings, the readers are more likely to accept the framings from conservative media. If the framings of the conservative newspapers influenced the South Korean public, they would evaluate the PyeongChang Olympics in a negative way (such as giving negative outlooks or assessments). To check this claim, some of Korea Gallup's weekly public opinion surveys on the public's outlooks and assessments about the PyeongChang Olympics would be helpful references. For the surveys, Korea Gallup hosted 1000 respondents each by conducting Random-Digit Dialing (RDD) surveys. According to the results of these surveys conducted before and after the PyeongChang Olympics, 73% of the respondents expected that the PyeongChang Olympics would be successfully held, and 84% of the respondents evaluated that the Olympics were successfully held [67]. After considering these results, we were able to find out that the conservative newspapers were not successful in influencing upon the public opinion about the Olympics. Korea Gallup also asked two questions before and after the event regarding North–South Korean joint marching and forming a North–South Korean unified female ice hockey team. However, before the Olympics, 53% of respondents perceived North–South Korean joint entrance was a valuable event for the Korean peninsula, 68% of the respondents evaluated this joint entrance optimistically after the Olympics. However, forming a North–South Korea unified female ice hockey team was evaluated somewhat negatively (50%) than positively (40%) before the Olympics. After the Olympics, however, this trend was overturned. Fifty percent of the respondents gave a positive evaluation, and 36% of the respondents gave a negative evaluation [67]. In the next section, we discuss more thoroughly on these contrasting results.

6. Discussion

This paper is now addressing our research questions and testing the hypotheses. For the first research question (Q1), different attitudes of conservative and liberal media from their evaluations towards the PyeongChang Olympics were found. As in Table 2, the current South Korean government has a compatible relationship with liberal media. Figures 10 and 11 support that the conservative media were inclined to show more unfavourable attitudes towards the government's performance of PyeongChang 2018. With this regard, the results prove that the first hypothesis (H1) is met. Echoing to the empirical studies on press partisanship, the conservative newspapers presented their resistance towards the current liberal government. By addressing the framing theory, each camp of newspapers had contrasting perspectives in framing their news agenda. Figure 12 shows that the conservative newspapers used the framing of "international sanctions against North Korea's nuclear possessions" in describing North Korea. *Chosun Ilbo*, in particular, showed more explicitly negative attitudes towards the South Korean government's invitation of North Korean officials and athletes to the PyeongChang Olympics. On the contrary, the liberal newspapers covered the PyeongChang Olympics positively. Figure 13 shows that the liberal newspapers used the framing of "Peace Promotion" to depict the Olympics.

For the second research question (Q2), this paper attempted to argue which type of newspapers were successfully delivering their messages to the Korean public. The second hypothesis was that conservative media would have been more successful in influencing the public opinion (H2). Comparing our data of media framing and Korea Gallup's public opinion surveys, the conservative newspapers failed to convey their messages to their audiences. The conservative newspapers covered more articles about PyeongChang Olympics, and their journalists employed more diverse framings and more effective words for their readers. In other words, the conservative media were more resourceful than the liberal media. However, the public opinion rejected the conservative media's salient issues and framings. Such resistance shows that our second hypothesis did not meet.

Our third research question (Q3) addressed how the public's attitudes were formed. According to the Korea Gallup's public opinion surveys on PyeongChang 2018, the public opinion was compatible with the liberal newspapers' attitudes towards PyeongChang 2018. Both the liberal newspapers and the South Korean public perceived PyeongChang 2018 as a peace promotion event, and they consider this event will contribute to peacebuilding process between North and South Koreas. This finding implies that the South Korean public would be more supportive to the current government. Considering the former President Park Geun-hye's impeachment, the general public has been disappointed with the conservative government. Korea Gallup conducted a public opinion survey on South Korean presidents' job performances. At the end of Park Geun-hye's term, 80% of the respondents evaluated her performance negatively. After her impeachment, 81% of the respondents evaluated positively about the following President Moon Jae-in's administration [67]. Since President Park's impeachment, the conservative parties seemed to have lost their public support and legitimacy in South Korean politics. In addition, the Korean public does not accept the information released by the conservative newspapers as much as they did before. Thus, our third hypothesis (H3) did not meet. Subsequently, it was expected that the conservative newspapers will continuously employ diverse ranges of framings and anti-government news coverage based on their resources for regaining their influence upon the public opinion.

In this study, the main point of argument was why the South Korean public perceptions are in the same line with the liberal newspapers' framings about PyeongChang 2018 (Q4). The conservative newspapers seem to be negligent in maintaining fair reporting attitudes. Rather, they attempt to bring their readers into the conflictual situations by building anti-North Korean sentiments. The media's core functions include monitoring the social environment, forming the public opinion and strengthening social norms [68]. In covering issues, media should maintain accurate, balanced and neutral reporting attitudes, so that they could establish a forum for public consensus [69]. The Korean public's support for the conservative government and media were significantly diminished. For the past few years, as the public has been disappointed with the wrongdoings of the conservative governments and the conservative media's overlooking attitudes to them. Our last hypothesis (H4) did not meet in this respect. The discordance between the conservative newspapers' framings and the South Korean public perceptions on PyeongChang 2018 imply that the readers proactively examine the genuineness of the news items before accepting the information from the news media.

7. Conclusions

This study emphasised that the public does not indiscriminately accept the messages from the media. In other words, the media cannot exercise their influence over their readers as much as they did before. This is not unrelated to the change in the media environment. Above all, the advent of the Internet brings us an emergence of convergence media, which transcends a simple merge of different types of media [70,71], and a new level media device that puts all types of media contents and broadcasting environment together [72] (p. 98). Such trend successfully led to the change of the general public. By using social media, a new communication network has been formed. It means that the public do not merely consume their messages and information from the media and the elites, but produce their discourses with other members of the society as part of a 'multitude' [73] (p. 52). The concept of the multitude, which came from Hardt and Negri's works, can be distinguished from 'masses', 'people', 'mob' and 'class' [74,75]. In other words, the multitude is not depicted as 'people' (which refers to someone challenges the sovereignty authority of the state) nor 'masses' (which refers to indiscriminate or uniform beings) but as a group of main agents who have diversity and distinctiveness. As they argued, the multitudes could be depicted as an image of the open and expandable network, which embraces the diverse ranges of expressions from the members of the society [73] (p. 54) [74,75].

Wellman and Morley explained the characteristics of the multitude by using the term of 'individualised mass' [76,77]. The multitudes can exist both as a group of independent individuals and a collective gathering [72] (p. 110). Although the multitudes do not collectively exist as the masses, they showed their strong interests to lead and improve the public opinion regarding social issues. Since the emergence of the individual media, their progressive tendencies are becoming more salient. In addition to the emergence of media convergence, the types of media services and recipients should be considered as well in order to understand the individualisation of the multitude's media-related activities [76,77].

Furthermore, interestingly, their 'individualised' behaviour in social communication enables the emergence of 'collective intelligence.' The collective intelligence is a concept specified by Pierre Levy [78]. He mentioned that the collective intelligence is universally distributed intelligence which can be occasionally improved, coordinated in real time, and mobilised effectively. He subsequently claimed that the current knowledge space could be formed due to the convergence of the Internet and the collective intelligence [78]. He summarised the collective intelligence in the following one sentence "no one knows everything, everyone knows something, all knowledge resides in humanity", and emphasised the starting point and final destination of the collective intelligence are 'mutual recognition' [78] (p. 20). In other words, "knowledge is simply the sum of what we know" [78] (p. 13). Furthermore, the collective intelligence is established in forms of a group of thoughts from the individual intelligence and collective knowledge, so it can be an appropriate way of solving the complicated problems of the modern society [78] (pp. 55–56). In addition, such collective intelligence can be realised in the form of diverse ranges of social media such as Facebook, Twitter, YouTube and of individual Internet-based broadcasters like a podcast.

Over the last nine years of the conservative power in South Korea, the collective intelligence has detected and substantiated not a few corruptions and scandals of the former conservative governments. Recently, this discovery was developed into the candlelight revolution and led to an unexpected and unprecedented regime change in South Korea. In light of Jenkin's account, the candlelight revolution can be interpreted as an impactful event which shows that the collective intelligence caused the change in the society [71]. In addition, this event can be interpreted as an evolution of communication structure, which led to social change [71]. As such, the conventional media should pay more attention to the sudden change of media environment and the evolution of the general public. Consequently, the South Korean media environment is facing the issue of maintaining its sustainability in terms of assuring the quality of being able to continue to exist as a reliable source of information. If conventional media neglect their roles supporting media sustainability, South Korean media ecology will face its further degeneration. Therefore, establishing a more sustainable media ecology will be an important prerequisite task for the South Korean society.

Author Contributions: S.-W.Y. conceptualised the theme of the paper and wrote the introduction, the literature review and conclusions of the paper. S.W.C. composed the theoretical framework, results and discussion section. He collected the data, designed the methodology of the article and undertook the analysis of using the software. Both authors contributed equally regarding preparation of the manuscript.

Funding: This research received no external funding.

Conflicts of Interest: The authors declare no conflict of interest.

References

1. Van Attelveldt, W. Introduction. In *Semantic Network Analysis: Techniques for Extracting, Representing and Querying Media Content*; BookSurge Publishing: North Charleston, SC, USA, 2008; ISBN 1439211361.
2. Kim, H.-T. A Study on the Strategy and SWOT Analysis of Hosting the 2018 Pyeongchang Winter Olympic Games. *J. Korea Entertain. Ind. Assoc.* **2011**, *5*, 14–24. (In Korean) [CrossRef]
3. Lee, S.-D.; Ji, W.-S.; Lee, S.; Park, S.-Y. Inter-Regional Development Projects for the Co-prosperity of Gyeonggi-Do and Gangwon-Do on the 2018 PyeongChang Winter Olympic Game, Gyeonggi Research Institute, Policy Research Series 2011-66. Available online: http://www.gri.re.kr/%EC%97%B0%EA%B5%AC%EB%B3%B4%EA%B3%A0%EC%84%9C/?brno=3838&prno=3242 (accessed on 21 October 2018). (In Korean)
4. Ryu, J.-H. 2018 PyeongChang Olympic Games and Regional Development Issues. *Korea Tour. Policy.* **2011**, *45*. Korea Culture & Tourism Institute. Available online: http://www.dbpia.co.kr/Journal/ArticleDetail/NODE01921445# (accessed on 21 October 2018). (In Korean)
5. Ahn, D. A Contemplation for the Success of 2018 PyeongChang Olympics. *Korea Tour. Policy* **2011**, *45*. Available online: http://www.dbpia.co.kr/Journal/ArticleDetail/NODE01921444 (accessed on 21 October 2018). (In Korean)
6. Cha, M. A Study on the Successful Hosting of the PyeongChang Winter Olympics through Regional Networking and Cooperation. *Plan. Policy* **2011**, *362*, 14–21. (In Korean)
7. Yeom, D. The Effect of Intention of the Local Residents for Cooperation for the 2018 Pyeongchang Winter Olympics on Regional Development and Sport Culture. *J. Korean Soc. Wellness* **2016**, *11*, 1–12. (In Korean) [CrossRef]
8. Kim, L.; Kim, N. Connecting Opinion, Belief and Value: Semantic Network Analysis of a UK Public Survey on Embryonic Stem Cell Research. *J. Sci. Commun.* **2015**, *14*. Available online: https://jcom.sissa.it/sites/default/files/documents/JCOM_1401_2015_A01.pdf (accessed on 19 October 2018).

9.   Kim, Y. Political Orientation of Korean Media and the Crisis in Social Communication. In *Communication Crisis in the Korean Society*; Korean Society for Journalism and Communication Studies, Ed.; Communication Books: Seoul, Korea, 2001; pp. 170–217. ISBN 9788964061862. (In Korean)

10. Kahn, K.; Kenney, P. The Slant of the News: How Editorial Endorsements Influence Campaign Coverage and Citizens' Views of Candidates. *Am. Political Sci. Rev.* **2002**, *96*, 381–394. [CrossRef]

11. Shoemaker, P.; Reese, S. *Mediating the Messages: Theories of Influences on Mass Media Content*, 2nd ed.; Longman Publication: New York, NY, USA, 1996; ISBN 0801312515.

12. Kang, J. Political Economy of Communication: A Study on Structural Barrier of Communication. In *Communication Crisis in the Korean Society*; Korean Society for Journalism and Communication Studies, Ed.; Communication Books: Seoul, Korea, 2011; pp. 65–89. ISBN 9788964061862. (In Korean)

13. Kang, M. Media War and the Crisis of Journalism Practices. *Korean J. Journal. Commun. Stud.* **2004**, *48*, 319–421. (In Korean)

14. Nam, S. Media Crisis and Professional Journalists' Responsibilities. *Kwanhun J.* **2009**, *113*, 89–109. (In Korean)

15. Patterson, T. *Out of Order*; Vintage Books: New York, NY, USA, 1994; ISBN 0679755101.

16. Kim, J. Editorial Tone of Major Korean Newspapers toward the Sunshine Policy during the Kim Dae Joong Government. *Korean Political Sci. Rev.* **2003**, *37*, 197–218. (In Korean)

17. Lee, W. Ideological Tendency and Assessment of the Government Policy through Reporting South-North Korea Issue: Comparative Analysis of Editorials under Kim Young-Sam and Kim Dae-Jung Administrations. *Korean J. Commun. Inf.* **2006**, *35*, 329–361. (In Korean)

18. Song, Y. A Study on the Korean Newspapers' Objectifying Strategies. *Korean J. Journal. Commun. Stud.* **2007**, *51*, 229–251. (In Korean)

19. Choi, H. A Study on the Diversity of Korean Newspapers: Analyzing the Tendencies of Covering Three Major Issues. *Korean J. Journal. Commun. Stud.* **2010**, *54*, 399–426. (In Korean)

20. Kim, K.; Noh, G. A Comparative Study of News Reporting about North Korea on Newspapers in South Korea. *Korean J. Journal. Commun. Stud.* **2011**, *55*, 361–387. (In Korean)

21. Koh, Y. An Analysis of News Reports about the Scandals of the Presidents' Relatives and In-laws': A News Frame Approach. *Commun. Theor.* **2007**, *3*, 156–195. (In Korean)

22. Kim, Y.; Lim, Y. An Analysis of News Reports about Government-Media Relationships and Media Policies: Comparison of News Contents under Noh-Lee Governments. *Korean J. Journal. Commun. Stud.* **2009**, *53*, 94–115. (In Korean)

23. Kim, I. A Study on the News Frame Analysis of 2008 Candlelight Protesting: Focusing on the Ideological Polarization of Main Newspapers. Ph.D. Thesis, Kyung Hee University, Seoul, Korea, 2010. (In Korean)

24. Kim, S.; Kim, C.; Kim, H. Discursive Politics of the Media and Economic Crisis: A Case Study about 'Korea's September Crisis in 2008'. *Korean J. Commun. Inf.* **2010**, *50*, 164–185. (In Korean)

25. Park, D.; Cho, Y.; Hong, J. A Qualitative Study of News Source-Reporter Relations: On the Problems of Beat Reporting System. *Korean J. Commun. Stud.* **2001**, *45*, 367–397. (In Korean)

26. Hong, J.; Kim, K. The Influence of the Ideological Tendency of the Press on the Theme and the Tone of the Press Related with New Media Policy. *J. Korea Contents Assoc.* **2017**, *17*, 162–177. (In Korean)

27. Entman, R. Framing: Toward Clarification of a Fractured Paradigm. *J. Commun.* **1993**, *43*, 51–58. [CrossRef]

28. Price, V.; Tewksbury, D. News Values and Public Opinion: A Theoretical Account of Media Priming and Framing. In *Progress in Communication Sciences*; Barnett, G., Boster, F., Eds.; Ablex: New York, NY, USA, 1997; Volume 13, pp. 173–212, ISBN 9781566402770.

29. Scheufele, D.; Tewksbury, D. Framing, Agenda Setting, and Priming: The Evolution of Three Media Effects Models. *J. Commun.* **2007**, *57*, 9–20. [CrossRef]

30. d'Angelo, P. News Framing as a Multiparadigmatic Research Program: A Response to Entman. *J. Commun.* **2002**, *52*, 870–888. [CrossRef]

31. Pan, Z.; Kosicki, G. Framing Analysis: An Approach to News Discourse. *Political Commun.* **2007**, *10*, 55–75. [CrossRef]

32. Fillmore, C.; Atkins, B. Toward a Frame-Based Lexicon: The Semantics of RISK and its Neighbors. In *Frames, Fields and Contrasts: New Essays in Semantic and Lexical Organization*; Lehrer, A., Kittay, E., Eds.; Lawrence Eribaum Associates: Hillsdale, NJ, USA, 1992; pp. 75–102, ISBN 9780805810899.

33. Nerlich, B.; Koteyko, N. Carbon Reduction Activism in the UK: Lexical Creativity and Lexical Framing in the Context of Climate Change. *Environ. Commun.* **2009**, *3*, 206–223. [CrossRef]

34. Degani, M. Values and Lexical Preferences in Obama's Speeches. In *Framing the Rhetoric of a Leader*; Palgrave Macmillan: London, UK, 2015; pp. 202–224, ISBN 9781349501014.
35. Newman, M.E.J.; Barabási, A.; Watts, D.J. *The Structure and Dynamics of Networks*; Princeton University Press: Princeton, NJ, USA, 2006; ISBN 9780691113579.
36. Barabási, A. Chapter 2 Graph Theory. In *Network Science*; Cambridge University Press: Cambridge, UK, 2016; ISBN 9781107076266.
37. Almgren, K.; Kim, M.; Lee, J. Extracting Knowledge from the Geometric Shape of Social Network Data Using Topological Data Analysis. *Entropy* **2017**, *19*, 360. [CrossRef]
38. De Saussure, F. *Course in General Linguistics*; Columbia University Press: New York, NY, USA, 2011; ISBN 9780231157261.
39. Sowa, J. *Principles of Semantic Networks*; Morgan Kaufmann: San Mateo, CA, USA, 1991; ISBN 1558600884.
40. Helbig, H. *Knowledge Representation and the Semantics of Natural Language*; Springer: Berlin, Germany, 2006; ISBN 9783540299660.
41. Drieger, P. Semantic Network Analysis as a Method for Visual Text Analytics. *Procedia-Soc. Behav. Sci.* **2013**, *79*, 4–17. [CrossRef]
42. Carley, K.; Pamquist, M. Extracting, Representing, and Analyzing Mental Models. *Soc. Forces* **1992**, *70*, 601–636. [CrossRef]
43. Doerfel, M. What constitutes semantic network analysis? A comparison of research and methodologies. *Connections* **1998**, *21*, 16–26.
44. Johnston, H. A Methodology for Frame Analysis: From Discourse to Cognitive Schemata. In *Social Movements and Culture*; Johnston, H., Klandermans, B., Eds.; UCL Press: London, UK, 1995; pp. 217–246, ISBN 0816625743.
45. Doerfel, M.; Barnett, G. A Semantic Network Analysis of the International Communication Association. *Hum. Commun. Res.* **1999**, *25*, 589–603. [CrossRef]
46. Jiang, K.; Barnett, G.; Taylor, L. Dynamics of Culture Frames in International News Coverage: A Semantic Network Analysis. *Int. J. Commun.* **2016**, *10*, 3710–3736.
47. Schultz, F.; Kleinnijenhuis, J.; Oegema, D.; Utz, S.; van Atteveldt, W. Strategic Framing in the BP Crisis: A Semantic Network Analysis of Associated Frames. *Public Relat. Rev.* **2012**, *38*, 97–107. [CrossRef]
48. Motta, G.; Baden, C. Evolutionary Factor Analysis of the Dynamics of Frames: Introducing a Method for Analyzing High-dimensional Semantic Data with Time-changing Structure. *Commun. Methods Meas.* **2013**, *7*, 48–82. [CrossRef]
49. David, C.; Legara, E.F.T.; Atun, J.M.L.; Monterola, C.P. News Frames of the Population Issue in the Philippines. *Int. J. Commun.* **2014**, *8*, 1247–1267.
50. Chosun Ilbo. Available online: www.chosun.com (accessed on 31 October 2018).
51. JoongAng Ilbo. Available online: http://joongang.joins.com (accessed on 31 October 2018).
52. The Hankyoreh. Available online: www.hani.co.kr (accessed on 31 October 2018).
53. The Kyunghyang Shinmun. Available online: www.khan.co.kr (accessed on 31 October 2018).
54. Modified from Korea Audit Bureau of Certification, The Number of Paid Circulation of Daily Newspapers in 2017. Available online: http://www.kabc.or.kr/about/notices/100000002623?param.page=¶m.category=¶m.keyword= (accessed on 26 September 2018). (In Korean)
55. Chung, S. External Images of the EU: Comparative Analysis of EU Representations in Three Major South Korean Newspapers and Their Internet Editions. Ph.D. Thesis, University of Canterbury, Christchurch, New Zealand, 2013.
56. Chosun.com. Are Choi Ryong-hae's Visit to PyeongChang and Supports to North Korean the Violation of Sanction against North Korea? 5 January. Available online: http://news.chosun.com/site/data/html_dir/2018/01/05/2018010500248.html (accessed on 27 August 2018).
57. Chosun.com. Violation against the Sanction in Case of Supporting North Korean Team's Stay, Ship or Airplanes to PyeongChang. 11 January. Available online: http://news.chosun.com/site/data/html_dir/2018/01/11/2018011100310.html (accessed on 27 August 2018).
58. Chosun.com. [Opinion] Can North-South Talks Reach the Disestablishment of North Korean Nuclear Weapons after PyeongChang Olympics. Available online: http://news.chosun.com/site/data/html_dir/2018/01/09/2018010903080.html (accessed on 1 October 2018).

59. Chosun.com. [Public Opinion] North-South Talks, We Should Welcome but should not be Moved. Available online: http://news.chosun.com/site/data/html_dir/2018/01/09/2018010903111.html (accessed on 1 October 2018).

60. Chosun.com. [Opinion] If there is no Taegukgi at the Opening Ceremony of PyeongChang Olympics. Available online: http://news.chosun.com/site/data/html_dir/2018/01/04/2018010403144.html (accessed on 27 August 2018).

61. Hani.co.kr. Around North Korea's New Year's Address, The Minjoo Party of Korea and People's Party "Welcomed" but the Liberty Korean Party Perceived it as "Mockery". Available online: http://www.hani.co.kr/arti/PRINT/825903.html (accessed on 1 October 2018).

62. Hani.co.kr. Joint March with Holding Korean Unification Flag . . . North and South open PyeongChang of Peace. Available online: http://www.hani.co.kr/arti/PRINT/827025.html (accessed on 1 October 2018).

63. Khan.co.kr. [Opinion] Trump, Utilise North-South Talks as an Entrance to North-US Talks. Available online: http://news.khan.co.kr/kh_news/khan_art_view.html?art_id=201801052117005 (accessed on 1 October 2018).

64. Hani.co.kr. Ri Son-gwon Began His Talk with Saying "the Present to Our People," Cho Myoung-gyon Responded with Saying "Well Begun is Half Done". Available online: http://www.hani.co.kr/arti/PRINT/827042.html (accessed on 1 October 2018).

65. Khan.co.kr. The Synergy Coming Out of North Korea's Participation in PyeongChang Olympics. Available online: http://sports.khan.co.kr/olympic/2018/pg_view.html?art_id=201801011837003&sec_id=530601 (accessed on 1 October 2018).

66. Hani.co.kr. Special Law on Supporting Unified Korean Team of PyeongChang Olympics, which was Supported by the Liberty Korean Party during MB Administration. Available online: http://www.hani.co.kr/arti/politics/assembly/828755.html (accessed on 1 October 2018).

67. Korea Gallup. Gallup Report. Available online: http://www.gallup.co.kr/gallupdb/report.asp?pagePos=4&selectyear=&search=&searchKeyword= (accessed on 12 September 2018).

68. Bang, J. The roles of News Reports for recovering the homogeneity after the unification. In *Korean Society for Journalism & Communication Studies*; Symposium & Seminar: Seoul, Korea, 1995; pp. 17–49. (In Korean)

69. Choi, J.; Ha, J. News Frames of Korean Unification Issues: Comparing Conservative and Progressive Newspapers. *Korean J. Commun. Stud.* **2016**, *24*, 121–145. (In Korean)

70. Castell, M. *The Rise of Network Society*; Blackwell Publishers: Oxford, UK, 2000; ISBN 0631221409.

71. Jenkins, H. *Convergence Culture: Where Old and New Media Collide*; New York University Press: New York, NY, USA, 2006; ISBN 9780814742815.

72. Lim, J. Candlelight and Media: The Mode of Convergence Media and the Emergence of Individualized Mass. *Korean J. Humanit. Soc. Sci.* **2011**, *35*, 97–122. (In Korean)

73. Song, T. The Multitude's Foreign Policy Debates and its Collective Behaviour through Social Media: The Impact of Changing Communication Environment on the Public's Foreign Policy Attitudes. *Korean J. Int. Stud.* **2013**, *53*, 41–87. (In Korean)

74. Hardt, M.; Negri, A. *Empire*; Harvard University Press: Cambridge, MA, USA, 2000; ISBN 0674006712.

75. Hardt, M.; Negri, A. *Multitude: War and Democracy in the Age of Empire*; Penguin Press: New York, NY, USA, 2004; ISBN 1594200246.

76. Wellman, B. Physical Place and Cyberspace: The Rise of Personalized Network. *Int. J. Urban Reg. Res.* **2001**, *25*, 227–252. [CrossRef]

77. Morley, D. *Media, Modernity, Technology: The Geography of the New*; Routledge: London, UK, 2007; ISBN 0415333423.

78. Levy, P. *Collective Intelligence: Mankind's Emerging World in Cyberspace*; Plenum Trade: New York, NY, USA, 1997; ISBN 0738202614.

 *sustainability*

Article

Skill Needs for Early Career Researchers—A Text Mining Approach

Monica Mihaela Maer-Matei [1,2], **Cristina Mocanu** [1], **Ana-Maria Zamfir** [1,*] **and Tiberiu Marian Georgescu** [2]

[1]   National Scientific Research Institute for Labour and Social Protection, 6-8 Povernei Street, 010643 Bucharest, Romania; matei.monicamihaela@gmail.com (M.M.M.-M.); mocanu@incsmps.ro (C.M.)

[2]   Department of Economic Informatics and Cybernetics, The Bucharest University of Economic Studies, 6 Piata Romana, 010552 Bucharest, Romania; tiberiugeorgescu@ase.ro

*   Correspondence: anazamfir2002@yahoo.com; Tel.: +40-21-3124069

Received: 9 April 2019; Accepted: 13 May 2019; Published: 15 May 2019

Abstract: Research and development activities are one of the main drivers for progress, economic growth and wellbeing in many societies. This article proposes a text mining approach applied to a large amount of data extracted from job vacancies advertisements, aiming to shed light on the main skills and demands that characterize first stage research positions in Europe. Results show that data handling and processing skills are essential for early career researchers, irrespective of their research field. Also, as many analyzed first stage research positions are connected to universities, they include teaching activities to a great extent. Management of time, risks, projects, and resources plays an important part in the job requirements included in the analyzed advertisements. Such information is relevant not only for early career researchers who perform job selection taking into account the match of possessed skills with the required ones, but also for educational institutions that are responsible for skills development of the future R&D professionals.

Keywords: skills; researchers; early career; text mining

---

1. Introduction

Research and development activities are one of the main drivers for progress, economic growth and wellbeing in many societies. Even if the investments made in research have been questioned by some stakeholders, technological developments are needed for more science advancements in order to push forward sustainable economic development, especially in emerging economies aiming to catch up with developed countries [1]. In the European Union (EU), the main policy instrument in this field, the European Research Area (ERA), promotes several priorities such as effective national research systems based on investments and national competition, transnational cooperation and competition, open labour market for researchers, gender equality, and optimal circulation of knowledge. Thus, current policies promote a Europe based on the freedom of movement of people and knowledge. In order to achieve this goal in R&D, a number of initiatives have been promoted such as the European Charter for Researchers and Code of Conduct for their Recruitment. The key purpose is to support researchers' movement across borders, sectors and disciplines. Such ambitious goals are supported by the EURAXESS platform which is a pan-European initiative that provides information on job opportunities and supports research careers. This initiative is endorsed by the European Union, member states and associated countries. The number of job vacancies advertised on the EURAXESS platform witnessed an increase of 7.8% in 2012–2014, followed by a decline of 5% in 2015–2016 [2]. However, the share of researchers expressing their satisfaction with the level of openness, transparency and merit-basis of the recruitment processes increased by 7.5% in the 2015–2016 period [2]. Concluding, the EURAXESS platform plays a key role in supporting an open labour market for researchers. The information provided via EURAXESS takes the form of job vacancies and funding opportunities advertisements, aiming to allow for the match between the supply and demand for researchers. The development of researchers' careers and their movement across borders, sectors and disciplines are influenced by the attractiveness and quality of information that are available for them. The main goal of this article is to explore and reveal key dimensions that characterize the entry level research labour market for selected scientific fields.

According to the European Framework for Research Careers, four professional categories exist among researchers, irrespective of their working context (universities, research institutions, NGOs, companies): (1) First Stage Researcher, (2) Recognized Researcher, (3) Established Researcher and (4) Leading Researcher. This study is focused on the entry level research labour market, namely on positions for first stage researchers. Usually, first stage researchers are PhD candidates who carry out research activities under supervision, have a good knowledge of their field of study, are able to collect data under supervision, as well as to analyze and assess complex ideas and to present their research outcomes.

Previous studies on career management of R&D professionals have showed that individuals respond and take decisions based on the structure of available opportunities [3], meaning that they consider the perceived rewarded activities when they develop their career orientation and strategies in order to reach expected career outcomes [4,5]. On the other hand, following the idea that career decisions and challenges vary significantly by career stage [6], we focus our analysis on first stage research positions. Previous studies on R&D professionals found that, during the exploration stage, career goals include the understanding of personal abilities and interests, evaluation of job requirements, and integration within research teams. Entry-level researchers face the need to develop their professional identity, to contribute with their knowledge and competences within the organization and team, as well as to cope with challenging tasks [7,8].

As opposed to other sectors, career systems in R&D have been extensively influenced by issues related to the level of professional competences of the researchers and other relevant skills such as team work and problem solving and less by the traditional advancement in the organizational hierarchy [9–11]. Thus, attracting and retaining researchers with the right mix of knowledge and competences became a key factor for more and more organizations [12] as individuals make job selections that are consistent with their personal orientations and profile [13,14]. The volume of researchers represents an important input for the innovation processes [15–17]. Many scholars consider that it is important to better understand the reactions of researchers to various career opportunities [4,18–20]. Career choices are made on the basis of career orientation, which represents a mix of self-perceived preferences, talents, needs and values [18,21,22]. Five types of career orientation have been identified among R&D professionals: technical orientation, manager orientation, project orientation, technical transfer orientation and entrepreneur orientation [20]. However, different career orientations share many common competences, values and professional roles [23]. One common challenge is that R&D professionals face rapidly changing demands determined by new technologies developments [24].

The way individuals respond to various job opportunities is explained by the person–organization fit theory which refers to the way the profile and skills of workers match with the needs, practices and expectations of the organizations [25,26]. From this point of view, information on job opportunities that are provided by organizations to first stage researchers shape the way individuals make job selections in the R&D sector. While many studies analyze career orientation and choices of R&D professionals by exploring data collected from researchers [18,21,22], this article is focused on information coming from organizations in the form of job vacancies advertisements. Such information is relevant not only for early career researchers who perform job selection taking into account the match of possessed skills with the required ones, but also for educational institutions that are responsible for skills development of the future R&D professionals. Various innovative approaches have been developed in order to better inform education and training institutions with respect to the nature and level of skills required from their graduates [27,28]. This article proposes a text mining approach applied to a large amount of data extracted from job vacancies advertisements, aiming to shed light on the main skills and demands that characterize first stage research positions in Europe.

2. Skills for RDI Sector: Some Hints from the Literature

Although the purpose of identifying skills relevant to research and innovation might seem appealing for decision makers in the area of education and training, and although there are several research endeavors aiming to provide some hints, finding the links between skills and RDI and understanding the policy relevance of the results are not easy tasks.

One of the most well-known measures of skill needs is the required level of education. As the share of higher education graduates as well as the share of doctoral and postdoctoral graduates increased in the population, the minimum level of education required for entry level positions in RDI and universities increased to doctorate level. PhD holders are among the most mobile populations, the international mobility often starting from the training period/program [29,30], so a better knowledge on required skills could improve the PhD holders' mobility, as well as the knowledge flows among European countries. Current RDI strategies aim to support the increase of PhD holders' numbers for a specific theme/research sector, being considered that usually a PhD veils some mix of skills that supports research and innovation [29,31].

Apart from the apparent consensus on the minimum required level of education in RDI, findings from the scientific literature are very heterogeneous, as a lot of skills and personal characteristics were under scrutiny and proved to influence research and innovation ideas and outputs [29,32]. The RDI sector is a very heterogeneous one, and studies carried out in the field used different typologies and focused on the role of different skills, not to mention the different conceptual approaches of skills and innovation used. Although the meaning of skills varies a lot through the literature, we use for this paper a broad sense of the concept, covering abilities, competences, knowledge, as well as personal attributes [29,33,34].

Studies addressing skills for the RDI sector are rarely comparable across industry [32], addressing mainly the corporate side of the sector and usually finding a mix of skills supporting research and/or innovation. The mix of required skills covers basic skills, technical skills, academic or methodologic skills, soft skills, etc. [29,32].

The mix of skills needed in RDI varies along sectors (business, university, NGOs), according to industry structure and competitiveness, type of RDI (fundamental, empirical, etc.) or type of innovation. Higher sectoral skills lead to higher sectoral productivity [35], as well as to higher investments in R&D [36], so the sector's characteristics, its structure and competitiveness, could influence the required mix of skills. Methodological limits in introducing sectoral and specific skills in comparative surveys also limit the possibility to identify specific skills supporting research and innovation and urge for more in-depth studies at the level of sub-sectors and occupations. Skill needs in RDI usually imply both theoretical and practical skills [32].

Big innovations and outputs are more likely to be produced by highly specialized companies [37], so technical and methodological skills remain at the core of job requirements in RDI, while communication, teamwork, sharing, etc. increase their importance.

Leadership, management and entrepreneurial skills are also addressed by the scientific literature, but are treated on a rather separate track. Management and entrepreneurial skills can be considered as transversal skills along the entire RDI sector, increasing self-regulation and adaptability, irrespective of sector specificities, but also fostering and mentoring the organizational space where innovation might appear. Entrepreneurial skills foster spill overs and contribute to increasing R&D returns [38]. Managerial and leadership skills are crucial not only for better positioning the company/organization on the market, but also to develop cooperation with other stakeholders and competitors in the field [39].

Globalization, ICT and the increasing importance of green skills are among the drivers of change for the future skill demands of RDI sector [28]. Globalization and ICT are changing the way economies work, increasing competitiveness and urging for collaboration. Soft skills such as communication,

communication in foreign languages, teamwork, working in multicultural teams and organizations, and working in multidisciplinary/interdisciplinary teams might become more and more important. Apart for the increasing importance of the so-called soft skills, globalization in RDI also leads to increasing levels of specific and technical skill needs. Large amounts of data available due to internet development call for new methods and skills to collect, organize and analyze them. Globalization also urges for skills that can support comparable studies, both quantitative and qualitative. Mass education is less probable to provide such a high level of skills, so self-learning and learning to learn are among the skills underpinning skills development in RDI.

Sustainability-oriented innovation changes the way economies operate, new green skills, green occupations and even green sectors emerging. Also, more responsible and ethical attitudes towards environment, culture, and communities becomes mandatory in RDI, although their future impact on the RDI is hard to be estimated [29,40].

The current findings point to a broader set of skills needed in RDI, with different sub-sectors needing different mixes of skills in different contexts and for different purposes [29,32]. This is why pulling out a core set of skills to substantiate teaching and learning policies in the field might be a tricky task, asking in fact for stakeholders' involvement in curriculum design, as well as for more specific studies undertaken for different sectors and occupations in order to provide more detailed findings.

The mix of skills needed in RDI is also in line with the mix of technical, communication, IT, ethical, legal and data science skills needed to support the objective of promoting and developing Open Science [41].

3. Data Gathering Process

Text mining represents a solution to the research challenge induced by new data sources such as text data posted on the web. This has led to an increase in the amount of data that can be extracted and analyzed in different domains. For example, the content analysis of job advertisements is one research topic based on voluminous textual data providing findings about training needs or technical skills required for specific jobs. This information could be useful for academic institutions in updating their curricula or for individuals in their career planning or for job analysts. The studies developed for identifying the skills valued by employers using online job vacancies focused on information regarding the activities associated with the job and on the attributes required from applicants. Some of the researches focused on specific jobs: big data jobs [42], IT jobs [43], information systems [44], and librarians [45], meanwhile other studies analyzed the communalities encountered for different professions [46,47]. The studies highlighted that constantly a combination of technical and soft skills is required. Previous researches on data analytics positions in the business sector found a common set of soft and transferable skills such as decision making, organization, communication and structure data management [48]. The findings were rather limited with respect to a similar set of technical skills, only statistics and programing skills being mentioned as common to the scrutinized jobs [48].

Big volumes of data were required as a source for quantitative research. In the planning stage of the study presented in this article, data were collected manually via a web browser. When the authors comprehended the potential of the research, they decided to develop tailor-made solutions to collect data automatically.

The source of data was EURAXESS web platform. EURAXESS is a pan-European initiative backed by the European Union which provides valuable information and resources to researchers. For this article, the authors were interested in collecting information about jobs offers in the research field [49]. Data collected is from public pages and it is used for research purposes only.

The authors chose five domains to analyze: Computer Science, Economics, Engineering, Environmental Science and Mathematics. Data about research job offers were gathered for the 1 May to 27 October 2018 timeframe. Data collected included the research field, researcher profile, date, description and requirements. Many of the job listings include more than one research field.

If at least one of them was among the six mentioned above, the data about the page were gathered. The authors decided to keep only the job listings that were accessible to early career researchers. Therefore, the specifications of the researcher profile had to include First Stage Researcher (R1), which according to EURAXESS, includes "individuals doing research under supervision in industry, research institutes or universities", including PhD candidates, but not PhD holders [49]. Some job listings were dedicated to First Stage Researchers, others were open for more experienced researchers as well. Date, description and requirements were gathered for all the pages that respected the criteria described above.

The solution was developed in Python 3 programming language, using Scrapy—An open-source framework [50]. Previous studies [51,52] have used similar technologies to collect big volumes of data automatically in the absence of an API (Application Programming Interface). Article [53] describes in detail the process of data acquisitions, difficulties encountered and solutions to solve them.

Although the solution is able to extract data much faster, a download rate of 36 pages/minute was set. The authors were very careful not to affect EURAXESS server performances. The solution is designed to collect from every scraped page only the relevant data. That is possible by finding the CSS selectors which contained the information required; 48,054 pages were automatically scraped. Out of all pages, 1571 were found for the Computer Science field, 1041 for Economics, 3004 for Engineering, 265 for Environmental Sciences, and 451 for Mathematics. Data were stored in JSON format and further processing was required to clean it before using it as an input in data analysis instruments.

4. Methodology

Consequently, the input in this investigation is represented by a considerable amount of textual data. To be more precise, a collection of documents representing descriptions of research job positions constitutes the source of information in this study. Usually, in the text mining literature, a collection of documents is labelled corpus. In order to extract information from it, as we know from big data theory, an interdisciplinary approach is recommended. Therefore, tools of informatics, programming, statistics, data analysis as well as the domain experts to evaluate and validate the outputs, are required.

As mentioned in the previous section, the findings of this study are based on the investigation of five different corpuses. Text mining was performed with tm library in R [54–56] and mainly used for text summarization. Other concepts employed in the text mining literature, besides the corpus, are document, token and lexicon or dictionary. In our investigation, the document is a research job offer; the tokens are the fundamentals units of analysis, represented by individual words. The lexicon or the dictionary consists of all unique words in a specific corpus, meanwhile the corpus size indicates the total number of words used [57]. Large texts are analyzed by computational methods based on statistical concepts. In order to use such methods, data transformation is required. This stage involves building structured representations similar to those employed in classical data mining such as matrices. The final results will consist of a matrix known as a document-term matrix whose elements are numerical, representing word frequency. The rows are the documents and the columns are the tokens [55,58]. At first, before completing data cleaning, this matrix is very large and extremely sparse. Table 1 emphasizes the vocabulary size and the corpus size before undertaking this pre-processing action. We have to mention that these numbers were computed after conducting some preliminary cleaning operations such as: eliminating extra white spaces, removing conjunctions and prepositions (stop words), removing punctuation, and converting to lower case.

There are universal regularities characterizing word frequency distribution, of which the best known is the theory of the minimum effort (Zipfs' law) [59,60]. This theory is used in the literature to compare and asses the quality of the informational content of a text. Zipfs' law states that the frequency of occurrence of a word is approximately an inverse power law function of its rank. The parameters appearing in this law will characterize the diversity of the vocabulary. In order to understand if the five corpuses investigated in this paper depict this universal law and moreover to see if there are significant differences between them in terms of vocabulary richness, we summarized the frequency distribution through a frequency spectrum. This involves computing Vm, representing the number of words occurring m times in the corpus. We have plotted the frequency spectrum for the first 50 elements [60,61]. The main conclusion we can draw from this representation is that the corpuses analyzed in this paper follows a typical frequency pattern suggested by the plot in Figure 1.

Table 1. Dimension of the corpus.

Field	Before Data Cleaning		After Data Cleaning	
	Vocabulary Size	**Corpus Size**	**Vocabulary Size**	**Corpus Size**
Engineering	21570	363158	608	108634
Economics	9280	126301	625	45295
Computer Science	16306	265725	851	84362
Environmental sciences	6939	41322	797	14199
Mathematics	7199	53841	587	17260

Source: authors' computation.

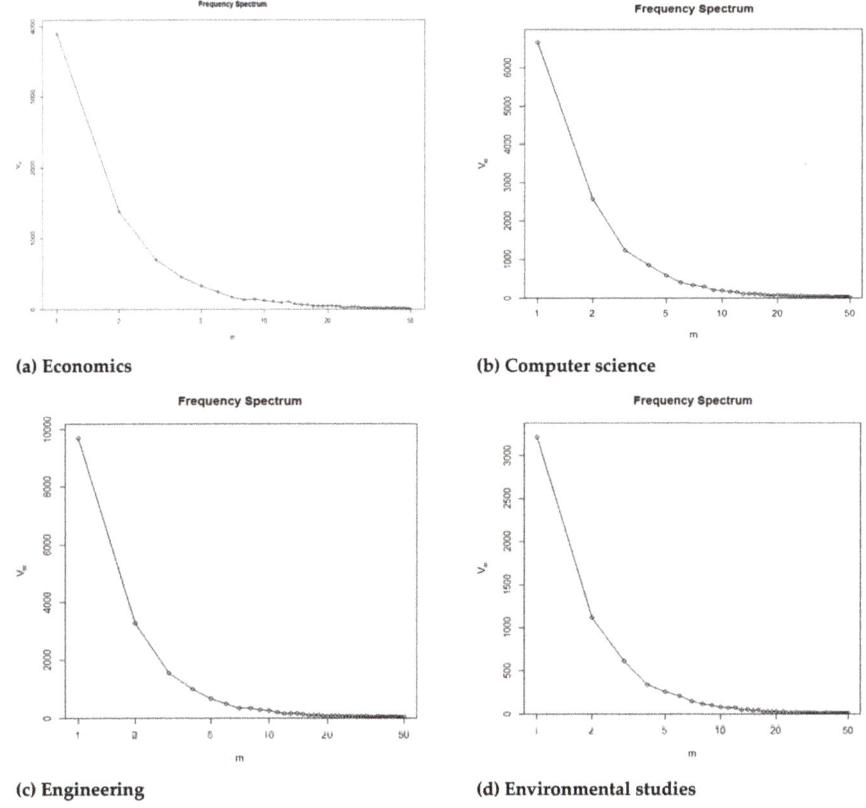

(a) Economics

(b) Computer science

(c) Engineering

(d) Environmental studies

Figure 1. Cont.

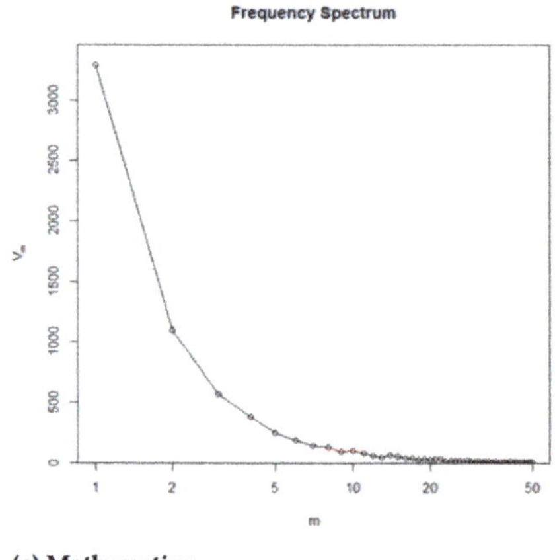

(e) Mathematics

Figure 1. Frequency spectrum.

Generally, in text mining, the data transformation process which leads to a frequency matrix is followed by data cleaning. Typically, this stage involves pre-processing the corpus through the following operations: eliminating extra white spaces, removing conjunctions and prepositions (stop words), removing punctuation, converting to lower case, and application of a stemming algorithm. As we mentioned before, we made use of all except the last one. We decided not to use a stemming algorithm, which removes word' suffixes with the purpose of dimension reduction. As we exemplified in the next section, retrieving the radicals of some words can lead to the loss of relevant information. We also defined and eliminated non-relevant words such as: applicant, required, email, and position, which are common in the textual data coming from job advertisements, with no informational value in the context of our investigation.

In order to reduce the size of the document-term matrix, we used two different procedures. The first one excludes the sparse terms and the second one eliminates the words appearing in almost all the documents.

The maximal allowed sparsity was set to 0.98. This means that those columns associated to very infrequent words were dropped. The sparsity was computed for each term by the formula:

$$sparsity_i = 1 - \frac{n_i}{N} \tag{1}$$

where n_i represents the number of occurrences for term i and N is the total number of documents in the corpus. In our analysis were kept only those words with a sparse factor of less than the threshold of 0.98. We consider that rare terms do not contribute to our findings given that we are interested in finding which are the most required skills.

On the opposite side, the document-term matrix contains words occurring in almost all the documents. Such words are not necessarily related to our topic, they are rather common words specific to the job posts. In this case, the elimination is made switching from a weighting system based on term frequency to a scheme known as inverse document frequency emphasizing the words with higher discriminative power [57,58,62].

The inverse document frequency for a specific term is computed by the formula:

$$idf_i = log_2\left(\frac{N}{d_i}\right) \tag{2}$$

where N represents the size of the corpus and d_i represents the number of documents where the term i appears [62].

Generally, in text mining, the statistical measure used to evaluate the importance of a certain term is given by the tf-idf which stands for term frequency-inverse document frequency. The importance increases proportionally to the number of occurrences in the document but is counterbalanced by the frequency of the word in the entire collection of documents. This implies normalization of a term frequency using the document length measured by the total number of words in the document (tf_i). Therefore, the tf-idf is computed by [57,58]:

$$tf\_idf_i = tf_i * idf_i \tag{3}$$

We have computed this statistic for each term and we discarded all the terms obtaining a value smaller than the first quantile. Following these procedures, we have significantly reduced the size of the term document matrix. The dimensions of each corpus and the summary statistics for the tf-idf values are given in Table 2.

Table 2. Term frequency—descriptive statistics.

Corpus	Length (No. of Documents)	Tf-Idf Statistics			
		1st Qu.	Median	Mean	3rd Qu.
Engineering	3004	0.052	0.066	0.069	0.082
Economics	1043	0.043	0.059	0.0616	0.074
Computer science	1571	0.034	0.042	0.044	0.052
Environmental sciences	265	0.033	0.043	0.05	0.06
Mathematics	451	0.045	0.056	0.064	0.074

Source: authors' computation.

The implications of these operations on the dimension of each corpus are revealed in Table 2 as well as in Appendix A. As illustrated in Table 1, the vocabulary size sharply decreased, and also the variance of this measure across the research fields significantly declined. Among all research fields, computer science depicts the largest vocabulary. As we will find in the next section, this could be explained by the diversity of the domains where computational methods are required. The representation included in Appendix A plots the top ten most frequent terms before and after undertaking the text cleaning, revealing that the final matrices do not include terms without informational value.

These collections of documents were used to identify the skills and knowledge required in the research sector for five different fields.

In the next section, the findings are represented via word clouds, a visual instrument highlighting the most frequently used terms in the advertisements of the vacancies or in the calls for applications. The frequency of a certain word is computed by the sum of the column it represents in the document-term matrices obtained after the cleaning and transformation process. The word clouds we inserted in the paper use the top 100 most frequent words [63,64].

For a better understanding of the word clouds, we have also extracted and represented the associations encountered for different terms. In essence, the correlations among those words are computed indicating the share of co-occurrences. This tool allows us to draw the context in which those terms are used.

5. Results

The main findings are extracted from the word cloud representation associated to each research field. The most obvious conclusion that can be drawn at first glance from the word frequency visualization is related to the interdisciplinary dimension of the research activity. This facet may also be a consequence of the fact that many of the job posts include more than one research field. Further research should deal with this issue using classification methods.

Without exception, the aspects related to "data" are very often specified in job descriptions and/or requirements. In order to understand the context, we analyzed the associations of this word and we found that it co-occurs with "protection", "statistics", "science", and "processing". In the Figure 2 inserted below are represented only the correlations exceeding a threshold of 0.5. For computer science for example, the highest correlation is of 0.35. In this case, it co-occurs with terms such as "analytics", "analysis", "processing", "machine" or "model". We can conclude that at least in mathematics, engineering and computer science, data mining or data processing skills are frequently required.

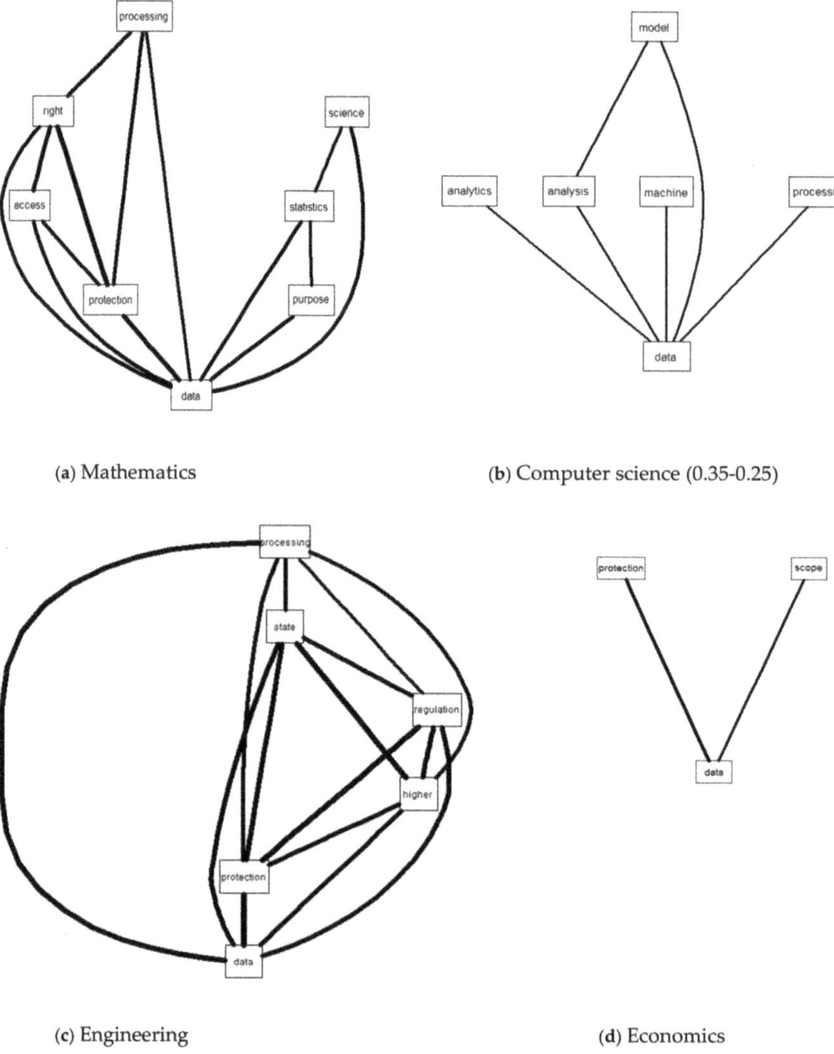

(a) Mathematics

(b) Computer science (0.35-0.25)

(c) Engineering

(d) Economics

Figure 2. Visualization of the correlations for term "data". Source: authors' computation.

Besides the term "data", another term common to all fields, appearing with a high frequency, is "model". In computer science, the highest correlations are found for "simulation" (0.39) and "scientist", pointing again towards data science skills. For economics, the most significant association was found with "energy" (0.39) and "analyses" (0.37). For the documents extracted from the engineering field, the first association (0.34) was found for "numerical", showing that the advertisements of the vacancies including the term "model" also contain "numerical". The corpus obtained for environmental sciences revealed many terms co-occurring with the term "model". We consider that the relevant ones could be "simulation" (0.67), "dynamics" (0.67) or "surveys" (0.64). For the mathematics field, the words that could explain the context in which it occurs are "simulation" (0.29) and "optimization" (0.24). Hence, the term "model" indicates requirements related to analytical skills. This conclusion is also supported by the frequency of the term "analysis" which is easily detectable in all five figures.

Additionally, the word "university" plays an important role in the documents we have analyzed, and this is due to the fact that most of the posts are coming from universities and implicitly the name of the university is mentioned in the job description. Regularly, a position in a university also implies teaching activity, which is reflected by our word cloud through terms such as "courses", "teaching", "assistant", and "professor".

The term "management" is common to all five-word clouds but it is difficult to summarize its correlations due to the fact it is related to a wide variety of aspects. For instance, no matter the field, it is associated with the following terms: "time", "risk", "financial", "supply chain", "organizational", "project", "industrial", "quality", "resources", "financial", "team", and "strategic". So, data processing and handling, teaching and management skills could be considered core competences for R&D professionals that transcend all the analyzed fields.

Specific skills required within the five analyzed fields are presented in the following paragraphs. For the vacancies associated with engineering field (Figure 3), words such as "physics", "energy", "materials", "mechanical", "electrical", "mathematics", and "electronics" are keywords for the technical knowledge required. On the other hand, the presence of the words "communication" or "language" unveil a different facet of the research activity which requires good oral and written communication skills. The term "language" is mostly associated with "English" (0.43) and "foreign" (0.4), emphasizing that "English" is used as a scientific and research language. The research topics within the project calls could be very heterogeneous and it is indicated by terms such as "medical", "sustainability" or "environmental". Another dimension that could be extracted from the representation is about IT-related competences. For instance, the term "software" acquired a significant frequency being correlated with "programming", "testing" or "computer".

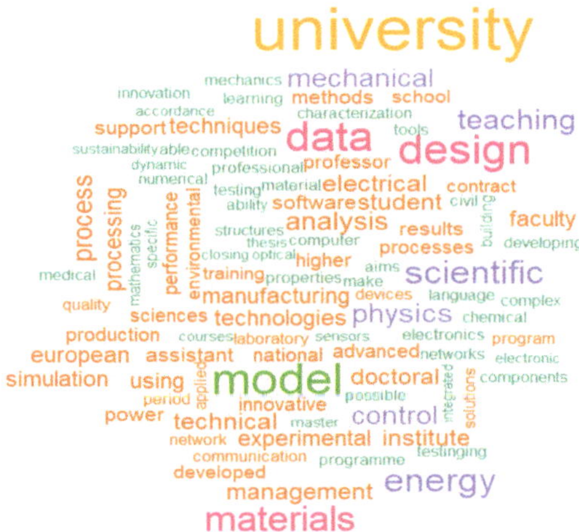

Figure 3. Engineering. Source: authors' computation.

The representation built for the environmental sciences field (Figure 4) highlights more keywords related to the research topics of the projects being advertised: "water", "climate", "ocean", "earth", "biology", "change", "ecosystem", and "sustainability". Therefore, this investigation also depicts the topical research areas.

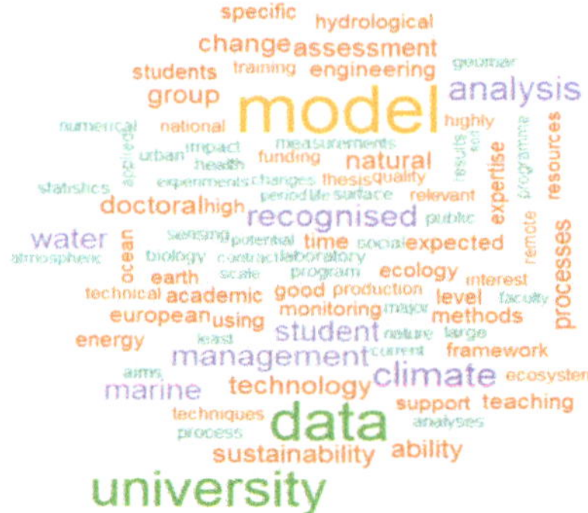

Figure 4. Environmental sciences. Source: authors' computation.

As mentioned in the methodology section, we did not use a stemming algorithm because we consider that relevant information could be lost. For example, in the word cloud associated to Mathematics field (Figure 5), one can distinguish the term "science" but also the plural form "sciences". Employing a stemming algorithm will lead to the elimination of the second one together with the absorption of its frequency by the radical of the word. However, if we analyze the associations for these words, we see that they are indicating different aspects. "Science" is rather related to data science meanwhile the prevalence of the word sciences comes from collocations like natural sciences or social sciences. Among technical skills, related to this research field are very well represented: "probability theory", "differential equations", and "physics". IT skills are now reflected by terms like "computer" or "program". Moreover, an interesting aspect also related to softwareengineering skill is highlighted by the associations found for the word "machine". The word "learning" is frequently associated with it (0.87), depicting that the candidates should be able to implement different machine learning algorithms. Among the top 100 terms, "team" is present, emphasizing that team-work is essential for research activity. The word "language" is associated with diplomas and certificates (0.4), showing that good communication in a foreign language is required.

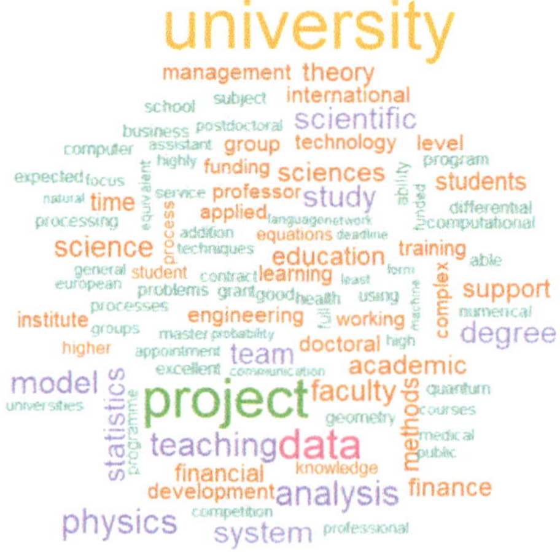

Figure 5. Mathematics. Source: authors' computation.

For the first time, the word cloud representing the Economics field (Figure 6) highlights an important dimension of the research activity, the publication of which is an outcome but also a selection criterion. This is summarized by "article", "publication" or "journals" words situated among first 100 most frequent. The magnitude of the term "business" is coming from the requirements related to the candidate studies: Master's degree or PhD or equivalent relevant experience in the field of finance or management/business administration are required. This is why the two words most correlated with "business" are "school" and "administration".

Figure 6. Economics. Source: authors' computation.

For the computer science field (Figure 7), terms such as "deep", "machine", "algorithms", "learning", "digital", "computing", and "intelligence", are keywords that could be anticipated. Therefore, successful candidates should have technical skills related to machine learning or its new area known as deep learning which is mostly based on artificial neural networks.

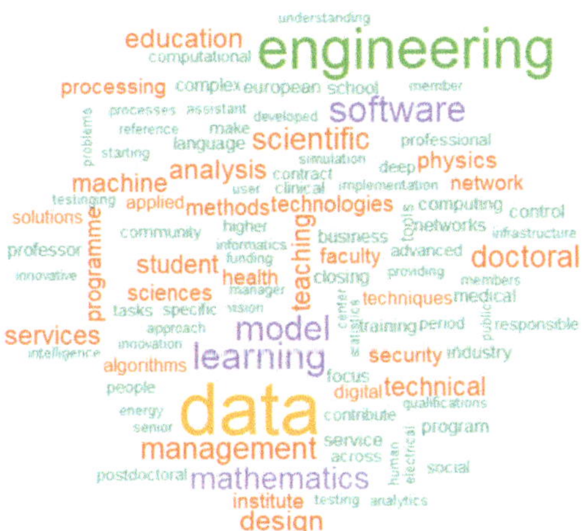

Figure 7. Computer science. Source: authors' computation.

The appearance of the words "medical", "health", "human", "social", and "clinical" could indicate topical research themes for the period we are analyzing. This is also due to the that fact that machine learning algorithms are often applied to medical data sets. The term innovative is pointing towards technologies and solutions that should be developed in the projects for which the research positions are opened.

6. Conclusions

Our article proposes a text mining approach in order to identify the mix of skills that are required from first stage researchers in Europe. The analysis was applied on job vacancies advertisements extracted from the EURAXESS platform for selected research fields: Computer Science, Economics, Engineering, Environmental Science and Mathematics.

First, the results of our analysis can be utilized by educational institutions that contribute to skill formation of future R&D professionals. Second, the results are relevant for early career researchers, PhD candidates and career guidance providers who can better understand the entry level research labour market in terms of skills and demands at the workplace. Third, the results can be useful for R&D companies themselves which can benefit from the overview of the main developments characterizing various research fields. They can better develop effective human resources policies in order to attract, develop and retain the right mix of skills.

Another important conclusion is that text mining analysis of job vacancies advertisements is very useful for identifying the mix of skills required by employers in R&D sector from first stage researchers. Our results show that data handling and processing skills are essential for early career researchers, irrespective of their research field. Also, first stage research positions are connected to universities and include teaching activities to a great extent. Management of time, risks, projects and resources plays an important part in the job requirements included in the analyzed advertisements. Considering the obtained word clouds, we can conclude that R&D professionals face rapidly changing demands determined by new technologies developments and environmental challenges. IT skills have also been highlighted by the word clouds in all research fields. In fact, one could see that nowadays first stage research positions include aspects that are embedded in all types of R&D career orientation (technical orientation, manager orientation, project orientation, technical transfer orientation and entrepreneur orientation). It indicates a diversification of job tasks for early career researchers in line with the increased interdisciplinary and transformations of the research sector.

Moreover, the proposed methodological approach is very helpful for exploring specific, technical skills which are much more complicated to be assessed and which are usually studied via in-depth sector level analysis. The main limitations of the study are related to the short time span covered by the gathered job advertisements, limited number of research fields that have been analyzed and the fact that the EURAXESS platform is used mostly by universities and less by research companies. In our future research, we plan to study the dynamics of the skill needs for R&D professionals, to compare entry level positions with more advanced ones, to perform cross-country comparisons, and to include more research fields in our analysis. Further research will extend the collection of textual data to other different research fields in order to extract specific latent variables known as topics. These represent a cluster of words with similar meanings and could lead to a classification of our documents according to the prevalence of topics that describe each document.

Author Contributions: Conceptualization, data processing and manuscript writing M.M.M.M.; manuscript writing and editing C.M and A.M.Z.; data gathering and manuscript writing T.M.G.

Funding: This paper has been elaborated in the NUCLEU Program 19N/2019, funded by Romanian Research and Innovation Ministry, projects PN 19130301 and PN 19130302. The APC was funded by PN 19130301.

Conflicts of Interest: The authors declare no conflict of interest. The funders had no role in the design of the study; in the collection, analyses, or interpretation of data; in the writing of the manuscript, or in the decision to publish the results.

Appendix A

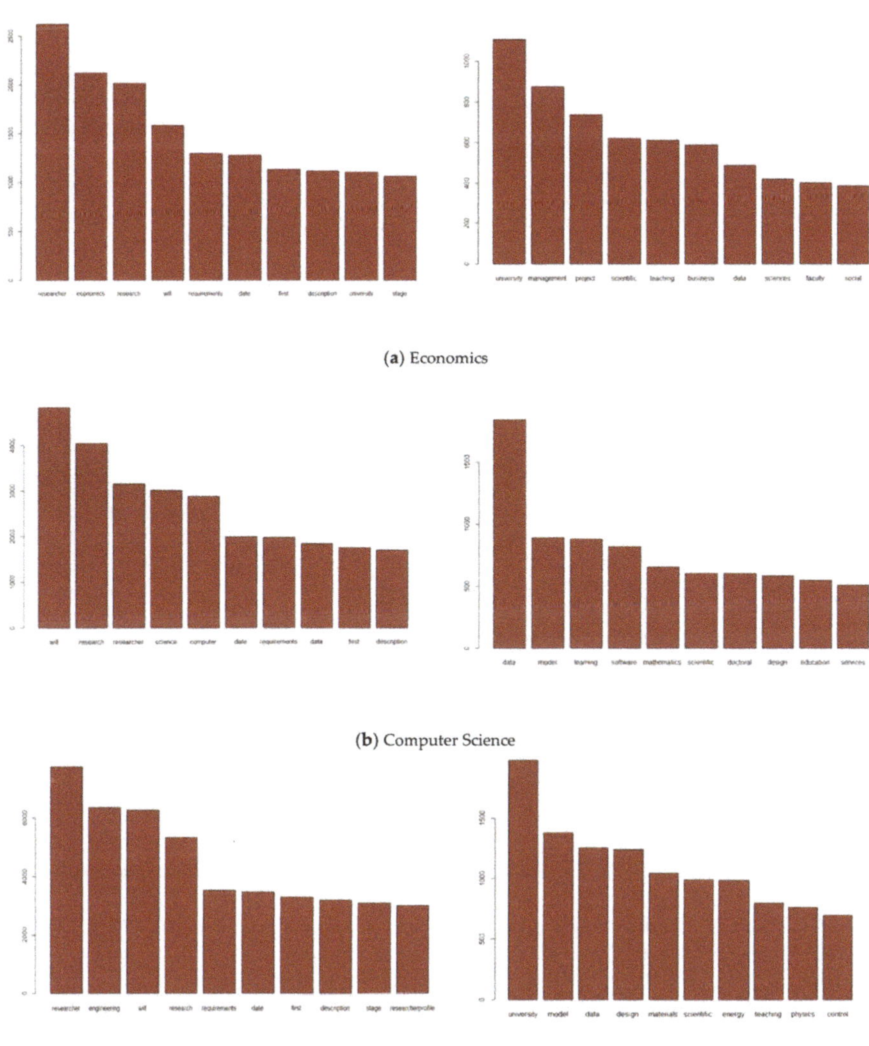

(**a**) Economics

(**b**) Computer Science

(**c**) Engineering

Figure A1. *Cont.*

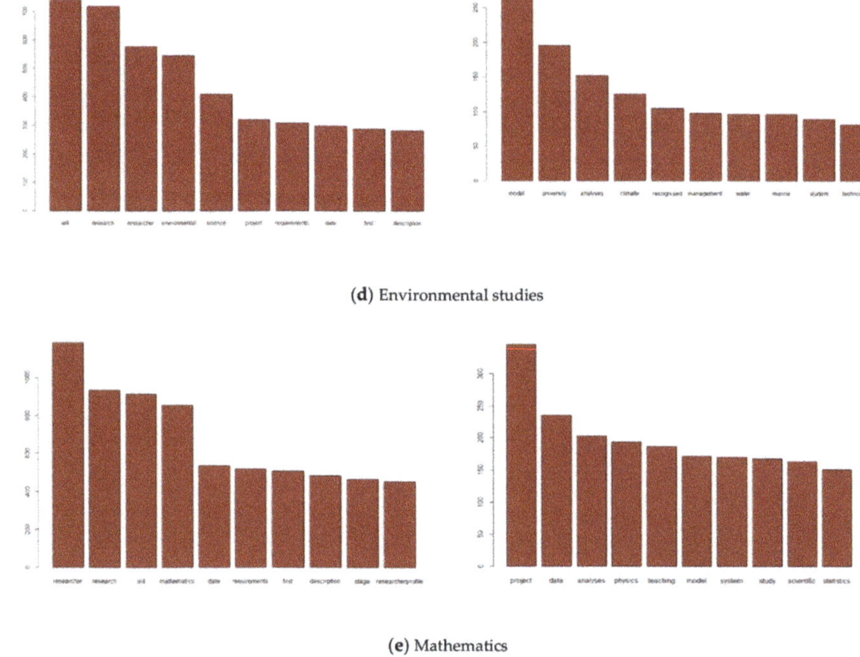

(**d**) Environmental studies

(**e**) Mathematics

Figure A1. Top ten most frequent words before and after data cleaning. Source: authors' computation.

References

1. Vuong, Q.H. The (ir)rational consideration of the cost of science in transition economies. *Nat. Hum. Behav.* **2018**, *2*, 5. [CrossRef]
2. European Commission. *European Research Area Progress Report 2018*; European Commission: Brussels, Belgium, 2019.
3. Roberts, K. The entry into employment: An approach towards a general theory. *Sociol. Rev.* **1968**, *16*, 165–184. [CrossRef]
4. Aryee, S. Career orientations, perceptions of rewarded activity, and career strategies among R&D professionals. *JET-M* **1992**, *9*, 61–82.
5. Allen, T.J.; Katz, R. The dual ladder: Motivational solution or managerial delusion? *R&D Manag.* **1986**, *16*, 185–197.
6. Dalton, G.W.; Thomson, P.H. *Novations: Strategies for Career Management*; Scott Foresman: Glenview, IL, USA, 1986.
7. Chen, T.Y.; Chang, P.L.; Yeh, C.W. Square of correspondence between career needs and career development programs for R&D personnel. *J. High Technol. Manag. Res.* **2003**, *14*, 189–211.
8. Chen, T.Y.; Chang, P.L.; Yeh, C.W. A study of career needs, career development programs, job satisfaction and the turnover intentions of R&D personnel. *Career Dev. Int.* **2004**, *9*, 424–437.
9. Debackere, K.; Buyens, D.; Vandenbossche, T. Strategic career development for R&D professionals: Lessons from field research. *Technovation* **1997**, *17*, 53–62.
10. Burack, E.H.; Burack, M.D.; Miller, D.M.; Morgan, K. New paradigm approaches in strategic human resource management. *Group Organ. Stud.* **1994**, *19*, 141–159. [CrossRef]
11. Wheelwright, S.C.; Clark, K.B. *Revolutionizing Product Development*; The Free Press: New York, NY, USA, 1992.
12. García-Sánchez, E.; Guerrero-Villegas, J.; Aguilera-Caracuel, J. How Do Technological Skills Improve Reverse Logistics? The Moderating Role of Top Management Support in Information Technology Use and Innovativeness. *Sustainability* **2019**, *11*, 58. [CrossRef]
13. Igbaria, M.; Kassicieh, S.K.; Silver, M. Career orientations and career success among research, and development and engineering professionals. *J. Eng. Technol. Manag.* **1999**, *16*, 29–54. [CrossRef]

14. Garden, A.M. Career orientations of software developers in a sample of high tech companies. *R&D Manag.* **1990**, *20*, 337–353.

15. OECD. *Workforce Skills and Innovation: An Overview of Major Themes in the Literature*; OECD: Paris, France, 2011.

16. National Science Board. *Research & Development, Innovation, and the Science and Engineering Workforce: A Companion to Science and Engineering Indicators 2012*; NSB-12-03; National Science Foundation: Arlington, VA, USA, 2012.

17. Davidescu, A.A.; Paul, A.M.V.; Gogonea, R.-M.; Zaharia, M. Evaluating Romanian Eco-Innovation Performances in European Context. *Sustainability* **2015**, *7*, 12723–12757. [CrossRef]

18. Gerport, T.J.; Domsch, M.; Keller, R.T. Career orientations on different countries and companies: An empirical investigation of West Germany, British and US industrial R&D professionals. *J. Manag. Stud.* **1988**, *25*, 439–462.

19. Aryee, S.; Leong, C.C. Career orientations and work outcomes among industrial R&D professionals. *Group Organ. Stud.* **1991**, *16*, 193–205.

20. Kim, Y.; Cha, J. Career orientations of R&D professionals in Korea. *R&D Manag.* **2000**, *30*(2), 121–137.

21. Igbaria, M.; Greenhaus, J.H.; Parasuraman, S. Career orientation of MIS employees: An empirical analysis. *MIS Q.* **1991**, *15*, 151–169. [CrossRef]

22. Schein, E.H. How career anchors hold executives to their career paths. *Personnel* **1975**, *52*, 11–24.

23. Turpin, T.; Deville, A. Occupational roles and expectations of research scientists and research managers in scientific research institutions. *R&D Manag.* **1995**, *25*, 141–157.

24. McCormick, K. Career paths, technological obsolescence and skills formation: R&D staff in Britain and Japan. *R&D Manag.* **1995**, *25*, 197–211.

25. Cable, D.M.; Judge, T.A. Person-organization fit, job choice decisions and organizational entry. *Organ. Behav. Hum. Decis. Process* **1996**, *67*, 294–311. [CrossRef]

26. Kristof, A. Person-organization fit: An integrative review of its conceptualizations, measurement, and implication. *PPSych* **1996**, *49*, 1–49. [CrossRef]

27. Yi, J.C.; Kang-Yi, C.D.; Burton, F.; Chen, H.D. Predictive Analytics Approach to Improve and Sustain College Students' Non-Cognitive Skills and Their Educational Outcome. *Sustainability* **2018**, *10*, 4012. [CrossRef]

28. Zhang, L.; Guo, X.; Lei, Z.; Lim, M.K. Social Network Analysis of Sustainable Human Resource Management from the Employee Training's Perspective. *Sustainability* **2019**, *11*, 380. [CrossRef]

29. OECD. *Skills for Innovation and Research*; OECD Publishing: Paris, France, 2011.

30. Auriol, L. *Careers of Doctorate Holders: Employment and Mobility Patterns*; OECD STI Working Paper 2010/4; OECD: Paris, France, 2010; 8p.

31. Forfas. *The Role of PhDs in the Smart Economy*; Advisory Council for Science, Technology and Innovation: Dublin, Ireland, 2009; Available online: http://www.sciencecouncil.ie/media/asc091215_role_of_phds.pdf (accessed on 3 April 2019).

32. Mietzner, D.; Kamprath, M. A Competence Portfolio for Professionals in Creative Industries. *Creat. Innov. Manag.* **2013**, *22*, 280–294. [CrossRef]

33. O*Net OnLine. Available online: https://www.onetonline.org/ (accessed on 30 March 2019).

34. Skills Panorama. Available online: https://skillspanorama.cedefop.europa.eu/en/glossary/s (accessed on 20 March 2019).

35. Sasso, S.; Ritzen, J. Sectoral cognitive skills, R&D, and productivity: A cross-country cross-sector analysis. *Educ. Econ.* **2019**, *27*, 35–51.

36. Piva, M.; Vivarelli, M. The role of skills as a major driver of corporate R&D. *Int. J. Manpow.* **2009**, *30*, 835–852.

37. Mariani, M. What determines technological hits? Geography versus firms competencies. *Res. Policy* **2004**, *33*, 1565–1582. [CrossRef]

38. Michelacci, C. Low returns in R&D due to the lack of entrepreneurial skills. *Econ. J.* **2003**, *113*, 207–225.

39. Quelin, B. Core Competencies, R&D Management and Partnerships. *Eur. Manag. J.* **2000**, *18*, 476–487.

40. CEDEFOP. *Future Skill Needs for the Green Economy*; Publications Office of the European Union: Luxembourg, Belgium, 2009.

41. European Commission. *Providing Researchers with the Skills and Competencies They Need to Practice Open Science*, Open Science Skills Working Group Report; European Commission: Brussels, Belgium, 2017.

42. Gardiner, A.; Aasheim, C.; Rutner, P.; Williams, S. Skill Requirements in Big Data: A Content Analysis of Job Advertisements. *J. Comput. Inf. Syst.* **2018**, *58*, 374–384. [CrossRef]

43. Wowczko, I.A. Skills and vacancy analysis with data mining techniques. *Informatics* **2015**, *2*, 31–49. [CrossRef]

44. Kennan, M.A.; Willard, P.; Cecez-Kecmanovic, D.; Wilson, C.S. IS knowledge and skills sought by employers: A content analysis of Australian IS early career online job advertisements. *Aust. J. Inf. Syst.* **2008**, *15*, 1168044.
45. Yang, Q.; Zhang, X.; Du, X.; Bielefield, A.; Liu, Y. Current market demand for core competencies of librarianship—A text mining study of American Library Association's advertisements from 2009 through 2014. *Appl. Sci.* **2016**, *6*, 48. [CrossRef]
46. Kobayashi, V.B.; Mol, S.T.; Berkers, H.A.; Kismihok, G.; Den Hartog, D.N. Text mining in organizational research. *Organ. Res. Methods* **2018**, *21*, 733–765. [CrossRef] [PubMed]
47. Karakatsanis, I.; AlKhader, W.; MacCrory, F.; Alibasic, A.; Omar, M.A.; Aung, Z.; Woon, W.L. Data mining approach to monitoring the requirements of the job market: A case study. *Inf. Syst.* **2017**, *65*, 1–6. [CrossRef]
48. Vermna, A.; Yurov, K.M.; Lane, P.L.; Yurova, Y.V. An investigation of skill requirements for business and data analytics positions: A content analysis of job advertisements. *J. Educ. Bus.* **2019**, *94*, 243–250. [CrossRef]
49. European Commission. Euraxess. 2019. Available online: https://euraxess.ec.europa.eu/ (accessed on 15 May 2019).
50. Scrapinghub Ltd. Scrapy. 2019. Available online: https://docs.scrapy.org/en/latest/intro/overview.html (accessed on 25 April 2018).
51. Zhang, S.; Feick, R. Understanding public opinions from geosocial media. *ISPRS Int. J. Geo-Inf.* **2016**, *5*, 74. [CrossRef]
52. Hernandez-Suarez, A.; Sanchez-Perez, G.; Toscano-Medina, K.; Martinez-Hernandez, V.; Perez-Meana, H.; Olivares-Mercado, J.; Sanchez, V. Social sentiment sensor in Twitter for predicting cyber-attacks using $\ell 1$ regularization. *Sensors* **2018**, *18*, 1380. [CrossRef] [PubMed]
53. Boja, C.E.; Herţeliu, C.; Dârdală, M.; Ileanu, B.V. Day of the week submission effect for accepted papers in Physica A, PLOS ONE, Nature and Cell. *Scientometrics* **2018**, *117*, 887–918. [CrossRef]
54. Feinerer, I. Introduction to the tm Package Text Mining in R. 2013. Available online: http://cran.r-project.org/web/packages/tm/vignettes/tm.pdf (accessed on 10 October 2018).
55. Meyer, D.; Hornik, K.; Feinerer, I. Text mining infrastructure in R. *J. Stat. Softw.* **2008**, *25*, 1–54.
56. TM Library R. Available online: https://cran.r-project.org/web/packages/tm/tm.pdf (accessed on 20 February 2019).
57. Solka, J.L. Text data mining: Theory and methods. *Stat. Surv.* **2008**, *2*, 94–112. [CrossRef]
58. Srivastava, A.N.; Sahami, M. *Text Mining: Classification, Clustering, and Applications*; Chapman and Hall/CRC: Boca Raton, FL, USA, 2009.
59. Ausloos, M.; Nedic, O.; Fronczak, A.; Fronczak, P. Quantifying the quality of peer reviewers through Zipf's law. *Scientometrics* **2016**, *106*, 347–368. [CrossRef]
60. Baayen, R.H. *Word Frequency Distributions*; Springer Science & Business Media: Berlin, Germany, 2002; Volume 18.
61. Baroni, M.; Evert, S. The zipfR Package for Lexical Statistics: A Tutorial Introduction. 2006. Available online: http://zipfr.r-forge.r-project.org (accessed on 5 June 2019).
62. Hornik, K.; Grün, B. Topicmodels: An R package for fitting topic models. *J. Stat. Softw.* **2011**, *40*, 1–30.
63. Hornik, K.; Meyer, D.; Buchta, C. Slam: Sparse Lightweight Arrays and Matrices. R Package Version 0.1-40. 2016. Available online: https://CRAN.R-project.org/package=slam (accessed on 8 February 2019).
64. Fellows, I. Wordcloud: Word Clouds. R Package Version 2.5. 2014. Available online: http://CRAN.R-project.org/package=wordcloud (accessed on 17 January 2019).

Article

Identifying Promising Research Frontiers of Pattern Recognition through Bibliometric Analysis

Inchae Park and Byungun Yoon *

Department of Industrial & Systems Engineering, School of Engineering, Dongguk University, 26, Pil-dong 3-ga, Chung gu, Seoul 100 715, Korea; inchae@dongguk.edu
* Correspondence: postman3@dongguk.edu; Tel.: +82-2-2260-8659

Received: 6 October 2018; Accepted: 1 November 2018; Published: 5 November 2018

Abstract: This paper aims at proposing a quantitative methodology to identify promising research frontiers (RFs) based on bibliographic information of scientific papers and patents. To achieve this, core technological documents are identified by suggesting several indices which measure paper impact, research impact, patent novelty, impact, marketability, and the right range to evaluate technological documents and which measure the research capability of research organizations (ROs) such as a RO's activity, productivity, market competitiveness, and publication impact. The RFs can be identified by clustering core technological documents, and promising indices of each RF which are from the perspectives of growth, impact, marketability, and science-based effect, are calculated to promising RFs. As an illustration, this paper selects the case of pattern recognition technology among various technologies in the information and communication technology sector. To validate the proposed method, emerging technologies on the hype cycle are utilized, allowing analysts to compare the results. Comparing the results derived from scientific papers and patents, the results from scientific papers are proper to suggest themes for research (R) in relatively long-term perspective, whereas the results from patents are appropriate for providing themes for development (D) in terms of relatively short-term view. This approach can assist research organizations and companies in devising a technology strategy for a future direction of research and development.

Keywords: promising technology; research frontier; bibliometric analysis; hype cycle

1. Introduction

As it is crucial to raise the competitiveness of scientific technology as a strategy for the future, the detection of promising technologies in an early stage is one of the most important challenges. If companies and countries cannot respond to rapidly changing technological trends in time and seize promising technological opportunities at an early stage, it is difficult for them to gain a competitive advantage in the market, and to lead technological innovation and social change. Thus, many developed countries recognize the importance of a promising technology discovery. Several research programs supporting the discovery of future technologies are conducted by Horizon 2020 of the European Union (EU), the Defense Advanced Research Projects Agency (DARPA) of the United States of America (US), and New Energy and Industrial Technology Development (NEDO) of Japan. In addition, many major companies and research institutes have attempted to explore promising technologies in diverse ways, in accordance with their own situations. Consequently, predicted promising technologies have been unveiled, such as the 10 breakthrough technologies from the Massachusetts Institute of Technology (MIT), the next 5 in 5 from International Business Machines Corporation (IBM), and the top 10 strategic technology trends from Gartner group.

In the previous studies related to promising technologies, relevant terms such as promising, emerging technologies, research front, frontier and so on have been utilized interchangeably. Many studies for detecting emerging technologies took a qualitative approach such as relevance-tree, Delphi method, and questionnaire survey analysis, which are based on domain expertise. These expert-based approaches have the advantage of easy validation; however, they also have the disadvantages of being expensive and time-consuming [1,2]. In contrast, quantitative approaches, such as computer-based methods and bibliometrics, can provide a complementary approach to handle massive data for exploring promising technologies [3]. In particular, bibliometrics has been widely utilized as a powerful tool for monitoring research trends [4] or technological trajectories [5–9] or analyzing technological changes [10,11] using various data, such as academic literature, patents, and other publications. Most of the previous studies on emerging technology using bibliometrics focused on the concept of fast growth among many perspectives on emerging technologies such as fast growth, radical novelty, prominent impact and so on. Moreover, the previous approaches also focused on the emerging research field using scientific publications and emerging technologies using patents, respectively. The background studies are summarized in Section 2. Under this theoretical background, two research propositions are suggested as follows:

Proposition 1. *Promising research frontiers (RFs) can be forecasted through a quantitative bibliometric approach using both scientific papers and patents by reflecting comprehensive views.*

Proposition 2. *The predicted results using scientific papers and patents can be shown to be different because of their characteristics.*

This paper proposes a data-driven model designed to identify promising RFs with comprehensive perspectives, which are technological growth, marketability, and the science-based effect. Several metrics are developed in this model to measure the quality of the technological documents, to evaluate research organizations (ROs), and to identify promising RFs quantitatively. Furthermore, the Girvan and Newman clustering algorithm and modularity concept are utilized in the model for grouping technological documents to identify RFs with quantitative approaches. It enables us to overcome the limitation of selecting the appropriate number of clusters through a qualitative approach because the algorithm can recommend the proper number of clusters automatically in conjunction with the modularity. In terms of data sourcing and collection, scientific paper and patent data are collected as technological document data from the Web of Science (WoS) and the United States Patents and Trademark Office (USPTO) database, respectively. The results derived from the proposed model are compared to the results of hype cycle in order to confirm the Proposition 1, and the results derived from scientific papers and patents are compared to observe the difference between them in order to confirm Proposition 2.

The Information and Communication Technology (ICT) industry has a complex and rapidly changing nature as technological convergence occurs and the technologies progress radically. ICT covers a wide spectrum of computing environments (e.g., laptop computers and smartphones) that carry out a broad range of communication and information functions. This connectivity is able to provide new opportunities that are changing the way that firms do business and transforming public service delivery. ICT has proven to be a key driver of economic growth through widespread diffusion of the Internet, mobile telephony, and broadband networks [12]. Due to this nature and environment of ICT, promising technology discovery is crucial in the ICT sector. Thus, the proposed methodology is applied to a pattern recognition technology field of the ICT sector because the pattern recognition technology area has experienced major growth due to the technological innovation of artificial intelligence and big data.

The results applied to pattern recognition technology are well-matched to the hype cycle [13] using both scientific papers and patents. The main finding is that the results from scientific papers are proper for suggesting themes for research (R) in a relatively long-term perspective while the results from patents are appropriate for providing themes for development (D) in terms of a relatively short-term view. It is partially supported by an R&D linear model that explains the seeds of innovation created by a research lab at the science level and companies develop technologies and products at technology and industry level [14]. From the results and implications, this research provides a brief guideline to differentiate the roles of scientific paper and patent data for strategic R&D planning by proposing priorities to utilize the proposed model in the discussion.

This research contributes in several ways. First, from the perspective of data utilization, promising technology is suggested by utilizing both scientific articles and patents. It is able to provide implications to a research organization for technology planning. Second, from the perspective of methodology, several indices are proposed using bibliographic information in respective steps to evaluate technological documents and research capability of the research organization, and measure comprehensive views of promising property. Finally, from the perspective of the utilization of the results, the results are well matched to hype cycle and provide distinctive implications derived from scientific papers and patent database.

The remainder of this paper is structured as follows. Section 2 introduces relevant previous literature. Section 3 describes the overall research concept of this study, database, data collection and quantitative methodology. Section 4 presents the results of the case study using the proposed methodology, which considers the pattern recognition technology field. Section 5 discusses the implications of the results. Lastly, Section 6 provides the contribution, limitation, and applications of the research.

2. Literature Review

2.1. Concept of Promising Technology

Promising technology can be defined differently from diverse viewpoints. Technical excellence can be considered as a factor for promising technology from the perspective of technology development. On the other hand, from the viewpoint of the market, the technology that is likely to make a great economic outcome after commercialization can be recognized as a promising technology. From the patent perspective, the technology that possesses core relevant patents can be regarded as a promising technology, as a patent is a legal means to protect the right of use of a technology. The term "promising technology" [1,15–17] is used interchangeably with other similar terms such as "emerging technology" [18–24], "research front" [2,25–28], and "research frontier" [1,29,30] etc. without it being defined clearly. Among the various related terms, Cozzens et al. [31] summarized the major concept of emerging technology by reviewing its definition in the literature: (i) fast recent growth [18,21]; (ii) transition or change to something new [19,20]; (iii) market or economic potential [19–21]; and (iv) science-based innovation [19]. Similarly, Rotolo et al. [32] identified five attributes of emerging technologies: (i) radical novelty [19,33]; (ii) relatively fast growth [31,34]; (iii) coherence [19,34–36]; (iv) prominent impact [18–21,31,34–37]; and (v) uncertainty and ambiguity [19–21,31,35,38,39]. However, Noh et al. [15] included four major concepts for promising technology in a broad sense: (i) technological vacancy; (ii) convergent technology; (iii) recent appearance and rapid growth of a technology regarding emerging technology; and (iv) customer-based technology. These perspectives on promising technology were not constructed to be mutually exclusive or collectively exhaustive, as they are affected by the purpose of the research, and the characteristics of technologies, respectively. To develop the conceptual model by reflecting comprehensive perspectives of the promising technology regarding Proposition 1, in this paper, the promising technology is identified as a highly growing, impactful and profitable technology, reflecting the major concepts of emerging technology from the works of Cozzens et al. [31] and Rotolo et al. [32], but the concepts of coherence and uncertainty from Rotolo et al. [32] are excluded, because it is difficult to measure and reflect them. The other concepts, such as technological vacancy, convergent technology, and customer-based technology, were also not considered, because they were too broad to deal with.

2.2. Detecting Promising Technology Using Bibliometrics

Bibliometrics is a method for analyzing publication data such as academic literature, patents, and other publications [40]. It can describe the research interests or the quantity of research, evaluate the impact of a technology or effectiveness of a research organization, and monitor research trends [41]. The approach can be used not only to understand the past by tracing the citation relation but also to forecast the future [42] because it is able to identify "hidden patterns" from large amounts of historical data [43]. Bibliometric analysis has been widely used to detect promising or emerging research areas or technologies as a quantitative approach. It can be exploited to provide an informative reference for forecasting promising technologies or research areas as the results are derived from the objective data-driven quantitative analysis. Table 1 shows the previous bibliometrics studies for promising technology from prior literature [31,32]. The previous researchers mostly focused on fast growth, among the several attributes of emerging technology. Other attributes, such as radical novelty, market impact, and science impact, were not reflected when detecting promising technologies. Terminologies such as research front, field, and frontier were utilized when they were using bibliographic information from scientific publications, whereas the studies using patent information utilized the term "emerging technology". Furthermore, a few studies utilized information from both patents and publications. To identify promising technologies using bibliometrics, various analysis techniques were employed such as co-citation analysis using bibliographic data, co-word analysis and text mining based on text information, network analysis for data visualization [44]. This summary shows a similar propensity to the summary suggested in Rotolo et al. [32] that effectively summarized the operational definitions, data, and methods of the previous literature. Many studies on emerging technology utilized publication and patent data respectively. Although some studies [22,30] utilized both forms of bibliographic data, they focused on the concept of fast growth. This research proposes promising research frontiers using both scientific papers and patents and the results are compared with regard to Proposition 2. Additionally, although there is an attempt [1] to identify promising research frontiers with consideration for not only fast growth but also market impact, it did not utilize scientific papers and consider science-based innovation perspective. Thus, this paper suggests promising research frontiers with comprehensive perspectives with both scientific papers and patents.

Table 1. Summary of previous studies on detection for promising technology using bibliometrics.

Concept of Emerging Technology		Literature	Terminology	Data	Method
from Rotolo et al. (2015) [32]	from Cozzens et al. (2010) [31]				
Relatively fast growth	Fast recent growth	Lee (2008) [16]	Promising/emerging research field	Publications	Co-word analysis
		Shibata et al. (2011) [2]	Emerging research front	Publications	Citation network; Clustering
		Iwami et al. (2014) [17]	Promising field	Publications	Citation network; Time transition analysis
		Toivanen (2014) [29]	Research frontier	Publications	Bibliometrics
		Corrocher et al. (2003) [18]	Emerging technology	Patents	Co-word analysis
		Breitzman and Thomas (2015) [23]	Emerging technology	Patents	Co-citation analysis; Clustering; Scoring
		Noh et al. (2016) [15]	Emerging technology	Patents	Network analysis; Textmining
		Park et al. (2016) [1]	Promising research frontier	Patents	Network analysis; Clustering; Index
		Park et al. (2015) [30]	Promising research frontier	Patents and publications	Network analysis; Clustering
		Visessonchok et al. (2014) [22]	Emerging technology	Patents and publications	Citation network; Clustering
Radical novelty	Transition/change to something new	Érdi et al. (2013) [24]	Emerging technology	Patents	Citation network; Clustering
Prominent impact	Market/economic potential	Park et al. (2016) [1]	Promising research frontier	Patents	Network analysis; Clustering; Index
	Science-based innovation	-	-	-	-
Coherence	-	-	-	-	-
Uncertainty and ambiguity	-	-	-	-	-

The conceptual model of the present research is related to the prior studies [1,30,45] in that the model derives core technological documents by the screening process and identifies research frontiers through a clustering method. The promising indices are updated based on the indices of prior research [1] and several indices are added because data source is extended and some analytic steps are added. To include newly emerging impactful technological documents, the model includes the step to evaluate leading research organizations and collects the technological documents of them. This conceptual model also proposes promising research frontiers by suggesting outliers as several previous studies [46–48] suggested technological opportunities as a weak signal.

3. Methodology

3.1. Research Concept and Overall Process

Figure 1 shows the overall research concept to detect promising technologies. In this research, data from both scientific papers and patents are firstly utilized as technological documents to identify promising technologies using bibliometrics. Second, the core technological documents are selected from the set of collected technological documents through the proposed screening methodology. Several quantitative indices are proposed by evaluating technological documents and the capacity of research organizations in the screening process. In particular, this paper considers the research capability of research organizations to include technological documents that need to be considered despite low scores in the suggested indices because top research organizations can lead the direction of technology development. Third, the finalized core documents are grouped into research frontiers (RFs) using clustering algorithm, or are otherwise determined as outlier documents. Finally, promising research frontiers and outlier documents are identified by calculating the proposed promising indices. The promising technologies are suggested with several types, and compared between those derived from scientific papers and patents.

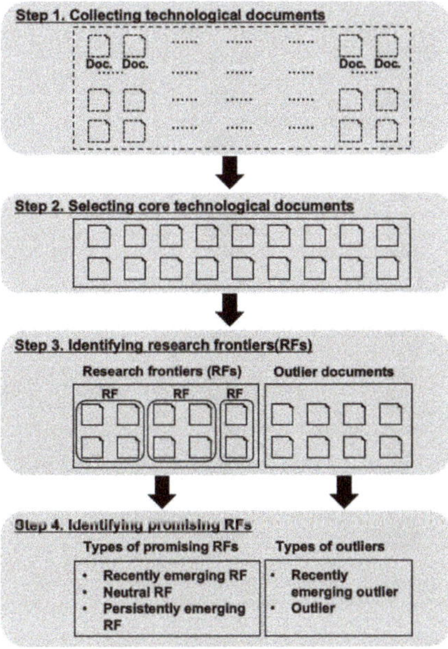

Figure 1. Research concept.

Figure 2 shows the detailed research process to identify the promising technologies. The promising technologies are identified with two perspectives, which are academic and technological, using scientific papers and patents. In the first step, scientific paper and patent data as technological document data are collected from the Web of Science (WoS) database and the United States Patents and Trademark Office (USPTO) database, respectively. In the second step, first of all, core technological documents are screened by evaluating the technological documents. An evaluation index is proposed in this research by reflecting the characteristics of the documents. Scientific papers are evaluated in terms of paper impact and academic research impact, whereas patents are evaluated from the viewpoints of novelty, impact, marketability, and the right range of patent, in order to derive core technological documents. Second, leading research organizations (ROs) are selected in the target technology area by evaluating the capacity of the RO. The RO capacity is evaluated in terms of the RO's activity for publications, RO's productivity for core publications, and impact of papers published from the ROs from the perspective of scientific paper. Meanwhile, the RO capacity in respect of patents is evaluated from the RO's activity for patent application, competitiveness of the patents registered from the RO, and the effect of patents registered from the RO. Third, the core technological document dataset is finally constructed by adding technological documents for the leading research organizations. This step is to include the technological documents that were underestimated using the evaluation index, because some recent technologies that have little chance to get high scores in the indices can be promising in the future. There is a presumption that the technological results from leading research groups had more potential to be promising technologies. In the third step, the research frontiers (RFs) are identified by clustering the core technological documents. In this step, RFs that have more than two documents, and outlier documents that are not grouped are extracted. In the final step, promising research frontiers for the academic perspective and the technology perspective are identified by calculating the promising indices. The promising indices for scientific papers and patents are proposed by considering the growth, impact, and science-based effects.

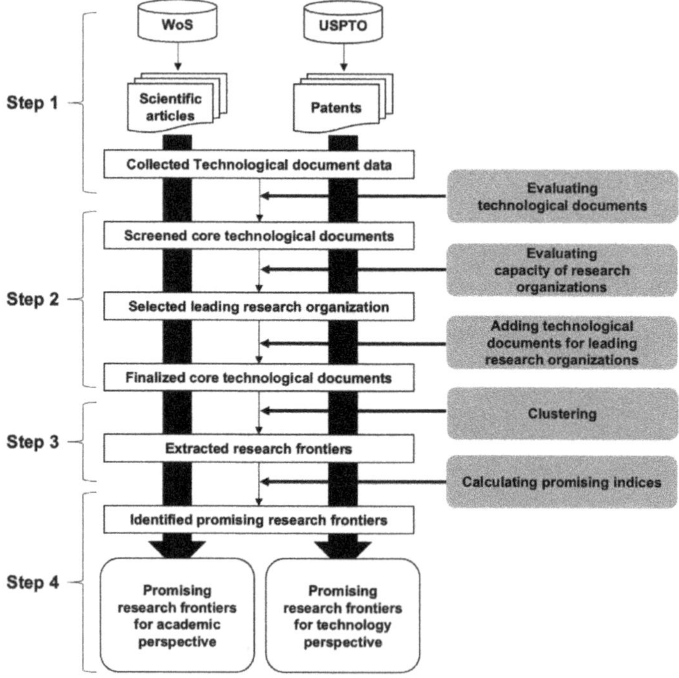

Figure 2. Research process.

3.2. Database, Data Collection, and Quantitative Methodology

3.2.1. Technological Documents Collection

In this step, the common process for both scientific papers and patents should be conducted: (1) target technology selection; (2) technology tree construction for the target technology; (3) searching keyword selection; (4) searching query construction; (5) data collection; and (6) noise removal. Data, including scientific journal papers and conference proceeding papers that had been published for 10 years, were collected from the WoS database. In addition, the registered patents for the first eight years, and the publicized and registered patents for the most recent two years were collected from the USPTO database. The proceeding papers and the publicized patents were collected to include more recent technological documents that would reflect the attribute of emerging technology, as those data represent more recent research themes.

We selected the technology field of pattern recognition as an illustration of the proposed method in this research. The technologies on pattern recognition have been widely utilized in character recognition, biometric recognition, human behavior pattern analysis, and medical image analysis. Furthermore, the technologies are fundamental to deep learning technology, which has recently received close attention. Thus, it is necessary to identify promising technologies in the relevant technologies in terms of academic and technological perspectives. Then, we built a technology tree for the pattern recognition technology and selected searching keywords and searching queries as shown in Table A1.

The bibliographic data on scientific papers, including core articles, journal and proceeding papers published between 2005 and 2014, were collected by searching in 'title' field of the WoS database using the searching queries of Table A1. Technology tree, which is a hierarchical structure of technology and structured as upper, middle, and lower classification in Table A1, and searching queries were constructed based on the literature survey and experts' opinion from a leading research institute of ICT field in Korea. The collected data includes bibliographic information on scientific papers on pattern recognition such as title, author, abstract, reference, citing reference and so on. After data collection and noise removal, 2421 scientific papers were collected, and 740 core scientific papers, which was the number of the papers published in Q1 journal, were extracted by the annual rate of total collected papers in a descending order, based on the criterion of the evaluated value. The noise data that are not relevant to pattern recognition technology were deleted by investigating the title and abstract of papers. The top 20 research organizations were extracted as leading ROs, using the proposed evaluation indices for ROs. The 76 scientific papers were those that had been published by the 20 leading ROs during the recent three years, and evaluated in the top 50% of the average value of the indices. Finally, 745 core scientific papers were extracted after adding the 76 papers published by 20 leading ROs, and deduplicating them. Table 2 shows the results of scientific paper data.

Table 2. Results of data collection on scientific paper.

Upper Classification	Middle Classification	Lower Classification	Collected Scientific Papers	Core Scientific Papers
Biometric recognition	Biometric recognition	DNA recognition	172	95
		Vein recognition	92	26
		Fingerprint recognition	233	81
		Iris recognition	189	47
Image recognition	Human recognition	Face recognition	361	120
		Action and gesture recognition	459	170
Voice recognition	Voice recognition	Voice recognition	915	206
	Total		2421	745

The data on patents registered from 2005 to 2014 and publicized from 2013 to 2014 in the USPTO database were collected. After data collection and noise removal, 5144 patents, which consisted of 3649 registered patents and 1495 publicized patents, were collected; and 648 patents, which was the number of patents whose family size was more than five, were extracted by the annual rate of the total collected patents in a descending order, based on the criteria of the evaluated value. The top 20 research organizations were extracted as leading ROs, using the proposed evaluation indices for ROs. The 922 patents were those that had been filed by 20 leading ROs during the most recent three years, and evaluated in the top 50% of the average value of the indices. Finally, 993 core patents were extracted after adding 922 patents filed by 20 leading ROs, and deduplicating them. Table 3 shows the results of patent data collection.

Table 3. Results of data collection on patent.

Upper Classification	Middle Classification	Lower Classification	Collected Patents	Core Patents
Biometric recognition	Biometric recognition	DNA recognition	141	20
		Vein recognition	141	19
		Fingerprint recognition	298	65
		Iris recognition	172	14
Image recognition	Object recognition	Object recognition	414	87
	Human recognition	Human detection and trace	561	93
		Face recognition	1390	334
		Action and gesture recognition	1203	237
Voice recognition	Utterance recognition	Isolated language recognition	416	76
		Continuous speech recognition	44	6
	Speaker recognition	Speaker recognition	264	42
	Total		5144	993

3.2.2. Core Technological Documents Selection by Evaluating Technological Documents

In this step, the common process for both scientific papers and patents should be conducted: Core technological documents are selected using the indices for each scientific paper and patent, as the two types of documents have different bibliographic information. The evaluation indices for each technological document are proposed that reflect their own characteristics. The number of core scientific papers is decided as the number of papers that are published in Q1 journal, which denotes the top 25% of the journal impact factors (JIFs), which are the yearly rankings of science and social science journals provided by Journal Citation Reports (JCR), published by Clarivate Analytics. The core scientific papers are selected based on the evaluation indices for scientific papers by the annual rate of total collected papers. The evaluation indices consist of the perspectives of paper impact and research impact. The paper impact index is proposed based on the number of forward citations for scientific paper as Dahlin and Behrens [49] utilized forward citations from the perspective of impact. The research impact index is suggested based on the JIFs and the number of forward citations because it would be potentially more impactful in terms of research impact perspective if the paper is published in journals with a high JIF. Both paper impact and research impact indices are transformed to a normalized value that is the value less the minimum value divided by the maximum value less the minimum value, as shown in (1) and (3). The research impact value is calculated by multiplying the journal impact factor for a scientific paper by the number of forward citations for the scientific paper, as shown in (2), and the calculated value is normalized as (3). The core scientific papers are extracted based on the average value of the two evaluation indices for scientific papers—paper impact, and research impact index—in a descending order, for as many as the calculated number by the annual rate of the total collected papers.

$$\text{Paper(Patent) Impact} = \frac{No.\ of\ forward\ citation - min(No.\ of\ forward\ citation)}{\max(No.\ of\ forward\ citation) - \min(No.\ of\ forward\ citation)} \tag{1}$$

$$\text{Research Impact} = \text{JIF} \times \text{No. of forward citation} \tag{2}$$

$$\text{Norm. Research Impact} = \frac{\text{Research Impact} - min(\text{Research Impact})}{\max(\text{Research Impact}) - \min(\text{Research Impact})} \tag{3}$$

Next, the number of core patents is decided as the number of patents that have more than five patent family countries. We utilized five patent families as standard to extract core patents, because the five patent offices—the United States of America (US), the European Union (EU), Japan (JP), China (CN), and Korea (KR)—are regarded as major patent offices. The core patents are selected based on the evaluation indices for patents by the annual rate of the total collected patents. The evaluation indices consist of the perspectives of patent novelty, impact, marketability, and right range. The novelty and impact indices are derived from the perspective of patent innovativeness, and these are developed as (4) and (1), simplifying the concept suggested in Dahlin and Behrens [49]. The patent that includes a lesser number of backward citations can be regarded as a novel patent, because the patent is dissimilar to past patents. That is, the patent that has a lesser number of references can be regarded as novel, in terms of the basis for innovation. Thus, the value is normalized and subtracted from one as (4). The patent impact index is proposed based on the number of forward citations as (1).

$$Patent\ novelty = 1 - \frac{No.\ of\ backward\ citation - min(No.\ of\ backward\ citation)}{max(No.\ of\ backward\ citation) - min(No.\ of\ backward\ citation)} \tag{4}$$

The patent marketability index is proposed based on the patent family size as (5), because the number of family patents can be perceived as the technology's potential market size [1]. The patent right range index is proposed based on the number of independent claims as (6). The number of independent claims in a patent can be considered as the right range of the patent, because each invention should be divided into claims, when a patent that includes more than two inventions is filed as one application [1]. The weighted sum of each value from the indices is calculated by deciding the weight using the analytic hierarchy process (AHP). Table 4 shows the evaluation indices for technological documents of both scientific papers and patents.

$$Patent\ Marketability = \frac{Patent\ family\ size - min(Patent\ family\ size)}{max(Patent\ family\ size) - min(Patent\ family\ size)} \tag{5}$$

$$Patent\ Right\ range = \frac{No.\ of\ independent\ claim - min(No.\ of\ independent\ claim)}{max(No.\ of\ independent\ claim) - min(No.\ of\ independent\ claim)} \tag{6}$$

Table 4. Evaluation indices for technological documents.

Source	Perspective	Bibliographic Information	Operational Definition
Scientific paper	Paper impact	Forward citation	The normalized number of forward citations for scientific papers
	Research impact	Journal impact factor (JIF), Forward citation	The normalized value that multiplies journal impact factor for the scientific paper by the number of forward citations for scientific papers
Patent	Patent novelty	Backward citation	The normalized number of backward citations for the patent that is subtracted from one
	Patent impact	Forward citation	The normalized number of forward citations for patents
	Patent marketability	Patent family	The normalized patent family size
	Patent right range	Claim	The normalized number of independent claims

3.2.3. Core Technological Documents Selection by Evaluating Research Organization

Although the core technological documents are selected by extracting those that have high values in the scoring model by year, there can be some potential core documents, because some indices are developed based on bibliographic information, such as the number of forward citations. For example, the number of forward citations can be increased as time goes by. Thus, the process of core document selection is redeemed by adding the leading research organization's documents in order to complement the recent research results by leading ROs, as the technological results from leading research groups have more potential to be promising technologies. To this end, the indices to evaluate

ROs in the technology field are proposed in this research, reflecting the characteristics of respective technological documents.

The leading ROs for a scientific paper are selected based on the evaluation indices for the leading ROs for scientific papers. The evaluation indices consist of the perspectives of RO's activity for publication, productivity for core publication, and impact of RO's publication. The index of RO's activity for publication is proposed based on the number of RO's scientific papers because the greater the number of publication by RO is, the more active the RO is in the technology field. The RO's activity is evaluated using (7) and it is normalized using (8).

$$\text{RO's activity index (AI)} \; = \; \frac{No. \; of \; papers(patents) \; of \; RO}{total \; No. \; of \; papers(patents)} \tag{7}$$

$$\text{Norm. AI} \; = \; \frac{\text{AI} - min(\text{AI})}{max(\text{AI}) - min(\text{AI})} \tag{8}$$

The index of the RO's productivity for core publication is proposed based on the number of RO's scientific papers, and journal impact factor (JIF) of the scientific paper. To this end, core journals in the technology field are defined as the journals whose JIF value is greater than the average JIF in the target technology area. The RO's productivity index (PI) is calculated as shown in (9), and normalized using (10), because the greater the number of the RO's scientific papers published in core journal is, the higher the RO's research productivity.

$$\text{RO's productivity index (PI)} \; = \; \frac{No. \; of \; RO's \; papers \; published \; in \; core \; journal}{total \; No. \; of \; RO's \; papers} \times 100 \tag{9}$$

$$\text{Norm. PI} \; = \; \frac{\text{PI} - min(\text{PI})}{max(\text{PI}) - min(\text{PI})} \tag{10}$$

The index for impact of RO's publication is proposed based on the number of RO's scientific papers, and forward citation of the scientific paper. The impact of RO's publication index (II) is calculated as shown (11) and normalized using (12).

$$\text{Impact of RO's publication index (II)} \; = \; \frac{\dfrac{Forward \; citation \; of \; papers \; by \; RO}{Forward \; citation \; of \; total \; papers}}{\dfrac{No. \; of \; papers \; published \; by \; RO}{Total \; No. \; of \; papers}} \tag{11}$$

$$\text{Norm. II} \; = \; \frac{\text{II} - min(\text{II})}{max(\text{II}) - min(\text{II})} \tag{12}$$

The top 20 leading ROs are extracted based on the average value of three evaluation indices for the RO using scientific papers. After domain experts reviewed the list of companies, the number of leading ROs was concluded to include most of influential and active ROs. The core scientific paper dataset is finalized by adding the scientific papers that are published by the top 20 leading ROs within the most recent three years, and positioned in the top 50%, based on the average score of three evaluation indices. Since the time duration of technology development is generally 2–3 years, we limited the time frame to the last three years to add recent papers. In addition, the criterion of scores in the indices (50%) was selected because the papers published in the Q1 and Q2 journals can be normally regarded as good quality papers. Although papers in Q2 journals might be not a high-quality paper, those that are published by leading ROs can have great potential for promising technology.

The leading ROs for patents are selected based on the evaluation indices for leading ROs for patents. The evaluation indices consist of the perspectives of RO's activity for patent application, market competitiveness of RO's patents, and effect of RO's patents. The index of RO's activity for patent application is calculated in the same way using (7), and it is normalized using (8). The index of market competitiveness of RO's patents is calculated in the same way using (5) but patent family size should be substituted by the value of RO's market competitiveness index (MCI) calculated by (13). Moreover, the index for the effect of RO's patents is calculated in the same way using (11) and it is normalized using (12); but forward citation of papers should be substituted by forward citation of patents. The top 20 leading ROs are extracted based on the average value of the three evaluation indices for ROs using patents. The core patent dataset is finalized by adding the patents that are publicized and registered by the top 20 leading ROs within the most recent three years, and positioned in the top 50%, based on the average score of the three evaluation indices. Table 5 shows the evaluation indices for research organizations from the perspective of scientific papers and patents.

$$RO's \text{ market competitiveness index (MCI)} = \frac{RO's \text{ patent family size}}{\text{the average patent family size}} \tag{13}$$

Table 5. Evaluation indices for research organizations.

Source	Perspective	Bibliographic Information	Operational Definition
Scientific paper	* RO's activity for publication	Frequency	The normalized value of the number of RO's papers divided by the total number of papers
	RO's productivity for core publication	Frequency, Journal impact factor	The normalized value of the percentage of the number of RO's papers published in the core journal among the number of RO's papers
	Impact of RO's publication	Frequency, Forward citation	The normalized value of the percentage of the number of forward citations for RO's papers among the number of forward citations for total papers divided by the percentage of the number of RO's papers among the number of total papers
Patent	RO's activity for patent application	Frequency	The normalized value of the number of RO's patents divided by the total number of patents
	Market competitiveness of RO's patents	Patent family	The normalized value of RO's patent family size divided by the average patent family size
	Effect of RO's patents	Forward citation	The normalized value of the number of forward citations for RO's patents divided by the number of forward citations for total patents

* RO: Research Organization.

3.2.4. Research Frontiers Extraction by Clustering

The core technological documents are grouped by a Girvan and Newman clustering algorithm [50], which is a hierarchical method to detect communities by removing edges from the original network. In this research, the original network is developed based on the normalized bibliographic coupling relation [51] that represents the degree of sharing references between technological documents. The normalized bibliographic coupling strength (NBCS) is defined as

$$NBCS_{ij} = \frac{r_{ij}}{\sqrt{n_i n_j}} \tag{14}$$

where $NBCS_{ij}$ is the normalized coupling strength between technological document i and j, r_{ij} is the number of sharing references between i and j, and $n_i(n_j)$ is the number of references in the reference list of document $i(j)$. The NBCS value is zero to one. After developing network based on normalized bibliographic coupling relation between documents, the edge betweenness centrality in the network, which is an extended concept of the vertex betweenness centrality [52], is calculated as [53]

$$C_{B_e}(e) = \sum_{s \neq t \in V} \frac{\sigma_{st}(e)}{\sigma_{st}} \tag{15}$$

where $C_{B_e}(e)$ is the edge betweenness centrality of edge e, σ_{st} is the number of shortest paths connecting node s to t, and $\sigma_{st}(e)$ is the number of shortest paths connecting node s to t passing through the edge e. Based on edge betweenness centrality value, Girvan and Newman clustering algorithm for discovering community structure in network were conducted. In the algorithm, the edge with the highest edge betweenness centrality is progressively removed. The edge betweenness is recalculated after removal of the edge with the highest value. The removal and calculation processes are repeated, until the modularity(Q) [50] is the highest, which means that the clustering process can provide the best set of groups in a way that maximizes the modularity. The modularity is defined as

$$Q = \sum_i (e_{ii} - a_i^2) = Tr\, e - e^2 \tag{16}$$

where e_{ij} is the fraction of all edges in the network that link vertices in community i to vertices in community j, the trace of the matric $Tr\, e = \sum_i e_{ii}$ gives the fraction of edges in the network that connect vertices in the same community, $a_i = \sum_j e_{ij}$ is the fraction of edges that connect to vertices in community i, and $\|x\|$ is the sum of the elements of the matrix x. The research frontiers (RFs) are identified by conducting this clustering process because the clusters are derived from the core technological documents. Moreover, the names of research frontiers are identified by reviewing the title and abstract of core technological documents.

3.2.5. Promising Research Frontiers Identification by Calculating Promising Indices

Promising research frontiers (RFs) are identified by using the promising indices, which are developed from the perspectives of growth, impact, marketability, and science-based effect. Those indices reflect the perspectives of rapid growth, market or economic potential, and scientific or technological change as attributes of promising technology introduced in the literature review section. The indices of growth and impact are common in scientific papers and patents, whereas the science-based effect index is for scientific papers, and the marketability index is for patents, because a paper includes rather academic and scientific information, whereas a patent includes technological information, which is likely to be commercialized. The growth and impact are defined as the growing potential of the RF and the applicability to other technologies, respectively, and the common indices—growth index (GI) and impact index (II)—are calculated using Equations (17) and (18), respectively.

$$\text{Growth Index (GI)} = \frac{A_i}{N} \times \left(\frac{\sum \left(\frac{P_t - P_{t-1}}{P_{t-1}} \right)}{n - 1} \times 100 \right) \tag{17}$$

where, A_i = the number of technological documents in RF i, P_t = the number of technological documents in RF i at time t, N = the total number of technological documents, and n = the data collection period.

$$\text{Impact Index (II)} = \frac{C_i}{P_i} \tag{18}$$

where, C_i = the number of forward citations in RF i, and P_i = the number of technological documents in RF i.

The science-based effect is defined as the effect of knowledge on science and technology. It is calculated with the journal impact factor using (19). The marketability index is defined as the potential for utilization as a product or service. It is calculated with the patent family size using (20). Table 6 shows the promising indices and the average score of the promising value from the three perspectives. However, the technological documents that are not grouped as RFs are considered as outliers, and the outlier documents are also evaluated by using impact, marketability, the science-based effect, and recentness, instead of growth, as the number of documents is just one, and the document does not belong to an RF.

$$\text{Sci} - \text{based Effect Index (SEI)} = \frac{IF_i}{P_i} \tag{19}$$

where, IF_i = sum of the impact factor of papers in RF i, and P_i = the number of technological documents in RF i.

$$\text{Marketability Index (MI)} = \frac{F_i}{P_i} \tag{20}$$

where, F_i = sum of the patent family size in RF i, and P_i = the number of technological documents in RF i. The equations for all indices in this paper are summarized in Table A2.

Table 6. Promising indices for promising research frontiers.

Source	Perspective	Bibliographic Information	Operational Definition
Scientific paper	Growth	Frequency	• Growing potential of research frontier (RF) • The value that multiplies the percentage of the papers in the RF among the total papers by the growth rate of papers in the RF
	Impact	Forward citation	• Applicability to other technologies • The sum of forward citations of papers in the RF divided by the number of papers in the RF
	Science-based effect	Journal impact factor	• Effect of knowledge on science and technology • The sum of JIFs of papers in the RF divided by the number of papers in the RF
Patent	Growth	Frequency	• Growing potential of the research frontier • The value that multiplies the percentage of the papers in the RF among the total patents by growth rate of patents in the RF
	Marketability	Patent family	• Potential for utilization as product and service • The family size of patents in the RF divided by the number of patents in the RF
	Impact	Forward citation	• Applicability to other technologies • The sum of forward citation of patents in the RF divided by the number of patents in the RF

The promising RFs are classified into four categories (recently emerging RFs, persistently emerging RFs, neutral RFs, and recently emerging outliers), by considering the level of technology development and the recentness of technological knowledge, based on the distribution of the publication year of technological documents in the RF, in order to suggest comprehensive interpretation of the results from scientific papers and patents. The recently emerging RF is defined as the cluster in which the technological documents published within the most recent three years account for more than 80 percent of all documents. The persistently emerging RF is defined as the cluster that includes technological documents that have emerged in more than five years among the total ten years. The neutral RF is defined as the cluster that includes technological documents that have emerged in less than five years among the total ten years, and in which the technological documents published within the most recent three years account for less than 80 percent of all documents. The recently emerging outlier is defined as the technological document itself that is not clustered, and that is published within the most recent three years. In addition, technological contents of promising research frontiers are presented to provide the practical information for technology development by conducting text mining.

4. Results

4.1. Results of the Analysis Using Scientific Papers

The research frontiers (RFs) shown in Table 7 were extracted by conducting Girvan and Newman clustering from the network based on the bibliographic coupling relation between the papers. The Girvan-Newman clustering was conducted at the upper classification level to derive the best clustering results using NetMiner, which is an application software for the visualization of large networks based on social network analysis. The modularity values were 28.85, 360.22, and 165.67 for each biometric, image, and voice recognition. As a result, 35 clusters that included at least two papers and 384 outliers were extracted. The clusters consisted of two recently emerging RFs, 22 neutral RFs, and 11 persistently emerging RFs. The promising RFs were extracted as the top 10 RFs in each type of cluster. Table 8 shows the title of the promising RF, the calculated values using the promising indices, and keywords derived through text mining. Vein and fingerprint recognition were included in recently emerging RF, biometric recognition, such as DNA and RNA recognition, was included in neutral RF, and gesture, RNA, and voice recognition were included in persistently emerging RF. Table 9 shows the title of the recently emerging outliers, the calculated values using promising indices, and keywords derived through text mining. The papers in the recently emerging outlier group can be considered as weak signals for promising research areas.

Table 7. Results of RFs on scientific papers.

Type	Title	No. of Scientific Papers (%)	No. of Clusters (%)
Cluster	Recently emerging RF	4 (1.11%)	2 (5.71%)
	Neutral RF	86 (23.82%)	22 (62.86%)
	Persistently emerging RF	271 (75.07%)	11 (31.43%)
Outlier	Recently emerging outlier	157 (40.89%)	-
	outlier	227 (59.11%)	-

4.2. Results of the Analysis Using Patents

The research frontiers (RFs) shown in Table 10 were extracted by conducting Girvan and Newman clustering from the network, based on the bibliographic coupling relation between patents. The Girvan-Newman clustering was conducted at the upper classification level using NetMiner, when the modularity values were 84.85, 28.71, and 13.81 for each biometric, image, and voice recognition. As a result, 64 clusters that included at least two papers, and 651 patents were extracted. The clusters consisted of 20 recently emerging RFs, 43 neutral RFs, and one persistently emerging RF. The promising RFs were extracted as the top 10 RFs in each type of cluster. Table 11 shows the title of the promising RF, the calculated values using promising indices, and keywords derived through text mining. Vein, face, and voice recognition were included in the recently emerging RFs, face, fingerprint, and biometric recognition were included in the neutral RFs, and vein recognition was included in the persistently emerging RF. Table 12 presents the title of the recently emerging outlier, the calculated values using the promising indices, and keywords derived through text-mining.

Table 8. Promising RF identification from scientific papers.

Type of Cluster	RF No.	Title of Promising RF	GI	II	SEI	Mean	Keywords
Recently emerging RF	RF 33	Sclera vein recognition	0.056	0	0.143	0.066	Iris, recognition, sclera, vein
	RF 35	Optimal extraction and fingerprint analysis	0	0.015	0.068	0.027	Extraction, spectrometry, determination
Neutral RF	RF 30	DNA Sequencing, and cancerous DNA recognition	0.011	1	1	0.670	DNA, mixture, synthetic, nanotube, recognition
	RF 16	The pattern of distribution of amino groups for RNA recognition	0.029	0.283	0.527	0.280	DNA, antibiotics, RNA. cleavage, molecular, genome
	RF 20	DNA microarray-based detection	0.010	0.336	0.429	0.258	DNA, detection, cell, microarray
	RF 410	Detection of actionable genomic alterations	0.028	0.357	0.328	0.238	Clinic, tumor, cancer, target, detection
	RF 10	RNA sequencing	0.215	0.089	0.247	0.184	RNA, gene, RNA-seq, cell, DNA, identify
	RF 272	Study on voice recognition	0.040	0.145	0.230	0.138	Voice, recognition, face, individual, speech
	RF 416	Face recognition method under lighting or color condition	0.065	0.242	0.044	0.117	Recognition, face, pattern, represent
	RF 13	Nanoscale DNA-polymer micelles	0.042	0.026	0.280	0.116	DNA, surfaces, micelles, individual, pattern, recognition
	RF 31	RNA recognition motif protein	0	0.066	0.260	0.108	RBM, RBP, MMA, transcription, pattern
	RF 29	HPV DNA detection	0.009	0.168	0.112	0.096	HPV, carcinoma, cervical, detect, DNA

Table 8. *Cont.*

Type of Cluster	RF No.	Title of Promising RF	GI	II	SEI	Mean	Keywords
	RF 92	Human action and gesture recognition	1	0.236	0.071	0.436	Action, recognition, motion, gesture, human, feature
	RF 1	RNA pattern recognition	0.339	0.335	0.601	0.425	RNA, immune, response, dsRNA, DNA, recognition, protein
	RF 2	Fingerprint recognition using model-based density map	0.970	0.118	0.055	0.381	Iris, recognition, detect, extract
	RF 415	Analytic techniques for face recognition	0.283	0.336	0.094	0.238	Face, recognition, discriminative, detect
Persistently emerging RF	RF 93	Cognition, action, and object manipulation	0.101	0.308	0.206	0.205	Action, activation, cognitive, recognition, inferior, demonstrate
	RF 254	Robust speech recognition algorithm	0.409	0.110	0.049	0.189	Speech, recognition, recognition, feature, signal, vector
	RF 257	Speech recognition by bilateral cochlear implant users	0.339	0.126	0.030	0.165	Speech, recognition, cochlear, hear, listen
	RF 417	Patterns of feature space, correlation, classification for face recognition	0.152	0.220	0.0821	0.151	Face, recognition, match, extract
	RF 17	DNA methylation patterns	0.057	0.099	0.258	0.138	Methylation, DNA, detect, cancer, hypermethylation
	RF 6	Detection of latent fingerprints	0.179	0.088	0.107	0.124	Fingerprint, detect, latent, contaminate, fluorescence, surfaces

Table 9. Recently emerging outlier identification from scientific papers.

Title of Recently Emerging Outlier (Paper)	II	SEI	Mean	Keywords
In-Situ Generation of Differential Sensors that Fingerprint Kinases and the Cellular Response to Their Expression	0.044	0.359	0.202	Kinases, protein, vitro
Fully Printed Flexible Fingerprint-like Three-Axis Tactile and Slip Force and Temperature Sensors for Artificial Skin	0.013	0.378	0.195	Tactile, skin, temperature, detect
Direct recognition of homology between double helices of DNA in Neurospora crassa	0.013	0.337	0.175	DNA, homology, identical, recognition
Fooling the Kickers but not the Goalkeepers: Behavioral and Neurophysiological Correlates of Fake Action Detection in Soccer	0.053	0.259	0.156	Action, predict, observe
The Negative Association of Childhood Obesity to Cognitive Control of Action Monitoring	0.049	0.259	0.154	Children, condition, amplitude, action
Human Parietofrontal Networks Related to Action Observation Detected at Rest	0.008	0.259	0.134	Observation, action, identified, correspondence
Detecting bacterial lung infections: in vivo evaluation of in vitro volatile fingerprints	0.129	0.108	0.119	Vitro, vivo, fingerprint, aeruginosa
Detection of a transient mitochondrial DNA heteroplasmy in the progeny of crossed genetically divergent isolates of arbuscular mycorrhizal fungi	0.031	0.197	0.114	Isolates, progeny, heteroplasmy, divergent
In Vivo Magnetization Transfer and Diffusion-Weighted Magnetic Resonance Imaging Detects Thrombus Composition in a Mouse Model of Deep Vein Thrombosis	0.017	0.209	0.113	Thrombus, histological, vein, detect
Interactions Between Visual and Motor Areas During the Recognition of Plausible Actions as Revealed by Magnetoencephalography	0	0.215	0.107	Action, activity, interact, recognition

Table 10. Results of RFs on patents.

Type	Title	No. of Patents (%)	No. of Clusters (%)
Cluster	Recently emerging RF	51 (14.91%)	20 (31.25%)
	Neutral RF	257 (72.15%)	43 (67.19%)
	Persistently emerging RF	34 (9.94%)	1 (1.56%)
Outlier	Recently emerging outlier	317 (48.69%)	-
	outlier	334 (51.30%)	-

Table 11. Promising RF identification from patent.

Type of Cluster	RF No.	Title of Promising RF	GI	II	MI	Mean	Keywords
Recently emerging RF	RF 45	Automatic face detection	0.148	0.724	0.375	0.415	Face, detect, measure, confidence, person, gesture
	RF 63	Displaying view for recognition	0.021	0.004	0.541	0.189	Recognition, detect, feature, synchronization
	RF 90	Facial decoding method	0.021	0.057	0.416	0.165	Motion, movement, contact, decoding, face, generate
	RF 237	Recursive motion recognition	0	0.139	0.333	0.157	Motion, region, detector, hand, gesture
	RF 8	Vein pattern detection	0	0	0.291	0.097	Determine, Vein Fistula, vessel, identified, atrium
	RF 12	Biometric sensor device for fingerprint	0.028	0	0.25	0.092	Sensor, encapsulation, biometric, fingerprint
	RF 154	Image discriminating method	0	0.059	0.208	0.089	Image, determine, voice, predetermine, recognition
	RF 175	Multi angle face recognition	0.084	0.013	0.166	0.088	face, detect, track, determine, facial, head
	RF 699	Voice control method	0.084	0	0.166	0.083	Voice, recognition, receive, language, speech
	RF 20	Blood vessel recognition for treat	0.080	0	0.166	0.082	Pressure, peripheral, hemodynamic, venous, vessel, configure
Neutral RF	RF 99	Human image recognition	1	0.075	0.791	0.622	Detect, face, image, gesture, eye, section, recognition
	RF 49	Image acquisition devices using face detection	0.009	1	0.708	0.572	Detect, magnification, gesture, face
	RF 52	Gesture image processing	0.309	0.149	0.708	0.389	Image, detection, face, motion, gesture, capturing, feature
	RF 51	Facial image processing	0.222	0.112	0.75	0.361	Image, detection, face, determine, gesture, feature
	RF 1	Detecting DNA	0.034	0.006	1	0.346	DNA, detecting, different, determine, molecule
	RF 74	Biometric authentication method	0.393	0.048	0.541	0.327	Image, detecting, face, feature, configure, apparatus, vector, signal
	RF 48	Image acquisition devices using face detection	0.014	0.238	0.708	0.320	Detecting, finger, gesture, determine, display
	RF 2	Fingerprint recognition using sensors	0.007	0.357	0.416	0.260	Fingerprint, sensor, finger, configure, capture
	RF 56	Automatic recognition by tracking method	0	0.304	0.416	0.240	Hand, focus, determine, face, track, human, autofocus
	RF 87	Facial feature selection	0.066	0	0.541	0.202	Search, face, detection, determine, configure, recognition
Persistently emerging RF	RF 7	Hand characteristic information	0.251	0.013	0.5	0.255	Fingerprint, sensor, substrate, detect, determine, finger

Table 12. Recently emerging outlier identification from patents.

Title of Recently Emerging Outlier (Patent)	II	MI	Mean	Keywords
Deletion gestures on a portable multifunction device	0.007	1	0.503	Deletable, gesture, detection, touch sensitive, multifunction
Architecture for controlling a computer using hand gestures	1	0	0.5	Gesture, image, control, recognition, hand
Illumination detection using classifier chains	0.363	0.529	0.446	Face, illumination, condition, correct
Image processing method using sensed eye position	0.003	0.823	0.413	Capture, detection, eye, face, graphic, capture
Fixed codebook searching apparatus and fixed codebook searching method	0	0.823	0.411	Impulse, codebook, processor, apparatus
Event recognition	0.146	0.588	0.367	Recognizes, gesture, determination
Real-time face tracking with reference images	0.169	0.529	0.349	Face, determination, relative, movement
Synchronization system and method for audiovisual programmes associated devices and methods	0.007	0.588	0.298	Recognition, synchronization, audiovisual, detection
Multi-dimensional disambiguation of voice commands	0.272	0.294	0.283	Action, audio, select, identifying
Systems and methods for interactively accessing hosted services using voice communications	0.003	0.529	0.266	Voice, convert, identified, recognition

4.3. Comparisons Results of the Analysis Using between Scientific Papers and Patents

Although there were several RFs that commonly emerged in both scientific paper and patent areas, the RFs for each technological document are classified into different categories and have different research themes. First, the fingerprint recognition-related research theme represented in the persistently emerging RF group and the recently emerging RF group were common in the scientific paper and patent areas. The RFs on the model for fingerprint recognition were distributed in terms of scientific papers (RF 35, RF 2, RF 6 in Table 8), whereas the RFs on fingerprint recognition using sensor in neutral RFs (RF 2 in Table 11), and RFs related to biometric sensor for fingerprint in recently emerging RFs (RF 12 in Table 11) were distributed in terms of patents. Second, the face detection research fields emerged in neutral and persistently emerging RFs for scientific papers, and in recently emerging and neutral RFs for patents. The research themes related to method and pattern for face detection were persistently emerged from the perspective of scientific papers (RF 415, RF 417 in Table 8), whereas the research themes on facial image processing and acquisition emerged in the neutral RFs group (RF 49, RF 51, RF 48, RF 87 in Table 11), and the themes on diverse methods were distributed in the recently emerging RFs group (RF 45, RF 90, RF 175 in Table 11) from the perspective of patents. Third, the gesture recognition research fields emerged in the persistently emerging RFs for scientific papers (RF 92 in Table 8), and in the neutral (RF 52, RF 56 in Table 11) and recently emerging RFs (RF 237 in Table 11) for patents. Fourth, the voice recognition research fields emerged in the persistently emerging and neutral RFs for scientific papers, and in the recently emerging and neutral RFs for patents. The research themes related to recognition algorithm persistently emerged from the perspective of scientific papers (RF 272, RF 254, RF 257 in Table 8), whereas the research themes on voice control method emerged in the recently emerging RF group from the perspective of patents (RF 699 in Table 11). Fifth, the DNA/RNA recognition research fields emerged in the persistently emerging, neutral, and recently emerging RFs for scientific papers, and in the recently emerging RFs for patents. The research themes related to DNA/RNA pattern recognition and sequencing were distributed from the perspective of scientific papers (RF 30, RF 16, RF 20, RF 410, RF 10, RF 13, RF 31, RF 29, RF 1, RF 17 in Table 8), whereas the research themes on DNA detection emerged in the recently emerging RFs group from the perspective of patents (RF 1 in Table 11). Finally, the vein recognition research fields emerged in the recently emerging RFs for both scientific papers and patents. From the perspective of scientific papers, the research theme was specified as sclera vein recognition (RF 33 in Table 8), whereas the research themes were rather general from the perspective of patents (RF 8, RF 20 in Table 11). In addition, the RFs of image recognition emerged in the neutral and recently emerging RFs groups from the perspective of only patents (RF 49, RF 154 in Table 11).

5. Discussion

5.1. Promising Research Frontiers with the Proposed Model and the Gartner's Hype Cycle

In terms of Proposition 1 on identifying promising research frontiers through a quantitative approach using technological documents, the predicted results based on data from 2005 to 2014 by the proposed model are compared to the results derived from the hype cycle for emerging technologies in 2015 [13], which is a graphical presentation developed by Gartner, the American IT research and advisory firm. The hype cycle provides five phases to present the maturity of emerging technologies, which are innovation trigger, peak of inflated expectations, trough of disillusionment, slope of enlightenment, and plateau of productivity. We matched the technologies related to facial expression recognition to affective computing technology on the hype cycle, biometric recognition relevant technologies to brain-computer interface (BCI) and biochips technology on the hype cycle, the voice recognition relevant technologies to speech-to-speech translation and natural language question answering on the hype cycle, and image recognition on human action to gesture control technology on the hype cycle. Tables 13 and 14 show the matched results. Both Tables suggested five phases of the hype cycle, matched technologies on the hype cycle, years to mainstream adoption that was proposed in the hype cycle, RF title, type of RF, and RF rank based on promising score among the total RFs.

Table 13. Results of matched promising RFs from scientific papers in Gartner's hype cycle.

5 Phases in Gartner's Hype Cycle	Matched Technologies	Years to Mainstream Adoption	RF No.	RF Title	Type of RF	Rank
	Affective computing	5 to 10 years	RF 415	Analytic techniques for face recognition	Persistently emerging RF	7
			RF 417	Patterns of feature space, correlation, classification for face recognition	Persistently emerging RF	13
			RF 416	Face recognition method under lighting or color condition	Neutral RF	17
Innovation trigger	Brain computer interface/Biochips	More than 10 years/5 to 10 years	RF 30	DNA Sequencing, and cancerous DNA Recognition	Neutral RF	1
			RF 1	RNA pattern recognition	Persistently emerging RF	3
			RF 16	The pattern of distribution of amino groups for RNA recognition	Neutral RF	5
			RF 20	DNA microarray-based detection	Neutral RF	6
			RF 410	Detection of actionable genomic alterations	Neutral RF	8
			RF 93	Cognition, action, and object manipulation	Persistently emerging RF	9
			RF 10	RNA sequencing	Neutral RF	11
			RF 17	DNA methylation patterns	Persistently emerging RF	15
			RF 13	Nanoscale DNA-polymer micelles	Neutral RF	18
			RF 31	RNA recognition motif protein	Neutral RF	19
			RF 29	HPV DNA detection	Neutral RF	20
Peak of inflated expectation/Trough of disillusionment	Speech-to-speech translation/Natural-language question answering	2 to 5 years/5 to 10 years	RF 254	Robust speech recognition algorithm	Persistently emerging RF	10
			RF 257	Speech recognition by bilateral cochlear implant users	Persistently emerging RF	12
			RF 272	Study on voice recognition	Neutral RF	14
Slope of enlightenment	Gesture control	2 to 5 years	RF 92	Human action and gesture recognition	Persistently emerging RF	2
-	-	-	RF 2	Fingerprint recognition using model-based density map	Persistently emerging RF	4
-	-	-	RF 6	Detection of latent fingerprints	Persistently emerging RF	16
-	-	-	RF 33	Sclera Vein Recognition	Recently emerging RF	27
-	-	-	RF 35	Optimal extraction and fingerprint analysis	Recently emerging RF	34

Table 14. Results of matched promising RFs from patents in Gartner's hype cycle.

5 Phases in Gartner's Hype Cycle	Matched Technologies	Years to Mainstream Adoption	RF No.	RF title	Type of RF	Rank
Innovation trigger	Affective computing	5 to 10 years	RF 49	Image acquisition devices using face detection	Neutral RF	2
			RF 45	Automatic face detection	Recently emerging RF	3
			RF 51	Facial image processing	Neutral RF	5
			RF 48	Image acquisition devices using face detection	Neutral RF	8
			RF 87	Facial feature selection	Neutral RF	12
			RF 90	Facial decoding method	Recently emerging RF	18
			RF 175	Multi angle face recognition	Recently emerging RF	39
	Brain computer interface/Biochips	More than 10 years/5 to 10 years	RF 1	Detecting DNA	Neutral RF	6
			RF 74	Biometric authentication method	Neutral RF	7
Peak of inflated expectation/Trough of disillusionment	Speech-to-speech translation/Natural-language question answering	2 to 5 years/5 to 10 years	RF 699	Voice control method	Recently emerging RF	42
Slope of enlightenment	Gesture control	2 to 5 years	RF 52	Gesture image processing	Neutral RF	4
			RF 56	Automatic recognition by tracking method	Neutral RF	11
			RF 237	Recursive motion recognition	Recently emerging RF	19
-	-	-	RF 99	Human image recognition	Neutral RF	1
-	-	-	RF 2	Fingerprint recognition using sensors	Neutral RF	9
-	-	-	RF 7	Hand characteristic information	Persistently emerging RF	10
-	-	-	RF 63	Displaying view for recognition	Recently emerging RF	14
-	-	-	RF 8	Vein pattern detection	Recently emerging RF	33
-	-	-	RF 12	Biometric sensor device for fingerprint	Recently emerging RF	36
-	-	-	RF 154	Image discriminating method	Recently emerging RF	37
-	-	-	RF 20	Blood vessel recognition for treat	Recently emerging RF	43

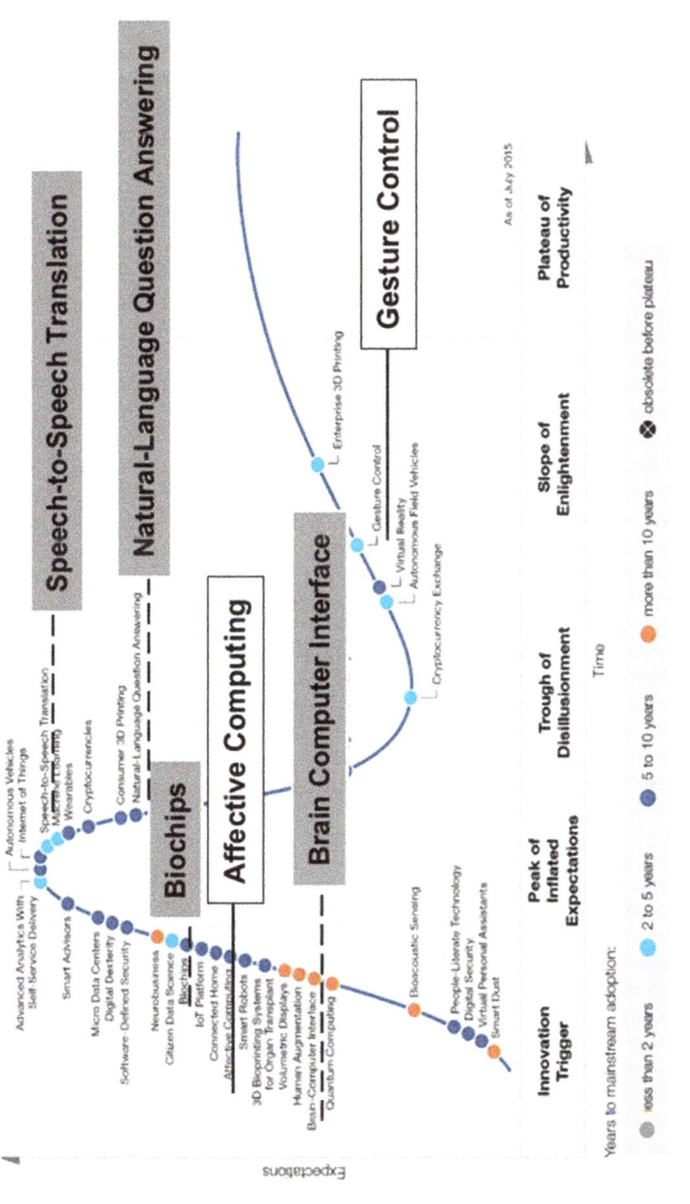

Figure 3. Pattern recognition relevant technologies on Gartner's hype cycle (Source: Burton and Walker, 2015).

The promising research frontiers predicted through the proposed method using data from 2005 to 2014 were well-matched to the emerging technologies for 2015 that were provided by Gartner's hype cycle, which can be considered as an expert-based quantitative approach, in both papers and patent perspectives. The 18 promising RFs were matched to technologies on the hype cycle among 22 promising RFs in terms of scientific papers. The four RFs that were not matched were the fingerprint and vein relevant research themes. The 13 promising RFs were matched to technologies on the hype cycle among 21 promising RFs in terms of patents. The eight RFs that were not matched included high ranked and neutral or persistently emerging RFs, such as fingerprint and hand characteristic recognition, and low ranked but recently emerging RFs, such as vein recognition and biometric sensor research themes. From the scientific paper perspective, the predicted 9 RFs among the top 10 RFs based on the promising score were matched, and from the patent perspective, 7 RFs among the top 10 RFs were matched. All matched RFs based on scientific papers were ranked in the top 20 promising score, whereas 11 RFs based on patents, which excepted 2 RFs among the 13 matched RFs, were ranked in the top 20. Most of the high ranked RFs had a tendency to be matched in the innovation trigger phase, DNA and RNA pattern recognition technology relevant RFs were matched to BCI and biochips, whose years to mainstream adoption were more than 10 year or 5 to 10 years from the scientific paper perspective, whereas the RFs related to affective computing technology whose years to mainstream adoption were 5 to 10 years were relatively more located in the innovation trigger phase from the patent perspective. However, they differed in that the RFs from scientific papers tend to be located in the innovation trigger and peak of the inflated expectation phases, whereas the RFs from patents tend to be located in the innovation trigger and slope of the enlightenment phases. Figure 3 compares the results of the predominant technologies in terms of the perspectives of papers and patent. The proposed promising research frontiers suggest the micro-level of research topics than the emerging technologies in Gartner's hype cycle shown in Tables 13 and 14. For example, there are many RFs with specific titles that are related to DNA or RNA sequencing and pattern recognition (relatively micro-level topics) are suggested in regard to BCI and biochips (macro-level topic) in the hype cycle. It can offer more micro-level information for strategic R&D planning for future promising technology because the suggested method is a bottom-up approach based on core technological documents.

5.2. Comparison of the Promising Research Frontiers from Scientific Papers and Patents

Regarding Proposition 2 on the difference between the results of promising research frontiers derived from scientific papers and patents, the academic papers account for high proportion in the order of persistently emerging RF, neutral RF, and recently emerging RF; whereas from the technological perspective, patents account for high proportion in the order of neutral RF, recently emerging RF, and persistently emerging RF shown in Tables 7 and 10. The rate of persistently emerging RFs from the results of scientific papers was 7.5 times higher than that from the results of patents, whereas the rate of recently emerging RFs from the results of patents was 15 times higher than that from the results of scientific papers. The differences can be interpreted by referring to the nature of scientific research and patents. Academic research has the characteristics of persistent momentum because collective efforts are invested to build a theoretical foundation for future research. However, since a new trial is critical in patents to develop a leading-edge technology and avoid the legal right of existing patents, the recently emerging RFs should be emphasized. It is consistent with the results of the previous research that analyzed scientific papers and patents in solar cell technology field in that scientific articles tended to include more basic research, whereas patents focused on applied and industrial technology [30,54].

For the comprehensive understanding with the results matched to Gartner's hype cycle, the results from scientific papers propose promising RFs that have relatively long years to mainstream adoption periods. The proposed method using scientific papers is appropriate to propose the promising research themes of research and development (R&D) with a long-term perspective. However, the results from patents suggest promising RFs that have relatively short years to mainstream adoption periods. Thus, the proposed method using patents is proper to suggest promising themes for the

R&D with a short-term perspective. The fact that scientific knowledge provides a fundamental basis for technology-oriented innovation, which consists of three main layers such as science, technology, and industry, is widely accepted [54]. This linear model explains that scientists and engineers in the research lab create the seeds of innovation, companies take up these seeds, develop technologies, and introduce them into production although this linear model is often criticized because there are many attempts to flexible technological collaborations between universities and firms in order to reduce uncertainty and risk of the R&D project [14]. The results of this research are partially supported by the linear model in that the RFs from scientific papers tend to play seeds of innovation with a long-term perspective of R&D whereas RFs from patents are related to applied technology in the short-term perspective. However, it is also partially supported by a flexible innovation model because the results are shown in results from both technological documents.

Several implications on the RFs of scientific papers and patents can be discussed in order to utilize the results. First, considerable RFs such as fingerprint recognition and face detection-related technology appear in both academic and practical worlds. Such commonly emerging areas should be regarded as a definitely promising technology category. Second, most RFs identified from the scientific papers are prior to RFs through patent analysis. However, an RF from the analysis of papers has not been realized by active patenting activities. Thus, a list of RFs which are in a persistently emerging RF for papers and simultaneously recently emerging RF for patents can be useful for research organizations to plan their technology investment. Third, a unique group of RFs that do not appear in the analysis for scientific papers but are involved in the recently emerging RFs category must be interesting to companies. Such RFs can be regarded as an emerging technology area that the academic papers related to the technology are not new. Thus, these implications can assist in implementing an effective technology strategy based on the analysis of both papers and patents.

To apply the proposed method to strategic R&D planning, the process using scientific papers should be considered in advance, rather than using patents. The process based on scientific papers is proper to propose the impactful emerging technology henceforth, because the promising RFs from papers are the technologies that have a time lag to be commercialized, whereas the promising RFs from patents are the technologies that actively are applied to a product, and have high technological maturity. Therefore, we suggest brief guidelines for using the method for strategic R&D planning in terms of priority. First, the promising area derived from scientific papers should be considered as the first priority. Second, recently emerging RFs should be preferentially taken into account, rather than neutral RFs and persistently emerging RFs. Finally, the promising area derived from patents can be considered when the RF is in the recently emerging RF group, and commonly emerged in the areas from the analysis of scientific papers.

6. Conclusions

A quantitative methodology for detecting promising research areas is proposed in this research, using bibliometric analysis based on both scientific papers and patents. The indices for evaluating technological documents, research organizations, and research frontiers are suggested using bibliographic information, by reflecting the characteristics of both scientific papers and patents. The proposed indices were developed by considering the attributes of promising technologies, such as fast recent growth, change to something new, market potential, and science-based innovation. The research frontiers are suggested by the Girvan and Newman clustering algorithm. The proposed method was applied to pattern recognition technology for illustration. The results of the proposed promising research frontiers are compared to the results of the hype cycle proposed by Gartner in order to confirm the Proposition 1 while the results of scientific papers and patents are compared in regard to the Proposition 2.

There are several findings from the results applying the model. First, the results derived from scientific papers can be utilized for suggesting themes for the research (R) of R&D, whereas the results derived from patents are proper to provide themes for the development (D) of R&D. Second, the rate of recently emerging RFs derived from patents is much higher than that derived from scientific papers, whereas the rate of persistently emerging RFs derived from scientific papers is much higher than that derived from patents. Third, the predicted promising RFs were well-matched to technologies on the Gartner's hype cycle. The RFs from scientific papers have a tendency to locate in the innovation trigger and peak of the inflated expectation phases, whereas the RFs from patents tend to be located in the innovation trigger and slope of the enlightenment phases.

The proposed method and results can be utilized in various ways. First, the results and method can be utilized to build strategies for collaborative R&D between universities and firms because it is the method considering both academic and industrial sides. Second, an R&D policy maker can utilize it as an objective reference data and a supporting tool for decision making on a policy of promising technology. Third, this method can be appropriate for small and medium-sized enterprises which have relatively lower capability to discover new technological opportunities by domain experts compared to large companies.

Overall, this study makes the following contributions. First, in the perspective of data utilization, a quantitative approach is suggested by using both scientific papers and patents as data for an academic and technology perspective respectively. In the process of data collection, several limitations were overcome. First, data was extracted by the annual rate of total data to prevent biased extraction of data. Second, the recent results of research by leading research groups are added to the extracted core technological document data, in order to include recent core documents. Second, from the perspective of methodology, several indices are proposed based on comprehensive understanding of the property of promising technology using bibliographic information in respective steps to evaluate technological documents and research capability of research organization, and to measure how promising the technology is. It is advantageous in that it is relatively simple to apply them to practice compared to using complicated data analytic methods such as citation-based analysis and network analysis. However, it has a limitation that the correlation check among indices was not thoroughly conducted, although the indices are developed based on different perspectives using different bibliographic information. In addition, in terms of clustering technological documents, the ambiguity of the number of clusters can be solved by using the modularity of Girvan-Newman clustering. Finally, in the perspective of the utilization of the results, the results show reliability because it was well matched to the hype cycle and consistency with the results and findings of the previous studies.

Although this research proposed a new approach to identifying promising technology, this paper has limitations. First, this paper briefly mentioned that recently emerging outlier documents can be considered as a weak signal for promising research themes in terms of novelty. However, although they can be a candidate for promising technology, we did not investigate the contents of outliers in detail. Second, in the process of adding technological documents of leading research organizations, the criteria to select the number of leading organizations and the cut-off value of paper quality are dependent on the domain experts. Even though this paper provided a rationale for the criteria, more robust criteria need to be suggested. Thus, future research can explore promising research themes based on outliers by extending the in-depth analysis. Furthermore, more sophisticated analysis such as sensitivity analysis on the criteria for the analysis on leading research organizations can improve the validity of the proposed approach.

Author Contributions: Research design, I.P. and B.Y.; Methodology, I.P. and B.Y.; Data analysis, I.P.; Investigation, I.P.; Funding acquisition, B.Y.; Writing–original draft, I.P.; Writing–review & editing, I.P. and B.Y.

Funding: This work was supported by Global Research Network program through the Ministry of Education of the Republic of Korea and the National Research Foundation of Korea (NRF-2016S1A2A2916222).

Conflicts of Interest: The authors declare no conflicts of interest.

Appendix A

Table A1. Searching queries for scientific paper and patent on pattern recognition.

Upper Classification	Middle Classification	Lower Classification	Patent Searching Query	Scientific Paper Searching Query
Biometric recognition	Biometric recognition	DNA recognition	TI = (Recogni* or Cogni* or Realiz* or Perce* or Sens* or detect*) and (DNA* or RNA*)) and AB = ((Recogni* or Cogni* or Realiz* or Perce* or Sens* or detect*) and (DNA* or RNA*)) and (RD >= 20050101 and RD <= 20141231)	((Recogni* or Cogni* or Realiz* or Perce* or Sens* or detect*) and (DNA* or RNA*)) and pattern*
		Vein recognition	TI = ((Recogni* or Cogni* or Realiz* or Perce* or Sens* or detect*) and vein) or AB = ((Recogni* or Cogni* or Realiz* or Perce* or Sens* or detect*) and vein) and (RD >= 20050101 and RD <= 20141231)	((((Recogni* or Cogni* or Realiz* or Perce* or Sens* or detect*) and vein))
		Fingerprint recognition	TI = ((Recogni* or Cogni* or Realiz* or Perce* or Sens* or detect*) and (fingerprint* or thumb*)) and (RD >= 20050101 and RD <= 20141231)	(((Recogni* or Cogni* or Realiz* or Perce* or Sens* or detect*) and (fingerprint* or thumb*)))
		Iris recognition	TI = ((Recogni* or Cogni* or Realiz* or Perce* or Sens* or detect*) and Iris) or AB = ((Recogni* or Cogni* or Realiz* or Perce* or Sens* or detect*) and Iris) and (RD >= 20050101 and RD <= 20141231)	(((Recogni* or Cogni* or Realiz* or Perce* or Sens* or detect*) and Iris))
Image recognition	Object recognition	Object recognition	TI = (("feature vector") or "SITF" or ("robust feature")) or AB = (("feature vector") or "SITF" or ("robust feature")) and (RD >= 20050101 and RD <= 20141231)	-
	Human recognition	Human detection and trace	TI = ((("Motion detection") or ("Multiple Threshold")) and (Recogni* or Cogni* or detect*)) or AB = ((("Motion detection") or ("Multiple Threshold")) near/2 (Recogni* or Cogni* or detect*)) and (RD >= 20050101 and RD <= 20141231)	(((("Motion detection") or ("Multiple Threshold")) and (Recogni* or Cogni* or detect*)))
		Face recognition	TI = ("HAAR" or ((Recogni* or detect*) near/2 (face*))) or AB = ("HAAR" or ((Recogni* or detect*) near/2 (face*))) and (RD >= 20050101 and RD <= 20141231)	(("HAAR" or ((Recogni* or detect*) near/2 (face*)))) and pattern*
		Action and gesture recognition	TI = (((Recogni* or Cogni* or detect*) near/2 (gesture* or action* or "Active Marker" or "Passive Marker"))) or AB = (((Recogni* or Cogni* or detect*) near/2 (gesture* or action* or "Active Marker" or "Passive Marker"))) and (RD >= 20050101 and RD <= 20141231)	((((Recogni* or Cogni* or detect*) near/2 (gesture* or action* or "Active Marker" or "Passive Marker"))))
Voice recognition	Utterance recognition	Isolated language recognition	TI = (isolat* or fix*) and (word* or voca* or speech* or language*) and ((VQ) or (Recogni* or Cogni* or Realiz* or Perce* or Sens*)) or AB = (isolat* or fix*) and (word* or voca* or speech* or language*) and ((VQ or LPC OR mfcc or vq or dtw) or (Recogni* or Cogni* or Realiz* or Perce* or Sens*)) and (RD >= 20050101 and RD <= 20141231)	("voice recognition" or "speech recognition" or "language recognition") and ("voice recognition" or "speech recognition" or "language recognition")
		Continuous speech recognition	TI = (connect* or continu* or flexi*) and (word* or voca* or speech*) and ((LPC or MFCC or VQ or DTW) or (Recogni* or Cogni* or Realiz* or Perce* or Sens*)) and (RD >= 20050101 and RD <= 20141231)	
	Speaker recognition	Speaker recognition	TI = ((((((voice or speach or sentence or pronounc*) and (Recogni* or Cogni* or Realiz* or Perce* or Sens* or detect*)) or "AVR" or "VAD" or "Automatic voice recognition"))) and identi*) or AB = ((((((voice or speach or sentence or pronounc*) and (Recogni* or Cogni* or Realiz* or Perce* or Sens* or detect*)) or "AVR" or "VAD" or "Automatic voice recognition"))) and identi*) and (RD >= 20050101 and RD <= 20141231)	

Note: The star * indicates any character string of zero or more characters. (e.g., 'Recogni*' can search for 'recognize', 'recognition' etc.)

Table A2. Summary of the equations for indices.

Indices	Source	Perspective	Bibliographic Information	Equations
Evaluation for technological documents	Scientific paper	Paper impact	Forward citation	**Equation (1)** Paper Impact $= \frac{No.\ of\ forward\ citation - min(No.\ of\ forward\ citation)}{max(No.\ of\ forward\ citation) - min(No.\ of\ forward\ citation)}$
		Research impact	Journal impact factor (JIF), Forward citation	**Equation (2)** Research Impact $=$ JIF $\times$ No. of forward citation **Equation (3)** Norm. Research Impact $= \frac{Research\ Impact - min(Research\ Impact)}{max(Research\ Impact) - min(Research\ Impact)}$
	Patent	Patent novelty	Backward citation	**Equation (4)** Patent novelty $= 1 - \frac{No.\ of\ backward\ citation - min(No.\ of\ backward\ citation)}{max(No.\ of\ backward\ citation) - min(No.\ of\ backward\ citation)}$
		Patent impact	Forward citation	**Equation (1)** Patent Impact $= \frac{No.\ of\ forward\ citation - min(No.\ of\ forward\ citation)}{max(No.\ of\ forward\ citation) - min(No.\ of\ forward\ citation)}$
		Patent marketability	Patent family	**Equation (5)** Patent Marketability $= \frac{Patent\ family\ size - min(Patent\ family\ size)}{max(Patent\ family\ size) - min(Patent\ family\ size)}$
		Patent right range	Claim	**Equation (6)** Patent Right range $= \frac{No.\ of\ independent\ claim - min(No.\ of\ independent\ claim)}{max(No.\ of\ independent\ claim) - min(No.\ of\ independent\ claim)}$
Evaluation for research organizations	Scientific paper	RO's activity for publication	Frequency	**Equation (7)** RO's activity index (AI) $= \frac{No.\ of\ papers\ of\ RO}{total\ No.\ of\ papers}$ **Equation (8)** Norm. AI $= \frac{AI - min(AI)}{max(AI) - min(AI)}$
		RO's productivity for core publication	Frequency, Journal impact factor	**Equation (9)** RO's productivity index (PI) $= \frac{No.\ of\ RO's\ papers\ published\ in\ core\ journal}{total\ No.\ of\ RO's\ papers} \times 100$ **Equation (10)** Norm. PI $= \frac{PI - min(PI)}{max(PI) - min(PI)}$
		Impact of RO's publication	Frequency, Forward citation	**Equation (11)** Impact of RO's publication index (II) $= \frac{\frac{Forward\ citation\ of\ papers\ by\ RO}{Forward\ citation\ of\ total\ papers}}{\frac{No.\ of\ papers\ published\ by\ RO}{Total\ No.\ of\ papers}}$ **Equation (12)** Norm. II $= \frac{II - min(II)}{max(II) - min(II)}$
	Patent	RO's activity for patent application	Frequency	**Equation (7)** RO's activity index (AI) $= \frac{No.\ of\ patents\ of\ RO}{total\ No.\ of\ patents}$ **Equation (8)** Norm. AI $= \frac{AI - min(AI)}{max(AI) - min(AI)}$
		Market competitiveness of RO's patents	Patent family	**Equation (13)** RO's market competitiveness index (MCI) $= \frac{RO's\ patent\ family\ size}{the\ average\ patent\ family\ size}$
		Effect of RO's patents	Forward citation	**Equation (11)** Impact of RO's publication index (II) $= \frac{\frac{Forward\ citation\ of\ patentss\ by\ RO}{Forward\ citation\ of\ total\ patents}}{\frac{No.\ of\ patents\ published\ by\ RO}{Total\ No.\ of\ patents}}$ **Equation (12)** Norm. II $= \frac{II - min(II)}{max(II) - min(II)}$

Table A2. *Cont.*

Indices	Source	Perspective	Bibliographic Information	Equations
Promising indices for promising research frontiers	Scientific paper	Growth	Frequency	**Equation (17)** Growth Index (GI) $= \frac{A_i}{N} \times (\frac{\sum(\frac{P_t - P_{t-1}}{P_{t-1}})}{n-1} \times 100)$
		Impact	Forward citation	**Equation (18)** Impact Index (II) $= \frac{C_i}{P_i}$
		Science-based effect	Journal impact factor	**Equation (19)** Sci − based Effect Index (SEI) $= \frac{IF_i}{P_i}$
	Patent	Growth	Frequency	**Equation (17)** Growth Index (GI) $= \frac{A_i}{N} \times (\frac{\sum(\frac{P_t - P_{t-1}}{P_{t-1}})}{n-1} \times 100)$
		Marketability	Patent family	**Equation (20)** Marketability Index (MI) $= \frac{F_i}{P_i}$
		Impact	Forward citation	**Equation (18)** Impact Index (II) $= \frac{C_i}{P_i}$

References

1. Park, I.; Park, G.; Yoon, B.; Koh, S. Exploring Promising Technology in ICT Sector Using Patent Network and Promising Index Based on Patent Information. *ETRI J.* **2016**, *38*, 405–415. [CrossRef]
2. Shibata, N.; Kajikawa, Y.; Takeda, Y.; Sakata, I.; Matsushima, K. Detecting emerging research fronts in regenerative medicine by the citation network analysis of scientific publications. *Technol. Forecast. Soc. Chang.* **2011**, *78*, 274–282. [CrossRef]
3. Ciarli, T.; Coad, A.; Rafols, I. Quantitative analysis of technology futures: A review of techniques, uses and characteristics. *Sci. Public Policy* **2016**, *43*, 630–645. [CrossRef]
4. Soranzo, B.; Nosella, A.; Filippini, R. Managing firm patents: A bibliometric investigation into the state of the art. *J. Eng. Technol. Manag.* **2016**, *42*, 15–30. [CrossRef]
5. Chen, N.; Liu, Y.; Cheng, Y.; Liu, L.; Yan, Z.; Tao, L.; Guo, X.; Luo, Y.; Yan, A. Technology resource, distribution, and development characteristics of global influenza virus vaccine: A patent bibliometric analysis. *PLoS ONE* **2015**, *10*, e0136953. [CrossRef] [PubMed]
6. Park, H.; Magee, C.L. Tracing technological development trajectories: A genetic knowledge persistence-based main path approach. *PLoS ONE* **2017**, *12*, e0170895. [CrossRef] [PubMed]
7. Youtie, J.; Porter, A.L.; Huang, Y. Early social science research about Big Data. *Sci. Public Policy* **2016**, *44*, 65–74. [CrossRef]
8. Zhou, Y.; Li, X.; Lema, R.; Urban, F. Comparing the knowledge bases of wind turbine firms in Asia and Europe: Patent trajectories, networks, and globalisation. *Sci. Public Policy* **2015**, *43*, 476–491. [CrossRef]
9. Roepke, S.; Moehrle, M.G. Sequencing the evolution of technologies in a system-oriented way: The concept of technology-DNA. *J. Eng. Technol. Manag.* **2014**, *32*, 110–128. [CrossRef]
10. Cho, Y.; Kim, M. Entropy and gravity concepts as new methodological indexes to investigate technological convergence: Patent network-based approach. *PLoS ONE* **2014**, *9*, e98009. [CrossRef] [PubMed]
11. Lee, W.J.; Lee, W.K.; Sohn, S.Y. Patent network analysis and quadratic assignment procedures to identify the convergence of robot technologies. *PLoS ONE* **2016**, *11*, e0165091. [CrossRef] [PubMed]
12. Van Reenen, J.; Bloom, N.; Draca, M.; Kretschmer, T.; Sadun, R.; Overman, H.; Schankerman, M. *The Economic Impact of ICT*; Final report; John Van Reenen London School of Economics: London, UK, 2010.
13. Burton, B.; Walker, M. *Hype Cycle for Emerging Technologies, 2015*; Gartner's Hype Cycle Special Report; Gartner: Stamford, CT, USA, 2015.
14. Niosi, J. Fourth-generation R&D: From linear models to flexible innovation. *J. Bus. Res.* **1999**, *45*, 111–117.
15. Noh, H.; Song, Y.-K.; Lee, S. Identifying emerging core technologies for the future: Case study of patents published by leading telecommunication organizations. *Telecommun. Policy* **2016**, *40*, 956–970. [CrossRef]
16. Lee, W.H. How to identify emerging research fields using scientometrics: An example in the field of Information Security. *Scientometrics* **2008**, *76*, 503–525. [CrossRef]
17. Iwami, S.; Mori, J.; Sakata, I.; Kajikawa, Y. Detection method of emerging leading papers using time transition. *Scientometrics* **2014**, *101*, 1515–1533. [CrossRef]
18. Corrocher, N.; Malerba, F.; Montobbio, F. *The Emergence of New Technologies in the ICT Field: Main Actors, Geographical Distribution and Knowledge Sources*; Department of Economics, University of Insubria: Varese, Italy, 2003.
19. Day, G.S.; Schoemaker, P.J. A different game. In *Wharton on Managing Emerging Technologies*; John Wiley & Sons Inc.: New York, NY, USA, 2000.
20. Hung, S.-C.; Chu, Y.-Y. Stimulating new industries from emerging technologies: Challenges for the public sector. *Technovation* **2006**, *26*, 104–110. [CrossRef]
21. Porter, A.L.; Roessner, J.D.; Jin, X.-Y.; Newman, N.C. Measuring national 'emerging technology' capabilities. *Sci. Public Policy* **2002**, *29*, 189–200. [CrossRef]
22. Visessonchok, T.; Sasaki, H.; Sakata, I. Detection and introduction of emerging technologies for green buildings in Thailand. In Proceedings of the 2014 Portland International Conference on Management of Engineering & Technology (PICMET), Kanazawa, Japan, 27–31 July 2014; pp. 620–631.
23. Breitzman, A.; Thomas, P. The Emerging Clusters Model: A tool for identifying emerging technologies across multiple patent systems. *Res. Policy* **2015**, *44*, 195–205. [CrossRef]

24. Érdi, P.; Makovi, K.; Somogyvári, Z.; Strandburg, K.; Tobochnik, J.; Volf, P.; Zalányi, L. Prediction of emerging technologies based on analysis of the US patent citation network. *Scientometrics* **2013**, *95*, 225–242. [CrossRef]

25. Lucio-Arias, D.; Leydesdorff, L. An indicator of research front activity: Measuring intellectual organization as uncertainty reduction in document sets. *J. Am. Soc. Inf. Sci. Technol.* **2009**, *60*, 2488–2498. [CrossRef]

26. Jarneving, B. Bibliographic coupling and its application to research-front and other core documents. *J. Inform.* **2007**, *1*, 287–307. [CrossRef]

27. Jarneving, B. A comparison of two bibliometric methods for mapping of the research front. *Scientometrics* **2005**, *65*, 245–263. [CrossRef]

28. Boyack, K.W.; Klavans, R. Co-citation analysis, bibliographic coupling, and direct citation: Which citation approach represents the research front most accurately? *J. Am. Soc. Inf. Sci. Technol.* **2010**, *61*, 2389–2404. [CrossRef]

29. Toivanen, H. The shift from theory to innovation: The evolution of Brazilian research frontiers 2005–2011. *Technol. Anal. Strateg. Manag.* **2014**, *26*, 105–119. [CrossRef]

30. Park, I.; Lee, K.; Yoon, B. Exploring Promising Research Frontiers Based on Knowledge Maps in the Solar Cell Technology Field. *Sustainability* **2015**, *7*, 13660–13689. [CrossRef]

31. Cozzens, S.; Gatchair, S.; Kang, J.; Kim, K.-S.; Lee, H.J.; Ordóñez, G.; Porter, A. Emerging technologies: Quantitative identification and measurement. *Technol. Anal. Strateg. Manag.* **2010**, *22*, 361–376. [CrossRef]

32. Rotolo, D.; Hicks, D.; Martin, B.R. What is an emerging technology? *Res. Policy* **2015**, *44*, 1827–1843. [CrossRef]

33. Small, H.; Boyack, K.W.; Klavans, R. Identifying emerging topics in science and technology. *Res. Policy* **2014**, *43*, 1450–1467. [CrossRef]

34. Srinivasan, R. Sources, characteristics and effects of emerging technologies: Research opportunities in innovation. *Ind. Mark. Manag.* **2008**, *37*, 633–640. [CrossRef]

35. Stahl, B.C. What does the future hold? A critical view of emerging information and communication technologies and their social consequences. In *Researching the Future in Information Systems*; Springer: Berlin, Germany, 2011; pp. 59–76.

36. Alexander, J.; Chase, J.; Newman, N.; Porter, A.; Roessner, J.D. Emergence as a conceptual framework for understanding scientific and technological progress. In Proceedings of the PICMET'12—Technology Management for Emerging Technologies (PICMET), Vancouver, BC, Canada, 29 July–2 August 2012; pp. 1286–1292.

37. Martin, B.R. Foresight in science and technology. *Technol. Anal. Strateg. Manag.* **1995**, *7*, 139–168. [CrossRef]

38. Boon, W.; Moors, E. Exploring emerging technologies using metaphors—A study of orphan drugs and pharmacogenomics. *Soc. Sci. Med.* **2008**, *66*, 1915–1927. [CrossRef] [PubMed]

39. Halaweh, M. Emerging technology: What is it. *J. Technol. Manag. Innov.* **2013**, *8*, 108–115. [CrossRef]

40. Setti, G. Bibliometric indicators: Why do we need more than one? *IEEE Access* **2013**, *1*, 232–246. [CrossRef]

41. Polanco, X. *Infométrie et Ingénierie de la Connaissance*; INIST: Vandœuvre-lès-Nancy, France, 1994.

42. Morris, S.; DeYong, C.; Wu, Z.; Salman, S.; Yemenu, D. DIVA: A visualization system for exploring document databases for technology forecasting. *Comput. Ind. Eng.* **2002**, *43*, 841–862. [CrossRef]

43. Daim, T.U.; Rueda, G.; Martin, H.; Gerdsri, P. Forecasting emerging technologies: Use of bibliometrics and patent analysis. *Technol. Forecast. Soc. Chang.* **2006**, *73*, 981–1012. [CrossRef]

44. Ki, W.; Kim, K. Generating Information Relation Matrix Using Semantic Patent Mining for Technology Planning: A Case of Nano-Sensor. *IEEE Access* **2017**, *5*, 26783–26797. [CrossRef]

45. Saka, A.; Igami, M.; Kuwahara, T. *Science Map 2008-Study on Hot Research Areas (2003–2008) by Bibliometric Method*; Institute of Science and Technology Policy Science and Technology Foundation Research laboratory: Washington, DC, USA, 2010.

46. GEUM, Y.; Jeon, J.; Seol, H. Identifying technological opportunities using the novelty detection technique: A case of laser technology in semiconductor manufacturing. *Technol. Anal. Strateg. Manag.* **2013**, *25*, 1–22. [CrossRef]

47. Lee, C.; Kang, B.; Shin, J. Novelty-focused patent mapping for technology opportunity analysis. *Technol. Forecast. Soc. Chang.* **2015**, *90*, 355–365. [CrossRef]

48. Kim, J.; Lee, C. Novelty-focused weak signal detection in futuristic data: Assessing the rarity and paradigm unrelatedness of signals. *Technol. Forecast. Soc. Chang.* **2017**, *120*, 59–76. [CrossRef]

49. Dahlin, K.B.; Behrens, D.M. When is an invention really radical?: Defining and measuring technological radicalness. *Res. Policy* **2005**, *34*, 717–737. [CrossRef]

50. Newman, M.E.; Girvan, M. Finding and evaluating community structure in networks. *Phys. Rev. E* **2004**, *69*, 026113. [CrossRef] [PubMed]
51. Glänzel, W.; Czerwon, H.-J. A new methodological approach to bibliographic coupling and its application to research-front and other core documents. In Proceedings of the International Society for Scientometrics and Informetrics, River Forest, IL, USA, 7–10 June 1995; pp. 167–176.
52. Freeman, L.C. A set of measures of centrality based on betweenness. *Sociometry* **1977**, *40*, 35–41. [CrossRef]
53. Girvan, M.; Newman, M.E. Community structure in social and biological networks. *Proc. Natl. Acad. Sci. USA* **2002**, *99*, 7821–7826. [CrossRef] [PubMed]
54. Shibata, N.; Kajikawa, Y.; Sakata, I. Extracting the commercialization gap between science and technology—Case study of a solar cell. *Technol. Forecast. Soc. Chang.* **2010**, *77*, 1147–1155. [CrossRef]

 *sustainability*

Article

A Conceptual Framework for Assessing an Organization's Readiness to Adopt Big Data

Celina M. Olszak * and Maria Mach-Król

Department of Business Informatics, University of Economics in Katowice, 1 Maja 50, 40-287 Katowice, Poland;
maria.mach-krol@ue.katowice.pl
* Correspondence: celina.olszak@ue.katowice.pl; Tel.: +48-501-260-509

Received: 11 September 2018; Accepted: 15 October 2018; Published: 17 October 2018

Abstract: The main aim of this paper is to provide a theoretically and empirically grounded discussion on big data and to propose a conceptual framework for big data based on a temporal dimension. This study adopts two research methods. The first research method is a critical assessment of the literature that aims to identify the concept of big data in organizations. This method is composed of a search for source materials, the selection of the source materials, and their analysis and synthesis. It has been used to develop a conceptual framework for assessing an organization's readiness to adopt big data. The purpose of the second research method is to provide an initial verification of the developed framework. This verification consisted of conducting qualitative research with the use of an in-depth interview in 15 selected organizations. The main contribution of this study is the Temporal Big Data Maturity Model (TBDMM) framework, which can help to measure the current state of an organization's big data assets, and to plan their future development with respect to sustainability issues. The proposed framework has been built over a time dimension as a fundamental internal structure with the goal of providing a complete means for assessing an organization's readiness to process the temporal data and knowledge that can be found in modern information sources. The proposed framework distinguishes five maturity levels: atemporal, pre-temporal, partly temporal, predominantly temporal, and temporal, which are used to evaluate data/knowledge, information technology (IT) solutions, functionalities offered by IT solutions, and the sustainable development context.

Keywords: big data; maturity model; temporal analytics; advanced business analytics

---

1. Introduction

In recent years, the source of power has shifted from land, finance, and capital to information and knowledge [1]. Organizations can use information: (1) to collect information about competitors, customers, suppliers, industries, and governments; (2) to collect and analyze background information, such as information about technology, politics, and society; (3) to observe the changing business environment and keep track of trends; and (4) to monitor the trends of suppliers of materials, exporting nations, and competitors. Historically, organizations have stored and explored internal information in order to better understand their business processes and to improve their decision-making on an operational and tactical level [2]. Recently, organizations have started to analyze the external information that comes from different dispersed information sources; e.g., from competitors, industries, government, administration, and healthcare [3,4]. In this respect, the study of big data is highly important. On the other hand, practice has shown questionable success from the application of big data. Organizations do not achieve an appropriate degree of benefit from big data usage. The reasons for this failure are not clear and have not been well-investigated. So, there is a need for a more systematic and deliberate study of tools for assessing organizations' readiness to adopt big data.

The first commonly accepted definitions of big data took into account three Vs: Volume, Velocity, and Variety [5–9]. Then, two other Vs (Veracity and Value) were added to form five Vs [10]. Recently, big data has been characterized by seven Vs: Volume, Velocity, Variety, Veracity, Variability, Visualization, and Value [11]. In the contemporary global economy, the velocity dimension of big data is particularly important, as the speed at which an organization takes action is critical to its gaining a sustainable competitive advantage. Organizations have to respond in almost real time to emergent challenges and opportunities [12]. For this reason, real-time big data analytics has gained importance. Two of the hottest trends for 2018 in big data analytics are real-time data analytics and data visualization [13]. All of the definitions of big data, such as those listed in [11], emphasize big data's volume, variety, and rate of change (velocity), and reveal at the same time the strong link between big data analytics and new business insights [14]. These analytical procedures may be even called temporal big data analytics; that is, analytics focused on data change and on a time dimension in the domain being analyzed. At the same time, the greatest benefits (as perceived by top management) for a competitive organization come from the use of all data available from inside and outside the organization at the same time; i.e., its integration and then real-time analysis from a holistic viewpoint [13]. A sustainable competitive advantage and sustainable development are becoming more and more dependent on an organization's ability to manage big data, information, and knowledge [15–17]. In this way, advanced data analytics and a sustainable competitive advantage are linked together. Research shows that any organization that is willing to be competitive and innovative must also be self-adjusting, sensitive, and responsive to have continued market relevance and viability/sustainability [18]. Real-time big data analytics is one of the ways to achieve this state.

It is widely recognized that new sources of insight need to be used, as big data analytics is a new enabler of a sustainable competitive advantage. It enables organizations to have improved business efficiency because of its high operational and strategic potential [19]. "Big data analytics (…) has emerged as the new frontier of innovation and competition" [20] (p. 190).

Although there have been numerous studies on big data over the last few years, they have not addressed the essence of temporal big data's maturity and the design of temporal big data maturity models for assessing an organization's readiness to adopt big data. They have been mainly focused on static and well-structured information. The issue of the design of a framework for assessing an organization's readiness to adopt big data based on the temporal dimension has been insufficiently investigated. The research studies are fragmentary and scattered. There is a lack of a comprehensive framework in this respect as well as a lack of examples on how to build and use such a framework in organizations. Moreover, as stated in [21], the existing frameworks are mainly technically focused. There is no maturity framework that addresses both big data adoption based on the temporal dimension and its implications for an organization's sustainable development issues. On the other hand, the existing maturity models, such as sustainability maturity models and stakeholders' relationships maturity models, do not address advanced analytics issues in a satisfactory manner. Hence, a comprehensive maturity framework aimed at combining these two perspectives is required.

The main aim of this study is to investigate the issue of big data and to propose a conceptual framework for big data based on the temporal dimension. Temporality in the proposed framework extends and enhances the existing maturity models for big data that have been developed, e.g., by [22–26]. The existing frameworks in fact only take into account the three classical Vs (Volume, Variety, and Veracity) or sometimes five Vs (Volume, Variety, Veracity, Visualization, and Value). The proposed framework also incorporates the sixth and the seventh Vs (Velocity and Variability); thus, it provides a complete toolkit for assessing not only the current state of an organization's big data assets and its current set of analytical tools, but also the organization's preparedness for processing this influx of temporal data and/or knowledge that can be found in modern sources of information.

This study adopts two research methods. The first research method is a critical assessment of the literature that aims to identify the concept of big data in organizations. This method is used to develop a conceptual framework for assessing an organization's readiness to adopt big data. The purpose of the second research method is to provide an initial verification of the developed framework. This verification consists of conducting qualitative research with the use of an in-depth interview in 15 selected organizations.

Our study makes several theoretical contributions to the relevant literature. First, big data (BD) is generally an unexplored field of research. Therefore, the current study contributes to the emerging literature on BD by investigating the issue of BD from a temporal perspective. Second, the current study is one of the rare studies that proposes a framework that would be helpful to measure the current state of an organization's big data assets and to plan for their future development. Third, this study demonstrates that the proposed framework distinguishes five maturity levels—atemporal, pre-temporal, partly temporal, predominantly temporal, and temporal—that are used to evaluate data/knowledge, information technology (IT) solutions, functionalities offered by IT solutions for organizations, and the sustainable development context. Finally, the current findings provide empirical evidence that the proposed framework may play an important role in the management of organizations.

The structure of the paper is as follows. In Section 2, related work on big data, sustainability, and maturity models is presented and discussed. Section 3 is devoted to the Temporal Big Data Maturity Model (TBDMM) framework. In Section 4, the reception of the TBDMM among managers is presented in detail. Section 5 contains a discussion of the research findings, and Section 6 contains our concluding remarks.

2. Related Work

2.1. Big Data and Sustainability Issues

The interest in analytics in the information systems domain has continued for many years. This is reflected in research that was conducted in the 1980s and 1990s and that referred mainly to Management Information Systems (MIS), Decision Support Systems (DSS), and Enterprise Information Systems (EIS) [2,27–29]. The 21st century has been characterized by the development of data warehousing, Online Analytical Processing (OLAP), Business Intelligence (BI), Competitive Intelligence (CI), and big data. The common assumption that underlies these analytical systems is that information that is appropriately acquired, collected, analyzed, integrated, and used may be a critical component in the success of an organization. Analytical systems may help organizations to reach strategic goals, make better decisions, improve business processes, increase profitability, and improve customer satisfaction [29–33]. At the same time, it is emphasized that this challenge becomes more difficult with the constantly increasing volume of information, both internal and external, that is available to an organization.

To better understand the concept of big data, it is worth recalling the basic assumptions that underlie the idea of BI and CI, which represent the most recent stages in the evolution of analytical systems. It is said that BI is focused mainly on acquiring and analyzing internal information; thus, it helps to improve internal business processes and decision-making on an operational and tactical level [2]. CI refers to the collection and exploration of external information that comes from an organization's environment [3,4]. Organizations believe that external information is as valuable to them (and sometimes even more) as information that comes from their own internal sources. According to many authors, organizations that are able to understand their environment and their competitors and establish competitive management strategies will win in this rapidly globalizing information society [1,28,34]. It has been highlighted [35] that BI cannot replace CI and vice versa. The main aim of BI is to manage an organization's well-structured internal data and processes. Typical BI tools include reporting, OLAP, data warehouse, and data mining and visualization toolkits. In contrast, data that originates from external sources, which is distinctive of CI, is usually semi-structured and unstructured. Therefore, a strong position in CI requires tools such as advanced data mining, predictive modeling, web mining, text mining, and opinion mining tools. Big data, compared to BI and CI, concerns an even broader spectrum of data and more complex and sophisticated data processing. In general, it is not possible to store and process this kind of data by conventional databases and data analysis techniques [36]. It requires tools and methods that are capable of analyzing and extracting patterns from large-scale data. The rise of big data has been caused by increased data storage capabilities, increased computational processing power, and the availability of huge volumes of data [37]. Big data mainly concerns unstructured information about competitors, customers, and other stakeholders of the organization (Table 1).

Table 1. The basic differences between Business Intelligence (BI), Competitive Intelligence (CI), and big data.

	Business Intelligence	**Competitive Intelligence**	**Big Data**
Purpose	Analysis of internal business processes, improvement of operational and tactical decisions	Analysis of external environment (mainly competitors)	Analysis of the whole environment of the organization: internal resources, customers, suppliers, users of the Internet, and communities of practices
Scope	Organization	Environment of organization (mainly competitors)	Whole environment of the organization
Content/data	Well-structured information, internal data originating from databases, Enterprise Resource Planning, transaction systems	Semi-structured or unstructured information, external data originating from competitors, customers, and the Internet	Unstructured content, external data that comes from public, open resources, the Internet, mobile devices, and social media
Used tools, technologies	Online Analytical Processing (OLAP), data mining, data warehouses	Advanced data mining, predictive modeling, web mining, text mining, and opinion mining	Advanced data mining, predictive modeling, web mining, opinion mining, text mining, exponential random graph models, search-based applications, dashboards, SOA (Service-Oriented Architecture), Hadoop, Spark, MapReduce, parallel processing, real-time processing, and machine learning techniques

A critical analysis of the literature on the subject shows that there is no consensus on the correct interpretation of the term "big data" (Table A1 [38–61] in Appendix A).

Ferguson [56] states that the term "big data" is "associated with the new types of workloads and underlying technologies needed to solve business problems that could not be previously supported due to technology limitations, prohibitive cost, or both". It has been highlighted that big data is therefore not just about data volumes but about analytical workloads that are associated with some combination of data volume, data velocity, and data variety that may include complex analytics and complex data types. It is believed that big data, when analyzed in combination with traditional organizational data, enables organizations not only to better understand their business, but first of all to change it and to have new sources of revenue, a more competitive position, and greater innovation. The most widely accepted definition of big data is the one by Gartner [5] (cf. Table A1). Gartner stresses the role of big data as a new type of asset, which—if properly addressed—may provide organizations with enhanced insights and better decision-making. In turn, the most extensive definition of big data has been given by [61], which focuses on data heterogeneity, the lack of structure, and the Internet as big data source. When analyzing such data, organizations gain a deeper understanding of the economic environment, which may lead to the creation of a sustainable business and a sustainable competitive advantage.

When analyzing the literature, it should be noticed that the concept of big data may be explained from a technical or an organizational perspective. From a technical point of view, big data means new technologies and tools that make it possible to process huge amounts of data that come from such new sources as sensors, social media, and real-time systems [9,38–50]. From an organizational point of view, big data means new ways of running a business, of decision making, and of understanding the customers, suppliers, and other stakeholders of organizations [7,8,24,51–54]. There are also some explanations that take both perspectives into account [5,6,55–60]. Some research shows that big data can completely change the functioning of an organization [62]. Recently, the notion of smart data— big data that has been initially preprocessed—has been gaining popularity [63,64]. "Smart data arises by expediently structuring information from big data which then can be used for knowledge advances and decision making" [65].

According to [24,53], big data is a new asset that creates valuable opportunities for organizations. The most outstanding one is creating multidimensional business insights that encompass, inter alia, insights into customers' interests, passions, affiliations, and associations, product performance recommendations, and location-based insights. In this way, an organization is able to quickly respond to external challenges/trends, to make faster decisions, and to obtain more precise answers from data [43].

It has been highlighted that big data, when explored and analyzed together with traditional organizational data, enables organizations not only to better understand their business, but first of all to change it and to have new sources of revenue, a more competitive position, and greater innovation. Nelson [51] says that the use of big data in organizations will contribute mainly to improving decision-making through crowdsourcing, a better visualization of data and complex relationships, making the reporting and monitoring of business decisions easier, making the detection of complex patterns in data faster, and making the search for information and results easier.

It has been stated that big data enables the creation of innovative business models, products, and services. It gives organizations a way to outperform their competition. This kind of data may be used to achieve a better understanding of an organization's customers, employees, partners, and operations [39]. Big data may complement BI systems, and provide in-depth insights and predictive analytics on unstructured massive data. Big data provides organizations with a completely new kind of insight with its analyses of social media, images, natural language, and so on. With big data, unstructured web content may be understood and used for predictive analysis. With such possibilities, big data sources may be used for analysis in such areas as decision-making, customer insights, competitive intelligence systems, marketing, and human resources. Customers and competitors are now understood better, and decisions are better, as they are fact-based [66]. Schmarzo [24] gives some examples of new insights that are possible with big data. These are, among others:

- resource scheduling based on purchase history, buying behaviors, and local weather and events;
- distribution and inventory optimization given current and predicted buying patterns and local demographic, weather, and events data;
- integrating analytics directly into products to create "intelligent" products; and
- insights about customers' usage patterns, product performance behaviors, and overall market trends.

These insights may be also accompanied by social media analysis when an organization obtains feedback on customers' needs and expectations concerning current and future products/services. As pointed out in [67], modern organizations face several new megatrends, one of which is globalization. It is, therefore, not sufficient to analyze only the immediate economic environment in order to be competitive; organizations have to also consider globalization processes. One of the current globalization challenges is to ensure the sustainable evolution of human existence in its social, environmental, and economic dimensions. Hence, business value creation must be geared towards sustainability [65]. According to [68], "sustainability is a paradigm for thinking about the future in which environmental, societal, and economic considerations are equitable in the pursuit of an improved lifestyle", and innovative solutions (such as business models, goods, and services) have to be developed according to this new paradigm. Big data has the potential to transform business processes [69], providing organizations with the opportunity to create new, sustainable business models, and to realize sustainable business value creation in the three sustainability dimensions: economic, social, and environmental [65]. As claimed in [70], these new business models can contribute to solving an environmental or a social problem. It has even been postulated that we treat the big data that is captured and analyzed by organizations as a knowledge commons that will help us to create innovative solutions to various socioeconomic problems [71]. It has also been argued that real-time big data analytics may provide the basis for a more efficient, sustainable, competitive, and productive organization [16]. In this way, temporal big data analytics may link together a successful organization and sustainability. However, there is a lack of a comprehensive framework that links the maturity of an organization's temporal big data analytics adoption with the sustainable development of an organization. There are some frameworks that address in a detailed manner such questions as sustainability maturity [72,73], stakeholders' relationships management maturity [74,75], or related concepts, such as learning maturity [76,77] and change management maturity [78]; however, they do not cover big data analytics issues in relation to management questions.

Along with the increasing number of data sources and types, one important challenge that emerges is how to ensure that the analytical results from big data are trustworthy. Research from IBM has shown that one in three managers do not trust the information used in their decision-making processes, and 27% of managers are unsure about the accuracy of the information [79]. The value of insights from big data depends not only on meeting the challenges from the seven Vs. The analytical process also has to ensure that there are solutions to such problems as data quality, data cleansing, real-time analysis and decision-making, parallel and distributed computing, exploratory analysis, and new models for big data computation (to name only a few) [37].

Despite the many potential benefits of big data analytics that were pointed out above, the current level of its take-up is not satisfactory [13]. This is most probably due to the fact that many organizations lack the knowledge that is required to organize and create big data insights and profit from them. The theory of maturity models may provide organizations with methods and guidelines to develop the big data idea.

2.2. Maturity Models

"Maturity" may be defined as "a state of being complete, perfect, or ready" [80]. Cooke-Davies [81] treats the notion of maturity more widely by pointing out that maturity may be graded from extreme immaturity to extreme maturity. Maturity is obtained in the process of improving certain needed tasks and activities. To assess the level of maturity in various domains, so-called maturity models have been proposed by several authors.

According to [80,82], a maturity model allows us to identify the strong and weak points of a domain. Hence, with such a model, it is possible to assess an organization (or part of it) and to describe its current state of development. Many maturity models originate from the popular Capability Maturity Model (CMM). It was proposed in 1991 for a software development process. This was then superseded by Capability Maturity Model Integration (CMMI), which integrates several sub-models concerning various areas of an organization. In CMMI—and also in many of the models that have followed it—the maturity of an area may be placed on one of several maturity levels. A typical set of maturity levels is presented in Figure 1. The main goal of integrating maturity models into an analysis is to codify knowledge on good processes/activities, codify knowledge on which assessment criteria can be brought to bear on the issue, and to elaborate upon a systematized way of assessing a domain [83].

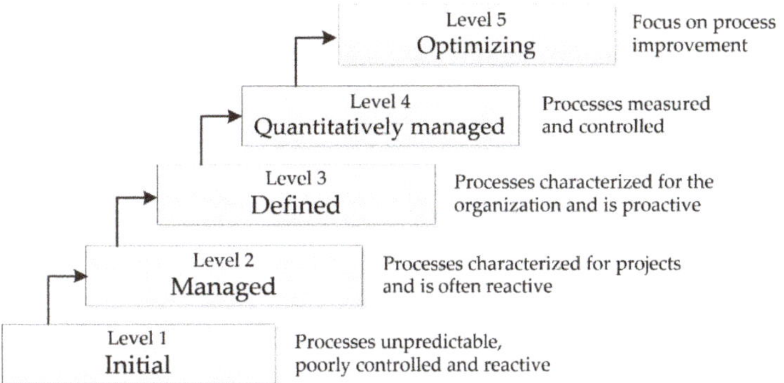

Figure 1. A description of maturity levels in the process approach (the Capability Maturity Model (CMM) model) [84].

Generally speaking, any maturity model falls into one of the following categories [85]:

- descriptive: used to determine an organization's level of maturity;
- prescriptive: describing a desired state and assessing an organization's distance to it;
- transitive: determining the steps that an organization must follow to reach the desired state.

As has already been highlighted, the current set of maturity models were first set up for process management and software development, but the ability of these models to be adapted for other domains has caused their popularity to spread. Hence, maturity models can also be used for the assessment of an organization's ability and capability to make use of big data.

2.3. Maturity Models for Big Data

Researchers have most commonly proposed to use the classical CMM and/or CMMI models to assess an organization's capability to adopt big data. Such an approach is presented in [86]. Spaletto [87] proposes to use the CMM to assess an organization's big data strategy and standards of data growth management. Such an approach would enable us to identify effective strategies for managing exponential data growth.

The authors in [88] propose the use of CMMI and/or Software Process Improvement and Capability Determination (SPICE) models to assess web engineering systems, because, when using them, one may consider also the latest trends, such as cloud computing, that are closely linked to big data. Hence, it would be possible to adapt this approach in order to investigate an organization's maturity in big data adoption.

The SaaS maturity model proposed in [89] may be used to measure the big data maturity of an organization. This is because any organization capable of using SaaS is also capable of using big data. These authors assess maturity in the context of service components and link them to the levels of maturity.

The first model designed to measure the maturity level of an organization in the context of big data analytics has been proposed by The Data Warehouse Institute (TDWI). TDWI's Big Data Maturity Model describes the steps that every organization has to follow while undertaking big data initiatives. The model shows how an organization can transform itself to fully profit from big data, and is composed of five levels [22]: nascent, pre-adoption, early adoption, corporate adoption, and mature/visionary.

Radcliffe Advisory Services [23] proposes a Big Data Maturity Model that aims to arrange the notions that are connected to big data, assess the current state of an organization, and build a vision for the use of big data in the future. The model is similar to other maturity models, and is composed of six levels: five main ones and an additional one, which is described as "level 0" ("In the Dark"), where organizations do not even realize that big data exists. The main levels are called Catching up, First Pilot(s), Tactical Value, Strategic Leverage, and Optimize and Extend. The Radcliffe model is rather general, and the company offers only a set of big data hints to help organizations pass through all of the maturity levels [90].

Schmarzo [24] proposed the Big Data Business Model Maturity Index as a tool for assessing the maturity of a business model in the context of big data usage. As Schmarzo claims, organizations may use this index to:

1. obtain information on their stage of use of advanced big data analytics, on their value creation process, and on their business models; that is, obtain information on their current state;
2. identify the desired target state.

Schmarzo's model is composed of five levels (stages): Business Monitoring, Business Insights, Business Optimization, Data Monetization, and Business Metamorphosis. The first three levels are focused on optimizing an organization's internal business processes. The last two levels are focused on the organization's environment: its clients and markets.

It has to be noted, however, that none of the abovementioned big data maturity models accommodate the important factor of time, even though the temporal dimension is essential due to big data's velocity. For this reason, in this paper, a new conceptual framework for assessing an organization's readiness to adopt big data is proposed. The framework, named the Temporal Big Data Maturity Model (TBDMM), explicitly incorporates the temporal dimension. Temporality in the proposed maturity framework is an element that complements the other models, and enables a full adoption of big data's seven Vs. The existing models in fact take into account only five Vs (Volume, Variety, Veracity, Visualization, and Value), while the framework proposed in this paper also incorporates the sixth and seventh Vs (Velocity and Variability). In this respect, the new maturity framework is more complete than the existing ones.

3. Framework of the Temporal Big Data Maturity Model (TBDMM)

3.1. Levels of Temporality

The main idea of the proposed framework—to incorporate the time dimension in it—results from several observations. First and foremost, all types of knowledge processed by organizations, such as internal knowledge and knowledge from big data, may be considered temporal. The "temporality" of knowledge is seen in knowledge changes; knowledge is mostly dynamic in nature and evolves in time. Hence, knowledge possesses an explicit time dimension that must not be omitted in order to not lose the temporal characteristics of a domain. In this way, time turns out to be one of the most important aspects of knowledge analytics in organizations.

Second, the time dimension is indispensable for making inferences about areas of interest that are dynamic, such as the economic and competition domains. Such inferences can be performed by intelligent computer systems that mimic human reasoning. Hence, data, knowledge, and reasoning may be considered at different levels of temporality.

Regarding the time dimension, there can be the following types of data:

- static data: this data does not contain any temporal context, nor can this context be inferred from it;
- sequences: ordered sequences of static data with no direct time stamps (relative ordering, such as "earlier" or "later");
- time-stamped sequences: sequences of static data stamped with time, which are collected in regular or irregular intervals; and
- fully temporal data: data that contains at least one time dimension; e.g., valid time or transaction time.

Knowledge temporality levels are similar, and may be characterized as follows:

- static knowledge: this knowledge does not contain any temporal context, nor can this context be inferred from it. An example of such knowledge is the sentence: "Every organization has to conform to legal rules";
- sequences: ordered sequences of events with no direct time stamps. These may be, e.g., events ordered by Allen's temporal relations [91]. An example of sequential knowledge may concern the legal domain; namely, the sequential knowledge about a law-creating process: passing a law $\rightarrow$ signing of the law by the President $\rightarrow$ publishing the law;
- time-stamped knowledge: static knowledge that has been extended with time stamps (an example of which is a description of a license issuing process: Application for a license $\rightarrow$ decision $\rightarrow$ valid period of the license); and
- fully temporal knowledge: knowledge that possesses at least one time dimension, e.g., knowledge on the varying prices of shares.

Depending on the type of data and knowledge, different reasoning rules may be applied:

- static rules: rules with no time context;
- temporally extended static rules; e.g., temporal descriptive rules; and
- rules that are proper to fully temporal knowledge; e.g., causal detection rules and temporal data mining rules.

As for the levels of temporality, all types of big data fall into the category of fully temporal data. However, as regards the various types of big data, it is also possible to discuss them in a temporal logic structure context. Big data differs depending on the source. These can be click streams, data from mobile and business transactions, user-generated content (UGC), social media data, sensor data, healthcare records, and management and finance data. The most intuitive temporal classification of big data is, therefore, the one that considers the time structure that is used to represent and process the data. Data that is generated periodically, such as financial or management data, may be formalized using point structures, while almost-continuous data flows, such as social media data, may need interval structures or point-interval ones. There is, of course, also the question of time granularity, which will not be discussed in this paper.

The above definitions of various atemporal and temporal types of data, knowledge, and reasoning have been used to develop the levels of the Temporal Big Data Maturity Model. The main assumption regarding the analytical maturity of an organization in the context of temporal big data states that the more mature the organization is, the more temporal solutions it uses in its analytics.

3.2. The Temporal Big Data Maturity Model (TBDMM)

In this paper, the Temporal Big Data Maturity Model (TBDMM) is advocated as a way to establish an organization's level of maturity in temporal big data analytics. It enables an organization to measure its big data assets and the effectiveness of its current analytical tools and to plan for their evolution. Furthermore, the model intentionally makes use of the time dimension. Therefore, it provides a complete toolkit for gauging an organization's fitness to process temporal data and/or knowledge. Hence, we propose the new framework as a guide for organizations that are willing to implement temporal big data analytics. With the TBDMM, organizations may be able to plan and execute the steps that are necessary to move from one level (their current state) to another (the desired one). However, to be fully functional, the framework needs to be accompanied by a self-assessment form. Such a form, originating from the characteristics of the subsequent maturity levels, will allow organizations to assess the state of temporal BD analytics in order to plan their subsequent activities.

The framework of the TBDMM comprises five maturity levels that are coordinated by the temporality levels that were discussed in the previous section. The names of the levels are: Atemporal, Pre-Temporal, Partly Temporal, Predominantly Temporal, and Temporal. At each level, the framework evaluates an organization's maturity against four areas: the data/knowledge from which insights are gained, the IT solutions used for analytics, the functionalities offered by these solutions, and the sustainable development context. Such a four-tier structure reflects the most significant points where an organization's preparedness to profit from temporal big data analytics is manifested. The numbering of the maturity levels starts with 0 to point out that, at the beginning, an organization does not gain insights from big data nor from temporal data/knowledge. The details of the maturity levels are provided below. The TBDMM model is composed of five levels in order to show the process of change in an organization in an adequately precise manner. Too few levels—e.g., three or four—would fail to capture all of the necessary features and characteristics of the various phases in the process of adopting temporal big data analytics. On the other hand, too many levels—e.g., more than five—would result in the dispersion of detail. The rationale that underlies using exactly five levels is to show the key trends in the path to an organization's full adoption of temporal big data.

Level 0 (Atemporal): as highlighted above, at Level 0 the data and knowledge that are used for analytics are atemporal in nature. These comprise multidimensional data and static knowledge (as defined in the previous section). Of course, in the OLAP model, a temporal dimension is included; however, at this maturity level, an organization does not make any temporal inferences with such data. What is more, at this level, an organization has no ability to make use of big data. The IT solutions that are enacted at Level 0 cover mainly a data warehouse, a BI system, and a knowledge base system. These allow various functions to be incorporated into an organization's functions, including performance monitoring, trend analysis, reporting, comparative analysis, benchmarking, and decision support with the use of static rules. Data warehouses and BI systems might not be classified as temporal, because, to date, questions on processes representation, persistence representation, temporal operators in queries, and temporal relations analysis have not been successfully solved. Trend analysis involves a similar picture. Although time series are time-stamped, they do not allow for temporal reasoning. They only record data in predefined time intervals. At Level 0, the organization is able to perform daily reporting from structured data and historical analyses on structured data, and may use some intelligent decision support for daily operations. However, these are only day-to-day insights, and do not provide the organization with the possibility to mine client data, to perform sentiment analysis, or to predict the decisions of an organization's market competitors. Such basic business analytics on structured data have very little potential to support a sustainable competitive advantage. At this level, any activities that aim to gain such an advantage are not supported properly, and managers mainly rely on their own beliefs to make decisions. At Level 0, sustainability issues are addressed in the organization in an ad-hoc manner. Processes for sustainability support are missing. The same concerns change management. A stakeholder network does not exist; stakeholders (e.g., suppliers and retailers) are selected ad-hoc on a financial basis only, with price as the main factor. Learning and communication issues are not perceived by senior staff to be important, and are not managed. Stakeholder management—if it exists—is performed with simple tools only.

Level 1 (Pre-Temporal): with respect to data/knowledge, an organization that is classified as being at this level makes use of similar structures to those that are listed in Level 0, but has begun making limited use of unstructured data sources (e.g., texts). The knowledge that is gained can now be said to be static or sequential. Additional solutions that are used by the organization to process the data/knowledge sources—apart from the solutions that are used at Level 0—also now include intelligent dashboards and sequential knowledge base systems. These increased abilities lead to the following functionalities becoming available: predictive analytics, advanced statistics, data mining on structured data, and text mining. The possibility to structure knowledge fragments qualitatively with temporal relations, such as "earlier" or "later", is now available. With the data/knowledge, IT solutions, and IT functionalities that are implemented at Level 1, an organization may gain some deeper insights into its customers by analyzing, e.g., customer relationships management (CRM) data. This provides an organization with, for example, profiles of clients, but they are based only on structured data. An organization is not able to utilize knowledge on clients' opinions or expectations. At this level, it is also possible to predict changes in the market environment, and consumers' and competitors' behavior. Decision support can be based on information about how the knowledge base changes with time. However, this support only reveals the directions of changes and the possible causes and effects of them. Such a level of analytics may give an organization a temporal competitive advantage; however, in the long run, it is not sufficient to maintain a sustainable advantage over competitors. At Level 1, sustainability management is supported by the planning of processes and activities as well as by predictions. Change management uses structured processes, predictions, event processing, and time-series analytics; however, it is still mostly reactive. The organization begins to establish a stakeholder network based on collaboration and BI and DM (Data Mining) insights. Learning occurs on an individual level. The organization may begin efforts to reduce environmental impacts.

Level 2 (Partly Temporal): at this maturity level, data/knowledge may partially be obtained from big data sources. Hence, data sequences, time-stamped data sequences, and time-stamped knowledge are in use. An organization gains insights from these kinds of data/knowledge by implementing business optimization software, time-stamped knowledge systems, and data mining systems. These IT solutions offer the following functionalities: embedded analytics, optimization, scheduling, pattern analysis, advanced data mining, and temporal descriptive reasoning rules. This kind of temporal rule is indispensable for describing the evolution of knowledge and knowledge sources in a reasoning system. At this level, through time-series analytics and business optimization software, it is possible to optimize business processes and market operations. Also possible is a temporal analysis of knowledge on the business environment, which provides an organization with the potential to predict changes in market conditions. Client data can be analyzed in an advanced manner (e.g., market basket analysis and natural groups based on demographic features and former buying decisions); however, these analyses do not contain the time component and are predominantly static. On the other hand, at this partly temporal level of maturity, organizations may implement some elements of temporal reasoning for decision support, and thus obtain insights into changes in clients' and competitors' behavior. Also, at this level, basic big data analytics may be implemented; hence, unstructured or semi-structured information on market trends may be added to the analytical landscape. With this form of business analytics, organizations are able to gain a stronger and more durable competitive advantage than on the previous level; however, as they cannot detect real-time changes in big data, it is difficult to obtain a sustainable advantage or sustainable business models. However, sustainability issues are integrated into business operations with the use of advanced IT solutions. Using big data analytics on customer data, the organization will begin to work on enhancing its image concerning the environmental context. Change management is built into other processes, and stakeholders are selected on the basis of quality. The main stakeholder network is based on advanced and structured collaboration processes, while stakeholder management is focused on the sharing of benefits. Customers' needs are analyzed using advanced DM and some big data analytics. Learning processes in an organization occur at both the individual and the team level. Processes are established to reduce environmental impacts.

Level 3 (Predominantly Temporal): at this level, the temporal dimension starts to dominate in the data/knowledge sources that an organization uses and in the processing of this data/knowledge. Apart from the data sources that are apparent at Level 2, an organization uses temporal big data from sensors and clickstreams. In addition, unstructured knowledge and data, such as legal texts, are brought together and used for analytics. This is possible with Hadoop (and presumably alternative big data tools, such as Spark) and partly temporal knowledge base (KB) systems; that is, KB systems within which only the structured knowledge is temporal, while the unstructured data is not. Text mining and web mining tools are also being taken advantage of. Such IT solutions offer functionalities concerning customer behavior analysis, personalized recommendations, market trends discovery, strategic analysis, temporal query processing, and temporal reasoning on the structured part of the knowledge. In the context of a sustainable competitive advantage or sustainable business models, the predominantly temporal level provides organizations with more knowledge on market conditions and on the competitive environment. Organizations will start to obtain insights from real-time data flow; e.g., clickstreams are analyzed.

Also, unstructured knowledge from opinion portals is used; however, organizations cannot track its evolution. Additionally, the knowledge base that is implemented in an organization is temporal. Temporal knowledge representation has many advantages with respect to environmental analysis. Some of these advantages are as follows: the representation of changes, their scope, and the resulting interactions among, e.g., market features; the representation of both discrete and continuous changes; and the representation of changes as processes with causal relationships that are explicitly defined. Temporal knowledge base systems collect experiences concerning the domain being depicted, and thus trace a domain's evolution and are able to draw new conclusions. Temporal reasoning on temporal knowledge may be qualitative; hence, it may concern complex relationships, descriptive information, or information that is only partially specified. Together with the possibility to model the persistence of notions, it is possible to encode in a temporal system the so-called erudite knowledge of experts. Such rich representation and reasoning possibilities provide organizations with valuable dynamic insights into the market and competitive environment situations. An organization may react to new challenges more quickly and accurately. At this level of maturity, big data analytics may also be implemented, providing broader predictive possibilities because of the intensive use of unstructured data. Decision support systems have a clear time component; e.g., when to bring a new product or service to the market. The competitive advantage of the organization is, therefore, becoming sustainable. Sustainability issues drive the business. The organization turns its attention to advanced and timely analyses of its customers. Sustainability and customer management are supported by temporal knowledge analytics and reasoning. Stakeholders are selected with respect to environmental concerns. The stakeholder network is built upon strong collaboration, consultation, and crowdsourcing. At the same time, stakeholder management is focused on stakeholder engagement. Innovations and creativity are fostered and supported by dynamic management. Communication and knowledge sharing are strongly promoted inside and outside the organization.

Level 4 (Temporal): as the level name suggests, an organization uses temporal big data and temporal knowledge in a mature way. Temporal big data maturity means making use of, e.g., social data; additionally, structured and unstructured temporal knowledge is now taken into account. An organization puts into practice such IT solutions as big data analytics toolkits, temporal knowledge bases, and multiagent systems that obtain social data. The functionalities offered by these IT solutions might include, *inter alia*, text and opinion mining, sentiment analysis, the discovery of customer usage patterns, a holistic analysis of clients, qualitative and quantitative temporal reasoning, and the representation and analysis of beliefs. At the temporal level, an organization fully incorporates big data in its business analytics, and pays special attention to social media data and real-time information on customers' opinions. This can be easily added to the temporal knowledge base system even though the real-time big data lack structure. The reach of the representation of data, information, and knowledge from the competitive environment enables advanced temporal analytics, and temporal reasoning, in every area of management. Through real-time insights into the competitive environment, through the early detection of customers' attitudes and expectations, and through temporal reasoning about the market's evolution and competitors' intentions, an organization is able to obtain the advantage of being the first to act on an issue, which may be easily transformed into a sustainable competitive advantage based on a sustainable business model. At the highest maturity level, sustainability is treated and understood as a strategic key concept. The organization advocates for continuous change and improvements. Innovative ideas are fostered and supported by unstructured big data analytics and by the use of the temporal reasoning system. An organization-wide learning environment is established. The stakeholder network is based on creative collaboration, trust, and on knowledge sharing. Hence, stakeholders actively participate in the organization's life and decision-making processes. Temporal big data analytics is used to explore the competitive environment and to support innovations and creativity. Learning and change management are continuously improved. Sustainability and environmental issues are built into the organization's strategy.

An illustration of the TBDMM model is presented in Figure 2.

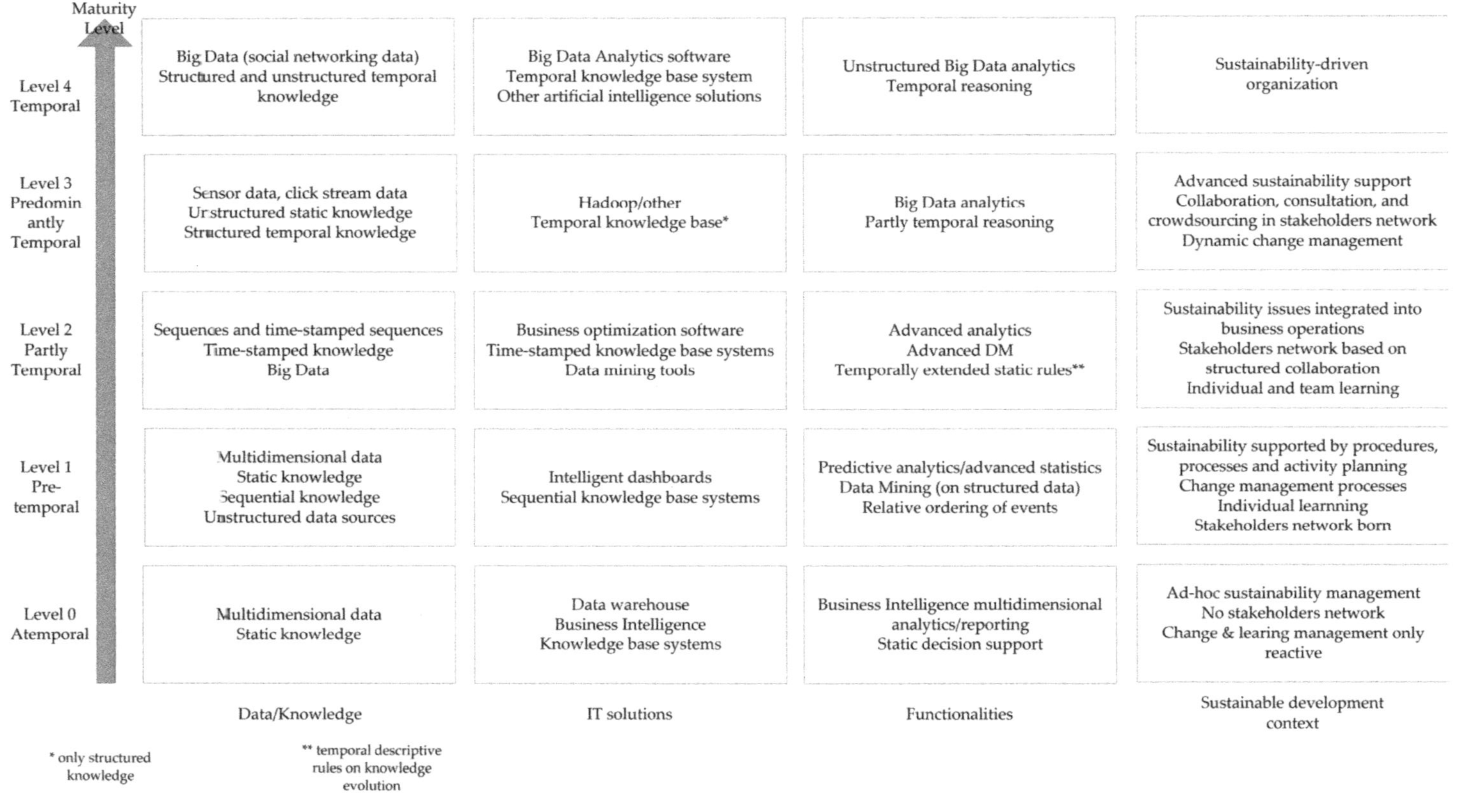

Maturity Level	Data/Knowledge	IT solutions	Functionalities	Sustainable development context
Level 4 Temporal	Big Data (social networking data) Structured and unstructured temporal knowledge	Big Data Analytics software Temporal knowledge base system Other artificial intelligence solutions	Unstructured Big Data analytics Temporal reasoning	Sustainability-driven organization
Level 3 Predominantly Temporal	Sensor data, click stream data Unstructured static knowledge Structured temporal knowledge	Hadoop/other Temporal knowledge base*	Big Data analytics Partly temporal reasoning	Advanced sustainability support Collaboration, consultation, and crowdsourcing in stakeholders network Dynamic change management
Level 2 Partly Temporal	Sequences and time-stamped sequences Time-stamped knowledge Big Data	Business optimization software Time-stamped knowledge base systems Data mining tools	Advanced analytics Advanced DM Temporally extended static rules**	Sustainability issues integrated into business operations Stakeholders network based on structured collaboration Individual and team learning
Level 1 Pre-temporal	Multidimensional data Static knowledge Sequential knowledge Unstructured data sources	Intelligent dashboards Sequential knowledge base systems	Predictive analytics/advanced statistics Data Mining (on structured data) Relative ordering of events	Sustainability supported by procedures, processes and activity planning Change management processes Individual learnning Stakeholders network born
Level 0 Atemporal	Multidimensional data Static knowledge	Data warehouse Business Intelligence Knowledge base systems	Business Intelligence multidimensional analytics/reporting Static decision support	Ad-hoc sustainability management No stakeholders network Change & learing management only reactive

* only structured knowledge

** temporal descriptive rules on knowledge evolution

Figure 2. The proposed Temporal Big Data Maturity Model (TBDMM).

The proposed framework extends the previous maturity models for big data adoption by suggesting that the time factor is to be taken as the primary factor in developing advanced business analytics that are focused on a sustainable competitive advantage. The TBDMM responds precisely to the modern challenges faced by organizations: the need to analyze an unstructured real-time data flow, the need to respond to consumers' expectations that are expressed in social media discussions, the need to understand changes in competitors' activities, and finally the need to incorporate the new insights into decision support systems.

4. Reception of the TBDMM

This section presents the results that were obtained from the survey conducted from April 2016 to October 2016. The survey has been aimed at identifying how organizations perform big data analytics (BDA), and hence how the proposed framework fits into their analytics needs. Specifically, the research questions were: what BDA activities are undertaken in the organizations under study, and do the solutions concerning data/knowledge, IT systems, and their functionalities that are incorporated in the new framework respond to these organizations' analytical needs?

4.1. Research Methodology and Research Sample

The proposed temporal framework for big data has been presented to managers during personal, semi-structured, Computer-Assisted Personal Interviewing (CAPI) interviews. In total, 25 organizations were invited to participate in the research, and 15 of them agreed. Hence, interviews with employees from 15 organizations were done. Formally, the research consisted of a case-study approach with multiple cases. Specifically, the interviews were performed in a two-stage process. At the first stage, all the respondents who agreed to participate in the research were sent the interview questionnaire and given as much time as they needed to fill it in. Then, they returned the questionnaires. During the second stage, personal meetings with each respondent were arranged. During these meetings, the interviews were performed on the basis of the completed questionnaires. The respondents were free to add comments to each question; also, some of the questions were extended and made more precise during the interview. The interviews were recorded using a recording application on a computer, and, after completing all the interviews, they were listened to and transcribed. In this way, the collected data and information set consisted of three subsets: the original completed questionnaires, the recordings, and the transcripts.

In the interview questionnaire, 15 closed and open-ended questions were present. Each respondent's particulars were covered by seven questions. The research questions referred to:

- the presence of temporal aspects in analysis and managerial decisions, and their significance;
- the maturity of the organization;
- the respondent's understanding of the term "big data";
- the importance of various business analytics types;
- the data/knowledge, IT solutions, and IT functionalities appropriate for BDA;
- the respondent's assessment of their employees' level of training in the context of BDA;
- the IT infrastructure and data quality in the organization; and
- the advantages of, and barriers to, BDA implementation.

The method of purposive sampling was used to select the research sample. Then, the sample was verified with the business activity of the organization, the number of staff members, the average annual turnover during the last three years, and the capital structure as independent variables. Also, such variables as the respondent's position in the organization, the organization's sector, and the period of market operation of the organization were used.

The group of organizations was composed of the following sub-groups: eight service organizations, one research and development (R&D) organization, four manufacturing organizations, one banking organization, and one commerce organization.

The research sample can also be divided into the sub-group of medium or large organizations (eight indications), and the sub-group of small organizations (seven indications). No organizations fell into the micro-organization sub-group. The next descriptive feature concerns the financial capital of the organizations under study. Ten organizations were financed with national capital, three with foreign capital, and two with mixed capital.

Taking into consideration the sector of operation, the research sample can be divided into the sub-group of ICT (Information and Communication Technology) production and ICT service/support (three indications), the sub-group of the professional, scientific, and technical sector (four indications), the sub-group of the financial sector (one indication) and the sub-group of other sectors (seven distinct indications).

The interviewed respondents were owners or management board members (eight persons), ICT managers/specialists (two persons), or persons that performed other functions in their organizations, such as technology department managers, advanced analytics managers, SEO (Search Engine Optimization) specialists, and risk assessment managers. The details of the backgrounds of the companies are given in Table A2 in Appendix B.

The research focused on the analytics needs of organizations concerning data/knowledge, IT solutions, and their functionalities. Due to the qualitative nature of the research, it was difficult to precisely define the model's reception indicators. Hence, attention has been paid to the question of whether respondents recognize the need to use solutions that are incorporated into the new temporal maturity framework (TBDMM).

4.2. Selected Survey Findings

4.2.1. Managers' Understanding of Temporal Big Data

The first two important questions of the survey concerned managers' understanding of big data, and the basic assumption of the TBDMM; namely, its explicit time dimension. First, respondents were asked to define the meaning of the "big data" notion. For almost all of the managers (14 persons), big data meant the flow and processing of huge amounts of data that is constantly changing, dispersed, and only loosely coupled. Five persons pointed out difficulties in the processing of such data, in using standard analytical methods, and in searching for dependencies among the data. These people understand big data to be vast datasets on which classical processing and storing methods are ineffective. Also, five persons linked the big data term with such notions as dynamics and variability. Similarly, five persons linked big data with its sources of origin: the internet, social media, cookies, web usage mining, the Internet of Things (IoT), and Google. Two managers pointed out big data's lack of structure and its irregular flows. Other remarks concerning the big data term were as follows:

- "people from organizations having no advanced information technology solutions do not know what they mean while using the big data notion";
- "big data does not exist, we focus on the analysis of a dataset's portion";
- "big data means creating and validating models based on machine learning tools, and using these models on complete datasets";
- "linked heterogeneous datasets owned by various organizations"; and
- Hadoop, Spark, Cassandra, HBase, and NoSQL.

To verify the time assumption of the TBDMM, respondents were asked to determine how important the time factor is in managerial analysis and decision-making. For 11 of them, time is important, and for 4 of them it is very important. Hence, all of the interviewed managers recognize the temporality of business analytics and decisions. This means that the assumption that the time factor is the key indicator of subsequent maturity levels in the presented framework has been fully justified.

4.2.2. The Data/Knowledge Aspect in the TBDMM

The next area of the survey concerned the data/knowledge that is used by organizations, and the data/knowledge that should be used for advanced business analytics. The majority of the studied organizations (10 indications) does not use big data to gain valuable insights. Five managers indicated the use of customer data, three of them indicated competitors' data, two managers pointed out social media data, one person indicated suppliers' data, and one person indicated other stakeholders' data. As for the quality of data used in the organizations for advanced analytics, three managers assessed it as high quality data, nine persons assessed it as average quality data, two persons assessed it as rather poor quality data, and one person declared that big data is not used for analytics, as there is no need to do it (Figure 3).

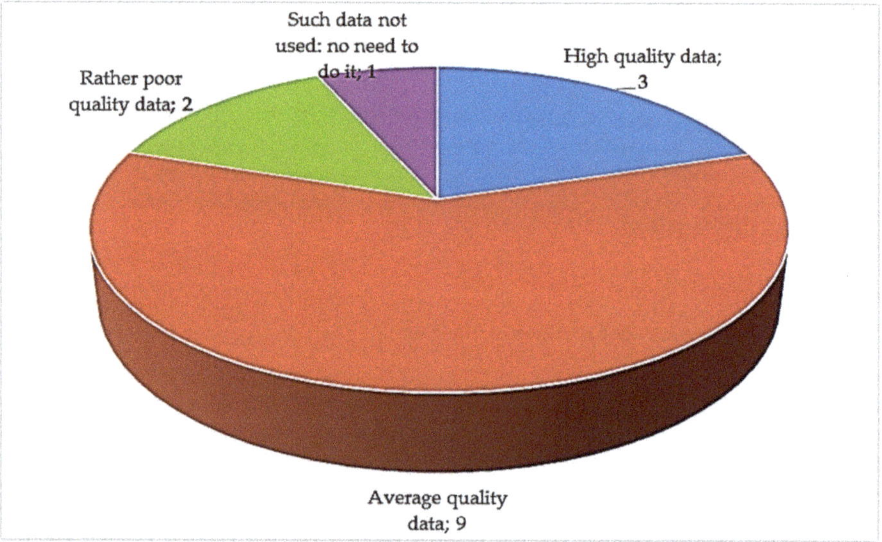

Figure 3. The quality of data for big data analytics (BDA) in the studied organizations.

It should be noted here that, while only one person who assessed the data quality declared big data to have no use, there were as many as 10 indications in the area of the data being used concern big data (as not being used at all). At first sight this may seem contradictory; however, most probably, this contradiction is due to a poor understanding of the big data notion and its features as discussed in Section 4.2.1.

In the area of the survey concerning the data/knowledge that should be used for BDA, the respondents were presented with a closed set of examples, and asked to express their opinion on each with respect to the solution's usability. They were also able to add their own remarks on this topic. Figure 4 illustrates how the interviewed managers responded.

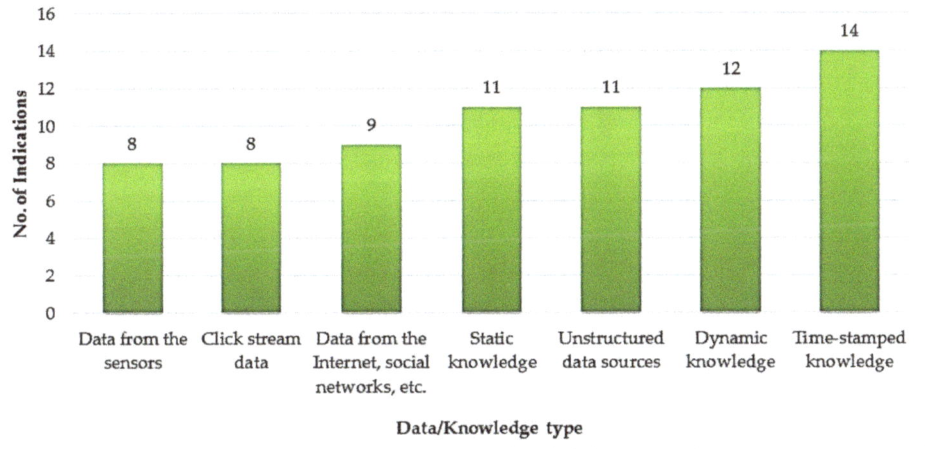

Figure 4. The most essential data/knowledge in BDA.

As expressed by the respondents, they understood static knowledge to be, e.g., managerial knowledge, models, and expert rules. Almost all of the interviewed persons (14) recognized time-stamped knowledge (such as time-series knowledge) to be indispensable to BDA. However, they indicated that this factor is strictly linked with the industry, and is highly project-dependent. For the majority of managers (12), dynamic (changing) knowledge is essential for BDA. Attention should be also paid to the high position of unstructured data, including social network data. Hence, managers recognize the need to make use of temporal data, including big data. The big data notion was not present in the managers' comments explicitly; however, as the respondents see the analytics potential of the web, social networks, sensors, click streams, etc., this indicates their interest in big data. Additionally, the respondents were asked to give their opinion on whether advanced BI and big data analytics should be unlimited across the organization, or conducted only in isolated fragments (departments) of the organization. For the majority of the managers (10 indications), such analytics should be limited to isolated fragments of the organization, and, according to five managers, it should be unlimited across the organization.

4.2.3. The IT Solutions Aspect in the TBDMM

The next part of the interview was devoted to the advanced analytic solutions in each organization, its IT infrastructure, and to the BDA solutions that, according to the respondents, should be implemented. The managers were asked about the types of business analytics solutions that are particularly important for their organization. The solutions that were most often indicated were dynamic analysis systems (12 indications) and reporting systems (11 indications). Also, real-time analytics systems were often chosen (nine indications). On the other end of the solutions spectrum, respondents pointed out scoring systems (one indication), risk analysis and decision support systems (three indications), and static analysis systems (understood as simple analytical systems with no time or dynamic aspects, four indications). Information on the respondents' opinions is given in Figure 5.

Each manager was also asked to evaluate the suitability of their organization's IT infrastructure for executing advanced business analytics. None of the managers assessed it as "very good" or "very bad". Seven managers assessed the IT infrastructure's suitability as "good"; four of them assessed it as "neither good nor bad"; and four of them assessed it as "bad".

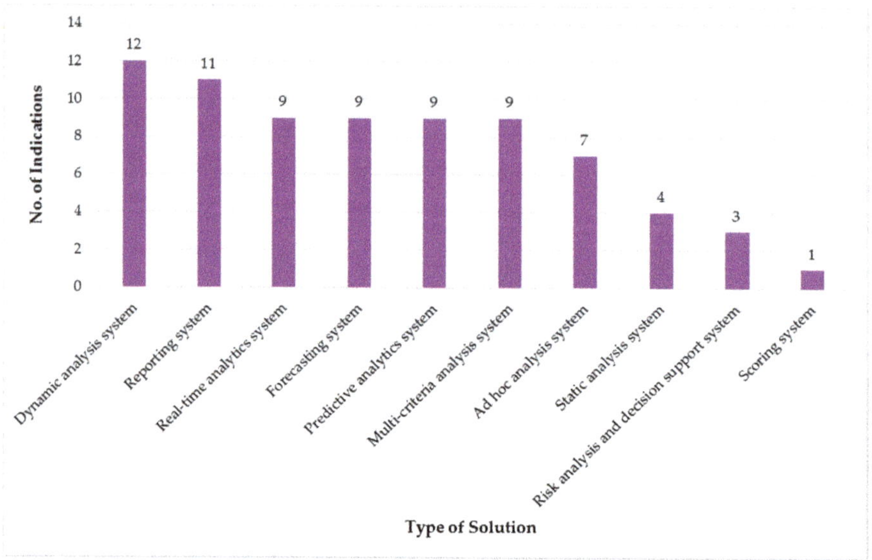

Figure 5. The business analytics solutions that were particularly important to the surveyed organizations.

Subsequently, the respondents were asked about what IT solutions their organization should use while preparing BDA. Again, they were presented with a closed set of examples, and asked to express their opinion on each with respect to the solution's usability for advanced business analytics, including big data analytics. As before, the respondents were able to add their own remarks during the interview. The results are presented in Figure 6.

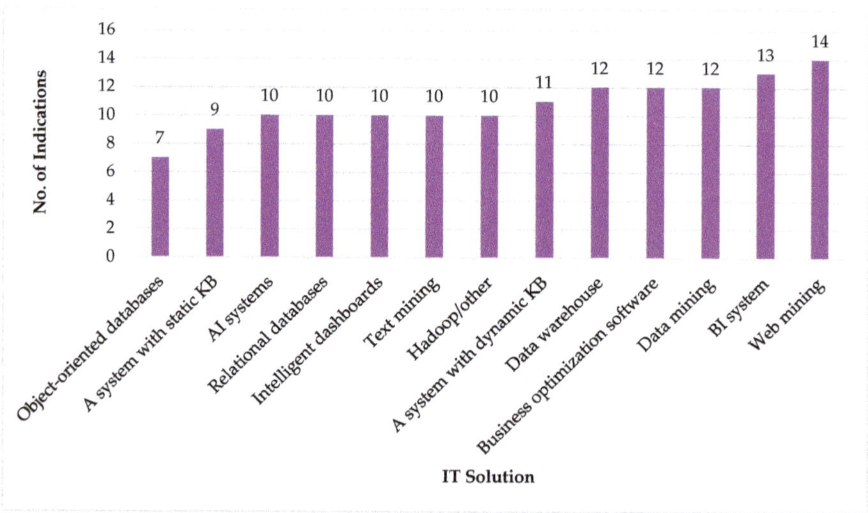

Figure 6. The most suitable information technology (IT) solutions for BDA in surveyed organizations. KB, knowledge base.

Almost all of the respondents (13) recognized BI systems as an indispensable tool for BDA. Similarly, almost all of them (14) admitted that using Web mining tools for BDA is very important for organizations. Ten managers recognized that Hadoop and other big data tools (such as Spark) support organizations in performing BDA.

As in the case of analytics sources, among the IT solutions the respondents brought to the forefront were those linked with time and big data. The respondents recognize—in the context of advanced business analytics—the need for and utility of Web/text mining tools, big data analytics tools (Hadoop, Spark), and temporal tools, such as temporal knowledge base systems. This is consistent with previous remarks showing that time and big data need to be taken into consideration in managerial analysis and decisions.

4.2.4. The IT Functionalities Aspect in the TBDMM

The subsequent interview topics concerned the IT functionalities used for advanced business analytics in the studied organizations, and those that should be offered by IT systems if they are to be useful in BDA. The functionalities that were used in the surveyed organizations correspond to the data/knowledge pointed out in the earlier part of the interviews (Section 4.2.2.) and are as follows: customer data analytics (five indications), competitors' data analytics (three indications), social media analytics (two indications), and suppliers' data and other stakeholders' data analytics (one indication each). As for the desired IT functionalities for BDA, the managers were presented again with a closed set of examples, and asked to express their opinion on each with respect to each solution's usability. The functionalities that were found to be necessary are as follows:

- multidimensional analytics/BI reporting;
- data mining and advanced data mining;
- big data analytics; and
- temporal inferences (with time as a distinctive feature).

"Advanced data mining" for the interviewed managers meant data mining procedures that go beyond typical data mining (DM) activities (such as clustering, classification, and associations); e.g., temporal data mining. Figure 7 presents the summarized results.

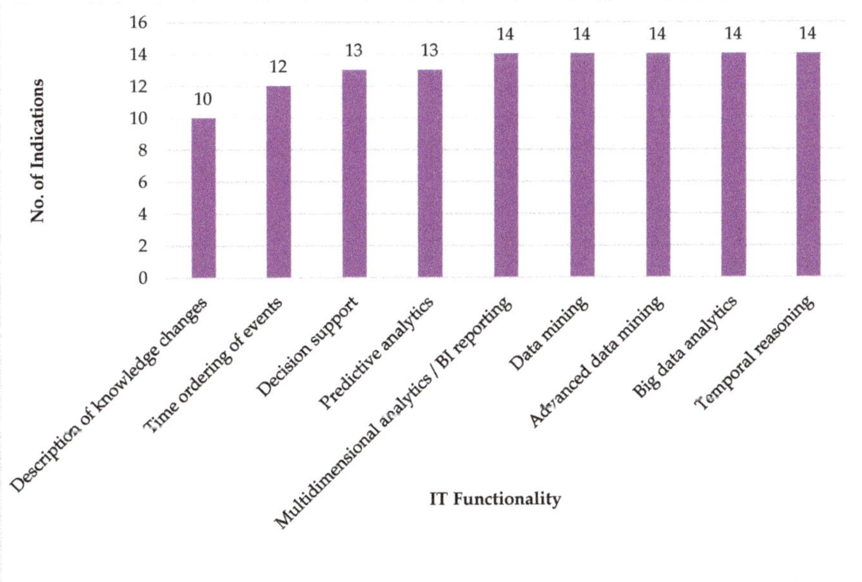

Figure 7. The most desired IT functionalities for BDA. BI, business intelligence.

Almost all of the respondents (14 indications) admitted that multidimensional analytics/BI reporting is of extreme importance as an IT functionality supporting advanced business analytics and BDA. The same number of respondents gave such an opinion on data mining. Twelve respondents stated that IT systems should also offer such functionality as event ordering in time (e.g., for events from the economic environment). The description of knowledge changes is an important IT functionality for 10 respondents. Fourteen respondents claimed that IT systems supporting advanced analytics and BDA should offer the functionality of temporal reasoning.

As in the previous areas of the interview, in the area of IT functionalities supporting advanced business analytics and BDA, the most acknowledged functionalities were solutions concerning data mining (including advanced data mining; e.g., temporal data mining) and functionalities concerning the temporal aspect of analytics (the description of changes in knowledge/data, causal analytics of event sequences, automatic temporal reasoning).

5. Discussion. The Sustainable Development Context

All of the respondents perceive an organization's maturity, in the context of advanced business analytics, to be characterized by identifying business processes, elaborating organizational strategy, and identifying analytics needs. Almost all of them (13 indications) claimed that organizational maturity depends strongly on the completeness and coherence of the data used for analytics. More than a half of the interviewed respondents (eight indications) are convinced that organizations need an appropriate IT infrastructure to perform advanced business analytics and to cooperate with their customers, suppliers, and all other stakeholders. According to four respondents, the analytics maturity of an organization is manifested in employing people with high analytics competency and skills. As these results show, the understanding of an organization's analytics maturity that the respondents have corresponds partially to the framework of the TBDMM that is proposed in this paper. This correspondence concerns the IT infrastructure, which is perceived as one of the maturity determinants, and at the same time is one of the three fundamental aspects (named IT solutions) in the TBDMM. On the other hand, however, the respondents did not consider big data processing and usage in organizations to be an important step towards analytics maturity (or they did not express this explicitly). This is probably due to the rather poor awareness among managers of big data's core characteristics as discussed in Section 4.2.1.

The majority of respondents (nine indications) assessed the level of training for employees (their competencies and skills) in their organizations to draw up BDA as good or very good. The rest of the respondents assessed these competencies and skills as unsatisfactory, or even very bad (two indications). In the area of interpretation of the prepared advanced analytical outputs, 10 respondents assessed their employees' level of training as good (9 indications) or very good (1 indication). However, for five respondents, this level was neither good nor bad (three indications) or "bad" (two indications), which means that this area is in need of special attention.

As for the benefits of big data analytics for organizations, the respondents most frequently pointed to the reliability of insights, allowing for: better decisions, a faster reaction to changes, better quality conclusions, an effective impact on the market, and a faster detection of trends or fluctuations in consumers' expectations (eight indications). Five respondents named such benefits as gaining detailed insights into market and consumer needs, and understanding clients' motivations and behaviors, which leads to a sustainable competitive advantage. The optimization of business processes resulting in a profitability increase was recognized as a benefit of BDA by three respondents. Two respondents were convinced that BDA allows for the targeting and accessing of selected consumer groups with a tailored offer. Similarly, two respondents linked BDA with the potential of e-commerce in social media. Other BDA benefits that were pointed out are as follows:

- the possibility of creating new products and entering new markets;
- technology cost optimization;
- the possibility of using new technologies;
- the prediction of client insolvency risk, and a better understanding of global risks; and

- easier credit scoring.

The respondents—even if they may not be sure what the term "big data" in fact means (Section 4.2.1.)—are able to sense the most outstanding benefits of BDA. They properly linked BDA with a sustainable competitive advantage, gained through enhanced insights into consumers' needs, opinions, and desires, that comes from (inter alia) social media and web analysis. This is in line with the IT functionalities aspect in the TBDMM, which encompasses, among other things, functionalities aimed at, e.g., text or web mining.

The respondents also spoke about barriers to the performance of big data analytics. The most often recognized barrier (10 indications) was a lack of qualified data scientists able to analyze and interpret big data. Five respondents pointed out financial barriers, high BDA implementation costs, and a lack of a big data strategy. Similarly, five respondents noticed no understanding or awareness of BDA among businesses. Four managers turned their attention to the low reliability, low quality, and randomness of big data. Three respondents recognized technological barriers. Other barriers to the successful performance of big data analytics that were pointed out by the respondents are as follows:

- the ROI (Return on Investment) is hard to estimate (two indications);
- data cleaning and storage issues (two indications);
- the economy is not mature enough to make use of big data insights;
- a lack of internal data access procedures;
- a lack of appropriate hardware; and
- communication and legal barriers.

The barriers to BDA implementation in organizations, as perceived by the respondents, most generally concern financial and educational questions. It may thus be assumed that organizations in which managers are aware of BDA's potential benefits would more easily overcome such obstacles as a lack of a big data strategy or a lack of financial support. Once BDA is well-understood and perceived as indispensable in an organization, the proposed TBDMM may be a helpful tool for the successful implementation of BDA.

During the interviews, special attention was paid to the areas (data/knowledge, IT tools, and IT functionalities) that are included in the proposed TBDMM. Special attention was paid to the temporal dimension, but it was also interesting to check whether the elements on the lower levels of the proposed model were perceived as useful and needed.

Comparing the model presented in Figure 2 to the respondents' answers, it can easily be seen that the managers used almost all of the notions and solutions that are included in all of the maturity levels of the proposed framework. Hence, this justifies the accommodation in the model of the structured, static solutions; i.e., the ones not designed for big data analytics. At the same time, it should be noted that, in the context of advanced business analytics (including big data analytics), the majority of answers concerned temporal solutions and unstructured data. In other words, the more mature (in the context of the effective use of big data analytics) is an organization, the more it is driven by temporal, unstructured sources of data/knowledge, temporal IT tools, and their temporal functionalities.

Summing up the respondents' statements on their knowledge and needs concerning big data analytics, and comparing them to the proposed temporal maturity model, it may be stated that the model's reception has been positive and its composition responds to the analytical scope of modern organizations. However, some general differences between industries could be noticed. The respondents from the IT and R&D sectors were the most aware of big data analytics' potential, its benefits, and its implementation barriers. Also, the respondent from the only light industry manufacturer was among those managers who understood the notion of big data, and the possible value that may be obtained for an organization from BDA. On the other end of the spectrum, the respondents representing the sector of services had the most numerous problems with defining the big data notion, and with assessing BDA's possible benefits. Surprisingly, one manager from a very large manufacturing organization declared that his organization uses only MS Excel for advanced analytics, and has no need to capture or analyze big data.

As mentioned earlier, the interviews were performed in 2016. This probably explains why the big data notion, and some of the notions linked to it, that are present in the new framework (e.g., unstructured big data analytics) were not understood properly by some of the respondents, especially the non-IT professionals [92]. Also, there were some concerns that were associated with the quality and reliability of the new data. Additionally, the respondents did not always see the direct link between advanced business analytics (including big data analytics) and a sustainable competitive advantage. It has also to be noted that some terms, e.g., "temporal knowledge base", "sequential knowledge", "temporal data mining", and other temporally associated ones are obviously not widely known, nor understood. The interviewed managers are not an exception. Often, during the interviews, they asked for explanations, and then acknowledged the temporal solutions to be very useful for advanced business analytics in a dynamic competitive environment. Hence, it would be desirable to popularize knowledge on temporal analytics, temporal IT solutions, and their functionalities among managerial staff. It may be assumed that with such knowledge managers would be more willing to implement temporal analytics solutions, including those for big data. This will give organizations better insight into, and understanding of, the dynamic environment of business operations.

6. Concluding Remarks

The usability of big data analytics for gaining a sustainable competitive advantage and for establishing successful sustainable business models has been drawing researchers' attention for several years [14,15,18,52,66,70]. Time is also widely accepted as an important aspect of business operations [13,16]. However, research on big data analytics in general, and on big data maturity models in particular, has not brought the temporal dimension to the forefront. The research is focused either on big data's operational and strategic potential [19] or on big data implementation and processing based on well-established IT solutions [22,24–26].

Our study attempts to formulate a new conceptual framework for assessing an organization's readiness to adopt big data. It argues that a time dimension is a critical driver in maturity models for big data adoption. The proposed framework consists of five maturity levels: atemporal, pre-temporal, partly temporal, predominantly temporal, and temporal. The maturity levels are used to evaluate the data/knowledge, IT solutions, functionalities offered by IT solutions, and sustainable development of organizations.

The proposed model was initially tested among 15 organizations. The survey that was conducted among respondents advocates that our framework extends the previous maturity models for big data adoption by suggesting that the time factor should be taken as the primary factor in the development of advanced analytics that are focused on the sustainable development of organizations. The proposed framework responds precisely to the modern challenges faced by organizations.

Our study makes several theoretical contributions to the relevant literature. First, big data adoption in organizations is a largely unexplored field of research. Therefore, the current study contributes to the emerging literature on assessing an organization's readiness to adopt big data by investigating temporal big data maturity models. Second, the current study is one of the rare studies that proposes a conceptual framework for assessing an organization's readiness to adopt big data based on the temporal dimension. This framework proposes a new perspective on the issue of assessing an organization's readiness to adopt big data. It responds to the current challenges that come from an organization's environment, such as the need to incorporate real-time big data analytics into decision support systems. Third, this study demonstrates how the proposed framework may be used to assess modern information sources in organizations. The interviews with managers that were discussed in this paper have proven that organizations are aware of the big data phenomenon and its analytics potential. However, they do need a tool—a kind of signpost—that will enable them to assess their organizations' preparedness to perform big data analytics and to implement in a coordinated manner solutions that are aimed at improving these analyses. This conclusion is drawn from the barriers to the successful performance of big data analytics that were listed by the managers. During the interviews, they described, e.g., a lack of qualified staff, a lack of a big data analytics implementation strategy, and the randomness, poor quality, and incredibility of such data. Using a big data maturity model may help organizations to overcome these barriers.

Finally, the current findings provide empirical evidence that the design of a conceptual framework for assessing organizations' readiness to adopt big data based on the temporal dimension may play an important role in their sustainable development. This framework offers guidance to managers who want to successfully incorporate temporal big data analytics into their organization's business, competition, and sustainability strategies. It highlights the role of real-time big data analytics and temporal decision support systems as critical tools for building a sustainable competitive advantage and sustainable business models. The proposed framework provides a comprehensive means to be used by managers in their efforts to respond to market challenges.

Although our results provide important contributions to theory and practice in the big data field, we discuss some limitations to keep in mind when interpreting the results of this study, while also suggesting some potential opportunities for future research.

The proposed model needs further validation and testing. This should lead to the development of best practice and contingency guidelines for managers, researchers, and software engineers concerned with big data usage. By understanding the idea of the proposed framework, organizations should be better able to profit from big data analytics. It seems that the next step in exploring the phenomenon of temporal big data maturity models should be quantitative tests that cover a much greater number of organizations and that compare the results obtained from different domains.

Author Contributions: C.M.O. conceived the idea, designed the study's outline, and proposed the interview questions; M.M.-K. and C.M.O. performed the literature review; M.M.-K. designed the TBDMM and conducted the interviews with managers; M.M.-K. analyzed the data from the interviews; and C.M.O. supervised the entire project, and reviewed and edited the manuscript.

Funding: This research received no external funding.

Conflicts of Interest: The authors declare no conflict of interest.

Appendix A

Table A1. Various definitions of the term "big data", listed in chronological order.

Author(s)	Definition
Technological Approach	
[38]	"Data sets are generally quite large, taxing the capacities of main memory, local disk, and even remote disk. We call this the problem of big data. When data sets do not fit in the main memory (in core), or when they do not fit even on the local disk, the most common solution is to acquire more resources."
[39]	The big data term "describes data sets that are growing exponentially and are too large, too raw or too unstructured for analysis using relational database techniques".
[40]	"Big data involves the data storage, management, analysis, and visualization of very large and complex datasets."
[36]	"Datasets whose size is beyond the ability of typical database software tools to capture, store, manage, and analyze."
[41]	The big data term is "used to describe data sets so large, so complex or that require so rapid processing (. . .) that they become difficult or impossible to work with using standard database management or analytical tools."
[42]	"Big data refers to datasets with sizes beyond the ability of common software tools to capture, curate, manage, and process the data within a specified elapsed time."

Table A1. *Cont.*

Author(s)	Definition
Technological Approach	
[43] (p. 89)	"Big data (…) means that the organization's need to handle, store, and analyze data (its volume, variety, velocity, variability, and complexity) exceeds its current capacity and has moved beyond the IT comfort zone."
[44]	"Big data involves the data storage, management, analysis, and visualization of very large and complex datasets. It focuses on new data-management techniques that supersede traditional relational systems, and are better suited to the management of large volumes of social media data."
[9]	"Big data consists of expansive collections of data (large volumes) that are updated quickly and frequently (high velocity) and that exhibit a huge range of different formats and content (wide variety)."
[45]	"The broad range of new and massive data types that have appeared over the last decade or so."
[46]	"Data of a very large size, typically to the extent that its manipulation and management present significant logistical challenges."
[47]	"Big data: the data-sets from heterogeneous and autonomous resources, with diversity in dimensions, complex and dynamic relationships, by size that is beyond the capacity of conventional processes or tools to effectively capture, store, manage, analyze, and exploit them."
[48]	"Big data typically refers to the following types of data: (1) traditional enterprise data; (2) machine-generated/sensor data (e.g., weblogs, smart meters, manufacturing sensors, equipment logs); and (3) social data."
[49]	"Big data is an emerging paradigm applied to datasets whose size is beyond the ability of commonly used software tools to capture, manage, and process the data within a tolerable elapsed time."
[50]	"«Big data» means enormous volumes of data. The data can be divided as structured data and unstructured data. Various methods are applied to collect these data."
Organizational approach	
[51] (p. 1)	"Consumers have unprecedented access to information and tools with which to consume information. Social media, mobile access, augmented reality and three-dimensional (3D) views of pictures and video have blurred the lines between our private and work personas and have fundamentally changed the way the consumer utilizes information."
[7]	"Big data, like analytics before it, seeks to glean intelligence from data and translate that into business advantage. However, there are three key differences: Velocity, variety, volume."
[52]	"Big data focuses on three main characteristics: the data itself, the analytics of the data, and presentation of the results of the analytics that allow the creation of business value in terms of new products or services."
[8]	"Big data is a combination of volume, variety, velocity and veracity that creates an opportunity for organizations to gain sustainable competitive advantage in today's digitized marketplace."
[53] (p. 21)	"Companies today are overgrown with information, including what many categorize as big data. The jungle includes information about customers, competition, media and channel performance, locations, products, and transactions, to name just a few (…)"
[24] (p. xxi)	"(…) big data is about leveraging the unique and actionable insights gleaned about your customers, products, and operations to rewire your value creation processes, optimize your key business initiatives, and uncover new monetization opportunities."
[54]	"A new attitude by businesses, non-profits, government agencies, and individuals that combining data from multiple sources could lead to better decisions."

Table A1. *Cont.*

Author(s)	Definition
Mixed (tech-organizational) approach	
[55]	"Big data involves more than simply the ability to handle large volumes of data; instead, it represents a wide range of new analytical technologies and business possibilities. These new systems handle a wide variety of data, from sensor data to Web and social media data, improved analytical capabilities, operational business intelligence that improves business agility by enabling automated real-time actions and intraday decision making, faster hardware and cloud computing including on-demand software-as-a service. Supporting big data involves combining these technologies to enable new solutions that can bring significant benefits to the business."
[56]	Big data is "associated with the new types of workloads and underlying technologies needed to solve business problems that could not be previously supported due to technology limitations, prohibitive cost or both."
[6]	"Big data: high-volume, high-velocity, and/or high-variety information assets that require new forms of processing to enable enhanced decision making, insight discovery, and process optimization."
[57]	"Big data: a cultural, technological, and scholarly phenomenon that rests on the interplay of (1) Technology: maximizing computation power and algorithmic accuracy to gather, analyze, link, and compare large datasets; (2) Analysis: drawing on large datasets to identify patterns in order to make economic, social, technical, and legal claims; (3) Mythology: the widespread belief that large datasets offer a higher form of intelligence and knowledge that can generate insights that were previously impossible, with the aura of truth, objectivity, and accuracy."
[58]	"Big data is massively generated by uncountable online interactions between people, by human–systems transactions, and by sensor devices."
[59]	"Big data is no subject to sampling, it is linked with building databases from electronic sources, with no intention of statistical inference."
[5]	"Big data is high-volume, high-velocity and high-variety information assets that demand cost-effective, innovative forms of information processing for enhanced insight and decision making."
[60]	"Big data is large-scale data with various sources and structures that cannot be processed by conventional methods and that is intended for organizational or societal problem solving."
[61]	"Big data often represents miscellaneous records of the whereabouts of large and shifting online crowds. It is frequently agnostic, in the sense of being produced for generic purposes or purposes different from those sought by big data crunching. It is based on varying formats and modes of communication (e.g., text, image, and sound), raising severe problems of semiotic translation and meaning compatibility. Big data is commonly deployed to refer to large data volumes generated and made available on the Internet and the current digital media ecosystems."

Appendix B

Table A2. The backgrounds of the companies that were selected for interviews.

Interview No	Business Activity	No of Employees	Annual Turnover for the Last 3 years	Capital Structure	Respondent's Position	Sector	Period of Existence on the Market
1	R&D	50–249	<EUR 50 million	National capital	Owner/Management Board	Professional, scientific, and technical	6–10 years
2	Manufacturing	>250	<EUR 50 million	Mixed (national and foreign)	Technology manager	Light industry	>10 years
3	Services	≤9	<EUR 2 million	National capital	Owner/Management Board	Professional, scientific, and technical	>10 years
4	Manufacturing	>250	>EUR 50 million	Foreign capital	Manager/specialist planning and production management	Automotive industry	>10 years
5	Services	>250	>EUR 50 million	Foreign capital	Advanced analytics manager	Telecommunications	1–5 years
6	Manufacturing	>250	>EUR 50 million	Foreign capital	ICT Manager/specialist	Consumer electronic	>10 years
7	Services	10–49	<EUR 2 million	National capital	Owner/Management Board	Professional, scientific, and technical	6–10 years
8	Services	≤9	<EUR 2 million	National capital	Owner/Management Board	Administration services and support	>10 years
9	Services	≤9	<EUR 2 million	National capital	Owner/Management Board	ICT service and support	>10 years
10	Services	50–249	<EUR 10 million	National capital	ICT Manager/specialist	Production of ICT	>10 years
11	Services	50–249	<EUR 2 million	National capital	Senior SEO specialist	Transportation and storage	6–10 years
12	Services	≤9	<EUR 2 million	National capital	Owner/Management Board	Construction	6–10 years
13	Manufacturing	10–49	<EUR 2 million	National capital	Owner/Management Board	Production of ICT	6–10 years
14	Banking	>250	>EUR 50 million	Mixed (national and foreign)	Other-risk assessment manager	Finance	>10 years
15	Commerce	10–49	<EUR 2 million	National capital	Owner/Management Board	Professional, scientific, and technical	>10 years

References

1. Ishikawa, A.; Nakagawa, J. *An Introduction to Knowledge Information Strategy. From Business Intelligence to Knowledge Sciences*; World Scientific: London, UK, 2013; ISBN 978-981-4324-42-7.

2. Negash, S.; Gray, P. Business Intelligence. In *Handbook on Decision Support Systems*; Burstein, F., Holsapple, C.W., Eds.; Springer: Berlin/Heidelberg, Germany, 2008; Volume 2, pp. 175–193, ISBN 978-3-540-48715-9.

3. Baaras, H.; Kemper, H.G. Management support with structured and unstructured data–an integrated Business Intelligence framework. *Manag. Inf. Syst.* **2008**, *25*, 132–148. [CrossRef]

4. Olszak, C.M. An Overview of Information Tools and Technologies for Competitive Intelligence Building: Theoretical Approach. *Issues Inform. Sci. Inf. Technol.* **2014**, *11*, 139–153. [CrossRef]

5. Gartner IT Glossary. Available online: http://www.gartner.com/it-glossary/big-data/ (accessed on 10 February 2015).

6. The Importance of "Big Data": A Definition. Available online: https://www.gartner.com/doc/2057415/importance-big-data-definition (accessed on 21 April 2018).

7. McAfee, A.; Brynjolfsson, E. Big Data: The Management Revolution. *Harvard Bus. Rev.* **2012**, *90*, 60–668.

8. Analytics: The Real-World Use of Big Data. Available online: https://www-01.ibm.com/common/ssi/cgi-bin/ssialias?htmlfid=GBE03519USEN (accessed on 10 February 2015).

9. Davis, C.K. Beyond data and analysis. *Commun. ACM* **2014**, *57*, 39–41. [CrossRef]

10. Erl, T.; Khattak, W.; Buhler, P. *Big Data Fundamentals: Concepts, Drivers & Techniques*; Prentice Hall: Upper Saddle River, NJ, USA, 2016; ISBN 978-0134291079.

11. Mikalef, P.; Pappas, I.O.; Krogstie, J.; Giannakos, M. Big data analytics capabilities: A systematic literature review and research agenda. *Inf. Syst. e-Bus. Manag.* **2017**, 1–32. [CrossRef]

12. Yang, W.; Meyer, K.E. Competitive dynamics in an emerging economy: Competitive pressures, resources, and the speed of action. *J. Bus. Res.* **2015**, *68*, 1176–1185. [CrossRef]

13. 2018 Big Data Trends: Liberate, Integrate & Trust. Available online: http://www.businessintelligenceinfo.com/business-intelligence/big-data/now-available-2018-big-data-trends-survey-report (accessed on 21 February 2018).

14. Rajaraman, V. Big Data Analytics. *Resonance* **2016**, *21*, 695–716. [CrossRef]

15. Ngai, E.W.; Gunasekaran, A.; Wamba, S.F.; Akter, S.; Dubey, R. Big data analytics in electronic markets. *Electron. Mark.* **2017**, *27*, 243–245. [CrossRef]

16. Kitchin, R. The Real-Time City? Big Data and Smart Urbanism. *GeoJournal* **2014**, *79*, 1–14. [CrossRef]

17. Townsend, A. *Smart Cities: Big Data, Civic Hackers, and the Quest for a New Utopia*; W.W. Norton & Co.: New York, NY, USA, 2013; ISBN 9780393082876.

18. Lusch, R.F.; Nambisan, S. Service innovation: A service-dominant logic perspective. *MIS Q.* **2015**, *39*, 155–175. [CrossRef]

19. Wamba, S.F.; Gunasekara, A.; Akter, S.; Ren, S.J.F.; Dubey, R.; Childe, S.J. Big data analytics and firm performance: Effects of dynamic capabilities. *J. Bus. Res.* **2017**, *70*, 356–365. [CrossRef]

20. Akter, S.; Wamba, S.F. Big data analytics in E-commerce: A systematic review and agenda for future research. *Electron. Mark.* **2016**, *26*, 173–194. [CrossRef]

21. Kayser, V.; Nehrke, B.; Zubovic, D. Data Science as an Innovation Challenge: From Big Data to Value Proposition. *Technol. Innovat. Manag. Rev.* **2018**, *8*, 16–25. [CrossRef]

22. TDWI Launches Big Data Maturity Model Assessment Tool. Available online: http://tdwi.org/Articles/2013/11/20/TDWI-Launches-Big-Data-Maturity-Model-Assessment-Tool.aspx?Page=2 (accessed on 17 June 2014).

23. Leverage a Big Data Maturity Model to Build Your Big Data Roadmap. Available online: http://www.radcliffeadvisory.com/research/download.php?file=RAS_BD_MatMod.pdf (accessed on 19 February 2014).

24. Schmarzo, B. *Big Data: Understanding How Data Powers Big Business*; John Wiley & Sons: Indianapolis, IN, USA, 2013; ISBN 978-1-118-73957-0.

25. Big Data & Analytics Maturity Model. Available online: http://www.ibmbigdatahub.com/blog/big-data-analytics-maturity-model (accessed on 18 December 2016).

26. Hortonworks Big Data Maturity Model White Paper. Available online: http://hortonworks.com/wp-content/uploads/2016/04/Hortonworks-Big-Data-Maturity-Assessment.pdf (accessed on 18 December 2016).

27. Liautaud, B.; Hammond, M. *E-Business Intelligence. Turning Information into Knowledge into Profit*; McGraw-Hill: New York, NY, USA, 2002; ISBN 978-0071364782.

28. Davenport, T.H.; Harris, J.G. *Competing on Analytics. The New Science on Winning*; Harvard Business School Press: Boston, MA, USA, 2007; ISBN 9781422103326.

29. Wixom, B.H.; Watson, H.J. The BI-based organization. *Int. J. Bus. Int. Res.* **2010**, *1*, 13–28. [CrossRef]

30. Weiss, A. A brief guide to competitive intelligence. *Bus. Inf. Rev.* **2002**, *19*, 39–47. [CrossRef]

31. Williams, S.; Williams, N. *The Profit Impact of Business Intelligence*; Morgan Kaufmann: San Francisco, CA, USA, 2007; ISBN 978-0-12-372499-1.

32. Howson, C. *Successful Business Intelligence: Secrets to Making BI a Killer Application*; McGraw-Hill: New York, NY, USA, 2008; ISBN 9780071498517.

33. Olszak, C.M. Toward better understanding and use of Business Intelligence in organizations. *Inf. Syst. Manag.* **2016**, *33*, 105–123. [CrossRef]

34. Paraschiv, D.; Pugna, I.; Albescu, F. Business intelligence & knowledge management—Technological support for strategic management in the knowledge based economy. *Inform. Econ. J.* **2008**, *12*, 5–12.

35. Steyl, J. Knowledge Management—BI vs. CI. Available online: http://it.toolbox.com/blogs/bi-ci/business-intelligence-vs-competitive-intelligence-32441 (accessed on 15 March 2017).

36. Manyika, J.; Chui, M.; Brown, B.; Bughin, J.; Dobbs, R.; Roxburgh, C.; Byers, A.H. Big Data: The Next Frontier for Innovation, Competition, and Productivity. McKinsey Global Institute, 2011. Available online: https://www.mckinsey.com/~/media/McKinsey/Business%20Functions/McKinsey%20Digital/Our%20Insights/Big%20data%20The%20next%20frontier%20for%20innovation/MGI_big_data_exec_summary.ashx (accessed on 15 March 2017).

37. Najafabadi, M.M.; Villanustre, F.; Khoshgoftaar, T.M.; Seliya, N.; Wald, R.; Muharemagic, E. Deep learning applications and challenges in big data analytics. *J. Big Data* **2015**, *2*, 1. [CrossRef]

38. Cox, M.; Ellsworth, D. Application-controlled demand paging for out-of-core visualization. In Proceedings of the 8th Conference on Visualization '97 (VIS '97), Phoenix, AZ, USA, 18–24 October 1997; Yagel, R., Hagen, H., Eds.; IEEE Computer Society Press: Los Alamitos, CA, USA, 1997.

39. Leadership Council for Information Advantage. Big Data: Big Opportunities to Create Business Value. Available online: http://www.emc.com/microsites/cio/articles/big-data-big-opportunities/LCIA-BigData-Opportunities-Value.pdf (accessed on 1 February 2015).

40. Big Data Analytics. TDWI Best Practices Report, 2011, Fourth Quarter. Available online: https://tdwi.org/research/2011/09/best-practices-report-q4-big-data-analytics.aspx (accessed on 21 April 2018).

41. NewVantage Partners. Big Data Executive Survey. Themes & Trends. Available online: http://newvantage.com/wp-content/uploads/2012/12/NVP-Big-Data-Survey-Themes-Trends.pdf (accessed on 1 February 2015).

42. Bharadwaj, A.; El Sawy, O.A.; Pavlou, P.A.; Venkatraman, N.V. Digital business strategy: Toward a next generation of insights. *MIS Q.* **2013**, *37*, 471–482. [CrossRef]

43. Kent, P.; Kulkarni, R.; Sglavo, U. Finding Big Value in Big Data: Unlocking the Power of High Performance Analytics. In *Big Data and Business Analytics*; Liebowitz, J., Ed.; CRC Press Taylor & Francis Group, LLC: Boca Raton, FL, USA, 2013; pp. 87–102, ISBN 9781466565784.

44. Bekmamedova, N.; Shanks, G. Social media analytics and business value: A theoretical framework and case study. In Proceedings of the 47th Hawaii International Conference on System Sciences (HICSS), Waikoloa, HI, USA, 6–9 January 2014.

45. Davenport, T. Big Data at Work. In *Dispelling the Myths, Uncovering the Opportunities*; Harvard Business Review Press: Boston, MA, USA, 2014; ISBN 9781422168165.

46. OED—Oxford English Dictionary. Available online: http://www.oed.com/view/Entry/18833#eid301162177 (accessed on 10 February 2015).

47. Sun, E.W.; Chen, Y.T.; Yu, M.T. Generalized optimal wavelet decomposition algorithm for big financial data. *Int. J. Prod. Econ.* **2015**, *165*, 194–214. [CrossRef]

48. Opresnik, D.; Taisch, M. The value of big data in servitization. *Int. J. Prod. Econ.* **2015**, *165*, 174–184. [CrossRef]

49. He, X.; Ai, Q.; Qiu, R.C.; Huang, W.; Piao, L.; Liu, H. A big data architecture design for smart grids based on random matrix theory. *IEEE Trans Smart Grid* **2017**, *8*, 674–686. [CrossRef]

50. Kusuma, S.; Kasi Viswanath, D. IOT And Big Data Analytics In E-Learning: A Technological Perspective and Review. *Int. J. Eng. Technol.* **2018**, *7*, 164–167. [CrossRef]

51. Nelson, G.S. Business Intelligence 2.0: Are We There Yet? SAS Global Forum. Business Intelligence/Analytics. Available online: http://support.sas.com/resources/papers/proceedings10/040-2010.pdf (accessed on 10 February 2015).

52. The Digital Universe in 2020: Big Data, Bigger Digital Shadows, and Biggest Growth in the Far East. IDC iView. Available online: https://www.emc.com/collateral/analyst-reports/idc-the-digital-universe-in-2020.pdf (accessed on 10 February 2015).

53. Suther, T.; Burkart, B.; Cheng, J. Jack and the Big Data Beanstalk: Capitalizing on a Growing Market Opportunity. In *Big Data and Business Analytics*; Liebowitz, J., Ed.; CRC Press Taylor & Francis Group, LLC: Boca Raton, FL, USA, 2013; pp. 21–42, ISBN 9781466565784.

54. 12 Big Data Definitions: What's Yours? Available online: http://www.forbes.com/sites/gilpress/2014/09/03/12-big-data-definitions-whats-yours/ (accessed on 10 February 2015).

55. White, C. Using Big Data for Smarter Decision Making. Available online: ftp://public.dhe.ibm.com/software/pdf/ro/Using-Big-Data-Smarter-Decision-Making.pdf (accessed on 6 August 2017).

56. Ferguson, M. Architecting a Big Data Platform for Analytics. IBM, White Paper, 2012. Available online: http://www.ibmbigdatahub.com/whitepaper/architecting-big-data-platform-analytics (accessed on 13 May 2018).

57. Boyd, D.; Crawford, K. Critical questions for big data: Provocations for a cultural, technological, and scholarly phenomenon. *Inf. Commun. Soc.* **2012**, *15*, 662–679. [CrossRef]

58. Big Data and Better Data. Available online: http://magazine.amstat.org/blog/2012/06/01/prescorner/ (accessed on 21 April 2018).

59. Big Data: A Perspective from the BLS. Available online: http://www.andrew.cmu.edu/user/jsmurray/teaching/303/files/bls_bigdata.pdf (accessed on 6 August 2017).

60. Kamioka, T.; Tapanainen, T. Organizational use of big data and competitive advantage—exploration of antecedents. In Proceedings of the 18th Pacific Asia Conference on Information Systems, PACIS 2014, Chengdu, China, 24–28 June 2014; p. 372.

61. Akter, S.; Wamba, S.F.; Gunasekaran, A.; Dubey, R.; Childe, S.J. How to improve firm performance using big data analytics capability and business strategy alignment? *Int. J. Prod. Econ.* **2016**, *182*, 113–131. [CrossRef]

62. Liebowitz, J. (Ed.) *Big Data and Business Analytics*; CRC Press Taylor & Francis Group, LLC: Boca Raton, FL, USA, 2013; ISBN 9781466565784.

63. Kaisler, S.; Money, W.; Cohen, S. Smart Objects: An Active Big Data Approach. In Proceedings of the 51st Hawaii International Conference on System Sciences, Waikoloa Village, HI, USA, 3–6 January 2018; Curran Associates, Inc.: Red Hook, NY, USA, 2018; pp. 809–818.

64. Edelenbos, J.; Hirzalla, F.; van Zoonen, L.; van Dalen, J.; Bouma, G.; Slob, A.; Woestenburg, A. Governing the Complexity of Smart Data Cities: Setting a Research Agenda. In *Smart Technologies for Smart Governments. Public Administration and Information Technology*; Rodríguez Bolívar, M., Ed.; Springer: Cham, Germany, 2018; Volume 24, pp. 35–54, ISBN 978-3-319-58576-5.

65. Stock, T.; Seliger, G. Opportunities of sustainable manufacturing in industry 4.0. *Procedia CIRP* **2016**, *40*, 536–541. [CrossRef]

66. Wielki, J. The Opportunities and Challenges Connected with Implementation of the Big Data Concept. In *Advances in ICT for Business, Industry and Public Sector*; Mach-Król, M., Olszak, C.M., Pełech-Pilichowski, T., Eds.; Springer: Cham, Germany, 2015; pp. 171–189, ISBN 978-3-319-11327-2.

67. Valacich, J.S.; Schneider, C. *Information Systems Today. Managing the Digital World*; Global Edition; Pearson Education Limited: Harlow, UK, 2017; ISBN 9781292215976.

68. Belaud, J.-P.; Negny, S.; Dupros, F.; Michéa, D.; Vautrin, B. Collaborative simulation and scientific big data analysis: Illustration for sustainability in natural hazards management and chemical process engineering. *Comput. Ind.* **2014**, *65*, 521–535. [CrossRef]

69. Papadopoulos, T.; Gunasekaran, A.; Dubey, R.; Altay, N.; Childe, S.J.; Fosso-Wamba, S. The role of Big Data in explaining disaster resilience in supply chains for sustainability. *J. Clean. Prod.* **2017**, *142*, 1108–1118. [CrossRef]

70. Schaltegger, S.; Wagner, M. Sustainable entrepreneurship and sustainability innovation: Categories and interactions. *Bus. Strategy Environ.* **2011**, *20*, 222–237. [CrossRef]

71. Pitt, J.; Bourazeri, A.; Nowak, A.; Roszczyńska-Kurasińska, M.; Rychwalska, A.; Santiago, I.R.; Sanchez, M.L.; Florea, M.; Sanduleac, M. Transforming Big Data into Collective Awareness For Transformative Impact On Society. *Computer* **2013**, *46*, 40–45. [CrossRef]

72. Cagnin, C.H.; Loveridge, D.; Butler, J. Business sustainability maturity model. In Proceedings of the Business Strategy and the Environment Conference, Devonshire Hall, Leeds, UK, 4–6 September 2005.

73. The Sustainability Management Maturity Model: Version 2.0. Available online: https://www.triplepundit. com/2009/12/the-sustainability-management-maturity-model-version-2-0/ (accessed on 1 October 2018).

74. Bourne, L. *SRMM®: Stakeholder Relationship Management Maturity*; Project Management Institute: Newtown Square, PA, USA, 2008.

75. Jawahar, I.M.; McLaughlin, G.L. Toward a descriptive stakeholder theory: An organizational life cycle approach. *Acad. Manag. Rev.* **2001**, *26*, 397–414. [CrossRef]

76. How Mature Is Your Learning Business? Available online: https://www.tagoras.com/learning-business-maturity-model/ (accessed on 1 October 2018).

77. The Evolution of The High-Impact Learning Organization. Available online: https://www.panopto.com/ blog/the-evolution-of-the-high-impact-learning-organization/ (accessed on 1 October 2018).

78. Five Levels of Change Management Maturity. Available online: https://www.prosci.com/resources/ articles/change-management-maturity-model (accessed on 1 October 2018).

79. Big Data at the Speed of Business. Available online: http://www-01.ibm.com/software/data/bigdata/ (accessed on 2 October 2014).

80. Lahrmann, G.; Marx, F.; Winter, R.; Wortmann, F. Business Intelligence Maturity Models: An Overview. In *Information Technology and Innovation Trends in Organizations, Proceedings of the VII Conference of the Italian Chapter of AIS (itAIS 2010), Naples, Italy, 8–9 October 2010*; D'Atri, A., Ferrara, M., George, J., Spagnoletti, P., Eds.; Italian Chapter of AIS: Naples, Italy, 2010.

81. Cooke-Davies, T.J. Measurement of Organizational Maturity. In *Innovations: Project Management Research*; Slevin, D.P., Cleland, D.I., Pinto, J.K., Eds.; Project Management Institute: Newtown Square, PA, USA, 2004; pp. 211–228, ISBN 193069959X.

82. Hribar Rajterič, I. Overview of Business Intelligence Maturity Models. *Management: J. Contemp. Manag. Issues* **2010**, *15*, 47–67.

83. Mircea, M. (Ed.) *Business Intelligence—Solution for Business Development*; InTechOpen: London, UK, 2012; ISBN 978-953-51-0019-5.

84. Wikimedia Commons. Available online: http://commons.wikimedia.org/wiki/File:Characteristics_of_ Capability_Maturity_Model.svg (accessed on 20 May 2014).

85. Röglinger, M.; Pöppelbuß, J.; Becker, J. Maturity models in business process management. *Business Process Manag. J.* **2012**, *18*, 328–346. [CrossRef]

86. Woo, J. Information Retrieval Architecture for Heterogeneous Big Data on Situation Awareness. *Int. J. Adv. Sci. Technol.* **2013**, *59*, 113–122. [CrossRef]

87. Spaletto, J. An Investigation of Strategies for Managing Exponential Data Growth in the Enterprise. *J. Leadersh. Org. Eff.* **2013**, *1*, 4–14.

88. Lee, H.-Y.; Wang, N.-J. The Construction and Investigation of Web Engineering for Extending the Enterprise's Core Competence Related Studies. *Int. J. Softw. Hardw. Res. Eng.* **2013**, *1*, 68–78.

89. Kang, S.; Myung, J.; Yeon, J.; Ha, S.W.; Cho, T.; Chung, J.M.; Lee, S.G. A General Maturity Model and Reference Architecture for SaaS Service. In *Database Systems for Advanced Applications, Part II*; Kitagawa, H., Ishikawa, Y., Li, Q., Watanabe, C., Eds.; Springer: Berlin/Heidelberg, Germany, 2010; LNCS 5982; pp. 337–346, ISBN 978-3-642-12097-8.

90. The Eight Building Blocks of Big Data: Management Summary. Available online: http://radcliffeadvisory. com/research/download.php?file=RAS_8BB_BD_MS.pdf (accessed on 17 May 2014).

91. Allen, J. Maintaining Knowledge about Temporal Intervals. *Commun. ACM* **1983**, *26*, 832–843. [CrossRef]

92. Mach-Król, M. Big Data analytics in Polish companies. Selected research results. In Proceedings of the International Conference on ICT Management for Global Competitiveness and Economic Growth in Emerging Economies, Wrocław, Poland, 23–24 October 2017; Kowal, J., Kuzio, A., Makio, J., Paliwoda-Pękosz, G., Soja, P., Sonntag, R., Eds.; University of Wrocław: Wrocław, Poland, 2017; pp. 64–77.

Article

Application of Machine Learning in Predicting Performance for Computer Engineering Students: A Case Study

Diego Buenaño-Fernández [1,*], David Gil [2] and Sergio Luján-Mora [3]

[1] Facultad de Ingeniería y Ciencias Aplicadas, Universidad de Las Américas, Av. de los Granados E12-41 y Colimes, Quito EC170125, Ecuador
[2] Departamento de Tecnología Informática y Computación, Universidad de Alicante, San Vicente del Raspeig, 03690 Alicante, Spain; david.gil@ua.es
[3] Departamento de Lenguajes y Sistemas Informáticos, Universidad de Alicante, San Vicente del Raspeig, 03690 Alicante, Spain; sergio.lujan@ua.es
* Correspondence: diego.buenano@udla.edu.ec; Tel.: +59-39-8449-8347

Received: 7 April 2019; Accepted: 14 May 2019; Published: 17 May 2019

Abstract: The present work proposes the application of machine learning techniques to predict the final grades (FGs) of students based on their historical performance of grades. The proposal was applied to the historical academic information available for students enrolled in the computer engineering degree at an Ecuadorian university. One of the aims of the university's strategic plan is the development of a quality education that is intimately linked with sustainable development goals (SDGs). The application of technology in teaching–learning processes (Technology-enhanced learning) must become a key element to achieve the objective of academic quality and, as a consequence, enhance or benefit the common good. Today, both virtual and face-to-face educational models promote the application of information and communication technologies (ICT) in both teaching–learning processes and academic management processes. This implementation has generated an overload of data that needs to be processed properly in order to transform it into valuable information useful for all those involved in the field of education. Predicting a student's performance from their historical grades is one of the most popular applications of educational data mining and, therefore, it has become a valuable source of information that has been used for different purposes. Nevertheless, several studies related to the prediction of academic grades have been developed exclusively for the benefit of teachers and educational administrators. Little or nothing has been done to show the results of the prediction of the grades to the students. Consequently, there is very little research related to solutions that help students make decisions based on their own historical grades. This paper proposes a methodology in which the process of data collection and pre-processing is initially carried out, and then in a second stage, the grouping of students with similar patterns of academic performance was carried out. In the next phase, based on the identified patterns, the most appropriate supervised learning algorithm was selected, and then the experimental process was carried out. Finally, the results were presented and analyzed. The results showed the effectiveness of machine learning techniques to predict the performance of students.

Keywords: educational data mining; learning analytics; machine learning; big data; prediction grades

---

1. Introduction

Quality education is one of the Sustainable Development Goals (SDGs) approved by the United Nations forum in 2015 [1] and is a fundamental challenge to support sustainable development worldwide. A key element that must be taken into account when talking about sustainable development is the principle of equal opportunities. In the educational field, this principle consists of guaranteeing every person the same possibilities in terms of access and completion of studies [2]. Student desertion in higher education is a critical issue that requires a global analysis. The dropout rates of university students generate a waste of resources for all actors in the education sector and even affect the evaluation processes of the institutions. In fact, the dropout rate is higher among engineering students [3]. In the present study, it is proposed to carry out predictive analysis of the final grades (FGs) of computer engineering students that will support the processes of academic quality and thus mitigate the student dropout rate. Efforts to transform our societies must prioritize education. Teachers and educational administrators must develop their understanding of sustainability and their ability to improve the curriculum and implement systems that allow for expanded learning opportunities [4].

In this sense, higher education institutions need to work on the development of educational models that emphasize the use of information and communication technologies (ICT), which could function as support tools for equal opportunities and social responsibility.

From this perspective, the application of ICT in educational environments is imperative because it can contribute significantly to the improvement of the teaching and learning process, as well as encourage the process of knowledge construction [5]. The application of technology in teaching-learning processes is known as Technology-enhanced learning (TEL). This term is used to describe the use of digital technology aimed at improving the teaching-learning experience. TEL has become relevant due to the emergence of a huge number of technological resources that help the development of critical thinking in students [6]. TEL incorporates many emerging technologies, including learning management systems (LMS), mobile learning applications, virtual and augmented reality interventions, cloud learning services, social networking applications for learning, video learning, robotics, data mining, and so forth [7].

According to the results of a study about the sustainability of higher education and the TEL [6], we must be very cautious when defining the necessary conditions for technology to serve as a benefit and not as an obstacle to teaching and learning. For instance, training teachers and educational administrators to develop predictive analytical competence is vital for measuring the potential results of the use of technology [8].

All the technologies mentioned above, which are being applied with ever greater impact on the educational field, generate and store a vast amount of data that is ubiquitously available [9]. This amount of data has exceeded the capacity for processing and analysis through conventional means. To fulfill the task of data analysis, it is necessary to work with new specific technologies, such as big data, intelligent data, data mining, and text mining, among others. The convergence of these technologies with educational systems will allow the analysis of these data and transform it into useful information for all stakeholders [10].

Educational data mining (EDM) and learning analytics are emerging disciplines that guide the process of analyzing educational data. This analysis is done through a variety of statistical methods, techniques, and tools, including machine learning and data mining. The objective of learning analytics is to provide an analysis of the data that originates in the educational repositories, as well as in the LMS, in order to understand and optimize the learning process and the environments in which it occurs [11].

There are several studies [9,11–14] that have proposed different classifications related to the use of data mining in educational environments. Among the most representative classifications are the following: Analysis and Visualization of Data; Providing Feedback for Supporting Instructors; Recommendations for Students; Predicting Student's Performance; Student Modeling; and Social Network Analysis. In the present work, we focused on the Predicting Student's Performance, one of the most popular EDM applications. The objective of the prediction is to estimate an unknown value of a variable from historical data related to it. In the present work, this variable is related to the grades and performance of students. That is, the estimation or prediction of student grades proposed is based on multiple historical academic characteristics that describe the student's behavior [15].

Based on these principles, the main objective of this work was to predict the grades of the students according to several characteristics of their academic performance. This was done by establishing dashboards to track the students individually, by subject, by area, etc. The expected consequence of this tracking is to decrease the dropout rate, as well as provide real-time student follow-up to improve the education system. The early identification of vulnerable students who are prone to drop out their courses is essential information for successfully implementing student retention strategies. The term student retention rate refers to the rate of students in a cohort who have not abandoned their studies for any situation. This rate is increasingly important for university administrators, as this directly affects graduation rates [16]. Once these students have been identified, through different prediction techniques, it will be easier to provide them with proper attention to prevent these students from abandoning their studies. Even early warning systems can be planned and designed to support student retention rates [17].

The case study analyzed in the present work will allow evaluating the effectiveness of the proposed method since educational administrators will obtain a validated alternative to replicate it in all the faculties of the university. By scaling the project for all the university's careers, the total data to be analyzed would be 16,000 students, each with an average of eight subjects and with three PGs (PGs) for each subject. This amount of data, together with the need for immediate visualization, puts us in front of two problems that are referenced when talking about big data issues: "volume" and "velocity" [18]. In other words, we are faced with such a large amount of data that traditional data processing applications cannot capture, process, and—finally—visualize the results in a reasonable amount of time. Big data emerged with the aim of covering the gaps and needs not met by traditional technologies [19]. In higher education, it is fundamental that both teachers and students have updated information, preferably in real time, to make timely decisions and corrective actions. The scaling up in the magnitude of data analyzed will lead in the future towards the design of a big data project.

The document is organized as follows. Section 2 presents the related studies that contribute to the conception of the problem and an evaluation of the techniques and methodologies used. Section 3 describes the materials and the method used. The first phase of the method emphasizes the data collection and the preprocessing process; the second phase presents the selection of the machine learning method; the third phase corresponds to the experimental process and results analysis; in the fourth phase the process of data visualization is described. Finally, Section 4, includes the discussion and conclusions of the contributions presented in this research.

2. Related Work

There is an extensive range of EDM-related work, where many interesting approaches and tools are presented that aim to fulfill the objectives of discovering knowledge, making decisions and providing recommendations. Below, we describe some of them that have served as a source of information for the present work.

In a study concerning the application of big data in the educational field [20], it can be seen that big data techniques can be used in various ways to support learning analytics, such as performance prediction, attrition risk detection, data visualization, intelligent feedback, course recommendation, student skill estimation, behavior detection, and grouping and collaboration of students, among others. In this study, the functionality of predictive analysis is emphasized, which is oriented to the prediction of student behavior, skill and performance.

In a study carried out at the university Northern Taiwan [21], the learning analytics and educational big data approaches were applied with the objective of making an early prediction of the final academic performance of the students in a course of calculation. This study applied principal component regression to predict students' final academic performance. In this work, variables external to the course, such as video-viewing behaviors, out-of-class practice behaviors, homework and quiz scores, and after-school tutoring, were included.

In a study about the factors that impact on the correctness of software [22], it is concluded that, when working with data mining in educational environments, two types of data analysis are generally used: approaches based on predictive models and approaches based on descriptive models. Predictive approaches generally employ supervised learning functions to estimate unknown values of dependent variables [23]. By contrast, descriptive models often use unsupervised learning functions in order to identify patterns that explain the structure of the extracted data [24].

The methods of collaborative filtering have become a novel technique to predict the performance of a student in future academic years, depending on their grades. In the educational field, collaborative filtering methods are based on the hypothesis that student performance can be predicted from grade history of all courses or modules successfully completed. An evaluation of grade prediction for future academic years is presented in Reference [25] using collaborative filtering methods based on probabilistic matrix factorization and Bayesian probabilistic models. The prediction model was evaluated in a simulated scenario based on a set of real data of student grades between the years 2011 to 2016 in a higher education institution in Macedonia.

In another work [26], the application of collaborative filtering methods was also identified, where the objective was to predict the performance of students at the beginning of an academic period, based on their academic record. The approach is based on representing student learning from a set of grades of their approved courses, in order to find students with similar characteristics. The research was conducted on historical data stored in the information system of Masaryk University. The results show that this approach is as effective as using commonly used machine learning methods, such as support vector machines.

In other research, the authors propose the development of methods that use historical datasets of student grades by courses, with the objective of estimating student performance [27]. Their proposal was based on the use of dispersed linear models and low-range matrix factorizations. The work evaluated the performance of the proposed techniques in a set of data obtained from the University of Minnesota that contained historical grades of a 12.5-year period. This work showed that focusing on course-specific data improves the accuracy of grade prediction.

In Reference [28], a novel approach is proposed that uses recommendation systems for the extraction of educational data, especially to predict the performance of students. To validate this approach, recommendation system techniques are compared with traditional regression methods, such as logistic or linear regression. An additional contribution of the work is the application of recommendation system techniques, such as matrix factorization in the educational context, in order to predict the future performance of students.

In one research study [29], academic data were collected from different secondary schools in the district of Kancheepuran, India. They used decision trees and naïve Bayes algorithms to run the classification of students. The study concluded the following:

- The parents' occupation, and not the type of school, played an important role in predicting the FG.
- The decision tree algorithm was best for student modeling.
- The FG for upper secondary students could be predicted from the students' previous data.

Regarding big data, the opportunities and benefits that it offers for education have recently been studied. An analysis of the relationship between big data and educational environments has been presented in Reference [30]. The work focuses on the different methods, techniques, tools, and big data algorithms that can be used in the educational context in order to understand the benefits and impact that can cause in the teaching and learning process. The discussion generated in this document suggests that the incorporation of an approach based on big data is of crucial importance. This approach can contribute significantly in the improvement of the learning process, for its implementation must be correctly aligned with the learning needs and the educational strategies.

A smart recommendation system based on big data for courses of e-learning is presented in Reference [31]. In this article, the method of rules of association is applied in order to discover the relationships between the academic activities carried out by the students. Based on the rules extracted, the most appropriate course catalog is defined according to the behavior and preferences of the student. Finally, in this work, a recommendation system was implemented using technologies and big data tools, such as: Spark Framework and Hadoop ecosystem. The results obtained show the scalability and effectiveness of the proposed recommendation system.

3. Materials and Methodology

In the present work, a methodology guided by the steps described in Figure 1 is used:

1. The collection and data cleansing of historical datasets of student grades takes place.
2. The methods of machine learning and data mining are selected.
3. The model for predicting student grades is generated from previously processed data.
4. The results obtained are analyzed and visualized.

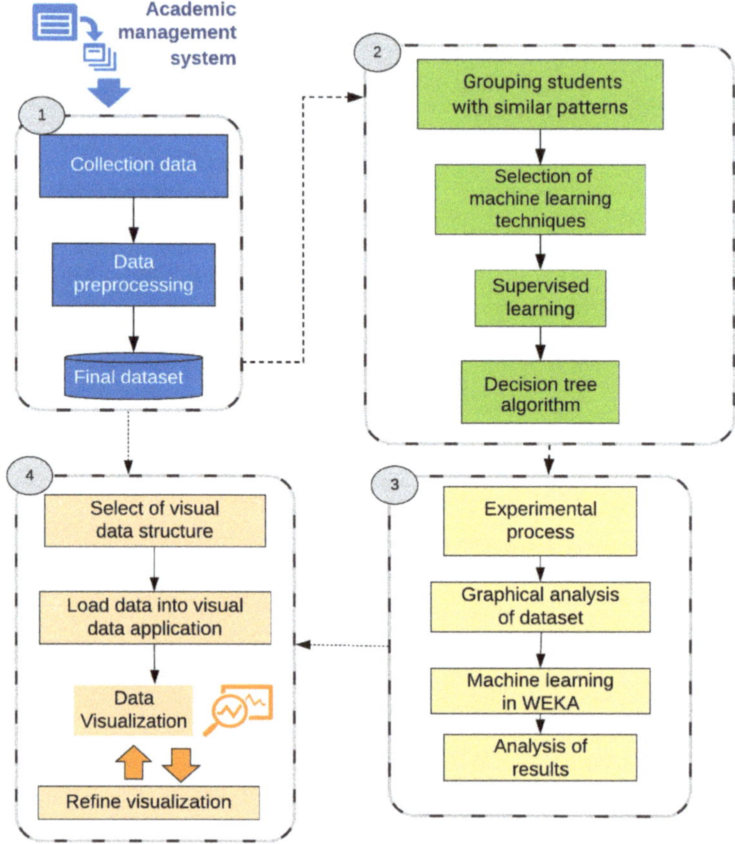

Figure 1. Methodology proposed.

3.1. Data Description

The dataset used for the present work is composed of the academic records of 335 students. The total number of historical records of students' grades was 6358, which corresponds to all the subjects taken by this group of students. The periods analyzed were from the semester 2016-1 to the semester

2018-2 in the Computer Systems Engineering Degree of a university in Ecuador. In addition, the dataset comprises a total of 68 subjects organized into seven knowledge areas (Programming and Software Development, Mathematics and Physics, Information Network Infrastructure, Electronics, Databases, Economy—Administration, General Education—Languages), as can be seen in Figure 2. In addition, Figure 2 shows the number of subjects by areas of knowledge. According to the educational model used by the university, curricular coherence is vertically aligned in each of the seven areas of knowledge, that is, what students learn in the course or module is used as the basis for the next academic course. However, it is important to point out an exception, since the transversal knowledge areas, such as Economics—Administration and General Education are more aligned horizontally, where there are no such strong dependencies in different subjects and academic years.

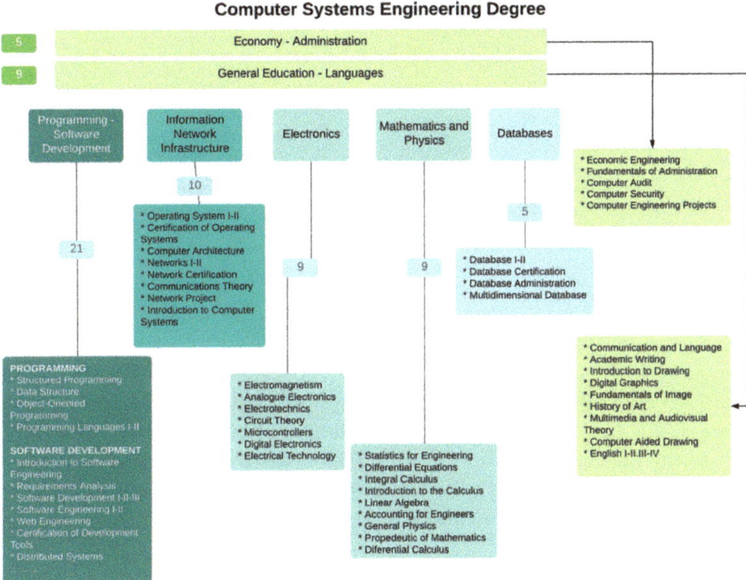

Figure 2. Subjects by area of knowledge.

The data were extracted from the institution's academic management system and stored in CSV format file. This information was periodically retrieved from the university's grades system and stored in an integrated data repository. From this repository, some dashboards useful for the stakeholders were built. Table 1 shows a sample of the dataset. In order to pass a subject, the student must obtain a FG (FG) equal to or higher than 6. The FG is composed of three partial components (i.e., PG) weighted differently: PG1 is 35%; PG2 is 35%; and PG3 is 30%. This formula applies equally to all subjects and is a curricular definition for the entire university.

Table 1. Sample of the dataset.

Academic Period	Subject Name	PG1	PG2	PG3	FG	Area	Situation
2016-1	General Physics	8.0	4.4	6.3	6.2	Mathematics and Physics	Pass
2017-1	Communication Theory	6.0	5.6	5.3	5.7	Infrastructure	Fail
2017-1	Digital Electronics	4.4	8.1	6.9	6.4	Electronics	Pass

In the data preprocessing phase, duplicate records and null value records in components PG1, PG2, and PG3 were eliminated. In addition, in this phase the subjects of the knowledge areas Economy—Administration and General Education—Languages were eliminated. Another important task was executing a process to anonymizing the data that was carried out to comply with international data protection standards. This process consisted of eliminating or substituting the personal data fields (identification number, names, and surnames) of both students and teachers.

Before the dataset was loaded into the WEKA (https://www.cs.waikato.ac.nz/ml/weka/) (Waikato Environment for Knowledge Analysis) machine learning software to carry out a series of experiments, it was of interest to observe and study the dataset in terms of visual graphs. Figure 3 shows the evolution of student grades from the first semester of 2016 to the last semester of 2018, showing the four-color lines for every grade PG1, PG2, PG3, and FG.

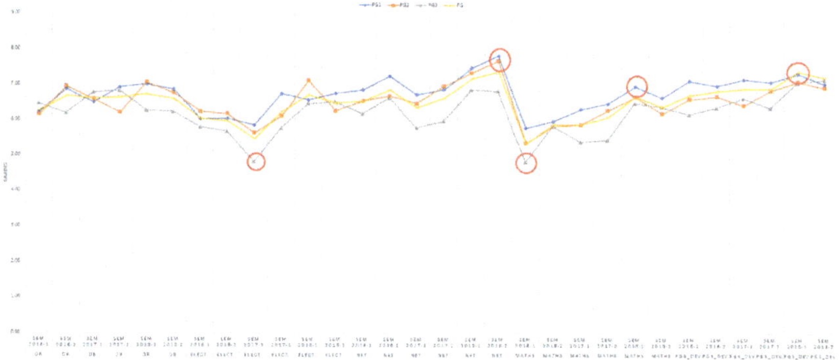

Figure 3. Trend of students' grades with greatest deviations highlighted with a red circle.

It is striking to verify that in general, there is a trend of similar grades by area. Inclusive, as can be seen in some interesting deviations that have been highlighted with a red circle. These peaks represent ascending and descending trends in grades by area of knowledge. It is possible to think that this could be due to virtual groupings (similar grades are obtained in the same area) by professors of subjects within the same area. Or, it could even be due to similar criteria in the evaluation of these professors who belong to the same area.

It is interesting to deepen the analysis, since, after consulting the course coordinators of the knowledge area, at first glance, it seems that these similar peaks of grades graphed in Figure 3 respond to a coincidence. For the analysis, it must be taken into account that a subject, in a certain area of knowledge, can be taught by different professors. In addition, in spite of the fact that the evaluation criteria are uniformly managed in the university, each teacher applies the academic freedom in their evaluation methods.

In Figure 3, some interesting deviations are highlighted with a red circle, with first highly descending peaks and then two others as highly ascending. It is worthwhile studying what these situations might be due to. At first, it seems the explanations could have to do with students attaining good grades in their first tests and then their grades deteriorating as the course advances. That might be the reason why PG3 decreased and vice versa with the last two red circles that show that the students at the end studied harder to get a better FG. In addition, there is an important factor that, since the semester 2018-1, the percentage weightings of each PG changed:

From 2016-1 to 2017-2, the FG was calculated as follows:

$$FG = PG1 \times 0.35 + PG2 \times 0.35 + PG3 \times 0.30$$

In these periods, students put their greatest interest (and effort) at the beginning of the course, PG1 and PG2. In many cases, just with these two PGs, they were able to pass the subject (although with the minimum mark required) and, therefore, neglected their academic performance in the PG3. For this reason, as of semester 2018-1, the FG is calculated as follows:

$$FG = PG1 \times 0.25 + PG2 \times 0.35 + PG3 \times 0.40$$

From this semester, it was observed that students improved their grades in PG3. Figure 4 shows all the data loaded graphically to more easily appreciate the correlation between all the columns with respect to the final result (pass or fail the course, column "Situation"; red = "fail", blue = "pass"). The aim of this figure is to show dashboards where it is possible to measure the influence of and relationship between every particular feature regarding the FG (Situation). Evidently, there are cases where that correlation is clearly identifiable. This FG, named "Situation", shown in Figure 4, clearly identifies (almost with a perfect line) that up to 5.6, the FG will be "fail", whereas over this value, the FG will be "pass".

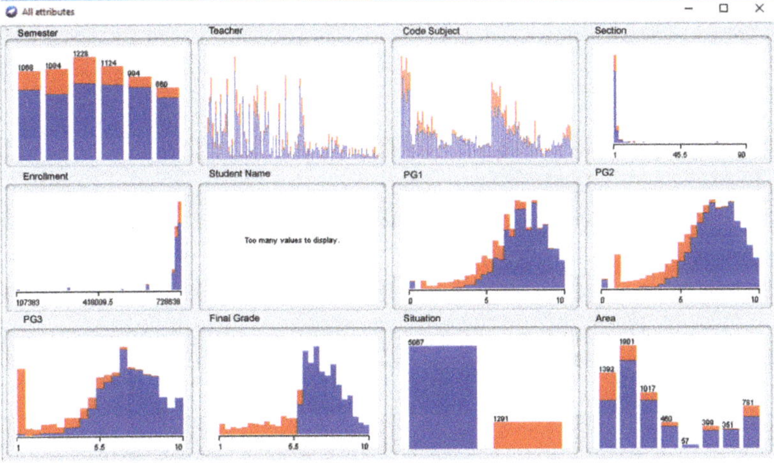

Figure 4. Visualization and correlation with all data after loading the dataset.

Most of the remaining dashboards are not as straightforward to interpret. They often show mixes of "red & blue" to confuse the correlation. Of course, there are general signs of these indicators, like the PGs (PG1–PG3), which indicate a trend to blue when the value increases, and they are red when the value is low. In fact, this is the clear objective of an indicator, obvious and concise.

It is also worthwhile mentioning the variables "Area" and "Code Subject", as it is widely believed that a particular area, as well as a specific subject, have a direct connection with the FG. The dashboard of "Code Subject" is harder to explain due to the high number of subjects. We could appreciate higher concentration of red in the central area, whereas at the beginning and just after the middle, there is a good proportion of blue. Nevertheless, there will be always a majority of blue as the classes (pass and fail) are totally unbalanced (5067 vs. 1291, respectively), as can be seen in Figure 5.

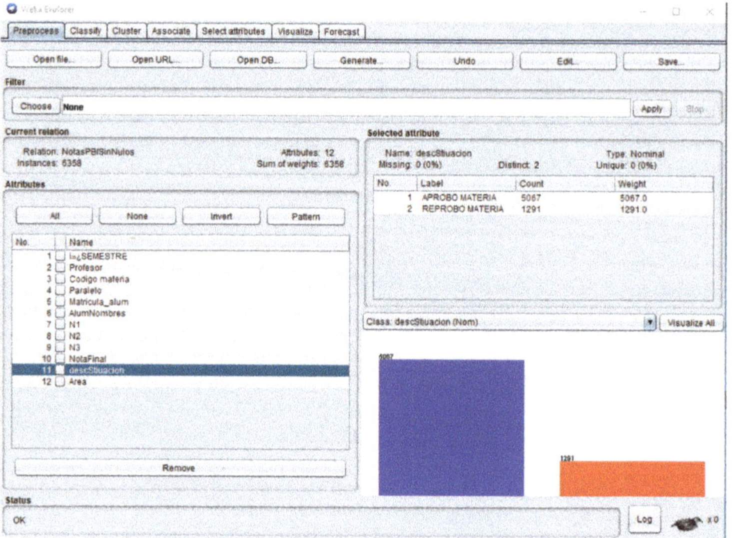

Figure 5. Initial dataset loaded in the system.

3.2. Selection of Machine Learning Techniques

In this research, we used data mining and machine learning techniques to provide an accurate prediction method for historical dataset of student grades. On the historical dataset of the student grades in the Computer Systems Engineering Degree, supervised learning techniques were applied to determine a predictive model that would lay the foundations for the future development of a system of recommendations for the students. Predicting the academic performance of students is considered one of the most common problems and, at the same time, represents a complex task of educational data mining.

Classification is the most widely used data mining technique, and this technique is applied over pre-classified data records in order to develop a predictive model that can be used to classify unclassified data records. This technique can be executed through the application of the decision tree algorithm. The process includes two steps: learning and classification [29]. In the learning step, the training dataset is analyzed using the chosen classification algorithm. The main benefit of applying the decision tree algorithm is that its results can be easily interpreted and explained, thanks to its graphical representation that summarizes a model of implicit decision rules.

3.3. Experimental Process

In the experimental phase, before applying machine learning tools, a study was carried out to group the information in order to identify groups of students with a certain pattern of behavior. The task of grouping data is particularly important since it is usually the first step in data mining processes. From this task, it is possible to identify groups with similar characteristics that can be used as a starting point to explore future relationships.

In a second phase, using the decision tree algorithm, some tests were done with the students' grades. For example, in a first test the grades of the (PG3) were eliminated in order to make a prediction of the (FG). With this test it was expected to identify the number of students who passed the subjects without this component. Then a prediction was attempted with the PG2 component eliminated. The results found are shown in the following section.

3.4. Data Visualization

The main purpose of data visualization is to present all the characteristics of the dataset through graphical representations. The visualization of data in a graphical format constitutes an element of support, so that the results of a process of data analysis are shown in an intuitive way for students, teachers or educational administrators. The data visualization process can be described in general terms in the following steps: obtain and debug the data; select the data visualization structure; load the data into the selected application; display the data in dashboards; and, finally, refine the process of visualization [32].

4. Results

In engineering degrees, it is not common to find regular students, that is to say that they pass consecutively all the subjects of various academic levels planned in the curriculum. With the historical dataset of student grades, a combination of variables was performed in order to obtain a group of students that have common attributes and on which some type of analysis can be carried out before applying machine learning algorithms. After combining student grades, subjects, and academic years, only four regular students were identified who have taken and passed the same subjects up to 6th semester; this is 37 subjects, which is equivalent to 62% of the total subjects (68) of the curriculum. As previously explained, 19 subjects were eliminated from certain knowledge areas of transversal training. Figure 6 shows the variation of the FG of the four students over six semesters and 19 subjects.

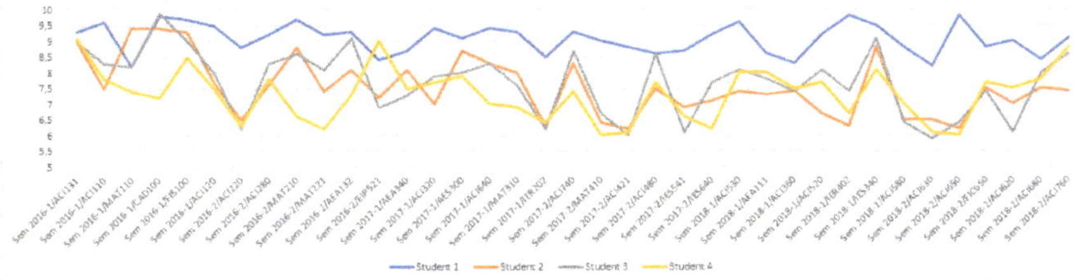

Figure 6. Variation of the final grades (FGs) (FG) of the four students.

These four students belong to the group that started the degree in the 2016-1 semester (2016-1 cohort). The number of students identified is very low, considering that in this cohort there was a new enrollment of 67 students, as can be seen in Table 2. That is, only 6% of students have managed to advance in the curriculum without failing any subject until the sixth semester (37 subjects). Table 2 shows the cumulative number of students who have dropped out of their studies corresponding to some cohorts, the attrition analysis is done at 6, 12, 18, 24, 30, and 36 months. The student dropout rate represents the number of students who drop out of their studies for different reasons. These reasons can be of an academic, economic, or personal nature. There are special cases in which students leave their studies for a certain time and then re-enroll. In these cases, the dropout rate takes atypical values, as can be seen in Table 2 in the academic period 2017-1, where the dropout number at 24 months (25) is lower than the dropout rate at 18 months (26).

Taking the 2016-2 cohort as a reference, an analysis was made of the peaks highlighted in Figure 6. The first subject observed with a low peak in the FG was Data Structures (ACI220). Figure 6 indicates that all the students lowered their FGs in this subject, with the FG near 6. Table 3 shows the statistical data of the subject (Data Structures) in the different periods of study. It was observed, in relation to the pass rates, that the subject has had a positive evolution throughout the semesters analyzed. The fail rate was reduced from 35% in the semester 2016-1 to 17% in the semester 2018-1.

Table 2. Dropout number per cohort.

Academic Period	Total New Enrolment	Dropout 6 Months	Dropout 12 Months	Dropout 18 Months	Dropout 24 Months	Dropout 30 Months	Dropout 36 Months
2016-1	67	12	21	29	32	36	38
2016-2	29	6	14	17	17	19	
2017-1	57	13	25	26	25		
2017-2	20	4	4	5			
2018-1	68	16	24				
2018-2	35	9					

Table 3. Statistics of subject Data Structures (ACI220).

Criteria	Statistics 2016-2	Statistics 2017-1	Statistics 2017-2	Statistics 2018-1
Total new enrolment	51	37	46	18
Average of grades	5.8	5.6	6.3	7.4
Pass rate	65%	65%	72%	83%
Fail rate	35%	35%	33%	17%

The second subject analyzed was Operating Systems II (ACI740) in the semester 2017-2; this subject has a peak of high FGs. Table 4 shows the statistical data of the subject in the different periods analyzed. It is interesting to consider some aspects identified around this subject. The subject has been taught by the same teachers in the three analyzed periods. The number of students per section is low in relation to other subjects. The average of pass rate of the subject is higher in relation to other subjects.

Table 4. Statistics of subject Operating Systems (ACI740).

Criteria	Statistics 2017-2	Statistics 2018-1	Statistics 2018-2
Total new enrolment	27	26	40
Average of grades	7.0	6.3	7.4
Pass rate	89%	70%	100%
Fail rate	11%	31%	0%

After the preliminary analysis, it became imperative to analyze the student retention and dropout values of the degree under study. Figure 7 shows the student retention and dropout rates accumulated for each cohort that began their studies in the academic periods we analyzed in this work. Figure 7 shows the retention and dropout rates at 6, 12, 18, 24, 30, and 36 months. When the rates are accumulated, it was observed that the cohort that began their studies in the semester 2016-1 had 29 students remaining after three years. The educational authorities must focus on these statistical data in order to implement actions that allow the dropout rate to be reduced.

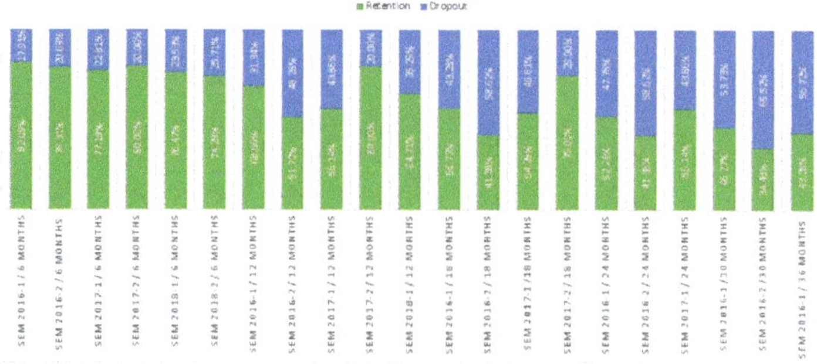

Figure 7. Retention and dropout rates by cohort.

4.1. Initial Situation: All Attributes

Figure 8 presents the first experiment carried out. The rule obtained by the decision tree is not very useful, as the tree in itself is very simple. However, the main feature retrieved, as we expected, was that a student needs to achieve over 5.9 grade to pass the subject.

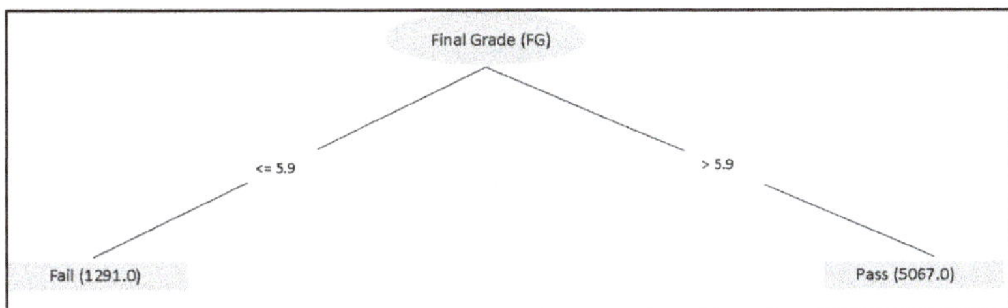

Figure 8. Decision tree with all the variables from the dataset.

Table 5 shows the accuracy, as well other measures, including the confusion matrix, obtained for this first experiment shown in Figure 8.

Table 5. Values for the accuracy of the decision tree and the confusion matrix using all attributes.

Correctly Classified Instances	6358	100%
Incorrectly Classified Instances	0	0%
=== Confusion Matrix ===		
a	b	<– classified as
1291	0	|a = Fail
0	5067	|b = Pass

4.2. Without Final Grade

As verified in the previous section, the first step is to run the decision tree with all the available input attributes. The analysis is that only the input variable of PGs are taken into account to predict whether the student will pass or not. Therefore, the next step is to remove this variable to check the incidence of the rest of the variables and their correlation in the final result. For this reason, in the following experiments, different tests were carried out, gradually eliminating some of these variables and assessing their weight in relation to the final prediction (if the student will pass or fail).

Figure 9 shows the confusion matrix obtained for this first experiment. On the other hand, Table 6 shows additional measures related to the results of the execution of the decision tree algorithm. The decision tree of Figure 9 offers a high accuracy, in spite of the FG being removed. Furthermore, the decision tree in itself provides good visual rules where is obvious to observe the influence of the input variables and their correlation with the FG. To go one step further, within the next subsection, we will explore the effect of PGs by removing some of them.

Table 6. Values for the accuracy of the decision tree and the confusion matrix without the FG.

Correctly Classified Instances	6139	96.5%
Incorrectly Classified Instances	219	3.5%
=== Confusion Matrix ===		
a	b	<– classified as
1122	169	|a = Fail
50	5017	|b = Pass

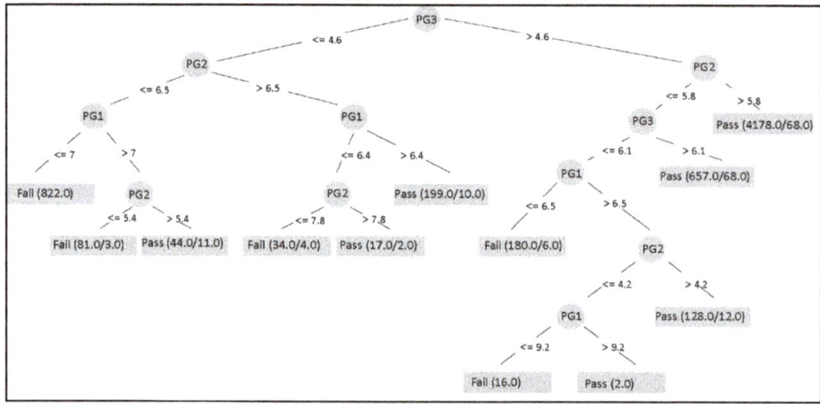

Figure 9. Decision tree without the FG.

4.3. Without PGs

In these experiments, we took out the PGs PG3 and PG2, respectively. Here, the objective with these tests was not only to build diverse decision trees—which in itself is great as it will provide us new rules and patterns—for every test, but most importantly, to weigh the significance of every PG, PG1–PG3. These results can be seen in Table 7.

Table 7. Values for the accuracy of the decision trees (without PG3 and PG2, respectively) and the confusion matrix.

Correctly Classified Instances	5815	91.5%
Incorrectly Classified Instances	543	8.5%
=== Confusion Matrix ===		
a	b	<– classified as
961	330	|a = Fail
213	4854	|b = Pass
Correctly Classified Instances	5915	93%
Incorrectly Classified Instances	443	7%
=== Confusion Matrix ===		
a	b	<–classified as
937	354	|a = Fail
89	4978	|b = Pass

What is really striking in these last experiments is the creation of clear and coherent decision trees and, consequently, the usefulness of the acquired decision rules. This allows a study on the PGs to determine which are the most decisive. For example, in Figure 10, the root of the decision tree shows that when PG2 is lower or equal than 5.7 and PG1 greater than 6.2, then the student will either fail if PG2 is lower than or equal to 4.0, or otherwise pass. With this information, teachers can build action plans of individualized learning for students classified under this rule. Figure 11 shows a similar decision tree, slightly more complex, where we are able to find patterns and rule analogously to the previous example of Figure 10; the difference in this test is that PG2 was removed, and we used the PGs PG1 and PG3.

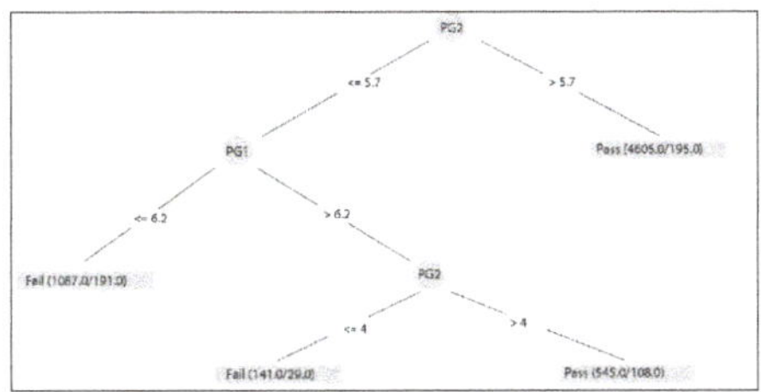

Figure 10. Decision tree without PG3.

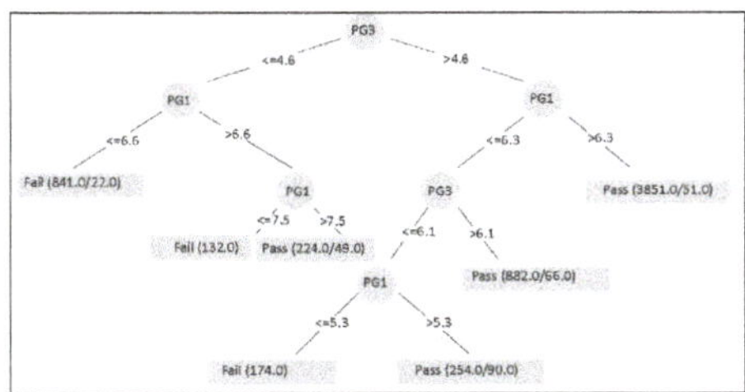

Figure 11. Decision tree without PG2.

4.4. Students Follow-Up

In this last subsection of the experimentation, we intend to address, possibly the most complex aspect, concerning student follow-up. For this challenge, we tried to predict the results of the students in the last year based on the results obtained in the previous academic courses (Figure 12).

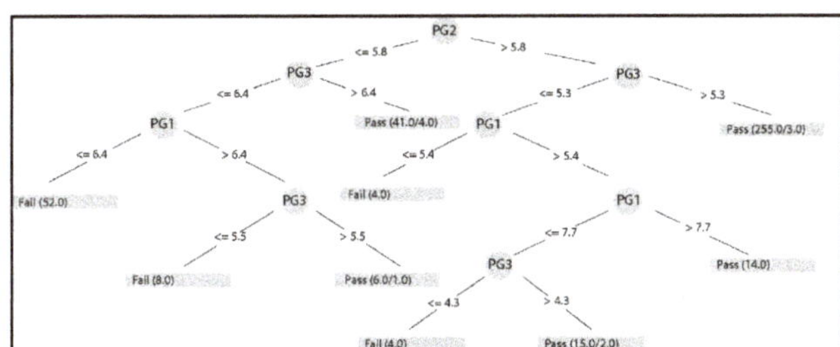

Figure 12. Decision tree to predict the results of the students in the last year.

We used a subset of the original dataset, including only students who belong to the database area, where the subjects are obviously similar. Figure 12 shows the decision tree we obtained with this experiment, and Table 8 shows the results achieved.

Table 8. Values for the accuracy of the decision tree and the confusion matrix.

Correctly Classified Instances	371	93%
Incorrectly Classified Instances	28	7%
=== Confusion Matrix ===		
a	b	<– classified as
59	19	|a = Fail
9	312	|b = Pass

Table 9 shows the classification errors for the last academic year. The decision tree predicts that a particular student (i.e., instance 166) will fail, whereas the current situation is that the student will pass.

Table 9. Classification errors for the last academic year.

Instance: 166			Instance: 182		
Academic period: Sem 2018-1			Academic period: Sem 2018-1		
Subject: ACI770- Multidimensional Database			Subject: ACI770- Multidimensional Database		
PG1: 6.0	PG2: 8.0	PG3: 4.9	PG1: 8.7	PG2: 4.7	PG3: 5.6
Predicted Situation: Fail			Predicted Situation: Fail		
Situation: Pass			Situation: Pass		
Instance: 187			Instance: 247		
Academic period: Sem 2018-2			Academic period: Sem 2018-2		
Subject: ACI530-Database I			Subject: ACI630- Database II		
PG1: 4.6	PG2: 7.8	PG3: 5.7	PG1: 5.8	PG2: 5.4	PG3: 6.6
Predicted Situation: Fail			Predicted Situation: Fail		
Situation: Pass			Situation: Pass		
Instance: 300					
Academic period: Sem 2018-2					
Subject: ACI810- Database Administration					
PG1: 6.1	PG2: 5.9	PG3: 5.9			
Predicted Situation: Fail					
Situation: Pass					

We have to find the reason among PGs. This instance has a lower qualification for a PG of 4.9. A similar explanation could be applied for instances 182 and 187, both of them with some PG below 5. Something more unknowable happens with instance 247, as well as 300, as they have PG not below 5 but still between 5 and 6. Therefore, these are the typical instances that belong to the outlier definition, between classes.

In Table 10, we can find the opposite perspective. Now, the decision tree predicts that a particular student (i.e., instance 160) will pass, whereas the current situation is that the student will fail. Again, it is important to depict and describe these results which depend on the PG. The system predicts that the student passes because some of the partial qualifications are high: instances 160, 233, and 238 are only low in PG3, although quite lower compared to the former PG1 and PG2.

Table 10. Classification errors for the last academic year.

Instance: 160			Instance: 233		
Academic period: Sem 2018-1			Academic period: Sem 2018-2		
Subject: ACI770- Multidimensional Database			Subject: ACI630-Database II		
PG1: 6.1	PG2: 6.0	PG3: 2.9	PG1: 7.7	PG2: 6.1	PG3: 4.0
Predicted Situation: Pass			Predicted Situation: Pass		
Situation: Fail			Situation: Fail		
Situation: Pass			Situation: Pass		
Instance: 235			Instance: 238		
Academic period: Sem 2018-2			Academic period: Sem 2018-1		
Subject: ACI530- Database I			Subject: ACI630- Database II		
PG1: 6.5	PG2: 5.7	PG3: 5.4	PG1: 7.0	PG2: 6.8	PG3: 4.3
Predicted Situation: Pass			Predicted Situation: Pass		
Situation: Fail			Situation: Fail		
Instance: 279			Instance: 358		
Academic period: Sem 2018-2			Academic period: Sem 2018-2		
Subject: Aci630- Database II			Subject: ACI040- Database certification		
PG1: 5.9	PG2: 6.5	PG3: 5.3	PG1: 5.4	PG2: 6.0	PG3: 5.0
Predicted Situation: Pass			Predicted Situation: Pass		
Situation: Fail			Situation: Fail		

5. Discussion and Conclusions

We carried out a complete series of experiments with the aim of establishing the best correlations between the input variables and the result, which is the prediction of whether the student will pass a certain subject or not.

The first and direct experiment was to use the FGs, but this fact did not represent a big step of our system (Figure 8). This was the reason why we used the PGs (Figures 9–11) that were the most influential variables. With all the PGs, we obtained a high accuracy for predicting the FG (or, to be more precise, the final situation, i.e., pass or fail) of 96.5%. If we removed PG3, the accuracy became 91.5%, whereas removing PG2 the precision became 93%. In addition, Figure 5 shows interesting correlations among the variables (e.g., how some areas influenced more than others).

These experiments have combined the selection of different PGs choice, as well as follow-up of the students. The results obtained by the experiments allow us to reach conclusions about the creation of action plans to avoid drop-out in the classrooms and to personalize the student follow-up as much as possible, as well as to make valuable information available to the student that allows them to evaluate their academic performance so that they take improvement actions in the subjects that have the highest risk of failing.

We need to continue collecting data to be able to do more tests and more follow-up to continue improving the prediction of the FGs. A future work that must be deepened is to group students according to different criteria—for example: FGs, affinities by area of knowledge, performance per semester, etc.

In this manuscript, we have proposed a methodology to monitor and predict grades in education. The objective of this approach was to obtain the best prediction results so that in a following work we can develop an individualized learning system. This approach led us to group students who meet certain common conditions—for example, those who have taken the same subjects and who have approved those subjects in the same academic period. This is not an easy task, since engineering students usually have very irregular behaviors when passing the required subjects of their curriculum. This is closely related to the fact that for engineering degrees, repetition rates are high, especially in subjects related to mathematics or engineering. For future research, it would be interesting to combine other variables, so that the prediction can be made based on similar academic patterns.

In the present study, an analysis of FGs was carried out by knowledge areas, such as database or network infrastructure areas. This is intended to justify that the grades in a subject can be predicted from student grades in the previous academic years of the subject. For example, the FGs of the course Database Certification can be predicted from the FGs of the subjects Databases I, Databases II, and Database Administration, while the FGs of the subject Certification of Networks can be predicted from the FGs of the subjects Networks I and Networks II.

As a result of the research carried out in the institution, the authorities of the university approved the change in the percentage assigned to each PG (PG), as we explained in the development of the work. In this way, it was possible to improve the grades and academic performance of students in the PG3, as well as reduce the rate of student absenteeism at the end of each academic period (PG3).

After we have verified the model proposed, the most imminent future work is to analyze and design a big data architecture that supports the processing of the large amount of academic data that the university generates periodically. This academic data should be also complemented with other data, such as personal and socio-economic information of the student and information on the student learning assessment system, among others. This large volume of data can be increased by scaling up the proposal of this paper for all the university's degrees. To define the project architecture, it is not recommended to use a traditional approach based on a data warehouse; rather, due to the nature of the proposed project, it will be necessary to create a documented, scalable, and flexible database that can support large indexing and data consultation by students, teachers, and educational administrators. Therefore, we plan to design an architecture that uses big data tools, such as Hadoop and MongoDB, in parallel.

Author Contributions: D.B.-F. made the following contributions to the study: performed literature review, the conception and design of the study, acquisition and pre-processing of data and drafting the article. S.L.-M. made the following contributions to the study: critical revision, drafting the article and approval of the submitted version. D.G. contributed to the following: acquisition of data, analysis, and interpretation of data, drafting the article and approval of the submitted version. All authors read and approved the final manuscript.

Funding: This work was supported in part by the Spanish Ministry of Science, Innovation and Universities through the ProjectECLIPSE-UA under Grant RTI2018-094283-B-C32.

Conflicts of Interest: The authors declare no conflict of interest.

References

1. United Nations. Sustainable Development Goals. Available online: http://www.undp.org/content/undp/en/home/sustainable-development-goals.html (accessed on 16 February 2019).
2. Shields, L.; Newman, A.; Satz, D. Equality of Educational Opportunity. In *Stanford Encyclopedia of Philosophy*; Zalta, E., Ed.; Stanford University: Stanford, CA, USA, 2017.
3. Paura, L.; Arhipova, I. Cause Analysis of Students' Dropout Rate in Higher Education Study Program. *Procedia Soc. Behav. Sci.* **2014**, *109*, 1282–1286. [CrossRef]
4. Mula, I.; Tilbury, D.; Ryan, A.; Mader, M.; Dlouha, J.; Mader, C.; Benayas, J.; Dlouhý, J.; Alba, D. Catalysing Change in Higher Education for Sustainable Development. *Int. J. Sustain. High. Educ.* **2017**, *18*, 798–820. [CrossRef]
5. Visvizi, A.; Lytras, M.D.; Daniela, L. Education, Innovation and the Prospect of Sustainable Growth and Development. In *The Future of Innovation and Technology in Education: Policies and Practices for Teaching and Learning Excellence*; Emerald Publishing Limited: Bingley, UK, 2018; pp. 297–305.
6. Casanova, D.; Moreira, A.; Costa, N. Technology Enhanced Learning in Higher Education: results from the design of a quality evaluation framework. *Procedia Soc.Behav. Sci.* **2011**, *29*, 893–902. [CrossRef]
7. Daniela, L.; Kalnina, D.; Strods, R. An Overview on Effectiveness of Technology Enhanced Learning (TEL). *Int. J. Knowl. Soc. Res.* **2017**, *8*, 79–91. [CrossRef]
8. Lee, J.; Choi, H. What affects learner's higher-order thinking in technology-enhanced learning environments? The effects of learner factors. *Comput. Educ.* **2017**, *115*, 143–152. [CrossRef]
9. Castro, F.; Vellido, A.; Nebot, À.; Mugica, F. Applying Data Mining Techniques to e-Learning Problems. In *Evolution of Teaching and Learning Paradigms in Intelligent Environment*; Springer: Berlin/Heidelberg, Germany, 2007; pp. 183–221.
10. Villegas-Ch, W.; Luján-Mora, S.; Buenaño-Fernandez, D.; Palacios-Pacheco, X. Big Data, the Next Step in the Evolution of Educational Data Analysis. In Proceedings of the International Conference on Information Technology & Systems (ICITS), Santa Elena, Ecuador, 10–12 January 2018; pp. 138–147.
11. Buenaño-Fernandez, D.; Villegas-CH, W.; Luján-Mora, S. The use of tools of data mining to decision making in engineering education—A systematic mapping study. *Comput. Appl. Eng. Educ.* **2019**, *27*, 744–758.
12. Romero, C.; Ventura, S. Educational Data Mining: A Review of the State of the Art. *IEEE Trans. Syst. Man Cybern. Part C (Appl. Rev.)* **2010**, *40*, 601–618. [CrossRef]
13. Baker, R.S.; Yacef, K. The State of Educational Data Mining in 2009: A Review and Future Visions, 1. *Int. Educ. Data Min. Soc.* **2009**, *1*, 3–17.
14. Baker, R.S. Data mining for education. *Int. Encycl. Educ.* **2010**, *7*, 112–118.
15. Elbadrawy, A.; Polyzou, A.; Ren, Z.; Sweeney, M.; Karypis, G.; Rangwala, H. Predicting Student Performance Using Personalized Analytics. *Computer* **2016**, *49*, 61–69. [CrossRef]
16. Piekarski, M.L. Student Retention - An issue, a discussion and a way forward. *Brittany Cotter Cobek Softw. Ltd.* **2013**, *1*, 29–35.
17. Márquez-Vera, C.; Cano, A.; Romero, C.; Noaman, A.Y.M.; Fardoun, H.M.; Ventura, S. Early dropout prediction using data mining: A case study with high school students. *Expert Syst.* **2016**, *33*, 107–124. [CrossRef]
18. Khalifa, S.; Elshater, Y.; Sundaravarathan, K.; Bhat, A.; Martin, P.; Imam, F.; Rope, D.; Mcroberts, M.; Statchuk, C. The Six Pillars for Building Big Data Analytics Ecosystems. *ACM Comput. Surv.* **2016**, *49*, 33. [CrossRef]
19. Provost, F.; Fawcett, T. Data Science and its Relationship to Big Data and Data-Driven Decision Making. *Big Data* **2013**, *1*, 51–59. [CrossRef]

20. Sin, K.; Muthu, L. Application of big data in educationDATA mining and learning analytics—A literature review. *ICTACT J. Soft Comput.* **2015**, *5*, 1035–1049. [CrossRef]
21. Lu, O.H.; Huang, A.Y.; Huang, J.C.; Lin, A.J.; Ogata, H.; Yang, S.J. Applying learning analytics for the early prediction of students' academic performance in blended learning. *Educ. Technol. Soc.* **2018**, *21*, 220–232.
22. Gil, D.; Fernández-Alemán, J.; Trujillo, J.; García-Mateos, G.; Luján-Mora, S.; Toval, A. The Effect of Green Software: A Study of Impact Factors on the Correctness of Software. *Sustainability* **2018**, *10*, 3471. [CrossRef]
23. Hong, S.J.; Weiss, S.M. Advances in predictive models for data mining. *Pattern Recognit. Lett.* **2001**, *22*, 55–61. [CrossRef]
24. Brooks, C.; Thompson, C. Predictive Modelling in Teaching and Learning. In *Handbook of Learning Analytics*; Lang, C., Siemens, G., Wise, A., Gasevic, D., Eds.; Society for Learning Analytics Research (SoLAR): Ann Arbor, MI, USA, 2017; pp. 61–68.
25. Rechkoski, L.; Ajanovski, V.V.; Mihova, M. Evaluation of grade prediction using model-based collaborative filtering methods. In Proceedings of the 2018 IEEE Global Engineering Education Conference (EDUCON), Tenerife, Spain, 17–20 April 2018; pp. 1096–1103.
26. Bydžovská, H. Are Collaborative Filtering Methods Suitable for Student Performance Prediction? In Proceedings of the Progress in Artificial Intelligence - 17th Portuguese Conference on Artificial Inteligence (EPIA), Coimbra, Portugal, 8–11 September 2015; pp. 425–430.
27. Polyzou, A.; Karypis, G. Grade prediction with models specific to students and courses. *Int. J. Data Sci. Anal.* **2016**, *2*, 159–171. [CrossRef]
28. Thai-Nghe, N.; Drumond, L.; Krohn-Grimberghe, A.; Schmidt-Thieme, L. Recommender system for predicting student performance. *Procedia Comput. Sci.* **2010**, *1*, 2811–2819. [CrossRef]
29. Khan, B.; Khiyal, M.S.H.; Khattak, M.D. Final Grade Prediction of Secondary School Student using Decision Tree. *Int. J. Comput. Appl.* **2015**, *115*, 32–36. [CrossRef]
30. Sedkaoui, S.; Khelfaoui, M. Understand, develop and enhance the learning process with big data. *Inf. Discov. Deliv.* **2019**, *47*, 2–16. [CrossRef]
31. Dahdouh, K.; Dakkak, A.; Oughdir, L.; Ibriz, A. Large-scale e-learning recommender system based on Spark and Hadoop. *J. Big Data* **2019**, *6*, 2. [CrossRef]
32. Godfrey, P.; Gryz, J.; Lasek, P. Interactive Visualization of Large Data Sets. *IEEE Trans. Knowl. Data Eng.* **2016**, *28*, 2142–2157. [CrossRef]

Article

Analyzing Online Car Reviews Using Text Mining

En-Gir Kim and Se-Hak Chun *

Department of Business Administration, Seoul National University of Science and Technology,
232 Gongreung-Ro, Nowon-Gu, Seoul 01811, Korea; engir@mail.ru
* Correspondence: shchun@seoultech.ac.kr; Tel.: +82-2-970-6487

Received: 27 February 2019; Accepted: 13 March 2019; Published: 17 March 2019

Abstract: Consumer reviews on the web have rapidly become an important information source through which consumers can share their experiences and opinions about products and services. It is a form of text-based communication that provides new possibilities and opens vast perspectives in terms of marketing. Reading consumer reviews gives marketers an opportunity to eavesdrop on their own consumers. This paper examines consumer reviews of three different competitive automobile brands and analyzes the advantages and disadvantages of each vehicle using text mining and association rule methods. The data were collected from an online resource for automotive information, Edmunds.com, with a scraping tool "ParseHub" and then processed in R software for statistical computing and graphics. The paper provides detailed insights into the superior and problematic sides of each brand and into consumers' perceptions of automobiles and highlights differences between satisfied and unsatisfied groups regarding the best and worst features of the brands.

Keywords: big data analytics; text mining; association rule; car review

---

1. Introduction

The rapid spread of the Internet has provided humanity with a new way to obtain information. It has now become the biggest source of information, with people conducting ever more searches on the Web. Alongside this, social media, another part of the Internet domain, has also captured the attention of netizens. Social media can take many different forms, one of which is product-review websites. These, along with other types of social media, provide a platform for consumers to share their experiences and opinions about the products they purchase and use, thereby providing other consumers with information about the pros and cons of these products. Such communication is also known as electronic word-of-mouth and has opened up new horizons in marketing. By reading consumer reviews, companies and marketers get to know their customers and thus obtain a better understanding of marketing opportunities, the competitive landscape, the market structure, and the features of their own and competitors' products that customers discuss.

Nowadays, with data spreading dramatically, many organizations as well as researchers strive to find patterns among data using datamining methods. Text mining is one of these methods and is used to analyze consumer reviews. For instance, we can take as an example a situation where consumers write comments or reviews about a mobile telephone they purchase and discuss their best or worst experiences. Consumer 1 purchases a mobile telephone from Company A while Consumer 2 purchases a mobile telephone from Company B. The autonomy of a mobile telephone from Company A is better than Company B, while the camera quality of the mobile phone from Company B is better than Company A. Hence, Consumer 2 is likely to write a review that will contain positive feedback about the camera. However, he is also likely to leave negative feedback about the autonomy of the camera. Thus, extracting meaningful information from consumer reviews such as the most frequent words and the relationship between them provides companies with an insight into the superior features of given brands as well as problematic features companies need to address and improve in future. This approach to gathering data and necessary information is sometimes much faster than administering a questionnaire survey.

Many studies have analyzed online reviews for various product categories such as books, movies, fashion, cosmetics, hotels, airlines, and restaurant services. However, there have been very few studies on car purchasing behavior, although some studies have identified the features of a car that affect purchasing behavior [1,2]. To the best of our knowledge, there has been no research to date on online car reviews using a text mining approach. This paper therefore analyzes consumer reviews for automobile brands and compares three competitive brands: Hyundai, Honda, and Ford. Car review websites such as Edmunds, Kelley Blue Book, and Motor Trend usually ask consumers to fill in two different subfields: "best features" and "worst features." Finding the most frequent words used for each subfield therefore gives us some information about the weak and strong aspects of given models. Using R programming, this paper will address three research questions using data from one of the best car review websites, Edmunds.com.

Firstly, this paper determines which terms occur most frequently in the "best features" and "worst features" reviews for each car and discusses their significance. Secondly, it analyzes eight essential car features by applying the association method for consumer reviews on each of three competitive automobile brands in 2012. The association method is then used to discuss the relationships between the most frequent terms and the eight different features of competing vehicles. This provides comprehensive information as to which features reviewers are most interested in and discuss. Finally, this paper compares the frequency and ratio of the terms for eight different features between satisfied and unsatisfied groups and presents relevant implications for the strategies used by car manufacturers.

The remainder of the paper is organized as follows. Section 2 reviews previous literature and presents the theoretical background to the study. Section 3 then presents research methodology and Section 4 presents the results. Section 5 concludes and discusses future research directions.

2. Research Background

2.1. Big Data Analytics and Business Value

With the rapid speed of the Internet and smart mobile devices, consumers can easily generate reviews and discussion on Web resources such as blogs, product review websites, chat rooms, and brand communities. These activities have exponentially grown in recent years and are increasing communication channels between consumers and firms. It has now become important to obtain meaningful information from Web resources such as overall product ratings and product reviews. Thus, organizations and companies are always looking for ways to use the power of big data analytics (BDA) to improve their decision making [3]. Big data analytics (BDA) improves data-driven decision making and provides ways to organize, learn, and innovate [3,4]. It is critical for organizational success that advantage is taken of all available information using big data analytics [4]. This will enable organizations to improve the management of operational risk, reinforce customer relationship management, enhance operational efficiency, and improve firm performance in general [5]. Liu [6], for instance, found that companies using analytics software could decrease customer acquisition costs by about 47% and increase their revenue by about 8%. Wamba et al. [7] also examined the effects of big data analytics capability on firm performance and found both direct and indirect impacts.

However, there is insufficient understanding of how organizations need to be structured and how they should utilize their big data initiatives to generate business value [8–10]. Grover et al. [11] described the value proposition of big data analytics by delineating its components and discussed constructs and relationships that focus on the creation and realization of such value. Dong and Yang [12] explained how and why social media analytics create super-additive value through synergies in functional complementarity between social media diversity and big data analytics. They found that social media diversity and big data analytics have a positive interaction effect on market performance, which is more salient for SMEs than for large firms [13]. Müller et al. [13] analyzed how firm performance is related to big data analytics and found that big data analytics (BDA) assets are associated with an average 3–7% improvement in firm productivity. It is necessary for organizations to understand the impact that big data quality has on firm performance [14]. Organizations should therefore establish specific processes and practices to realize value from their big data investments where different factors are emphasized depending on the context of examination [15].

2.2. Text Mining and Association Rule

Text mining is a data mining technique that obtains structured information from unstructured text information in order to summarize and classify textual data generated through traditional data mining and statistical techniques [16]. Text mining tasks are usually classified into five types of task: information extraction, text categorization, text clustering, document summarization, and association analysis [17–19]. The information extraction technique involves finding important words from social media and brand names, such as those relating to product features. These are then sorted into defined categories or topics in the categorization task. Using computer programs, text categorization treats texts as a bag of words and counts word frequencies while text clustering combines similar documents in groups without predefined categories [16]. Document summarization summarizes the most important concepts from a large collection of texts, enabling data analysts to identify changes in consumer preferences over time and market trends in general. Association analysis is an association rule which aims to find associations for a certain term and is based on counting co-occurrence frequencies [16].

The association rule refers to discovering unpredicted and unique rules from large datasets, finding correlations between elements in transactional databases, and then linking the information by discovering common relationships between the different factors [20]. It is an important data mining technique that is used to identify attribute-value conditions that frequently appear together in datasets [21–25]. There are two types of association rule methods: classical and relational rule mining. Classical association rules only consider co-occurrences between the attribute values while relational association rules are able to depict various types of relations between attribute values. From this perspective, relational association rule mining is an effective unsupervised learning model that can discover hidden patterns in data [25].

As text mining has gained increasing momentum in recent years, comment mining has much attention being given to sentiment analysis and opinion mining and becomes an important technique to obtain information from user-generated contents [13]. User-generated content can be applied to very different types of media where reviews have been researched and have been found to influence consumers to buy products. Consumers are likely to use reviews if they perceive the credibility of the source to be high [26,27]. Online social networks now enable users to share their own lives, generate and interact with vast amounts of multimedia content (text, audio, video, and images), and supplement these with feedback, comments, or feelings. The role of big data technologies is therefore becoming more important in obtaining meaningful information from users' interactive activities [28]. Amato et al. [29] developed a more effective and efficient mechanism for a text pre-processing task where each linguistic term is assigned with a weight that is computed using the well-known tf-idf formula. Yahav et al. [30] proposed an adjustment to tf-idf that accounts for this bias introduced by between-participant discourse to the study of comments in social media and illustrated the effects of both the bias and correction through data from seven Facebook fan pages.

2.3. Consumer Car Purchasing Behavior

Online consumer reviews have a significant impact on consumers and businesses. They are more reliable than information provided by sellers because they offer personalized advice as well as ratings of products or services [31]. Many studies have analyzed the relationship between online reviews and product sales for well-known websites such as eBay [32], Amazon [33–35], and Airbnb [36,37] and various product categories such as books [33–35,38], movies [39–41], fashion [42], cosmetics [43], hotels [44–47], washing machines [48], online lectures [49], restaurants [50,51], and airlines [52].

However, there have been very few studies on consumer car reviews. Kulkarni et al. [53] examined whether Internet use is associated with different choice patterns for cars and found that Internet users rely more on ratings while non-Internet users rely more on recommendations. Sagar et al. [1] considered whether factors affecting car choice behavior such as competition, consumer preferences, and government policies are salient features. Kaushal [2] identified car purchasing behavior through 39 items and validated the usefulness of five factors: safety & security, quality, performance, value, and technology. In this paper, we use mining methods to analyze eight different features: performance, comfort, value, interior, reliability, safety, technology, and exterior for three competitive brands in the automobile market in 2012. Such an approach gives marketers the opportunity to eavesdrop on their own consumers and on consumers in the automotive market in general. In particular, engineers and marketers from automotive companies, based on the results of the review analysis, can obtain structured information about both the superior and problematic aspects of their vehicles and those of their rivals, thereby gaining a competitive advantage in the market. Consequently, the implementation of such an approach can trigger sales growth and improve firm performance.

3. Research Methodology

3.1. Text Mining Approach to Car Reviews

To obtain adequate and proper data, reviews were collected from one of the biggest online resources for automotive information, Edmunds.com. The process of mining is divided into several steps.

3.1.1. Scraping Data from Websites

Data was collected using the scraping tool "ParseHub." This is a useful instrument when dealing with information of any kind. It can be adapted to any website, and scholars can extract any piece of information, be it a text or an image (e.g., Figure 1). In this study, units such as title, model of vehicle, best features, worst features, ratings for 8 different features, and total ratings were collected. The process of scraping is illustrated in Figure 1.

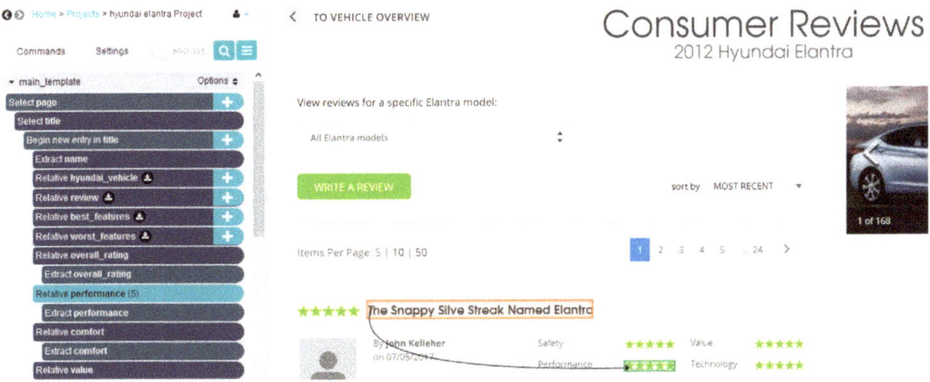

Figure 1. Consumer reviews.

The results can be saved in JSON or CSV format. In this study, they were saved to CSV format, slightly corrected in an Excel program, and then changed to XSLX format for further usage. The output appears as shown in Figure 2.

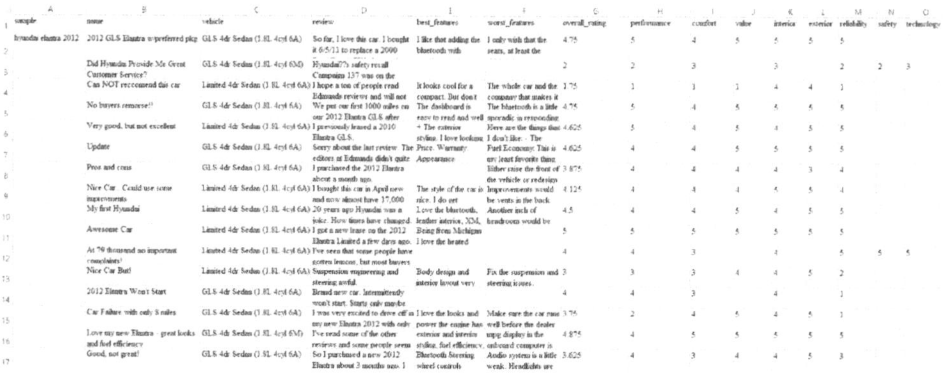

sample	name	vehicle	review	best_features	worst_features	overall_rating	performance	comfort	value	interior	exterior	reliability	safety	technology
hyundai elantra 2012	2012 GLS Elantra w/preferred pkg	GLS 4dr Sedan (1.8L 4cyl 6A)	So far, I love this car. I bought it 6/5/11 to replace a 2000	I like that adding the bluetooth with	I only wish that the seats, at least the	4.75	5	4	5	5	5	5		
	Did Hyundai Provide Me Great Customer Service?	GLS 4dr Sedan (1.8L 4cyl 6M)	Hyundai?'s safety recall Campaign 137 was on the			2	2	3		3		2	2	3
	Can NOT reccomend this car	Limited 4dr Sedan (1.8L 4cyl 6A)	I hope a ton of people read Edmunds reviews and will not	It looks cool for a compact. But don't	The whole car and the company that makes it	1.75	1	3	1	4	4	1		
	No buyers remorse!!	GLS 4dr Sedan (1.8L 4cyl 6A)	We put our first 1000 miles on our 2012 Elantra GLS after	The dashboard is easy to read and well	The bluetooth is a little sporadic in responding	4.75	5	4	5	5	5	5		
	Very good, but not excellent	Limited 4dr Sedan (1.8L 4cyl 6A)	I previously leased a 2010 Elantra GLS.	+ The exterior styling. I love looking	Here are the things that I don't like. - The	4.625	5	4	5	4	5	5		
	Update	GLS 4dr Sedan (1.8L 4cyl 6A)	Sorry about the last review. The editors at Edmunds didn't quite	The Price. Warranty. Appearance	Fuel Economy. This is my least favorite thing	4.625	4	4	5	5	5	5		
	Pros and cons	GLS 4dr Sedan (1.8L 4cyl 6A)	I purchased the 2012 Elantra about a month ago.		Either raise the front of the vehicle or redesign	3.875	4	4	4	4	3	4		
	Nice Car. Could use some improvements	Limited 4dr Sedan (1.8L 4cyl 6A)	I bought this car in April new and now almost have 17,000	The style of the car is nice. I do get	Improvements would be vents in the back	4.125	4	4	4	5	5	4		
	My first Hyundai	Limited 4dr Sedan (1.8L 4cyl 6A)	20 years ago Hyundai was a joke. How times have changed.	Love the bluetooth, leather interior, XM,	Another inch of headroom would be	4.5	4	4	5	4	5	5		
	Awesome Car	Limited 4dr Sedan (1.8L 4cyl 6A)	I got a new lease on the 2012 Elantra Limited a few days ago.	Being from Michigan I love the heated		5	5	5	5	5	5	5		
	At 79 thousand no important complaints!	Limited 4dr Sedan (1.8L 4cyl 6A)	I've seen that some people have gotten lemons, but most buyers			4	4	3		4		5	5	5
	Nice Car But!	Limited 4dr Sedan (1.8L 4cyl 6A)	Suspension engineering and steering awful.	Body design and interior layout very	Fix the suspension and steering issues.	3	3	3	4	4	5	2		
	2012 Elantra Won't Start	GLS 4dr Sedan (1.8L 4cyl 6A)	Brand new car. Intermittently won't start. Starts only maybe			4	4	3		4		1		
	Car Failure with only 8 miles	GLS 4dr Sedan (1.8L 4cyl 6A)	I was very excited to drive off in my new Elantra 2012 with only	I love the looks and power the engine has	Make sure the car runs well before the dealer	3.75	2	4	5	4	5	1		
	Love my new Elantra - great looks and fuel efficiency	GLS 4dr Sedan (1.8L 4cyl 6M)	I've read some of the other reviews and some people seem	exterior and interior styling, fuel efficiency,	mpg display in the onboard computer is	4.875	4	5	5	5	5	5		
	Good, not great!	GLS 4dr Sedan (1.8L 4cyl 6A)	So I purchased a new 2012 Elantra about 3 months ago. I	Bluetooth Steering wheel controls	Audio system is a little weak. Headlights are	3.625	4	3	4	4	5	3		

Figure 2. Extracted data.

All reviews were divided into two groups: satisfied and unsatisfied. The overall rating by consumers was used as the basis for this division. The average mean of overall ratings thus became a separating point for all three car samples and the next condition was set as follows:

$$\text{If ((car} = 1) \text{ and (overall\_rating} >= 4.30)) \text{ GD} = 1$$
$$\text{If ((car} = 1) \text{ and (overall\_rating} < 4.30)) \text{ GD} = 0$$
$$\text{If ((car} = 2) \text{ and (overall\_rating} >= 4.60)) \text{ GD} = 1$$
$$\text{If ((car} = 2) \text{ and (overall\_rating} < 4.60)) \text{ GD} = 0$$
$$\text{If ((car} = 3) \text{ and (overall\_rating} >= 3.70)) \text{ GD} = 1$$
$$\text{If ((car} = 3) \text{ and (overall\_rating} < 3.70)) \text{ GD} = 0$$

where Car #1 is a Hyundai Elantra, Car #2 is a Honda Civic, and Car #3 is a Ford Focus. GD is a variable for group diversity where GD = 1 relates to the satisfied group and GD = 0 relates to the unsatisfied group. The result of the data split is presented in Table 1.

Table 1. Frequency of the two groups.

Car Brands	Unsatisfied Group (GD = 0)	Satisfied Group (GD = 1)
Elantra (#1)	58	58
Civic (#2)	69	87
Focus (#3)	132	135

3.1.2. Input Data

For further analysis, data was input into the R program using the "xlsx" package. This enables the R user to read, write, and format Excel files. Best features, worst features, satisfied and unsatisfied variables, reviewers' id, name of automobile brand, and car id relative to automobile brand were included as input. Part of the input data is displayed in Figure 3.

In the output we acquired 539 reviews, each of which referred to either the satisfied group or unsatisfied group in accordance with the sat/unsat variable obtained from the average mean of the overall rating.

car_var	car	bestdf	worstdf	sat_unsat_var	id_var
1	hyundai	I like that adding the bluetooth with steering wh>	I only wish that the seats, at least the driver's>	1	1
1	hyundai	NA	NA	0	2
1	hyundai	It looks cool for a compact. But don't judge this>	The whole car and the company that makes it is no>	0	3
1	hyundai	The dashboard is easy to read and well designed. >	The bluetooth is a little sporadic in responding >	1	4
1	hyundai	+ The exterior styling. I love looking at my 201?>	Here are the things that I don't like: - The glov>	1	5
1	hyundai	Price. Warranty. Appearance compared to Civic and>	Fuel Economy: This is my least favorite thing abo>	1	6
1	hyundai	NA	Either raise the front of the vehicle or redesign>	0	7
1	hyundai	The style of the car is nice. I do get compliment>	Improvements would be vents in the back seat. Als>	0	8
1	hyundai	Love the bluetooth, leather interior, XM, sunroof>	Another inch of headroom would be nice. My hair b>	1	9
1	hyundai	Being from Michigan I love the heated front and r>	NA	1	10
1	hyundai	NA	NA	0	11
1	hyundai	Body design and interior layout very favorable.	Fix the suspension and steering issues.	0	12
1	hyundai	NA	NA	0	13
1	hyundai	I love the looks and power the engine has conside>	Make sure the car runs well before the dealer put>	0	14
1	hyundai	exterior and interior styling, fuel efficiency, r>	mpg display in the onboard computer is about 2 - >	1	15
1	hyundai	Bluetooth Steering wheel controls Styling Ride	Audio system is a little weak. Headlights are ver>	0	16
1	hyundai	Back up camera (once you have one you cant imagi>	Maybe a bit more power in the low end would be ni>	1	17
1	hyundai	If you call space a feature, then it's on my list>	A few things -- the storage compartment where the>	0	18
1	hyundai	Seats were comfortable. Drove nice and smooth. Ex>	NA	0	19
1	hyundai	MPG!!!! The Bluetooth is a great feature, and wor>	This is my first car with the slanted windshield.>	1	20
1	hyundai	keyless entry, smart key, nav system, backup came>	Let's add climate controlled AC/heat and put a sp>	1	21
1	hyundai	Nice interior and exterior. The seats are comfort>	Improve sound control. Fix the shifter gear noise>	1	22
1	hyundai	Great looks inside and out Awesome plush interior>	Get rid of the base model with no options and hav>	1	23
1	hyundai	How it drives and the interior. I bought it for t>	Scratching my head on this one. I guess I'd sugge>	1	24
1	hyundai	exterior and interior design, interior room (feel>	AC vents in the center are positioned too low. Th>	1	25

Figure 3. Input data.

3.1.3. Data Manipulation

After the data were superficially arranged, R program tools were used to process the data and divide the reviews into four different groups: satisfied best features, satisfied worst features, unsatisfied best features, and unsatisfied worst features. Such segregation can be deciphered as meaning that, even though consumers may be satisfied or unsatisfied, they still identify some best and worst features of the car they are reviewing. An example of the output is presented in Figure 4, where rows are organized randomly to increase clarity.

car_var	car	best_and_worst_features	sat_unsat	sat_unsat_var	id_var	group	group_var
2	honda	6-speed Torquey at low rpm Handling Efficient thr>	unsatisfied	0	186	UNsatisfied-BEST	3
2	honda	NA	satisfied	1	220	satisfied-WORST	2
2	honda	The only complaint I have is the rattles on the i>	satisfied	1	149	satisfied-WORST	2
3	ford	Sporty exterior design.	unsatisfied	0	505	UNsatisfied-BEST	3
1	hyundai	NA	unsatisfied	0	27	UNsatisfied-BEST	3
3	ford	Honesty from FMC; customer service that means som>	unsatisfied	0	490	UNsatisfied-WORST	4
2	honda	NA	satisfied	1	119	satisfied-WORST	2
3	ford	NA	unsatisfied	0	423	UNsatisfied-WORST	4
2	honda	I had my windows tinted, which helps a lot due to>	satisfied	1	121	satisfied-WORST	2
3	ford	NA	unsatisfied	0	359	UNsatisfied-BEST	3
3	ford	We had a little trouble figuring out some of the >	satisfied	1	374	satisfied-WORST	2
2	honda	Econ mode-great idea! I was also able to add an e>	unsatisfied	0	173	UNsatisfied-BEST	3
3	ford	Ford Manufacturing will need to either recall thi>	unsatisfied	0	348	UNsatisfied-WORST	4
2	honda	NA	satisfied	1	188	satisfied-WORST	2
2	honda	The EV only mode; its the first Honda IMA hybrid >	unsatisfied	0	226	UNsatisfied-BEST	3
1	hyundai	- You can feel the road bumps more than in the Cr>	satisfied	1	43	satisfied-WORST	2
3	ford	NA	satisfied	1	338	satisfied-WORST	2
2	honda	Better mileage, more room than previous generatio>	satisfied	1	159	satisfied-BEST	1

Figure 4. Output data.

3.1.4. Data Cleansing

To make it possible to count words and identify their co-occurrence in reviews, text must first undergo the cleaning process. Some features of text are thus removed, such as numbers, white spaces, punctuation, and common English words that have no semantic meaning. As well as converting text to lower case, fixing contractions and text stemming is also essential to obtain accurate and valuable data. We therefore received all the words occurred in all 1078 reviews, which totaled 2299 words. A sample of the output is presented in Figure 5.

	row.names	1	2	3	4	5
1	ability	0	0	0	0	0
2	able	0	0	0	0	0
3	abs	0	0	0	0	0
4	absolutely	0	0	0	0	0
5	accelerate	0	0	0	0	0
6	accent	0	0	0	0	0
7	acceptable	0	0	0	0	0
8	access	0	0	0	0	0
9	accessories	0	0	0	0	0
10	accident	0	0	0	0	0
11	accord	0	0	0	0	0
12	accumulated	0	0	0	0	0
13	accurate	0	0	0	0	0
14	acheat	0	0	0	0	0
15	achieve	0	0	0	0	0
16	across	0	0	0	0	0
17	action	0	0	0	0	0
18	active	0	0	0	0	0
19	acts	0	0	0	0	0
20	actual	0	0	0	0	0
21	acura	0	0	0	0	0
22	add	1	0	0	0	0
23	addition	0	0	0	0	0
24	addons	0	0	0	0	0
25	address	0	0	0	0	0
26	adequate	1	0	0	0	0

Figure 5. Terms that occurred after review cleaning.

3.1.5. Data Mastering

For further data analysis, the final step was to combine our primary database and database, which consisted of occurring words. Thus, with the help of R tools, a master table was created, as shown in Figure 6.

car_var	car	best_and_worst_features	sat_unsat	sat_unsat_var	id_var	group	group_var	ability	able	abs	absolutely	accelerate	accent
1	hyundai	I like that adding the bluetooth with steering wheel con...	satisfied	1	1	satisfied-BEST	1	0	0	0	0	0	0
1	hyundai	The dashboard is easy to read and well designed. Gas m...	satisfied	1	4	satisfied-BEST	1	0	0	0	0	0	0
1	hyundai	- The exterior styling. I love looking at my 2012 Elantra. ...	satisfied	1	5	satisfied-BEST	1	0	0	0	0	0	0
1	hyundai	Price. Warranty. Appearance compared to Civic and othe...	satisfied	1	6	satisfied-BEST	1	0	0	0	0	0	0
1	hyundai	Love the bluetooth, leather interior, XM, sunroof and I'm...	satisfied	1	9	satisfied-BEST	1	0	0	0	0	0	0
1	hyundai	Being from Michigan I love the heated front and rear se...	satisfied	1	10	satisfied-BEST	1	0	0	0	0	0	0
1	hyundai	exterior and interior styling, fuel efficiency, ride comfort	satisfied	1	15	satisfied-BEST	1	0	0	0	0	0	0
1	hyundai	Back up camera (once you have one you cant imagine lif...	satisfied	1	17	satisfied-BEST	1	0	0	0	0	0	0
1	hyundai	MPG!!!! The Bluetooth is a great feature, and works reall...	satisfied	1	20	satisfied-BEST	1	0	0	0	0	0	0
1	hyundai	keyless entry, smart key, nav system, backup camera, gre...	satisfied	1	21	satisfied-BEST	1	0	0	0	0	0	0
1	hyundai	Nice interior and exterior. The seats are comfortable and ...	satisfied	1	22	satisfied-BEST	1	0	0	0	0	0	0
1	hyundai	Great looks inside and out Awesome plush interior w/pr...	satisfied	1	23	satisfied-BEST	1	0	0	0	0	0	0
1	hyundai	How it drives and the interior. I bought it for the mix of f...	satisfied	1	24	satisfied-BEST	1	0	0	0	0	0	0
1	hyundai	exterior and interior design, interior room (feels like a la...	satisfied	1	25	satisfied-BEST	1	0	0	0	0	0	0
1	hyundai	Everything for the money.	satisfied	1	28	satisfied-BEST	1	0	0	0	0	0	0
1	hyundai	Style, layout, many standard features, larger trunk, goo...	satisfied	1	31	satisfied-BEST	1	0	0	0	0	0	0
1	hyundai	My favorite features are the comfortable seats, amazing ...	satisfied	1	34	satisfied-BEST	1	0	0	0	0	0	0
1	hyundai	Mileage, warranty, inside room, looks, and price	satisfied	1	37	satisfied-BEST	1	0	0	0	0	0	0
1	hyundai	Love the blue tooth, interior design, good on gas 535 b...	satisfied	1	39	satisfied-BEST	1	0	0	0	0	0	0
1	hyundai	-The exterior is very stylish. -The seats are comfortable. -...	satisfied	1	40	satisfied-BEST	1	0	0	0	0	0	0

Figure 6. Master table.

4. Results of the Study

4.1. Frequency Analysis

Once the master table was created, the actual analysis could be conducted. For this we utilized "WordCloud," one of R program's utilities. This is a visualization method that displays how frequently words appear in a given sample of text, and the way it works is quite simple. The more frequently a specific word appears in a database, the bigger and bolder it appears in the word cloud. The results of our cases will now be discussed.

4.1.1. Best Features

Figure 7 shows the best features of three cars. Words with the highest co-occurrence are represented in this word cloud, with the most frequent and important words located in the center and the least frequent words located on the edges. Hence, the closer the words are to edges, the less frequent they are. In the case of Hyundai, "seat" is the most frequent word, followed by "interior," then "style," and then the rest. In the case of Honda, the most frequent words are, in order, "mpg," "gas," "seat," "comfort," "mileage," "dash," "control," "display," "smooth," "steering," "wheel," "econ," "fun," and so on. In contrast with Hyundai, where the main advantage was design and style, consumers mostly emphasize characteristics related to value, technology, and movement on a road. In the case of Ford, the most frequent words are "handles," "seat," "interior," "system," "sync," "style," "comfort," "gas," "transmission," "exterior," "mileage," and so on.

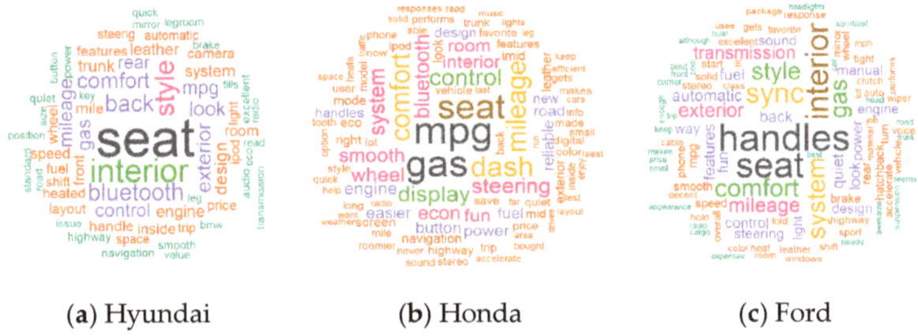

(a) Hyundai   (b) Honda   (c) Ford

Figure 7. Best features (Hyundai, Honda, and Ford) using word cloud.

In the case of Hyundai, after filtration only 14 words remain. As shown in the barplot in Figure 8, there are many words relating to appearance. These are "interior" (24), "style" (19), "exterior" (13), "look" (13), and "design" (10). The interpretation of this result is that consumers mostly liked the car design. The most frequent word is "seat," which occurs 43 times. Because this word has such high frequency, association analysis was conducted to determine its significance. We performed correlation analysis on the most frequent word, as shown in Figure 9. For example, in the case of Hyundai, the term "seat" has a high correlation with words such as "position," "front," and "back." Hence, we can assume that this word refers to the convenience and comfort consumers felt when they sat in a Hyundai Elantra. Consistent with this interpretation, words such as "back," "comfort," and "rear" might also refer to comfort, which was one of the best features for consumers. Similarly, the occurrence of words such as "mpg" and "gas" means that consumers were satisfied with Hyundai's fuel consumption. For the word "control," the most closely associated word was "steering" with a correlation of 0.78. This means consumers were likely to be satisfied with their control over the movement of a vehicle.

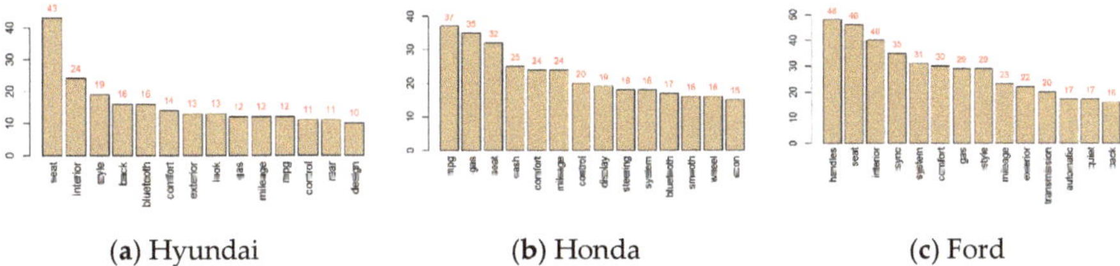

(a) Hyundai   (b) Honda   (c) Ford

Figure 8. Best features (Hyundai, Honda, and Ford) (barplot).

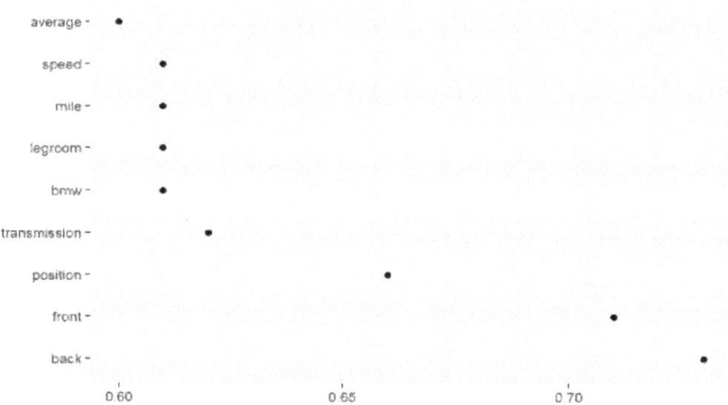

Figure 9. Correlation graph for the term "seat" (Hyundai).

In the case of Honda, the first two words are "mpg" and "gas" which means this car has very low fuel consumption. Additionally, for some consumers, seats seem to be very comfortable. The rest of the words are related to the dashboard and technological features such as "dash," "control," "display," "steering," "system," "bluetooth," and "econ." The word "steering" refers more to technology than holding the road because the correlated words were "wheel," "control," "electronic assist," and "dash." The word "econ" was correlated with words such as "mode" and "feature." This is explained by the fact that the Honda Civic has an econ button as a special function, which has become one of its most favored features.

In the case of Ford, the most frequent word is "handles," which occurred 48 times. Because it is quite difficult to interpret this word, an association analysis was conducted as shown in Figure 10. We can assume that the word "handles" does not refer to the means by which a thing is held, carried, or controlled, but how easily a car is to handle on a road. Such words as "turn," "directions," "balance," "quiet," "turn," and others can help to precisely interpret the meaning of this word. Another frequent word was "sync," which are correlated with some words such as voice, system, phone, ipod, control, navigation, and so on.

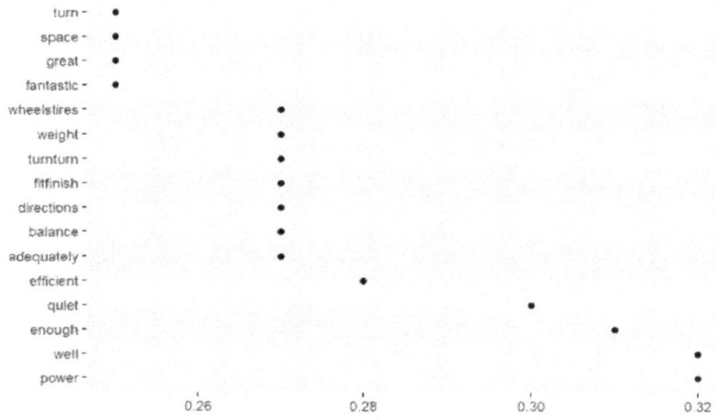

Figure 10. Correlation graph for the term "handles" (Ford).

4.1.2. Worst Features

The same analysis was then conducted using reviews that contained the worst features of three brands. For a Hyundai Elantra, the most frequent word is "mpg," which occurred 28 times, as shown in Figure 11. Although "mpg" also occurred in the results for best features, it is not impossible for the same term to appear in worst features. Here, we can assume that many consumers were not satisfied with fuel consumption and that these consumers outnumber those who were satisfied.

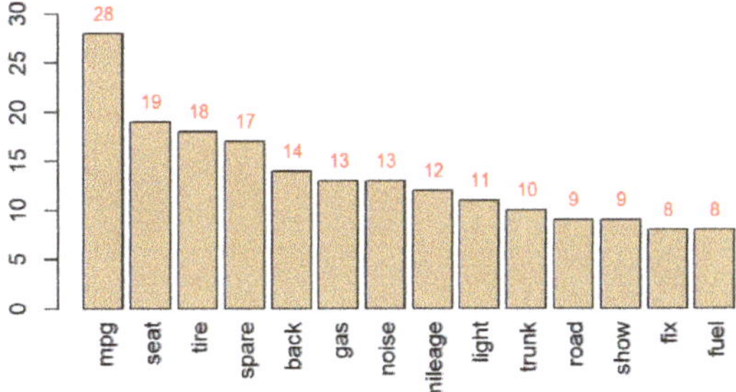

Figure 11. Hyundai's worst features—barplot.

Moreover, if we consider the correlation analysis for "mpg" in Figure 12, we can see that there are highly correlated words such as "show," "computer," "onboard," and "display." Therefore, we assume there might be some problem related to displaying the mpg on the onboard computer. This hypothesis was checked manually, and it was found that many consumers were complaining about an incorrect mpg display.

```
$mpg
       show    computer efficiency         fil      fillup     onboard         got
       0.90        0.68       0.68        0.68        0.68        0.68        0.64
    average     display
       0.61        0.61
```

Figure 12. Correlation for the term "mpg" (Hyundai).

Consistent with this finding, the correlation for the word "gas" yielded a similar result as shown in Figure 13, so we can assume that the words "estimates" and "misleading" are referring to the same problem.

```
$gas
     mileage                    eco   estimates  misleading
        0.73                   0.52        0.52        0.52
```

Figure 13. Correlation for the term "gas" (Hyundai).

The barplot for worst features shows that consumers were unsatisfied with spare tire ("spare" and "tire" were the most highly correlated with = 0.89), noise on the road, fog lights, trunk, mpg efficiency, as well as the mpg display and seats. As mentioned previously, "mpg," "gas," "fuel," as well as "seat," occurred inconsistently in the results for best features. Such phenomena could be accounted for by the differing preferences of every individual. Furthermore, based on the proportion of words for both groups, mpg and mileage are more likely to be considered poor rather than superior features because the sum of occurrences in best features is 36; in worst features it is 61.

In the case of Honda, we can see that one of the most frequent words in terms of worst features is "interior," as shown in Figure 14.

The correlation graphic shows that this is highly correlated with "cheap," as shown in Figure 15. Therefore, we can assume that some customers did not like the quality and appearance of their interior, viewing it as a drawback rather than an advantage. Additionally, "fabric" is correlated with "interior." Although the correlation is quite low, we can still assume that reviewers were unsatisfied with the material of their interior. Furthermore, some consumers were not satisfied with back or front seats. Moreover, fog lights, mirrors, and noise on roads became some of the worst features in the Honda Civic as shown in Figure 14.

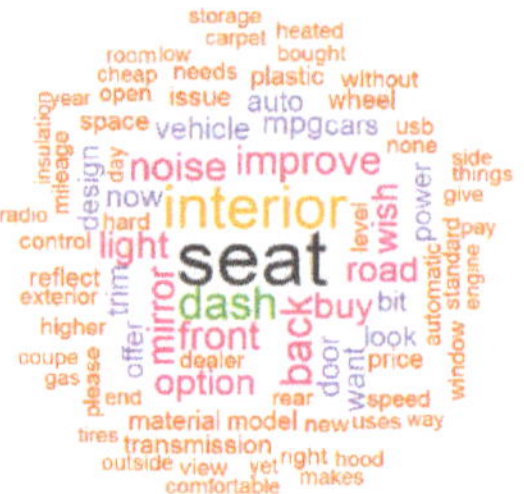

Figure 14. Honda's worst features using word cloud.

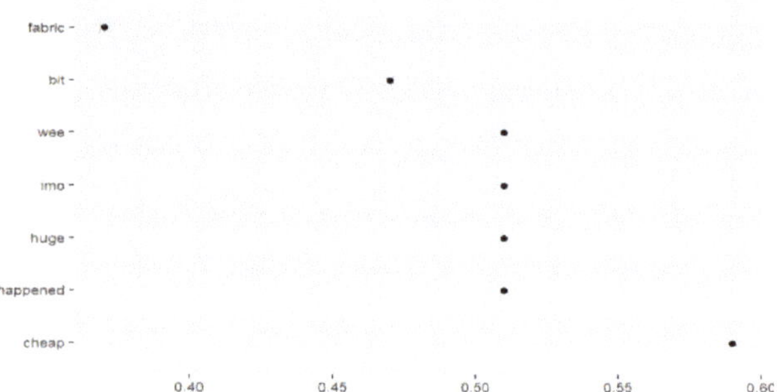

Figure 15. Correlation for the term "interior" (Honda).

In the case of Ford, the worst features were "transmission," "seat," "back," "control," "fix," "issue," "shift," "rear," "wheel," "system," and so on. The most frequent word "transmission" occurred 57 times, comparatively larger than other terms. To understand the transmission flaw in the Ford Focus, an association graph was built, as shown in Figure 16.

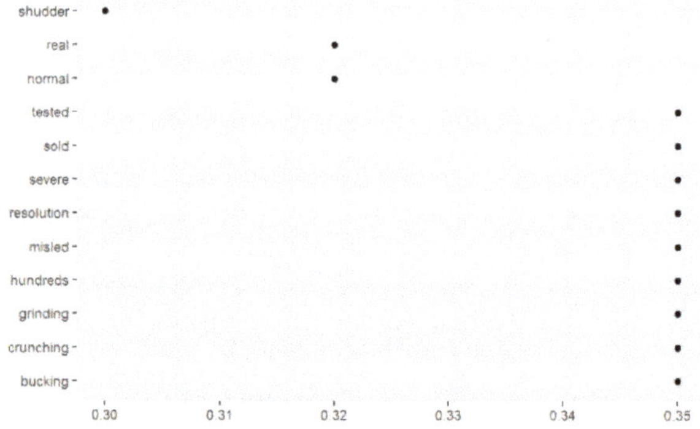

Figure 16. Correlation for the term "transmission" (Ford).

Among the terms highly correlated with "transmission" were "severe," "grinding," "crunching," and "bucking." Therefore, we can assume there is a problem with the transmission, as it is perceived as making strange sounds and being inconvenient to use. In addition, the terms "issue," "fix," "manual," "problem," and "shift" also correlated with "transmission," which means that it is probably the most significant problem with the Ford Focus. Ford also seems to have problems in terms of technology. For instance, the terms "control" and "wheel" were highly correlated with "steering," "device," "equipment," "aux," "cruise," "dashboard," and other words, which can be interpreted as

Ford exhibiting a deficiency in equipment. There are also consumers who are certainly not satisfied with the seats and space in a cab, both front and back (rear).

4.1.3. Implications and Discussion

Based on the results, we can assume that the biggest strength of the Hyundai Elantra car is its design. This is supported by the fact that the most frequent words are related to car appearance. These include "interior," "style," exterior," "look," and "design." The worst features for Hyundai appear to be gas consumption and some problems with technology, such as the mpg display on the onboard computer and problems with a spare tire. In contrast to the Hyundai car, the biggest strengths of the Honda car are low gas consumption, dashboard, controls on the steering wheel, and the "econ" mode, which improves fuel efficiency. The worst feature for Honda appears to be its interior, which reviewers emphasized as cheap. Furthermore, they were unsatisfied with the material it was made of. In the case of Ford, the biggest strength appeared to be manipulation of the car. This is consistent with the high frequency of the word "handles." The, other best features for Ford were the interior and exterior, noiselessness during the ride, and the "Ford Sync" system which allowed users to control automotive functions using their voice. The biggest disadvantage for Ford was found to be transmission. This was supported by the high frequency of the word "transmission" and other frequent yet negative words, such as "issue," "fix," "manual," "problem," and "shift." Another important disadvantage relates to technology, specifically a problem with the controls on the steering wheel. In addition, in all three cases one of the most frequent words was "seat," which appeared in both "best features" and "worst features" categories, suggesting that reviewers are divided in their opinions. Hence, it can be concluded that it is difficult for all three car brands to find favor in the eyes of all consumers.

4.2. Analysis of Car Features Using the Association Rule

According to Edmunds.com, eight different consumers rate features whereby each one of eight features refers to certain terms and involves some form standard conception. Otherwise, every individual might have a different conception about each of the features relative to other individuals.

- Performance involves terms such as acceleration, braking, road holding, and shifting.
- Comfort relates to front seats, rear seats, getting in/out, and noise/vibration.
- Value involves fuel economy, maintenance cost, purchase cost, and resale value.
- Interior implies cargo/storage, instrumentation, interior design, and the logic of controls.
- Reliability relates to repair frequency, dealership support, engine, and transmission.
- Safety consists of headlights, outward visibility, parking aids, and rain/snow traction.
- Technology stands for entertainment, navigation, Bluetooth, and USB ports.
- Exterior stands for exterior design.

The question that arises is: how are the most frequent terms for each car related to the eight different features and what is their frequency? To answer this question, both groups of reviews, which contain best features and worst features, were combined and analyzed using text mining tools. The aim was to determine the frequency of every word that occurs in reviews. The process was conducted for all three car brands: Hyundai Elantra 2012, Honda Civic 2012, and Ford Focus 2012. The 24 most frequent terms were chosen as a sample and, using the same association approach as in previous research, the relationship between the terms and eight different features were found. The results are presented in Table 2.

Table 2. Relationship between terms and eight features.

Hyundai Elantra			Honda Civic			Ford Focus		
Terms	Freq	Features	Terms	Freq	Features	Terms	Freq	Features
seat	62	Comfort	seat	61	comfort	seat	82	comfort
mpg	40	Value	mpg	46	value	transmission	77	reliability
back	30	Comfort	dash	40	tech	sync	56	Tech
interior	26	Interior	gas	40	value	interior	52	interior
gas	25	Value	interior	35	interior	handle	51	Perform/comfort
mileage	24	Value	mileage	30	value	back	48	comfort
bluetooth	20	Tech	comfort	29	comfort	system	48	perform/tech
style	20	interior/exterior	control	26	perform/comfort/tech	control	44	perform/comfort/tech
tire	19	Interior	display	23	interior/tech	gas	40	Value
trunk	19	Interior	wheel	23	perform/tech	comfort	36	comfort
rear	18	comfort/interior	road	22	perform/comfort	manual	34	reliability
light	17	safety/tech	steering	22	perform/tech	mileage	33	Value
spare	17	Interior	back	21	comfort	rear	32	comfort/interior
control	16	Perform/comfort/tech	bluetooth	21	tech	automatic	31	reliability
engine	16	Reliability	system	21	perform/tech	style	31	interior/exterior
look	16	interior/exterior	power	20	perform	wheel	28	perform/tech
system	16	perform/tech	room	19	comfort/interior	issue	27	reliability
comfort	15	Comfort	design	18	interior/exterior	look	27	interior/exterior
design	15	interior/exterior	engine	18	reliability	exterior	26	exterior
front	15	Comfort	econ	17	value/tech	front	26	comfort
fuel	15	Value	front	17	comfort	shift	26	reliability
noise	14	comfort	light	17	safety/tech	speed	26	perform
speed	14	perform	look	17	interior/exterior	light	25	safety/tech
exterior	13	exterior	noise	17	comfort	steering	25	perform/tech

Although the frequency for the most frequent terms was found, the total number of all reviews for each vehicle brand was different; specifically, 116 for the Hyundai Elantra, 156 for the Honda Civic, and 267 for the Ford Focus, respectively. Thus, we have to adjust the numbers to the common denominator to interpret the comparison between three brands more clearly. To do this, the following formula was applied:

$$F = \frac{frequency\ of\ a\ given\ term}{total\ quantity\ of\ reviews\ for\ a\ given\ car}$$

where F is approximate occurrence of a given term in one review.

Thus, all words according to a specific feature were summed, and their frequency before and after adjustment was determined.

4.2.1. Analysis Results for Eight Features

As shown in Table 3, the highest frequency was for terms related to comfort and interior features in Hyundai, comfort and technological features in Honda, and comfort in Ford (Criteria: $F \geq 1$). Therefore, reviewers were mostly interested in these features and discussed these most heatedly. These can now be scrutinized more closely for each case.

(1) Hyundai

If we compare the frequency of terms for interior and the average rating score of consumers for interior we see that terms related to interior appear more often in the satisfied group and in best features because the score for interior is quite high. Therefore, the Hyundai company has won the favor of consumers in respect to its interior. The comfort score is 3.97, which is neither high nor low. This suggests there might be some factors reviewers were not satisfied with. Thus, a more precise analysis is needed. Another feature worth considering is the exterior feature. Compared to other automobile brands, the frequency of terms related to the exterior is significantly higher than for Honda and Ford. It can therefore be assumed that the exterior is also the strongest feature for Hyundai. Its score is highest among the scores for all features and the likelihood that terms related to an exterior would mostly occur in the satisfied group and best features is very high.

Table 3. Frequency before and after adjustment.

Features	Hyundai			Honda			Ford		
	Freq (before)	Freq (after)	Score	Freq (before)	Freq (after)	Score	Freq (before)	Freq (after)	Score
performance	46	0.40	3.72	134	0.86	4.29	222	0.83	3.33
comfort	170	1.47	3.97	212	1.36	4.26	319	1.19	3.70
value	104	0.90	4.13	133	0.85	4.52	73	0.27	3.63
interior	150	1.29	4.39	112	0.72	4.14	142	0.53	3.80
exterior	64	0.55	4.70	35	0.22	4.40	84	0.31	4.47
reliability	16	0.14	3.90	18	0.12	4.66	195	0.73	3.19
safety	17	0.15	4.05	17	0.11	4.10	25	0.09	2.95
technology	69	0.59	3.70	210	1.35	4.10	226	0.85	2.84

(2) Honda

Comparing the frequencies and scores for comfort and technology, it is very likely that terms related to these will occur mostly in the satisfied group and best features because the score for both groups is pretty high. Furthermore, the frequency and score for technology is the highest among all three automobile brands, which can be interpreted as Honda being a technological leader.

(3) Ford case

Although the frequency of terms related to comfort is high, the score is quite low. Although it is not the lowest score compared to other features, the result suggests that such terms would appear in both satisfied and unsatisfied groups and in both the best and worst features groups. Furthermore, reliability is also worth mentioning, because the frequency of terms related to this feature is much higher than for Hyundai and Honda. With a very low score for reliability, we can assume that terms will mostly appear in worst features for both satisfied and unsatisfied groups.

4.2.2. Comparison of Two Groups' Reviews

In this section, we compare the reviews of both groups and find terms whose influence is greater than others. We also compare the differences between satisfied and unsatisfied groups of reviewers. What, therefore, are the frequency and ratio of terms for eight different features between satisfied and unsatisfied groups and what are the implications of this?

Performance

As shown in Table 4, in the case of Hyundai consumers rarely mentioned words related to performance in comparison to Honda and Ford consumers.

The satisfied group mentioned the words "control," "system," and "speed" more often than the unsatisfied group, although it is difficult to say they were definitely satisfied with these factors because these terms appeared in both best features and worst features. The frequency of these terms in the unsatisfied group is very low. Along with a fairly low score for performance, we can assume that reviewers were unsatisfied due to factors other than "control," "system," and "speed." In the case of Honda, consumers were generally satisfied with the performance because the most frequent words for this feature mostly appeared in best features, and, looking at the performance score of 4.29 in Table 3, we can assume they were significant to a certain degree. The result for Ford is ambiguous, but we can say with confidence that reviewers like how the car handles as the term "handle" appeared much more frequently than other terms and, in 95% of cases, appeared in best features for both satisfied and unsatisfied groups.

Table 4. Performance (best/worst features of each group).

Words	Hyundai			Honda			Ford		
	Satisfied Best/Worst	Unsatisfied Best/Worst	Sum	Satisfied Best/Worst	Unsatisfied Best/Worst	Sum	Satisfied Best/Worst	Unsatisfied Best/Worst	Sum
control	0.09/0.04	0.01/0.00	0.14	0.08/0.01	0.04/0.03	0.17	0.04/0.07	0.01/0.04	0.16
system	0.07/0.03	0.01/0.03	0.14	0.06/0.01	0.05/0.01	0.13	0.07/0.04	0.04/0.02	0.18
speed	0.06/0.05	0.01/0.00	0.12				0.04/0.04	0.00/0.02	0.10
wheel				0.06/0.01	0.04/0.01	0.15	0.03/0.05	0.00/0.01	0.10
road				0.06/0.03	0.01/0.04	0.14			
steering				0.06/0.01	0.06/0.02	0.14	0.04/0.03	0.01/0.01	0.09
power				0.05/0.02	0.03/0.03	0.13			
handle							0.15/0.01	0.03/0.00	0.19
Sum	0.22/0.12	0.03/0.03	0.40	0.37/0.08	0.24/0.17	0.86	0.39/0.25	0.10/0.10	0.83

Comfort

As shown in Table 5, Hyundai drivers felt comfortable in the car, and most were very satisfied with the seats. However, it seems that it has some problems with noise as the average score for comfort is low. Looking at the results for Honda, consumers who were satisfied with the Honda Civic purchase felt very comfortable in a cab and were pleased with the space provided, but it is likely that both satisfied and unsatisfied groups were unsatisfied with the back seats. In the case of Ford, there were people who found it comfortable and people who did not. The unsatisfied group did not discuss the comfort feature as much as the satisfied group, and it seems that individuals from the satisfied group liked neither the front nor the back seats. The most positive aspect of Ford mentioned by reviewers is the manipulation of the car.

Table 5. Comfort (best/worst features of each group).

Words	Hyundai			Honda			Ford		
	Satisfied Best/Worst	Unsatisfied Best/Worst	Sum	Satisfied Best/Worst	Unsatisfied Best/Worst	Sum	Satisfied Best/Worst	Unsatisfied Best/Worst	Sum
seat	0.27/0.13	0.10/0.03	0.53	0.12/0.09	0.08/0.10	0.39	0.13/0.10	0.04/0.03	0.31
back	0.09/0.08	0.04/0.04	0.26	0.04/0.05	0.01/0.04	0.13	0.04/0.09	0.01/0.03	0.18
control	0.09/0.04	0.01/0.00	0.14	0.08/0.01	0.04/0.03	0.17	0.04/0.07	0.01/0.04	0.16
comfort	0.09/0.01	0.03/0.00	0.13	0.12/0.01	0.03/0.02	0.19	0.09/0.00	0.02/0.02	0.13
rear	0.08/0.03	0.02/0.00	0.16				0.03/0.05	0.01/0.03	0.12
front	0.06/0.03	0.01/0.03	0.13	0.02/0.05	0.01/0.03	0.11	0.01/0.07	0.01/0.01	0.10
noise	0.01/0.07	0.00/0.04	0.12	0.00/0.03	0.03/0.05	0.11			
road				0.06/0.03	0.01/0.04	0.14			
handle							0.15/0.01	0.03/0.00	0.19
room				0.08/0.02	0.01/0.01	0.12			
Sum	0.69/0.39	0.21/0.18	1.47	0.52/0.29	0.22/0.33	1.36	0.52/0.39	0.13/0.15	1.19

Value

As shown in Table 6, most Hyundai holders were not satisfied with the fuel consumption of this car. Nevertheless, some of the satisfied group felt that Hyundai's mpg was not bad. The reason for this might depend on individual satisfaction levels in relation to mpg assessment. In the case of Honda, all terms related to fuel consumption constantly appeared in best features for both satisfied and unsatisfied groups. Occurring several times in worst features was the word minimal, which means Honda's mpg is very high and probably best among the three brands. Ford holders did not mention fuel consumption as much as owners of the other cars, but it is likely that Ford does not have any problems with fuel consumption and is credibly even better than Hyundai's mpg. Hence, there are other factors that resulted in the low score for value.

Table 6. Value (best/worst features of each group).

Words	Hyundai			Honda			Ford		
	Satisfied Best/Worst	Unsatisfied Best/Worst	Sum	Satisfied Best/Worst	Unsatisfied Best/Worst	Sum	Satisfied Best/Worst	Unsatisfied Best/Worst	Sum
gas	0.09/0.03	0.01/0.09	0.22	0.17/0.02	0.06/0.01	0.26	0.08/0.04	0.03/0.00	0.15
mpg	0.09/0.22	0.02/0.03	0.34	0.12/0.04	0.12/0.02	0.29			
mileage	0.09/0.04	0.13/0.02	0.21	0.08/0.02	0.08/0.02	0.19	0.06/0.02	0.03/0/01	0.12
fuel	0.06/0.07	0.00/0.00	0.13						
econ				0.07/0.01	0.03/0.01	0.11			
Sum	0.33/0.35	0.04/0.17	0.90	0.44/0.08	0.28/0.06	0.85	0.14/0.06	0.06/0.02	0.27

Interior

As shown in Table 7, the interior was most often discussed in the Hyundai case, and it is clear that reviewers from both satisfied and unsatisfied groups were greatly satisfied with this attribute. However, there is also a problem with a spare tire, which often appeared in worst features for both satisfied and unsatisfied groups. Opinions about the interior for Honda were divided among the satisfactory group, but the common element for both groups is that they liked the dashboard display. Additionally, the satisfied group often mentioned "room," which means they were satisfied with this feature. The interior results for Ford holders were good rather than bad, but it seems this was not the most important feature for reviewers.

Table 7. Interior (best/worst features of each group).

Words	Hyundai			Honda			Ford		
	Satisfied Best/Worst	Unsatisfied Best/Worst	Sum	Satisfied Best/Worst	Unsatisfied Best/Worst	Sum	Satisfied Best/Worst	Unsatisfied Best/Worst	Sum
interior	0.15/0.00	0.06/0.02	0.22	0.08/0.04	0.01/0.10	0.22	0.11/0.03	0.04/0.02	0.19
style	0.09/0.01	0.08/0.00	0.17				0.09/0.00	0.02/0.00	0.12
tire	0.01/0.09	0.00/0.07	0.16						
trunk	0.07/0.07	0.01/0.02	0.16						
rear	0.08/0.03	0.02/0.03	0.16				0.03/0.05	0.01/0.03	0.12
spare	0.00/0.10	0.00/0.04	0.15						
look	0.09/0.02	0.02/0.01	0.14	0.04/0.02	0.01/0.03	0.11	0.03/0.03	0.03/0.01	0.10
design	0.08/0.03	0.01/0.02	0.13	0.05/0.02	0.01/0.03	0.12			
display				0.08/0.01	0.04/0.01	0.15			
room				0.08/0.02	0.01/0.01	0.12			
Sum	0.56/0.34	0.19/0.20	1.29	0.33/0.11	0.10/0.19	0.72	0.26/0.11	0.10/0.06	0.53

Exterio

As shown in Table 8, in all three cases, customers were plenty satisfied with the exterior, especially Hyundai and Ford users. For instance, the occurrence of terms related to the exterior was very high for the Hyundai Elantra and the term "exterior" never appeared in worst features in either of the two groups. In the case of Ford, relative to other features, the exterior was the only factor with which consumers were satisfied. It is also the only factor which has a high score in the Ford sample. However, looking at the frequency of the term, we can assume that it was not the hottest topic for discussion compared to Hyundai. For the Honda Civic, the term "exterior" did not appear at all, which means it was not the main factor in determining whether customers purchased this car.

Table 8. Exterior (best/worst features of each group).

Words	Hyundai			Honda			Ford		
	Satisfied Best/Worst	Unsatisfied Best/Worst	Sum	Satisfied Best/Worst	Unsatisfied Best/Worst	Sum	Satisfied Best/Worst	Unsatisfied Best/Worst	Sum
style	0.09/0.01	0.08/0.00	0.17				0.09/0.00	0.02/0.00	0.12
look	0.09/0.02	0.02/0.01	0.14	0.04/0.02	0.01/0.03	0.11	0.03/0.03	0.03/0.01	0.10
design	0.08/0.03	0.01/0.02	0.13	0.05/0.02	0.01/0.03	0.12			
exterior	0.09/0.00	0.02/0.00	0.11				0.03/0.01	0.05/0.01	0.10
Sum	0.35/0.05	0.12/0.03	0.55	0.10/0.04	0.03/0.06	0.23	0.15/0.04	0.10/0.02	0.32

Reliability and Safety

As shown in Table 9, among the 24 most frequent terms for Hyundai and Honda, only one word, "engine," was related to reliability. There might be other words such as "engine" that could influence a decision to rate reliability, but these did not appear among the most frequent words. Therefore, it is hard to determine the extent to which the term "engine" affected the reliability score, but many people from the satisfied group for Honda were satisfied with its engine and mentioned it a few times. In the Hyundai group, the term "engine" occurred almost equally in best features and worst features. In the satisfied group it occurred more often in best features while in the unsatisfied group it appeared more frequently in worst features, which makes sense. Thus, we can assume there was an approximately equal number of people who were satisfied and unsatisfied with the engine.

Table 9. Reliability and safety (best/worst features of each group).

Words	Hyundai			Honda			Ford		
	Satisfied Best/Worst	Unsatisfied Best/Worst	Sum	Satisfied Best/Worst	Unsatisfied Best/Worst	Sum	Satisfied Best/Worst	Unsatisfied Best/Worst	Sum
	Reliability feature								
engine	0.06/0.02	0.02/0.04	0.14	0.05/0.01	0.03/0.03	0.12			
transmission							0.06/0.09	0.01/0.12	0.29
manual							0.05/0.06	0.00/0.01	0.13
automatic							0.06/0.03	0.01/0/02	0.12
issue							0.00/0.04	0.01/0.04	0.10
shift							0.03/0.04	0.00/0.03	0.10
Sum	0.06/0.02	0.02/0.04	0.14	0.05/0.01	0.03/0.03	0.12	0.19/0.27	0.04/0.23	0.73
	Safety feature								
light	0.04/0.05	0.01/0.04	0.15	0.03/0.04	0.01/0.03	0.11	0.04/0.04	0.01/0.01	0.09

Reliability was actively discussed in the Ford group. It is clear that Ford has serious problems in this field, mostly to do with transmission. The frequency of the term "transmission" was 0.29, the highest among all terms in the reliability group and more than three times higher than the frequency of the term "engine" in the Hyundai and Honda groups. Furthermore, the frequency of terms "manual," "automatic," "issue," and "shift" was also high. Based on frequency analysis, where a link was found between these terms and "transmission," we can say that both satisfied and unsatisfied groups criticized transmission, and the occurrence of these terms in total was 0.73, which is very high for just one specific part of a car. The low score of 3.19 in Table 3 is consistent with the results for frequency, so Ford must solve this problem in order to secure clients' trust. In the case of safety, it is difficult to interpret the results as there is only one word, "light," that, after association analysis, was correlated with several features such as "safety" and "technology." Hence, it would be a mistake to judge the significance of the relationship between safety scores for all three car brands and the term "light" as well as its frequency in the satisfied and unsatisfied groups.

Technology

Among the three brands, technology was the most frequently discussed by Honda owners, quite frequently by Ford owners, and least often in the Hyundai group as shown in Table 10. Along with comfort, technology was the hottest topic for discussion in the Honda group. We can say with confidence that Honda holders from both satisfactory groups greatly enjoyed using the steering wheel, Bluetooth, econ function, and inward system. Opinions about the dashboard varied, as there were reviewers in both satisfied and satisfied groups who liked or did not like this feature. Hence, Honda consumers were very satisfied with the technological side, and the technological level is probably the highest among the three brands. In the case of Ford, satisfied and unsatisfied groups mentioned terms related to technology in both best features and worst features, and the ratio was quite similar. Indeed, one of the most frequent terms was "sync," which refers to Ford's special feature. Looking at the results, it seems that, regardless of the satisfactory group, some reviewers enjoyed using this system and some did not. Therefore, the technological side of the Ford Focus was worth paying attention to, but it is unclear whether this is beneficial or disadvantageous for an automotive company. Given the very low score in Table 3 for the technology, we can presume that reviewers who mentioned these words in worst features evaluated it very negatively, while reviewers who mentioned these words in best features did not evaluate it highly. If we look at the results for Hyundai, we can say that consumers liked the Bluetooth system, but we cannot say the same about other terms. Therefore, we suppose there are other factors that resulted in the low score for technology.

Table 10. Technology (best/worst features of each group).

Words	Hyundai			Honda			Ford		
	Satisfied Best/Worst	**Unsatisfied Best/Worst**	**Sum**	**Satisfied Best/Worst**	**Unsatisfied Best/Worst**	**Sum**	**Satisfied Best/Worst**	**Unsatisfied Best/Worst**	**Sum**
system	0.07/0.03	0.01/0.03	0.14	0.06/0.01	0.05/0.01	0.13	0.07/0.04	0.04/0.02	0.18
control	0.09/0.04	0.01/0.00	0.14	0.08/0.01	0.04/0.03	0.17	0.04/0.07	0.01/0.04	0.16
bluetooth	0.09/0.03	0.04/0.01	0.17	0.07/0.02	0.04/0.01	0.13			
light	0.04/0.05	0.01/0.04	0.15	0.03/0.04	0.01/0.03	0.11	0.04/0.04	0.01/0.01	0.09
dash				0.09/0.04	0.07/0.05	0.26			
display				0.08/0.01	0.04/0.01	0.15			
wheel				0.06/0.01	0.04/0.03	0.15	0.03/0.05	0.00/0.01	0.10
steering				0.06/0.01	0.06/0.02	0.14	0.04/0.03	0.01/0.01	0.09
econ				0.07/0.01	0.03/0.01	0.11			
sync							0.10/0.06	0.03/0.02	0.21
um	0.29/0.16	0.07/0.08	0.59	0.60/0.16	0.38/0.21	1.35	0.34/0.29	0.10/0.11	0.85

4.2.3. Implications and Discussion

Based on the results of this research, the following propositions can be stated. Firstly, among the three car brands, the Hyundai Elantra car has the best marks in relation to "interior" and "exterior," but, in terms of the interior, there is a problem with a spare tire that needs to be solved. A few people were also unsatisfied with gas consumption, so it would be better for engineers from Hyundai to improve the mpg index. Furthermore, Hyundai has problems in terms of technology, one of which is incorrect mpg displays. Moreover, despite consumers' satisfaction with comfort, there was a problem with noise on roads. This seems to be the main reason for a relatively low score for comfort. Hence, Hyundai should reconsider the value and technological particularities of the Elantra car to make it more competitive on the market. Secondly, among the three car brands, the Honda Civic received positive feedback for all features and was found to be best in terms of value and technology. It has the best mpg index and the best technological equipment compared to Hyundai and Ford. Despite satisfaction with all features, Honda engineers should pay attention to the interior, because many consumers criticized it for its cheapness. In addition, Honda should consider the issue of comfort, because back seats were also found to be a weak point. Thirdly, the Ford car was evaluated very poorly regarding all features with the exception of the exterior, where it received a high score. However, according to the results, this was not the most discussed topic among reviewers. Among the 24 most frequent terms, negative terms were in the majority. These were related to the topic of reliability, where reviewers severely criticized transmission and found this to be the biggest problem in the Ford automobile. Apart from problems with transmission, most consumers were quite unsatisfied with both back and front seats. The only feature reviewers were truly pleased with, according to the results, was manipulation of the car. In previous research, reliability was found to be one of the most significant factors for buyers. Therefore, looking at the poor evaluation of this car, reviewers were greatly disappointed with its reliability and, for this reason, the scores for other features were slightly biased. Hence, Ford marketers and engineers must completely reconsider and reassess their car from all sides, starting with the reliability feature.

In addition, the results for several features, such as safety, were unclear and ambiguous, thereby making interpretation difficult. This can be explained by the lack of terms chosen for the analysis. To fill such blind spots, a more extended analysis is needed.

5. Conclusions

In this paper, consumer reviews of three different competitive automobiles—the Hyundai Elantra, the Honda Civic, and the Ford Focus—were examined. The results can be summarized as follows:

Firstly, each car model was analyzed in terms of its best and worst features, thereby underlining the superior features of a given car as well as its problematic features. It was reached by virtue of finding the words appearing most frequently in corresponding reviews. In terms of best features, the Hyundai Elantra has car design, seats, interior, bluetooth, and steering control; the Honda Civic has low gas consumption, seats, dashboard, and technological equipment such as Bluetooth and "econ" mode; and the Ford Focus has car manipulation, exterior, interior, a quiet ride, and the "Ford Sync" function. In terms of the worst features, the Hyundai Elantra has an incorrect mpg display, problems with a spare tire, noise on the road, fog lights, trunk, and mpg inefficiency; the Honda Civic has a cheap interior, seats, fog lights, mirrors, and noise on the road; and the Ford Focus has problems with transmission, seats, space in the cab, and controls on the steering wheel. For the Ford Focus, some features such as seats, gas consumption, dashboard, and "Ford Sync" systems were found in both best and worst features, which can be interpreted as a difference of opinion among reviewers.

Secondly, eight specific yet different features were analyzed using consumers' reviews of best and worst features. The results showed that consumers actively discussed the comfort feature for all three brands. In particular, Hyundai reviewers emphasized the interior and exterior, Honda reviewers were interested in technology, and Ford reviewers paid attention to reliability.

Thirdly, the ways in which the views of both satisfied and unsatisfied groups differed were analyzed. The results showed that Hyundai received the best marks in terms of design and interior but needs to reconsider the value and technology features to make the Elantra more competitive on market. The Honda Civic does not have any critical issues relating to any factors. It has the best mpg index and technological equipment compared to Hyundai and Ford. However, it should consider its cheap interior and comfort feature. In contrast, Ford should completely reconsider and reassess its car, starting with its reliability.

This paper has the following limitations. Firstly, even though the Edmunds website is one of the biggest online resources for automotive information, there was still a general lack of reviews. To obtain more significant results, a larger amount of reviews will be needed. Furthermore, it would be useful to obtain data on sales figures for all three car models to make a comparison between sales and the results of this study. In addition, because only 24 of the most frequent terms were chosen for the analysis, the results for some features, such as safety, were unclear and ambiguous, thereby making interpretation difficult. Hence, it is necessary to increase the number of terms to find more related to safety features.

Author Contributions: Conceptualization, E.-G.K. and S.-H.C.; formal analysis, E.-G.K. and S.-H.C.; methodology, S.-H.C.; software, E.-G.K. and S.-H.C.; visualization, E.-G.K.

Funding: This research received no external funding.

Conflicts of Interest: The author declares no conflict of interest.

References

1. Sagar, A.D.; Chandra, P. *Technological Change in the Indian Passenger Car Industry, Energy Technology Innovation Policy Discussion Paper: BCSIA Discussion Paper 2004-05*; Energy Technology Innovation Project; Kennedy School of Government, Harvard University: Cambridge, MA, USA, 2004.
2. Kaushal, S.K. Confirmatory factor analysis: An empirical study of the fourwheeler car buyer's purchasing behavior. *Int. J. Glob. Bus. Manag. Res.* **2014**, *2*, 90–104.
3. Janssen, M.; van der Voort, H.; Wahyudi, A. Factors influencing big data decision-making quality. *J. Bus. Res.* **2017**, *70*, 338–345. [CrossRef]
4. Olszak, C.M. Toward better understanding and use of business intelligence in organizations. *Inf. Syst. Manag.* **2016**, *33*, 105–123. [CrossRef]
5. Kiron, D. Organizational alignment is key to big data success. *MIT Sloan Manag. Rev.* **2013**, *54*, 54307.
6. Liu, Y. Big data and predictive business analytics. *J. Bus. Forecast.* **2014**, *33*, 40–42.
7. Wamba, S.F.; Gunasekaran, A.; Akter, S.; Ren, S.J.; Dubey, R.; Childe, S.J. Big data analytics and firm performance: Effects of dynamic capabilities. *J. Bus. Res.* **2017**, *70*, 356–365. [CrossRef]
8. Vidgen, R.; Shaw, S.; Grant, D.B. Management challenges in creating value from business analytics. *Eur. J. Oper. Res.* **2017**, *261*, 626–639. [CrossRef]
9. Günther, W.A.; Mehrizi, M.H.R.; Huysman, M.; Feldberg, F. Debating big data: A literature review on realizing value from big data. *J. Strateg. Inf. Syst.* **2017**, *26*, 191–209. [CrossRef]
10. Mikalef, P.; Pappas, I.O.; Krogstie, J.; Giannakos, M. Big data analytics capabilities: A systematic literature review and research agenda. *Inf. Syst. e-Bus. Manag.* **2018**, *16*, 547–578. [CrossRef]
11. Grover, V.; Chiang, R.H.L.; Liang, T.; Zhang, D. Creating Strategic Business Value from Big Data Analytics: A Research Framework. *J. Manag. Inf. Syst.* **2018**, *35*, 388–423. [CrossRef]
12. Dong, J.Q.; Yang, C.-H. Business value of big data analytics: A systems-theoretic approach and empirical test. *Inf. Manag.* **2018**, in press. [CrossRef]
13. Müller, O.; Fay, M.; Brocke, J.V. The Effect of Big Data and Analytics on Firm Performance: An Econometric Analysis Considering Industry Characteristics. *J. Manag. Inf. Syst.* **2018**, *35*, 488–509. [CrossRef]
14. Côrte-Real, N.; Ruivo, P.; Oliveira, T. Leveraging internet of things and big data analytics initiatives in European and American firms: Is data quality a way to extract business value? *Inf. Manag.* **2019**, in press.
15. Mikalef, P.; Boura, M.; Lekakos, G.; Krogstie, J. Big data analytics and firm performance: Findings from a mixed-method approach. *J. Bus. Res.* **2019**, *98*, 261–276. [CrossRef]
16. Tang, C.; Guo, L. Digging for Gold with a Simple Tool: Validating Text Mining in Studying Electronic Word-of-Mouth (eWOM) Communication. *Mark. Lett.* **2015**, *26*, 67–80. [CrossRef]
17. Pennebaker, J.; Mehl, M.; Niederhoffer, K. Psychological aspects of natural language: Our words, our selves. *Annu. Rev. Psychol.* **2003**, *54*, 547–577. [CrossRef]
18. Gupta, V.; Lehal, G.S. A survey of text mining techniques and applications. *J. Emerg. Technol. Web Intell.* **2009**, *1*, 60–76. [CrossRef]
19. Ramanathan, V.; Meyyappan, T. Survey of text mining. *Proc. Int. Conf. Technol. Business Manag.* **2013**, 508–514.

20. Khader, N.; Lashier, A.; Yoon, S.W. Pharmacy robotic dispensing and planogram analysis using association rule mining with prescription data. *Expert Syst. Appl.* **2016**, *57*, 296–310. [CrossRef]

21. Ren, J.; Li, W.; Wang, Y.; Zhou, L. Graph-mine: A key behavior path mining algorithm in complex software executing network. *Int. J. Innov. Comput. Inf. Control* **2015**, *11*, 541–553.

22. Calders, T.; Dexters, N.; Gillis, J.J.; Goethals, B. Mining frequent itemsets in a stream. *Inf. Syst.* **2014**, *39*, 233–255. [CrossRef]

23. Han, J. *Data Mining: Concepts and Techniques*; Morgan Kaufmann Publishers Inc.: San Francisco, CA, USA, 2005.

24. Tan, P.N.; Steinbach, M.; Kumar, V. *Introduction to Data Mining*, 1st ed.; Addison-Wesley Longman Publishing Co.; Inc.: Boston, MA, USA, 2005.

25. Czibula, G.; Czibula, I.G.; Miholca, D.; Crivei, L.M. A novel concurrent relational association rule mining approach. *Expert Syst. Appl.* **2019**, *125*, 142–156. [CrossRef]

26. Müller, J.; Christandl, F. Content is king–But who is the king of kings? The effect of content marketing, sponsored content & user-generated content on brand responses. *Comput. Hum. Behav.* **2019**, *96*, 46–55.

27. Ayeh, J.; Au, N.; Law, R. Do we believe in TripAdvisor? Examining credibility perceptions and online travelers' attitude toward using user-generated content. *J. Travel Res.* **2013**, *52*, 437–452. [CrossRef]

28. Amato, F.; Moscato, V.; Picariello, A.; Sperlí, G. Multimedia social network modeling: A proposal. In Proceedings of the 2016 IEEE Tenth International Conference on, IEEE Semantic Computing, ICSC, Laguna Hills, CA, USA, 3–5 February 2016; pp. 448–453.

29. Amato, F.; Moscato, V.; Picariello, A.; Sperl, G. Diffusion Algorithms in Multimedia Social Networks: A preliminary model. In Proceedings of the IEEE/ACM International Conference on Advances in Social Networks Analysis and Mining, Sydney, Australia, 31 July–3 August 2017; pp. 844–851.

30. Yahav, I.; Shehory, O.; Schwartz, D. Comments Mining With TF-IDF: The Inherent Bias and Its Removal. *IEEE Trans. Knowl. Data Eng.* **2019**, *31*, 437–450. [CrossRef]

31. Bickart, B.; Schindler, R.M. Internet forums as influential sources of consumer information. *J. Interact. Mark.* **2001**, *15*, 31–40. [CrossRef]

32. Resnick, P.; Zeckhauser, R. *Trust among Strangers in Internet Transactions: Empirical Analysis of eBay's Reputation System*; Baye, M.R., Ed.; The Economics of the Internet and Ecommerce; Emerald Group Publishing Limited: Bingley, UK, 2002; pp. 127–157.

33. Chevalier, J.A.; Mayzlin, D. The effect of word of mouth on sales: Online book reviews. *J. Mark. Res.* **2006**, *43*, 345–354. [CrossRef]

34. Schneider, M.J.; Gupta, S. Forecasting sales of new and existing products using consumer reviews: A random projections approach. *Int. J. Forecast.* **2016**, *32*, 243–256. [CrossRef]

35. Mudambi, S.M.; Schuff, D. What makes a helpful review? A study of customer reviews on amazon.com. *MIS Q.* **2010**, *34*, 185–200. [CrossRef]

36. Lawani, A.; Reed, M.R.; Mark, T.; Zheng, Y. Reviews and price on online platforms: Evidence from sentiment analysis of Airbnb reviews in Boston. *Reg. Sci. Urban Econ.* **2019**, *75*, 22–34. [CrossRef]

37. Cheng, M.; Jin, X. What do Airbnb users care about? An analysis of online review comments. *Int. J. Hosp. Manag.* **2019**, *76*, 58–70. [CrossRef]

38. Sohail, S.S.; Siddiqui, J.; Ali, R. Feature extraction and analysis of online reviews for the recommendation of books using opinion mining technique. *Perspect. Sci.* **2016**, *8*, 754–756. [CrossRef]

39. Zhang, X.; Dellarocas, C. The lord of the ratings: Is a movie's fate is influenced by reviews? In Proceedings of the 2006 ICIS, Milwaukee, WI, USA, 10–13 December 2006; pp. 1959–1978.

40. Reinstein, D.A.; Snyder, C.M. The influence of expert reviews on consumer demand for experience goods: A case study of movie critics. *J. Ind. Econ.* **2005**, *53*, 27–51. [CrossRef]

41. Lee, J.H.; Jung, S.H.; Park, J.H. The role of entropy of review text sentiments on online WOM and movie box office sales. *Electron. Commer. Res. Appl.* **2017**, *22*, 42–52. [CrossRef]

42. Kawaf, F.; Istanbulluoglu, D. Online fashion shopping paradox: The role of customer reviews and facebook marketing. *J. Retail. Consum. Serv.* **2019**, *48*, 144–153. [CrossRef]

43. Kim, S.G.; Kang, J. Analyzing the discriminative attributes of products using text mining focused on cosmetic reviews. *Inf. Process. Manag.* **2018**, *54*, 938–957. [CrossRef]

44. Xu, X.; Li, Y. The antecedents of customer satisfaction and dissatisfaction toward various types of hotels: A text mining approach. *Int. J. Hosp. Manag.* **2016**, *55*, 57–69. [CrossRef]

45. Hu, Y.; Chen, Y.; Chou, H. Opinion mining from online hotel reviews—A text summarization approach. *Inf. Process. Manag.* **2017**, *53*, 436–449. [CrossRef]

46. Geetha, M.; Singha, P.; Sinha, S. Relationship between customer sentiment and online customer ratings for hotels—An empirical analysis. *Tour. Manag.* **2017**, *61*, 43–54. [CrossRef]

47. Lee, P.; Hu, Y.; Lu, K. Assessing the helpfulness of online hotel reviews: A classification-based approach. *Telemat. Inf.* **2018**, *35*, 436–445. [CrossRef]

48. Wang, Y.; Lu, X.; Tan, Y. Impact of product attributes on customer satisfaction: An analysis of online reviews for washing machines. *Electron. Commer. Res. Appl.* **2018**, *29*, 1–11. [CrossRef]

49. Oza, K.S.; Naik, P.G. Prediction of Online Lectures Popularity: A Text Mining Approach. *Procedia Comput. Sci.* **2016**, *92*, 468–474. [CrossRef]

50. Nakayama, M.; Wan, Y. The cultural impact on social commerce: A sentiment analysis on Yelp ethnic restaurant reviews. *Inf. Manag.* **2019**, *56*, 271–279. [CrossRef]

51. Gao, S.; Tang, O.; Wang, H.; Yin, P. Identifying competitors through comparative relation mining of online reviews in the restaurant industry. *Int. J. Hosp. Manag.* **2018**, *71*, 19–32. [CrossRef]

52. Korfiatis, N.; Stamolampros, P.; Kourouthanassis, P.; Sagiadinos, V. Measuring service quality from unstructured data: A topic modeling application on airline passengers' online reviews. *Expert Syst. Appl.* **2019**, *116*, 472–486. [CrossRef]

53. Kulkarni, G.; Ratchford, B.T.; Kannan, P.K. The Impact of Online and Offline Information Sources on Automobile Choice Behavior. *J. Interact. Mark.* **2012**, *26*, 167–175. [CrossRef]

Article

Context–Problem Network and Quantitative Method of Patent Analysis: A Case Study of Wireless Energy Transmission Technology

Jason Jihoon Ree [1], Cheolhyun Jeong [1], Hyunseok Park [2] and Kwangsoo Kim [1,*]

[1] Department of Industrial and Management Engineering, Pohang University of Science and Technology, Pohang 37673, Korea; jjree@postech.ac.kr or jasonree32@gmail.com (J.J.R.); inbass@postech.ac.kr (C.J.)
[2] Department of Information System, Hanyang University, Seoul 04763, Korea; hp@hanyang.ac.kr
* Correspondence: kskim@postech.ac.kr; Tel.: +82-54-279-8234

Received: 13 February 2019; Accepted: 5 March 2019; Published: 11 March 2019

Abstract: Identification of prevalent problems is an important process of strategic innovation for stakeholders of trending technologies. This paper proposes a systematic and replicable method of patent analysis to identify problems to be solved requisite for sustainable technology planning and development, by implementing the concept of 'context' to facilitate problem identification. The main concept of the method entails the importance of the connections between contextual information and problems to provide more focused, relevant, and constructive insights essential for instating goals for research and development activities. These context–problem entities and their entwined connections are discovered using keyword pattern matching, grammar-based text mining, and co-word analysis techniques. The intermediary outputs are then utilized to generate the proposed context–problem network (CP net) for social network, grammar, and quantitative data analysis. For verification, our method was applied to 737 patents in the wireless energy transmission technology domain, successfully yielding CP net data. The detailed analysis of the resulting CP net data delivered meaningful information in the wireless charging technology field: The main contexts, "batteries", "power transmission coils", and "cores", are found to be most relevant to the main problems, "maximizing coupling efficiency", "minimizing DC signal components", and "charging batteries". The results provide a wide range of informative perspectives for individuals, the scientific community, corporate, and market-level stakeholders. Furthermore, the method of this study can be applicable to various technologies since it is independent of specific subject domains. Future research directions aim to improve this method for better quality and modeling of contexts and problems.

Keywords: systematic and replicable patent analysis method; problem-solved concept; context–problem network; network data analysis; sustainable wireless energy transmission technology

---

1. Introduction

Innovation commonly initiates when prevalent problems are accurately identified. For effective innovative endeavors, an innovator must define the currently existing problems and select which of them to prioritize and focus on. Patents are practical sources from which to identify problems to be solved because they possess solutions to specific problems in the form of the most state-of-the-art innovations. Thus, patent analysis is an effective way to identify problems to be solved for sustainable technology development and management. To obtain valuable information from patents, the analysis is generally performed manually by technology domain experts. However, manual analysis requires extensive time and effort. Moreover, in the contexts of identifying problems, a simple bibliographic analysis is not applicable, since problems are described in natural language in the description sections of patents.

Hence, to alleviate the time and labor for manual patent analysis, a plethora of research has dedicated much effort for the advancements of natural language processing methods in the contexts of patents. As a result, various influential and effective methods of computer-assisted patent analysis utilizing natural language processing and text mining have been reported [1–17]. The majority of these methods are founded on keyword-based patent analysis, which extracts keywords from the patent descriptions and ascertains patterns with technical significance [1–10]. These keyword-based methods, however, reveal severe limitations in finding exact correlations between the extracted keywords and how the keywords are exactly related to the technologies (i.e., problems to be solved) achieved in patents. Later, to overcome such limitations, grammar-based methods of patent analysis have been proposed [11–17]. Compared to the keyword-based analyses, the grammar-based methods are more complex because of the inclusion of syntactic and semantic analysis, which introduces the concept of 'function'. Here, 'function' is defined as the action changing a feature of any object that may provide valid information on the purposes of a technology [18]. Despite the analysis complexity, the grammar-based methods have rarely considered the identification of problems, which are directly related to the technology innovated in patents. Since patent descriptions encompass heterogeneous sections such as technology field, prior art, existing problems, summary of invention, and components of inventions, each sentence delineates a different purpose and meaning, even within the same section. Thus, identifying the sentences that describe the problems is essential before conducting grammar-based analyses. To extract the problems solved in patents, the 'problem-solved concept' (PSC), a statement of the problem that a patent solves, was first proposed by Phelps [19] and was further explored by other researchers [20,21], who introduced heuristic rules to automatically extract the PSC from a patent description. However, the limitation of this research is the absence of a method for analyses of the extracted PSC results. Moreover, the identification of the problem without information regarding the contexts or circumstances (e.g., where, how, when) of the problem limits the ability to provide perspective and significance of analysis results. For example, when the extracted problem after extraction using text mining takes a form similar to 'increase sensitivity', it is difficult to comprehend the meaning, situation, and significance of the problem. On the contrary, with the introduction of context (e.g., 'of touchpad'), the extracted problem accommodates more meaning and a stronger foundation for subsequent analyses.

In this study, we report a systematic and replicable method of patent analysis to identify key contexts and problems to be solved for sustainable technology development and management. This method is composed of context–problem network (CP net) generation and quantitative analysis (Figure 1). A CP net is generated in a sequential manner (Figure 2). In the first stage, sentences derived from patent parsing are selected by matching keyword patterns. Contexts and problems are extracted from the selected sentences by grammar-based text mining. In the second stage, the extracted contexts and problems are further normalized via stop word elimination, stemming, and semantic conversion. In the final stage, the normalized contexts and problems are used to determine co-occurrences by co-word analysis to generate a network, CP net. Then, the resulting CP net data are quantitatively analyzed through a combination of three different approaches (centrality, neighbor, and grammar analyses), providing constructive key information for technology innovation. Our method has been applied to the wireless energy transmission technology domain as a case study. As a result, meaningful information of the field is determined: The contexts, 'batteries', 'power transmission coils', and 'cores', are observed to be most relevant to the problems, 'maximizing coupling efficiency', 'minimizing DC signal components', and 'charging batteries'. As shown in the case study, the contexts and problems selected through our proposed method can be meticulously investigated to recognize associated problems and contexts directly related to the specific area of interest. Particularly, the outcomes of the case study are applied and analyzed to correspond to the current status quo of the wireless energy transmission technological domain from the perspectives of individuals, scientific communities, corporations, and market-level stakeholders to demonstrate the feasibility and exploitation of our method and results. Above all, the environmental and sustainability issues pertaining to wireless energy transfer technology and its developments are discussed. The proficiency to cover wide-ranging perspectives for strategic innovation indicates the suitability of this method to many other high-impact technologies, independent of specific subject domains.

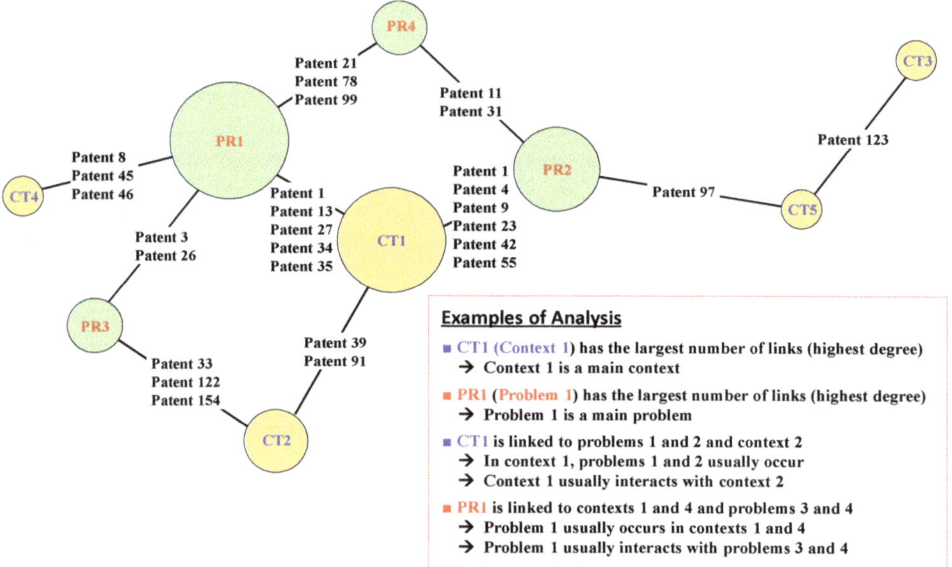

Figure 1. Example of context–problem network (CP net) generation and analysis; CT: Context; PR: Problem.

The rest of this paper consists of the sections as follows: Section 2 describes related works, Section 3 depicts the procedure and methods for generation of the CP net, Section 4 presents the method for analyzing the CP net, Section 5 portrays the application of the proposed method in a case study of wireless energy transmission technology, and Section 6 describes the conclusion.

2. Related Works

In this section, methods for extracting information by patent analysis and text mining are reviewed. Then, representative methods for analyzing extracted information are described in a brief manner.

2.1. Patent Analysis and Text Mining

In general, keyword-based text mining methods have been frequently used for the analysis of unstructured information. Notably, Yoon and Park [4] reported a method of generating a keyword-based patent map by a step-wise method to extract keywords, transform patents into keyword vectors, calculate similarities between the vectors, visualize the vectors into a patent map, analyze some indices, and cluster patents. This method was used to identify technology opportunities [5,6] and to forecast technology trends [6]. Similar methods have been adopted for patent analysis in various commercial software products, such as ClearForest, VantagePoint, Aureka®, STN® AnaVist™, ATMS/Analyzer, and VxInsight™. However, such methods of generating a keyword-based patent map were reported to reveal certain limitations in achieving exact relationships between keywords, as well as between keywords and problems solved [12].

As an alternative approach to keyword-based text mining methods, a subject–action–object (SAO) network and analysis method was suggested [12]. This method extracted SAOs, evaluated associations among subjects, actions, and objects, visualized a SAO network, and analyzed various network indices.

In follow-up studies, similar concepts were used to build a technology tree [13], to develop technology roadmaps [14], to identify technology trends [15], and to identify competition trends [16]. In addition to SAO-based methods, other researchers have modeled and extracted technological implications. An invention property–function network analysis, which analyzes relationships between adjectives and verbs, was reported [17]. Furthermore, a method for extracting function requirements and design parameters of patents by using prepositions ('to' and 'for') was proposed [22].

However, such grammar-based methods of patent analysis have limitations when applied to problem identification and analysis. Descriptions of a patent include heterogeneous sections, such as technology field, prior art, existing problems, summary of invention, and components of invention. Due to the eclectic natures of these segments of patents, each sentence has a different purpose and meaning, despite being in the same section. This underlying fact alone provides adequate reason for identifying the sentences that describe the problem, prior to conducting grammar-based methods to the text.

Despite the advancements of keyword-based or grammar-based patent analysis methods, a systematic and quantitative method for analyzing problems has yet to be explored. Although the most relatable research developed thus far involved the extraction methods of problems [19,20], the contexts of the problems were not considered.

2.2. Co-Word Analysis and Social Network Analysis

Co-word analysis (CWA) exploits associations between keywords [23,24]. The process of CWA is generally composed of definition, association, and analysis. In the definition step, interesting keywords are defined. In the association step, co-occurrences of the keywords are recorded when they appear in the same document. After recording co-occurrences, the analysis step applies statistical association rules to discover interesting associations between keywords. Such a process can be done automatically via computer and thus, is efficient in analysis of large datasets. Due to these advantages, CWA is used in many applications, like identifying technology trends [3], discovering new knowledge [25], searching information [26], and investigating inventor connections [27].

Recently, social network analysis (SNA) has been frequently used with CWA. SNA explores the relationships ('ties', 'arcs', or 'edges') between actors ('nodes' or 'vertices') [28,29]. SNA focuses on relationships among people, but its applications have been extended to include relationships between any connected entities, such as organizations, computers, documents, information, and knowledge. Several indices have been established to evaluate relationships between actors in a social network and to reveal which actor or cluster of actors are more central (i.e., more significant) relative to other actors [28,30]. Numerous measures can be exploited to assess centrality, such as degree, closeness, betweenness, and eigenvector. Degree centrality indicates the weights of adjacent nodes associated with the target node. High degree centrality can be interpreted as a node that is directly linked to many different nodes relative to other nodes and their respective links. Closeness centrality indicates the distance of a node from all other nodes in a network. A node with short geodesic distances to other nodes has high closeness. This measure considers indirect links to all nodes, so it represents the global importance of a node.

Leading research has found SNA particularly advantageous when extracting unknown and prospective internal relationships from large quantities of text [31–35]. SNA has also been applied to analyze bibliographic information of patents. For example, it was used to analyze relationships among inventors, applicants, and citations [29,36–39]. In more recent research, SNA and CWA have been used collectively to analyze unstructured information, such as technology keywords and research topics [4,40]. Additionally, SNA has been used to analyze relationships among components of grammar structures [12,17].

3. Generation of CP Net

In this study, the procedure for generating a CP net is composed of three stages: Extraction, normalization, and generation (Figure 2). Specific methods of the stages are described in the following sections.

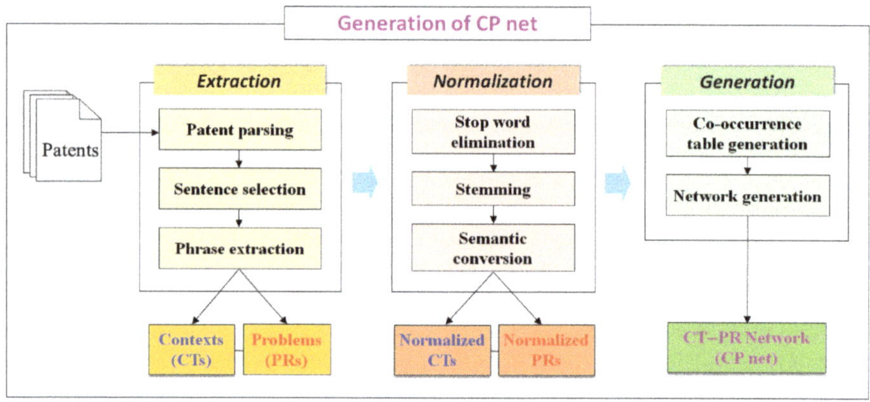

Figure 2. Procedure of generating a context–problem network (CP net).

3.1. Extraction of Contexts and Problems

Full-text patent documents have various formats, according to the patent office, database, and file type. This research utilizes US patents retrieved from the United States Patent and Trademark Office (USPTO). From the website, full-text patents are retrieved in html format. By utilizing Python-based natural language processing libraries, such as BeautifulSoup, unstructured texts can be extracted and categorized into subsections. The subsection 'background of the invention' is particularly significant due to the high likelihood of containing the description of the problem.

In our study, 'problem' is defined as a minimum expression of the problem (i.e., invention or invented one) which a patent solved. The 'context' is defined as a modifier of the problem, which includes interacting products, services, humans, tasks, and environments. Such components are described in the research of context-aware computing [41–46]. The context provides keys to understand the problem exactly, because they are direct or indirect causes of the problem.

In general, a patent is documented with numerous sentences that consist of contexts and further describe problems to be solved. In this study, sentences derived from patent parsing are selected by matching to keyword patterns. Sentences from patents can typically be categorized into four types: (i) Description of problems, (ii) description of the needs of users, (iii) description of objectives, and (iv) description of functions. In addition to these patterns, an enormous number of expressions may exist. However, this study uses four representative types of patterns to select sentences, as listed in Table 1 [19]. Utilizing these keyword patterns, the sentences from the full texts of patents are matched, selected, and categorized using code implemented in Python. If a sentence results in a positive keyword match with one or more of the keyword patterns (Table 1), it is placed in a collective list of sentences categorized by the patent of origin. Sentences that describe problems straightforwardly depict the problem precisely (e.g., "the existing problem is inefficiency of searching" or "the matter is that the speed of the transference is too slow"). Moreover, particular sentences describe the needs of users, such as "there is a need for improving efficiency" and "accurate methods are required". There also exist sentences that describe objectives of the patents, like "objective of this invention is to reduce human errors" and "purpose is to increase convenience". Certain sentences include verbs which have an increasing, decreasing, or solving connotation that denote functionality, like "to maximize heat conductivity" and "to improve mechanical strength".

Table 1. Keyword patterns for selecting sentences.

Types	Keyword Patterns
Problems	problem, drawback, matter, trouble, defect, weak, flaw, fault, shortcoming, demerit, fail, wrong, error, harm, complain, disadvantage, bad, too, low, loss, slow, complex, complicate, frustrate, difficult, hard to, restrict, limit, disable, uneasy, uneasiness, unpleasant, inconvenient, uncomfortable, discomfort, usability, throughput, expensive, awkward, danger, pervert, fussy, fastidious, refractor, stress, distress, hurt, painful, pain, suffer, anxiety, strain, burden, tense, injury, stuck, undermine, ruin
Needs	need for, demand for, requirement for, want for, desire for, be needed, be demanded, be required, be desired, be desirable
Objectives	objective of . . . invention, purpose of . . . invention, object is to, purpose is to
Functions	to minimize, to reduce, to lessen, to weaken, to diminish, to eliminate, to remove, to decrease, to maximize, to increase, to improve, to enhance, to promote, to multiply, to enlarge, to escalate, to raise, to strengthen, to expand, to address, to solve, to resolve, to settle, to remedy

After sentence selection, grammar-based text mining is applied to extract contexts and problems by employing extraction rules (Table 2). Most problems can be modeled as an action–object structure, i.e., a combination of a verb (VB) and an object of the verb (OV) [19]. Such modeling is also valid for the four types of problems, for example, "problem is interrupting (VB) communications (OV)", "there is a need for transferring (VB) electricity (OV)", "objective of this invention is to increase (VB) efficiency (OV)", and "purpose is to minimize (VB) interruption (OV)".

Table 2. Rules and examples of extraction.

Types	Rules	Example
Context (CT)	Preposition (PP) + Object (OP)	• 'In the wireless communication network' • 'Of mobile device'
Problem (PR)	Verb (VB) + Object (OV)	• 'Minimize energy consumption' • 'Transfer energy'

In contrast to the problem identification, modeling of the context has yet been studied. Thus, a key strength of our research is the extraction of the context in addition to the problem. In fact, the context is a surrounding situation when problems arise. Examples include a product, a service, a component, a human, a task, or an environment. Therefore, the context typically explains what is connected to the problem, and how it is associated. In this study, the preposition–object structure is used for modeling the context. The object, usually a noun phrase, refers to an object (OP) of the preposition and explains the identity or circumstance connected to the problem. The preposition (PP) explains how the OP is connected to the problem. For example, a context 'in the wireless communication network' means a problem occurs 'in' (PP) the environment of the wireless communication network (OP).

Because the problem and the context are modeled as grammatical structures, the part-of-speech (POS) of each word must be identified. In this research, the Stanford Log-linear Part-Of-Speech Tagger is used [47]. It tags all words with one of 36 POS tags or 12 punctuation tags. After tagging, the grammatical structures are identified by matching to sequences of POS tags. Specifically, the context is identified by a sequence of the preceding preposition and the following noun phrase, whereas the problem is identified by a sequence of the preceding verb phrase and the following noun phrase. Overall, CPs are extracted by patent parsing, sentence selection, and phrase extraction (Figure 2).

3.2. Normalization of Extracted Contexts and Problems

The contexts and the problems, which are extracted in the first stage, are further normalized for statistical analysis. First of all, 'stop words' such as 'the', 'a', 'what', 'they', and 'himself' [48] are eliminated. Additionally, some POSs, such as article determiners and personal pronouns, provide insignificant value for analysis, and are excluded.

Every word has various inflected or derived words. Specifically, a word takes different forms depending on the tense, person, plurality, gender, and case. Humans can easily recognize that the words are closely related, but computers recognize them as different entities. Thus, to enable a computer to process inflected or derived words accurately, they must be converted to a common form. For this purpose, the morphological approach is usually applied. In our study, Porter's stemmer was used for such purposes [49]. Porter's stemmer reduces words to their stem. For example, the different words 'argue', 'argued', 'arguing', and 'argues' are reduced to their stem 'argu'. Even though the suffixes are removed, the fundamental meaning is still clear and does not affect the interpretation of the original word.

In addition, the semantic approach was used for understanding human expression through language. This approach guides a computer-based system to understand the meanings of expressions. Semantic dictionaries such as WORDNET [50] have been used in the semantic approach of this study. By using the dictionary, words can be converted into their real meanings, and this process reduces problems of synonymy and homonymy. However, in certain cases, word conversions may be inaccurate because the dictionary is not customized to fit specific technological domains, thereby causing confusion especially when analyzing technology-specific patents. For this reason, only a few general words, such as words signifying 'maximize', 'minimize', 'transfer', and 'provide', have been converted in this study.

3.3. Context–Problem Network Generation

In this stage, CWA is performed for the normalized contexts and problems in order to generate a co-occurrence table, which records co-occurrences of contexts and problems. The co-occurrence table is for allowing initial observation and assessment of the relationships between all contexts and problems by organizing them in a matrix, and further providing the basis for generating a context–problem network. Importantly, co-occurrences between a single context and a problem are most common, but in alignment with the goals of this research to identify key contexts and problems, instances of context–context and problem–problem co-occurrences are also considered when applicable. The inclusion of context–context and problem–problem pairs prevents important information from loss or distortion for sentences with multiple problems or contexts, while also providing vital relationships of pairs of contexts or problems that may exist in the same patent. Henceforward, 'co-occurrence' refers to the link or relationship discovered through CWA between context and problem, context and context, and problem and problem.

In generating the co-occurrence table, co-occurrences of contexts and problems are recorded following three main rules: (i) Add one point for each sentence of a patent when the co-occurrence exists in the same sentence, (ii) add one point when the co-occurrence exists in the same patent, but (iii) a maximum of one point per patent is allowed in the case of multiple co-occurrences. The logic behind this scoring method of the co-occurrence table is (a) to add importance to context and problem co-occurrences for each existing sentence, (b) to add significance to context and problem co-occurrences in the same patent, and (c) to avoid over-exaggeration of the importance of the co-occurrence due to multiple co-occurrences within the same patent. Redundant points awarded to each co-occurrence within the same patent can affect the resulting network generated from the co-occurrence table negatively by misleadingly emphasizing their importance. The reasoning for acknowledging the co-occurrences within the same patent is to account for co-occurrences that transpire in separate sentences. For example, when considering the following two sentences, 'new features of the touchpad provide the following functionalities' and '4. increased sensitivity and precision', the context 'of the touch pad' is related to the problems 'increase sensitivity' and 'increase precision'. However, this relationship will be unnoticed if only co-occurrences in the same sentence are considered.

To further demonstrate how the co-occurrence table is generated, the process of recording the five co-occurrences of CT2 and PR2 in Figure 3a,b is examined as follows. When considering Sentence-2 ('the purpose is to increase transmission speed in the wireless network') in Patent-2, one point for the co-occurrence between the normalized context CT2: 'in wireless network' and the normalized problem PR2: 'maxim transmiss speed' is recorded. Furthermore, if Sentence-3 of Patent-2 states 'apparatus provides power and increases transmission speed in all wireless networks', another point for the co-occurrence between the context CT2: 'in wireless network' and problem PR2: 'maxim transmiss speed' is documented. Furthermore, this CT2–PR2 co-occurrence takes place in the same patent (i.e., Patent-2), registering one additional count. Thus, a resulting total of three co-occurrences are registered for the pair from Patent-2. Likewise, when considering Sentence-1 ('transfer power and amplify energy transmission speed in wireless network by primary coil') in Patent-1, the CT2–PR2 co-occurrence is recorded with a value of two: One count for co-occurring in the same sentence (i.e., Sentence-1) and one count for co-occurring in the same patent (i.e., Patent-1). Hence, the CT2–PR2 co-occurrence total in Figure 3a,b equates to five, with three counts from Patent-2 and two counts from Patent-1. Additionally, self-co-occurrences (i.e., CT1–CT1; PR1–PR1) and the directions of the co-occurrences (i.e., CT1 → PR1; PR2 ← CT2) imply insignificant meaning to the context–problem pair relationships and are not considered for our research, and thus, are represented as '-' in the co-occurrence table. In such a manner, all possible co-occurrences can be counted and accumulated by analyzing the remaining normalized contexts and problems.

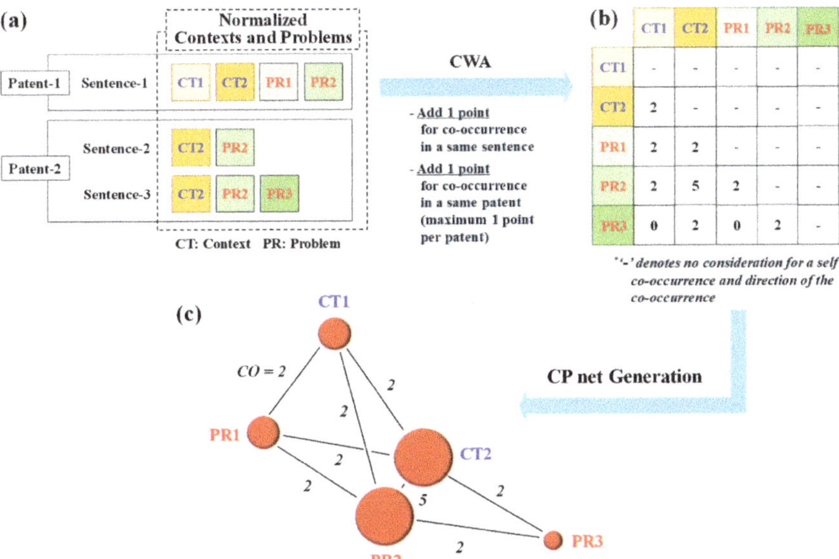

(b)	CT1	CT2	PR1	PR2	PR3
CT1	-	-	-	-	-
CT2	2	-	-	-	-
PR1	2	2	-	-	-
PR2	2	5	2	-	-
PR3	0	2	0	2	-

Figure 3. Example process for the generation of a co-occurrence table and network from normalized contexts and problems: (**a**) Contexts and problems which are extracted and normalized from the sentences of patents; (**b**) co-occurrence (*CO*) table which is constructed from the normalized contexts and problems in (**a**) by co-word analysis (CWA), where the numbers in the individual boxes denote *CO* values; (**c**) network of contexts and problems (CP net) which is generated from the co-occurrence table in (**b**) by using the NetMiner software. In (**c**), *CO* denotes the value of the co-occurrences. A higher *CO* value indicates higher degree centrality between the nodes (i.e., contexts and problems). Larger-sized nodes represent higher numbers of links between the adjacent nodes. The relative positional locations of the contexts and problems represent their closeness centralities in part.

With the three main rules described above, the total number ($TCO_{x,y}$) of the co-occurrences of each pair of interest among all the normalized contexts and problems can be formulated as an equation in the following:

$$TCO_{x,y} = \sum_{i=1}^{m} CO(x,y)_{S_i} + \sum_{j=1}^{n} CO(x,y)_{P_j} \tag{1}$$

where x is a node (i.e., a context or a problem) and y is another node (i.e., a context or a problem) where $x \neq y$, S_i is the ith of m sentences, P_j is the jth of n patents, $CO(x,y)_{S_i}$ is a co-occurrence of x and y in the ith sentence (1 if they co-occur; 0 otherwise), and $CO(x,y)_{P_j}$ is a co-occurrence of x and y in the jth patent (1 if they co-occur; 0 otherwise). Consequently, the values of the co-occurrences of contexts and problems are assessed for all patents to generate the co-occurrence table.

The co-occurrence table often leads to a complex network of contexts and problems. Thus, applying a cut-off value (CV) to $TCO_{x,y}$ is quite beneficial to transfigure the co-occurrence table into a network in a simpler but more valuable form. As CV increases, the efficiency of network generation and the simplicity of the resulting network increase, so that analysis can be performed intuitively. However, if the cut-off value is too high, important connections may be omitted. Thus, the CV should be determined case-by-case. Use of CV binarizes the weight ($WL_{x,y}$) of the links between the paired nodes (i.e., x and y in Equation (1)):

$$WL_{x,y} = \begin{cases} TCO_{x,y}, & if \; TCO_{x,y} \geq CV \\ 0, & otherwise \end{cases} . \tag{2}$$

Here, the co-occurrence weight ($WL_{x,y}$) determines the unique strength, which is illustrated by the different numbers and distances of the links between the x and y node pairs. Overall, adjustments to the CV are performed to reduce network complexity while preserving important co-occurrences prior to proceeding to the analysis of the generated context and problem network.

After completion of the co-occurrence table, the relational data in the form of a matrix can be suitably transfigured into a network of contexts and problems (CP net) to conduct SNA. Figure 3b,c illustrates the generation of the CP net representing the data collected from the co-occurrence table. The CP net can be generated from the co-occurrence table obtained above by using the NetMiner software (Cyram Company, Seongnam, Korea) as follows. The co-occurrence table is imported into the software, then the 'Analyze >> Centrality >> Degree/Closeness' module is utilized to generate the summary statistics, degree centrality vector, spring visualization, and concentric visualization of the data. Furthermore, through this process, vital information of the degree centrality and closeness centrality of the original co-occurrence data is obtained. The degree centrality and closeness centrality indices are formulated by the software.

Here, the degree centrality (D_x) of a node x (i.e., a context or a problem) is computed simply by the sum of co-occurrences of its adjacent nodes and then normalized by the total number of nodes, N_{node}, in the entire network as:

$$D_x = \frac{1}{N_{node} - 1} \cdot \sum_{y=1}^{p} TCO_{x,y} \tag{3}$$

where y is the yth of the p nodes co-occurring with the node x in which $x \neq y$ and the subtraction of '1' from N_{node} means the exclusion of the node x itself. In the degree centrality analysis, it is additionally noted that p represents the total link number of the node x to the other nodes (illustrated as node size in the NetMiner software).

The closeness centrality (C_x) of a node x is measured by the inverse of the sum of distances, d, from the node x to all other nodes, which is then normalized by multiplying the total number of nodes (except the x itself) in the entire network:

$$C_x = (N_{node} - 1) \left(\sum_{y=1}^{q} d(x,y) \right)^{-1} \tag{4}$$

where y is the yth of q nodes directly or indirectly linked to the node x, and d is the shortest number of hops between the node x and y.

4. Analysis of CP Net

The resulting network, CP net, is one-mode data with undirected links. Thus, in this study, the CP net data could be quantitatively characterized in three types of analyses: (i) Centrality analysis, (ii) neighbor analysis, and (iii) grammar analysis (Figure 4).

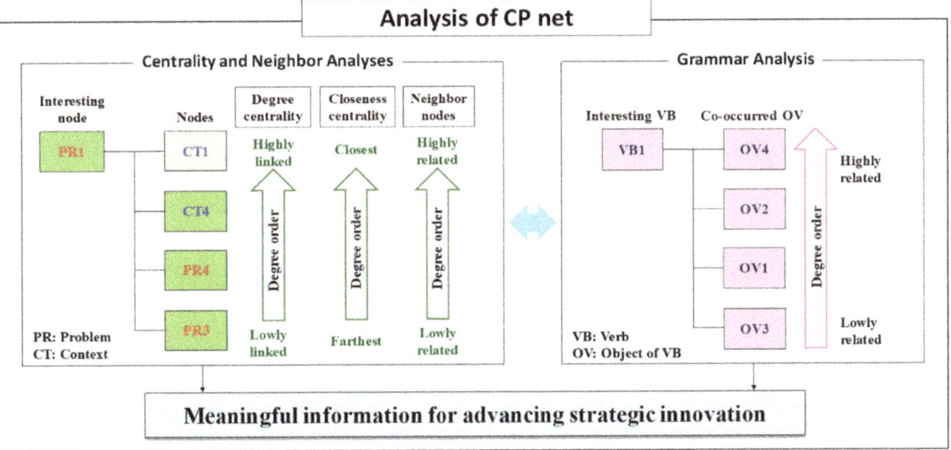

Figure 4. Centrality, neighbor, and grammar analyses of CP net.

Centrality analysis measures the activities or influences of nodes to find a central node. The centrality of a node determines its relative importance in a social network [28]. Thus, a context with high centrality indicates important products, services, components, places, times, tasks, or people. On the same principle, a problem with high centrality may indicate important shortages, errors, functions, tasks, objectives, or needs. In general, centrality analysis adopts three measures: Degree, betweenness, and closeness.

This research uses degree centrality and closeness centrality as representative indices of an importance (Table 3; Figure 4). Degree centrality can be calculated by assessing the co-occurrence weights between adjacent nodes. In the perspective of contexts and problems, high degree centrality signifies that the context or problem is directly linked to many nodes. Because only direct links are considered, this measure represents local importance. The closeness measure indicates the distance of a node from all other nodes in a network. A context or problem with short geodesic distances to all (direct and indirect) other nodes has high closeness, thus depicting its overall global importance.

Table 3. Centrality analysis of CP net.

	Context	Problem
High degree centrality	• Local context which directly affects many other contexts or problems	• Local problem which is directly affected by many other contexts or problems
High closeness centrality	• Overall context which widely affects all other contexts or problems	• Overall problem which is widely affected by all other contexts or problems

If users are interested in specific contexts or problems, they can select the node of interest and analyze its neighboring nodes (Figure 4). The highly-related neighbor nodes are identified by referring to the weights of the links. For example, if users are interested in the context 'of mobil phone' and neighbor nodes, like the context 'of convers', and the problem 'maxim qualiti' is identified, they may learn that the important problem is to maximize the quality of conversation in the context of the mobile phone.

The grammar analysis is more specific, which can provide relationships among grammatical components of a context or a problem (Figure 4). The context usually consists of PP and OP, and the problem usually consists of VB and OV. By separating the grammatical components from the construction, users can focus on a specific component. Especially, the VBs 'maximize' and 'minimize' are very useful for analyzing problems. For example, if we only consider problems that include the VB 'maximize' and sort them by degree order, we can learn which OVs other inventors are trying to maximize. The OV can be related to criteria such as efficiency, convenience, and cost. In a similar way, users can analyze the CP net in depth according to their interest.

5. Case Study: Wireless Energy Transmission Technology

The transfer of electricity over a tiny air gap was first demonstrated in 1888 by using an oscillator of Hertz which was connected with induction coils. Furthermore, long-distance transfer of microwaves was also demonstrated in 1896 by Tesla [51]. Thereafter, no further introduction of a method converting microwaves back to electricity has been documented for an extensive period. However, sixty-eight years later, the conversion of microwaves to electricity was realized by Brown and coworkers using a rectenna. The rectenna, a rectifying antenna, was invented in 1964 and disclosed as a patent in 1969 [52]. This invention opened up the possibilities for practical wireless charging technology, namely wireless energy transmission (WET) technology. A mid-range non-radiative wireless charging method was additionally reported in 2007 [53,54].

Currently, WET technology has been highly regarded as a next-generation charging technology, which can be used in diverse applications including mobile electronic devices, electric vehicles, and biomedical devices. In particular, the rise in interest for WET technology can be mainly explained by the adoption of the technology in portable electronic devices [55].

Thus, in this research, the WET technology domain has been selected for the case study. We have constructed and analyzed a CP net using the systematic and replicable method developed above. We have retrieved 737 patents disclosed for WET technology between 2004 and 2013 from the USPTO official website. The patents were full text documents in html file format. Through patent parsing, keyword matching, and grammar-based text mining, a total of 1734 context–problem pairs were extracted, with at least one from each of 515 (69.9%) among 737 patents. To increase extraction efficiency, linguistic research could be beneficial for modeling the contexts and problems in future studies. Following the normalization process via stop word elimination, stemming, and semantic conversion, 1248 contexts and 1261 problems were remained to comprise 2509 nodes and 13,619 undirected links for use in SNA.

Due to the high complexity of the resulting network, links with $TCO_{CT,PR} < 7$ were eliminated, and then the CP net was re-generated (Figure 5). Then, centrality analysis was applied (Figure 6), and top nodes with high centrality were identified (Table 4). Even though stop words had been eliminated, the result included some meaningless expressions, which were eliminated manually. The significant context included components such as 'a battery', 'a coil', and 'a core'. This illustrates that such components are highly related to major problems. In terms of problems, naturally, problems related to 'providing power' showed high scores. For example, problems like 'maximizing coupling efficiency', 'minimizing DC signal components', 'charging battery', and 'discharging system' are nominated, as shown in Table 4.

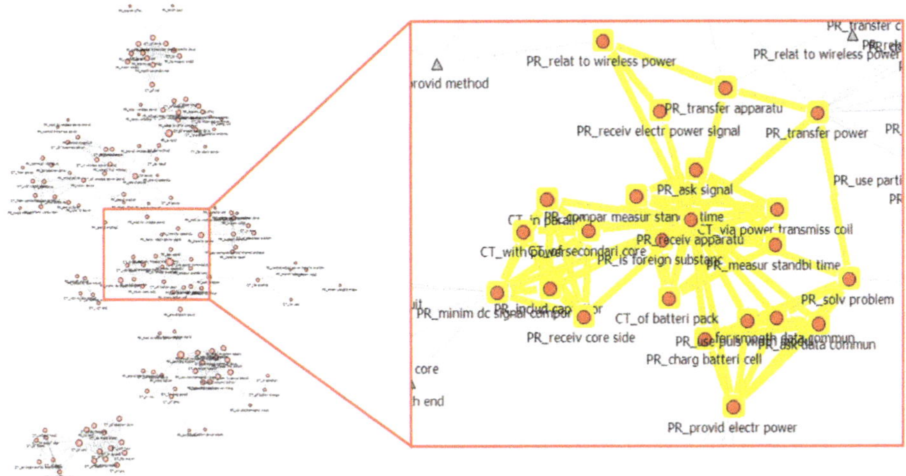

Figure 5. Spring visualization of the CP net for wireless energy transmission (WET) technology; CT: Context; PR: Problem.

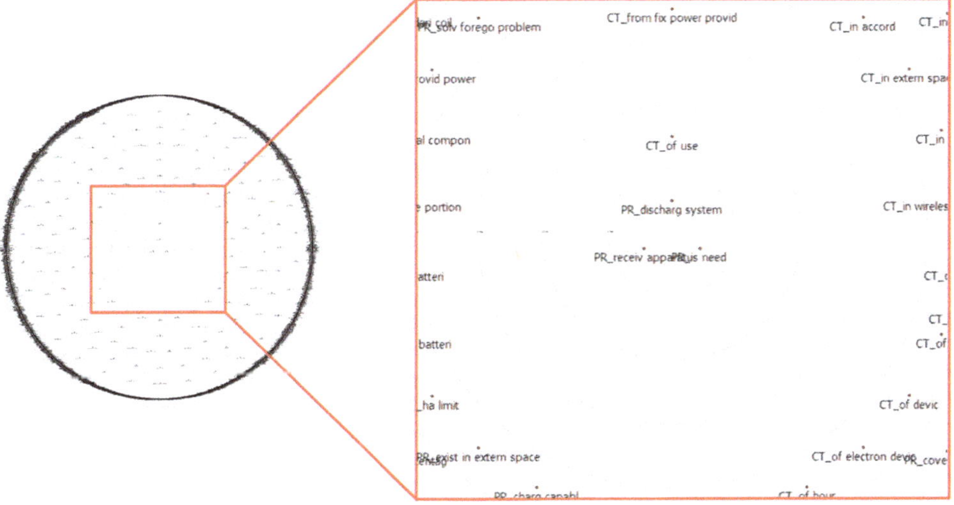

Figure 6. Concentric visualization of the CP net for WET technology analyzed by the degree centrality; CT: Context; PR: Problem.

Table 4. Centrality analysis results of the CP net for WET technology.

	Degree Centrality	**Closeness Centrality**
Contexts	CT_of batteri	CT_of power
	CT_of reson length	CT_for smooth data commun
	CT_from fix power provid	CT_with power
	CT_of electron devic	CT_of secondari core
	CT_by primari coil	CT_in parallel
	CT_with hour	CT_via power transmiss coil
	CT_while frequenc	CT_of batteri pack
	CT_of power	CT_in wireless power provid
	CT_of hour	CT_of wireless power provid
	CT_of coil	CT_for power
Problems	PR_provid power	PR_provid electr power
	PR_discharg system	PR_provid power
	PR_transfer power	PR_transfer power
	PR_maxim coupl effici	PR_minim dc signal compon
	PR_minim dc signal compon	PR_receiv electr power signal
	PR_use electromagnet wave	PR_maxim coupl effici
	PR_includ wireless drive portion	PR_use puls width modul
	PR_includ second batteri	PR_charg batteri cell
	PR_includ first batteri	PR_discharg system
	PR_minim sensit	PR_measur standbi time

(CT: Context; PR: Problem).

Once the contexts and problems were identified by the centrality analysis, neighbor analysis was performed for specific interests. In this case study, we focused on analyzing the context 'of coil' and its neighbors as a trial (Table 5). According to the result, the coil context was found to have strongest relations with problems such as 'maximizing coupling efficiency', 'having magnetic shield', 'disclosing energy transfer devices', 'covering primary coil', 'maximizing percentage', 'getting wind of devices', 'reaching secondary coil', and 'minimizing sensitivity'. Thus, when considering the component 'coil' as a target of improvement or invention, information regarding these relations would be imperative for establishing the purpose and direction.

Table 5. Neighbor analysis results of the context 'of coil' for WET technology.

Source Node	**Target Nodes with the Highest Degree**
CT_of coil	PR_maxim coupl efficien, PR_ha magnet shield, PR_disclos energi transfer devic, PR_cover primary coil, PR_maxim percentage, PR_wind of devic, PR_reach secondari coil, PR_minim sensit …

(CT: Context; PR: Problem).

Finally, grammar analysis was performed for in-depth analysis. For illustration purposes, we focused on problems with the VBs 'maximize' or 'minimize' (Table 6). By analyzing the VBs 'maximize' and 'minimize', we were able to observe that prior inventions have aimed to solve or optimize existing problems in the wireless charging technology field. In other words, the most significant objectives that inventors have directed their efforts toward could be identified. It was discovered that many inventors focused their advancements in maximizing 'coupling efficiency', 'overall efficiency', 'safety', 'accuracy', 'battery life', and 'ease of use'. On the other hand, inventors were also trying to minimize 'DC signal component', 'sensitivity', 'size', 'cost', and 'unsightly mess'.

Table 6. Results of grammar analysis of CP net with verbs (VBs), 'maxim' and 'minim' for WET technology.

Maximize	Minimize
PR_maxim coupl effici	PR_minim dc signal compon
PR_maxim percentag	PR_minim sensit
PR_maxim effici	PR_minim need
PR_maxim safeti	PR_minim size
PR_maxim accuraci	PR_minim cost
PR_maxim batteri life	PR_minim amount
PR_maxim eas	PR_minim unsightli mess

(PR: Problem).

From the results of our patent analysis method, an overview of the existing problems and contexts of the technological domain can be obtained. This is beneficial for inventors to gain extensive insights of the domain at large. Furthermore, the results can also provide valuable, in-depth information of a preselected specific area of interest within the technological domain, whether the initial search begins from the context or problem.

As an overview of the case study regarding the WET technology domain, we conclude that three main contexts, (i) battery, (ii) power transmission coil, and (iii) core, are the components most relevantly related to the three main problems, (1) maximizing coupling efficiency, (2) minimizing DC signal components, and (3) charging battery. For this technology, it can be observed that these contexts, in accordance to the problems, have been the most imperative features to be satisfied. Once the comprehension of the significant contexts and problems of the entire domain is performed, one can look more meticulously into selected problems (i.e., 'maximizing of coupling efficiency'/'minimizing DC signal component') or contexts (i.e., 'of coil') to detect respective contexts or problems directly related to the particular area of interest. For example, when considering the problem 'maximizing of coupling efficiency', one could find that this problem pertains mainly to the coil. After identifying the problem and which component to innovate, the user can concentrate on these areas for constructive and effective innovation.

The outcomes of our research provide valuable information that can be utilized at an individual level and also applied to strategic innovative planning from a larger perspective. From the point of view of an individual, user, or inventor, we believe that these findings are beneficial for anyone who is performing innovative research and development (R&D) in the WET technology domain. By accurately identifying the key problematic areas that require solutions for new product development, an individual or small group of inventors can exploit and focus on the outcomes of this research by identifying the contexts referring to the fundamental components (i.e., coil) to accomplish incremental improvements of the component based on the discovered problems and contexts.

Moreover, the yielded results from this research can be applied to a larger technological domain or market-level perspective. Through the outcomes of our method, a scientific or industrial community aimed at advancing and standardizing the WET technology can identify that the major problems of inefficiency and DC signal noise relate to short- and long-ranged (i.e., 'resonance length') wireless energy transfer. This holds true for WET technology due to its existing inefficiencies to transfer power as compared to direct, wired energy transfer, both for short-, and more so for long-range applications. As a community, joint efforts and collaborations can be formed among research groups and corporations to develop solutions and set requirements to meet user and industry standards collectively by identification of the most critical problems ('maximize coupling efficiency' and 'minimize DC signal component') found through our method.

From a market standpoint, the results of our method isolate key contexts (i.e., 'resonance length') requiring attention in order to fulfill the needs of consumers, thus increasing the rate of adoption. Presently, short-distance wireless energy transfer is generally utilized for mobile and medical devices. With the inconveniences of being unable to displace the device from the charger, the WET technology domain remains unable to utterly meet consumer and market needs and satisfaction. Additionally, long-range WET technology, usually aimed for larger WET systems (e.g., electric vehicle wireless charging), can also be found in its early development stage of R&D. Moreover, based on these results, industry leaders of WET technology can consolidate their corporate R&D plans around these findings for more cost-effective and revenue-generating directions. Through the developments to increase the efficient range or wireless energy transfer based on the results of our method, we expect rises in market adoption, consumer satisfaction, and new product applications.

Furthermore, these outcomes directly address environmental and sustainability issues in WET technology. At the current development state of WET technology, evolving environmental concerns stem mainly from the inefficiencies (propagation loss and low transfer efficiency) of wireless charging (i.e., more energy is needed to charge battery via wireless charging, compared to traditional wire charging, and thus more conventional fuel consumption is required for WET). Our research points attention toward the problems of maximizing coupling efficiency and minimizing DC signal (i.e., noise) in the contexts of the core and power transmission coil components of WET devices, and helps emphasize the importance of innovation in improving these problems. Fortunately, the innovative goals for these contexts–problems coincide from both technological and environmental perspectives. Supplementary investigation of the direct sustainability and environmental issues caused by WET technology, such as the adverse effects of radiative wireless charging, may be required for additional perspectives and directions for ecofriendly WET innovation.

In summary, the informative results acquired through our method can be widely applied and exploited by individuals, corporations, and research societies to aid in addressing key problems and components to focus strategic innovation in generating satisfactory technological and environmental advancements. Most importantly, this method can be applied universally to different technological domains and applications.

6. Conclusions

In this study, we developed the systematic and replicable method of patent analysis as an identification tool of contexts and problems to be solved in patents. This method generates and further analyzes CP net data in a quantitative manner. The CP net is constructed via the extraction, normalization, and network generation from patent analysis and text mining as follows. Contexts and problems were extracted from full-text patents after sentence selection. The extracted contexts and problems were further normalized by eliminating stop words, stemming, and semantically converting words. From the normalized contexts and problems, a co-occurrence table was generated to create a CP net, which can be analyzed in a comprehensive manner by using the combined centrality, neighbor, and grammar analyses.

This systematic and sustainable method was applied to analyze a total of 737 patents disclosed for advanced wireless energy transfer technology. In this case study, a set of CP net data was successfully generated. The CP net data were analyzed, providing essential information in the wireless charging technology field. The centrality analysis revealed components like 'battery', 'coil', and 'core' as important information. Problems related to providing energy, like 'maximizing coupling efficiency', 'minimizing DC signal components', and 'charging battery' were recognized. When the context 'a coil' is focused in the neighbor analysis, several problems like 'having magnetic shield', 'disclosing energy transfer devices', 'covering primary coil', 'getting wind of devices', 'reaching secondary coil', and 'minimizing sensitivity' were found to be highly related. The grammar analysis showed that many inventors are trying to maximize 'coupling efficiency', 'safety', 'accuracy', and 'battery life', or to minimize 'sensitivity', 'size', and 'cost'. The results were then deliberated and tailored to explain current situations of WET technology R&D. Furthermore, the discussion of strategic innovative planning and implications were further dissected into individual, scientific community, corporate, and market levels. In short, the case study concludes with effective examples in which a wide variety of domain stakeholders can utilize the results acquired through our method for the advancements of the WET technology and efforts required to alleviate its environmental impacts.

Overall, the systematic and sustainable method of patent analysis has been demonstrated as a very powerful tool to identify problems to be solved in patents. Our research intends to contribute to the advancements of patent mining and analysis with focuses in syntactic and semantic analyses, as well as to help provide managerial implications of strategic technological planning and innovation, granted the valuable insights obtained through our method. Furthermore, this method is domain-independent and is applicable in various technologies for technology planning and innovation.

However, the method has a limitation. Namely, despite elimination of stop words, expressions that were overly general remained, and these interfered with clear interpretation. To overcome this limitation, the need for a customized stop word dictionary for contexts and problems identification is necessary. In spite of the limitation, the generation process performed well, and the CP net was useful to identify problems to be solved. The authors' future research endeavors will aim to improve this method for higher quality of contexts and problems extracted, to incorporate modern, state-of-the-art machine learning and text mining techniques for CP net generation efficiency, and to cope with larger data sets.

Author Contributions: Conceptualization, J.J.R., C.J., H.P., and K.K.; data curation, J.J.R.; investigation, J.J.R.; methods, J.J.R and C.J.; discussions, J.J.R., C.J., H.P., and K.K.; writing—original draft, J.J.R.; writing—review and editing, J.J.R.; supervision, K.K.

Funding: This research was supported by the National Research Foundation of Korea (NRF) grant funded by the Ministry of Science and ICT (Grant No.: NRF-2016R1A2B4008381) and the Business for University Entrepreneurship Center funded by the Ministry of SMEs and Startups in 2014 (Grant No. 10016911).

Acknowledgments: The authors appreciate the financial support from the NRF, the Business for University Entrepreneurship Center, and the Ministry of SMEs and Startups allowing for this research to be accomplished.

Conflicts of Interest: The authors have no conflict of interest.

References

1. Kim, J.; Choi, J.; Park, S.; Jang, D. Patent keyword extraction for sustainable technology management. *Sustainability* **2018**, *10*, 1287. [CrossRef]
2. Joung, J.; Kim, K. Monitoring emerging technologies for technology planning using technical keyword based analysis from patent data. *Technol. Forecast. Soc. Chang.* **2017**, *114*, 281–292. [CrossRef]
3. Lee, S.; Lee, S.; Seol, H.; Park, Y. Using patent information for designing new product and technology: Keyword based technology roadmapping. *R D Manag.* **2008**, *38*, 169–188. [CrossRef]
4. Yoon, B.; Park, Y. A text-mining-based patent network: Analytical tool for high-technology trend. *J. High Technol. Manag. Res.* **2004**, *15*, 37–50. [CrossRef]
5. Lee, S.; Yoon, B.; Park, Y. An approach to discovering new technology opportunities: Keyword-based patent map approach. *Technovation* **2009**, *29*, 481–497. [CrossRef]
6. Yoon, B.; Park, Y. A systematic approach for identifying technology opportunities: Keyword-based morphology analysis. *Technol. Forecast. Soc. Chang.* **2005**, *72*, 145–160. [CrossRef]
7. Hasan, M.A.; Spangler, W.S.; Griffin, T.; Alba, A. *COA:* Finding novel patents through text analysis. In Proceeding of the 15th ACM SIGKDD International Conference on Knowledge Discovery and Data Mining, Paris, France, 28 June–1 July 2009.
8. Park, I.; Yoon, B. Identifying promising research frontiers of pattern recognition through bibliometric analysis. *Sustainability* **2018**, *10*, 4055. [CrossRef]
9. Tseng, Y.-H.; Lin, C.-J.; Lin, Y.-I. Text mining techniques for patent analysis. *Inform. Proc. Manag.* **2007**, *43*, 1216–1247. [CrossRef]
10. Yang, Y.; Akers, L.; Klose, T.; Barcelon Yang, C. Text mining and visualization tools – Impressions of emerging capabilities. *World Pat. Inf.* **2008**, *30*, 280–293. [CrossRef]
11. Lim, J.; Choi, S.; Lim, C.; Kim, K. SAO-based semantic mining of patents for semi-automatic construction of a customer job map. *Sustainability* **2017**, *9*, 1386. [CrossRef]
12. Choi, S.; Yoon, J.; Kim, K.; Lee, J.; Kim, C.-H. SAO network analysis of patents for technology trends identification: A case study of polymer electrolyte membrane technology in proton exchange membrane fuel cells. *Scientometrics* **2011**, *88*, 863–883. [CrossRef]

13. Choi, S.; Park, H.; Kang, D.; Lee, J.; Kim, K. An SAO-based text mining approach to building a technology tree for technology planning. *Expert Syst. Appl.* **2012**, *39*, 11443–11455. [CrossRef]

14. Choi, S.; Kim, H.; Yoon, J.; Kim, K.; Lee, J.Y. An SAO-based text-mining approach for technology roadmapping using patent information. *R D Manag.* **2013**, *43*, 52–74. [CrossRef]

15. Yoon, J.; Kim, K. Identifying rapidly evolving technological trends for R&D planning using SAO-based semantic patent networks. *Scientometrics* **2011**, *88*, 213–228.

16. Yoon, J.; Park, H.; Kim, K. Identifying technological competition trends for R&D planning using dynamic patent maps: SAO-based content analysis. *Scientometrics* **2013**, *94*, 313–331.

17. Yoon, J.; Choi, S.; Kim, K. Invention property-function network analysis of patents: A case of silicon-based thin film solar cells. *Scientometrics* **2011**, *86*, 687–703. [CrossRef]

18. Savransky, S.D. *Engineering of Creativity: Introduction to TRIZ Methodology of Inventiveproblem Solving*; CRC Press: Boca Raton, FL, USA, 2000.

19. Phelps, D.J. Automatic concept identification: Extracting problem solved concepts from patent documents. presented at the Information Retrieval Facility Symposium, Vienna, Austria, 8–9 November 2007.

20. Tiwana, S.; Horowitz, E. Extracting problem solved concepts from patent documents. In Proceedings of the 2nd International Workshop on Patent Information Retrieval, Hong Kong, China, 6 November 2009.

21. Jeong, C.; Kim, K. Creating patents on the new technology using analogy-based patent mining. *Expert Syst. Appl.* **2014**, *41*, 3605–3614. [CrossRef]

22. Li, Z.; Tate, D. Patent analysis for systematic innovation: Automatic function interpretation and automatic classification of level of invention using natural language processing and artificial neural networks. *Int. J. Syst. Innov.* **2010**, *1*, 10–26.

23. Callon, M.; Courtial, J.-P.; Laville, F. Co-word analysis as a tool for describing the network of interactions between basic and technological research: The case of polymer chemsitry. *Scientometrics* **1991**, *22*, 155–205. [CrossRef]

24. Callon, M.; Courtial, J.-P.; Turner, W.A.; Bauin, S. From translations to problematic networks: An introduction to co-word analysis. *Soc. Sci. Inf.* **1983**, *22*, 191–235. [CrossRef]

25. He, Q. Knowledge discovery through co-word analysis. *Libr. Trends* **1999**, *48*, 133–159.

26. Ding, Y.; Chowdhury, G.G.; Foo, S. Bibliometric cartography of information retrieval research by using co-word analysis. *Inf. Proc. Manag.* **2001**, *37*, 817–842. [CrossRef]

27. Moehrle, M.G.; Walter, L.; Geritz, A.; Müller, S. Patent-based inventor profiles as a basis for human resource decisions in research and development. *R D Manag.* **2005**, *35*, 513–524. [CrossRef]

28. Freeman, L.C. Centrality in social network concept clarification. *Soc. Netw.* **1978**, *1*, 215–239. [CrossRef]

29. Sternitzke, C.; Bartkowski, A.; Schramm, R. Visualizing patent statistics by means of social network analysis tools. *World Pat. Inf.* **2008**, *30*, 115–131. [CrossRef]

30. Lee, S.; Cha, Y.; Han, S.; Hyun, C. Application of association rule mining and social network analysis for understanding causality of construction defects. *Sustainability* **2019**, *11*, 618. [CrossRef]

31. Su, W.; Wang, Y.; Qian, L.; Zeng, S.; Baležentis, T.; Streimikiene, D. Creating a sustainable policy framework for cross-border e-commerce in China. *Sustainability* **2019**, *11*, 943. [CrossRef]

32. Kan, Z.; Zhang, G. Study on the text mining and Chinese text mining framework. *Inf. Sci.* **2007**, *25*, 1046–1051.

33. Aggarwal, C.C.; Zhai, C.X. *Mining Text Data*; Springer Press: New York, NY, USA, 2012.

34. Sakaki, T.; Okazaki, M.; Matsuo, Y. Tweet analysis for real-time event detection and earthquake reporting system development. *IEEE Trans. Knowl. Data Eng.* **2013**, *25*, 919–931. [CrossRef]

35. Ngai, E.W.T.; Lee, P.T.Y. A review of the literature on applications of text mining in policy making. In Proceedings of the Pacific Asia Conference on Information Systems (PACIS), Chiayi, Taiwan, 27 June–1 July 2016.

36. Cantner, U.; Graf, H. The network of innovators in Jena: An application of social network analysis. *Res. Policy* **2006**, *35*, 463–480. [CrossRef]

37. Balconi, M.; Breschi, S.; Lissoni, F. Networks of inventors and the role of academia: An exploration of Italian patent data. *Res. Policy* **2004**, *33*, 127–145. [CrossRef]

38. Von Wartburg, I.; Teichert, T.; Rost, K. Inventive progress measured by multi-stage patent citation analysis. *Res. Policy* **2005**, *34*, 1591–1607. [CrossRef]

39. Chang, S.-B.; Lai, K.-K.; Chang, S.-M. Exploring technology diffusion and classification of business methods: Using the patent citation network. *Technol. Forecast. Soc. Chang.* **2009**, *76*, 107–117. [CrossRef]

40. Lee, B.; Jeong, Y.-I. Mapping Korea's national R&D domain of robot technology by using the co-word analysis. *Scientometrics* **2008**, *77*, 3–19.

41. Hartson, R.; Pyla, P.S. *The UX Book: Process and Guidelines for Ensuring a Quality User Experience*; Elsevier: Amsterdam, The Netherlands, 2012.

42. Helander, M.G.; Landauer, T.K.; Prabhu, P.V. *Handbook of Human-Computer Interaction*; Elsevier: Amsterdam, The Netherlands, 1997.

43. Schilit, B.N.; Theimer, M.M. Disseminating active map information to mobile hosts. *IEEE Netw.* **1994**, *8*, 22–32. [CrossRef]

44. Brown, P.J.; Bovey, J.D.; Chen, X. Context-aware applications: From the laboratory to the marketplace. *IEEE Pers. Commun.* **1997**, *4*, 58–64. [CrossRef]

45. Ryan, N.S.; Pascoe, J.; Morse, D.R. Enhanced reality fieldwork: The context-aware archaeological assistant. In *Computer Applications & Quantitative Methods in Archaeology*; Tempus Reparatum: Birmingham, UK, 10–17 April 1997.

46. Dey, A.K. Context-aware computing: The CyberDesk project. In Proceedings of the AAAI 1998 Spring Symposium on Intelligent Environments, Stanford, CA, USA, 23–25 March 1998; pp. 51–54.

47. Toutanova, K.; Klein, D.; Manning, C.D.; Singer, Y. Feature-rich part-of-speech tagging with a cyclic dependency network. In Proceedings of the 2003 Conference of the North American Chapter of the Association for Computational Linguistics on Human Language Technology, Edmonton, AB, Canada, 27 May–1 June 2003.

48. STOPWORDS English Stopwords. 2018. Available online: http://www.ranks.nl/stopwords (accessed on 23 June 2018).

49. Porter, M.F. An algorithm for suffix stripping. *Program* **1980**, *14*, 130–137. [CrossRef]

50. Miller, G.A. WordNet: A lexical database for English. *Commun. ACM* **1995**, *38*, 39–41. [CrossRef]

51. Tesla, N. Apparatus for Transmitting Electrical Energy. U.S. Patent 1,119,732, 1 December 1914.

52. Brown, W.C.; George, R.H.; Heenan, N.I.; Wonson, R.C. Microwave to DC converter. U.S. Patent 3,434,678, 25 March 1969.

53. Kurs, A.; Karalis, A.; Moffatt, R.; Joannopoulos, J.D.; Fisher, P.; Soljacic, M. Wireless power transfer via strongly coupled magnetic resonances. *Science* **2008**, *317*, 83–86. [CrossRef] [PubMed]

54. Karalis, A.; Joannopoulos, J.D.; Soljacic, M. Efficient wireless non-radiative mid-range energy transfer. *Ann. Phys.* **2008**, *323*, 34–48. [CrossRef]

55. Vanderelli, T.A.; Shearer, J.G.; Shearer, J.R. Method and Apparatus for a Wireless Power Supply. U.S. Patent 7,027,311, 11 April 2006.

MDPI

St. Alban-Anlage 66

4052 Basel

Switzerland

Tel. +41 61 683 77 34

Fax +41 61 302 89 18

www.mdpi.com

Sustainability Editorial Office

E-mail: sustainability@mdpi.com

www.mdpi.com/journal/sustainability